Methods in Enzymology

Volume LI

Purine and Pyrimidine Nucleotide Metabolism

EDITED BY

Patricia A. Hoffee

DEPARTMENT OF MICROBIOLOGY
UNIVERSITY OF PITTSBURGH
SCHOOL OF MEDICINE
PITTSBURGH, PENNSYLVANIA

Mary Ellen Jones

DEPARTMENT OF BIOCHEMISTRY
UNIVERSITY OF SOUTHERN CALIFORNIA
SCHOOL OF MEDICINE
LOS ANGELES, CALIFORNIA

1978

ACADEMIC PRESS   New York   San Francisco   London

A Subsidiary of Harcourt Brace Jovanovich, Publishers

ACADEMIC PRESS, INC.
111 Fifth Avenue, New York, New York 10003

United Kingdom Edition published by
ACADEMIC PRESS, INC. (LONDON) LTD.
24/28 Oval Road, London NW1   7DX

LIBRARY OF CONGRESS CATALOG CARD NUMBER: 54–9110

ISBN 0–12–181951–5 (v. 51)

PRINTED IN THE UNITED STATES OF AMERICA

Table of Contents

Biosynthetic Enzymes

Section I. Activation of Ribose Phosphate

Section II. *De Novo* Pyrimidine Biosynthesis

A. Single Enzymes

Section IV. Deoxynucleotide Synthesis

Degradative and Salvage Enzymes

Section V. Nucleotidases and Nucleosidases

Section VI. Pyrimidine Metabolizing Enzymes

A. Kinases

B. Deaminases

C. Phosphorylases and *Trans*-Deoxyribosylase

E. General Methods

Contributors to Volume LI

Article numbers are in parentheses following the names of contributors.
Affiliations listed are current.

AHMED T. H. ABDELAL (4), *Biology Department, Georgia State University, Atlanta, Georgia*

LINDA B. ADAIR (7), *Department of Chemistry, Gulf Coast Community College, Panama City, Florida*

K. C. AGARWAL (64, 72, 79), *Division of Biological and Medical Sciences, Brown University, Providence, Rhode Island*

R. P. AGARWAL (49, 67, 72, 79), *Division of Biological and Medical Sciences, Brown University, Providence, Rhode Island*

T. AMAYA (10), *Fujisawa Pharmaceutical Co., Ltd., Osaka, Japan*

ELIZABETH P. ANDERSON (40, 42), *Nucleic Acid Section, Lab of Pathophysiology, National Cancer Institute, Bethesda, Maryland*

WILLIAM J. ARNOLD (77), *Department of Medicine, Abraham Lincoln School of Medicine, University of Illinois at the Medical Center, Chicago, Illinois*

CHARALAMPOS ARSENIS (34), *Department of Molecular Biology, Vanderbilt University, Nashville, Tennessee*

W. L. BELSER (19), *Department of Biology, University of California, Riverside, California*

STEPHEN H. BISHOP (27), *Department of Zoology, Iowa State University, Ames, Iowa*

RAYMOND L. BLAKLEY (32), *Department of Biochemistry, University of Iowa, Iowa City, Iowa*

JAMES BLANK (58), *Department of Microbiology, University of Pittsburgh School of Medicine, Pittsburgh, Pennsylvania*

EDWARD BRESNICK (46), *Department of Biochemistry, The University of Vermont, College of Medicine, Burlington, Vermont*

ELINOR F. BRUNNGRABER (50), *120 East 88 Street, New York, New York*

JOHN M. BUCHANAN (24, 25), *Department of Biology, Massachusetts Institute of Technology, Cambridge, Massachusetts*

MICHAEL E. BURT (3), *Department of Surgery, Cornell University Medical College, New York Hospital, New York, New York*

R. CARDINAUD (60), *Service de Biophysique, Department de Biologie, Centre d'Etudes Nucleaires de Saclay, Gif-sur-Yvette, France*

GIOVANNI CERCIGNANI (51), *Laboratory of Biochemistry, Faculty of Sciences, University of Pisa, Pisa, Italy*

DONALD CHANG (66), *Department of Medicine, Washington University, and The Jewish Hospital of St. Louis, St. Louis, Missouri*

TA-YUAN CHANG (6), *Department of Biochemistry, Dartmouth Medical School, Hanover, New Hampshire*

MING S. CHEN (45), *Department of Pharmacology, Yale University, School of Medicine, New Haven, Connecticut*

Y. -C. CHENG (47), *Department of Experimental Therapeutics, Roswell Park Memorial Institute, New York State Department of Health, Buffalo, New York*

CAROLE J. COFFEE (65), *Department of Biochemistry, University of Pittsburgh, School of Medicine, Pittsburgh, Pennsylvania*

PATRICK F. COLEMAN (18), *Hyland Laboratories, Costa Mesa, California*

WAYNE E. CRISS (61), *Department of Oncology, Howard University, Cancer Research Center, Washington, D. C.*

ROWLAND H. DAVIS (16), *Department of Molecular Biology and Biochemistry, University of California, Irvine, California*

MARTIN R. DEIBEL, JR. (44), *Department of Biochemistry, The Ohio State University, Columbus, Ohio*

THOMAS F. DEUEL (2, 66), *Departments of Medicine and Biological Chemistry, Washington University, and The Jewish Hospital of St. Louis, St. Louis, Missouri*

R. BRUCE DUNLAP (14), *Department of Chemistry, University of South Carolina, Columbia, South Carolina*

STAFFAN ERIKSSON (30), *Medical Nobel Institute, Department of Biochemistry I, Karolinska Institute, Stockholm, Sweden*

LEONARD F. ESTIS (3), *McArdle Laboratory, University of Wisconsin, Madison, Wisconsin*

HARALD E. FISCHER (27), *M. D. Anderson Hospital and Tumor Research Institute, University of Texas Cancer Center, Texas Medical Center, Houston, Texas*

KATHARINE J. GIBSON (1), *Department of Biochemistry, University of Illinois, Urbana, Illinois*

CHARLES L. GINTHER (48), *Department of Bacteriology, University of California, Davis, California*

MORTON D. GLANTZ (69, 71), *Department of Chemistry, Brooklyn College of the City University of New York, Brooklyn, New York*

DONALD P. GROTH (78), *Department of Biochemistry, Emory University, Atlanta, Georgia*

STANDISH C. HARTMAN (22), *Department of Chemistry, Boston University, Boston, Massachusetts*

RUDY H. HASCHEMEYER (3), *Department of Biochemistry, Cornell University Medical College, New York, New York*

DOLPH HATFIELD (20), *Laboratory of Chemical Carcinogenesis, National Cancer Institute, National Institutes of Health, Bethesda, Maryland*

JOY HOCHSTADT (75, 76), *Department of Microbiology, New York Medical College, Basic Science Building, Valhalla, New York*

PATRICIA A. HOFFEE (58, 70), *Department of Microbiology, University of Pittsburgh, School of Medicine, Pittsburgh, Pennsylvania*

BOR SHYUE HONG (25), *Boston Biomedical Research Institute, Boston, Massachusetts*

SARAH HOPPER (31), *Department of Pharmacology-Physiology, School of Dental Medicine, University of Pittsburgh, Pittsburgh, Pennsylvania*

JOHN L. INGRAHAM (4, 48), *Bacteriology Department, University of California, Davis, California*

PIER LUIGI IPATA (51), *Laboratory of Biochemistry, Faculty of Sciences, University of Pisa, Pisa, Italy*

DAVID H. IVES (43, 44), *Department of Biochemistry, The Ohio State University, Columbus, Ohio*

MARY ELLEN JONES (6, 7, 21), *Department of Biochemistry, University of Southern California, School of Medicine, Los Angeles, California*

DORIS KARIBIAN (8), *Laboratoire Physiologie Microbienne, U.E.R. Luminy, Marseille, France*

PRABHAKAR RAO KAVIPURAPU (21), *Senior Scientist, Jet Propulsion Laboratory, Pasadena, California*

WILLIAM N. KELLEY (77), *Department of Internal Medicine and Department of Biological Chemistry, University of Michigan Medical Center, Ann Arbor, Michigan*

JAMES G. KENIMER (78), *National Heart and Lung Institute, National Institutes of Health, Bethesda, Maryland*

MARC W. KIRSCHNER (5), *Department of Biochemical Sciences, Princeton University, Princeton, New Jersey*

K. KOBAYASHI (10, 11), *Department of Health Chemistry, Faculty of Pharmaceutical Sciences, Kyoto University, Kyoto, Japan*

SHU-HEI KOBAYASHI (38), *Department of Biochemistry, Kanazawa Medical University, Uchinada, Kahoku-gun Ishikawa, Japan*

DANIEL E. KOSHLAND, JR. (12), *Department of Biochemistry, University of California, Berkeley, California*

HAZEL B. LEUNG (33), *Biochemistry Department, Temple University School of Medicine, Philadelphia, Pennsylvania*

ARTHUR S. LEWIS (69, 71), *Department of Biochemistry, Albert Einstein College of Medicine, Bronx, New York*

JEROME M. LEWIS (22), *Department of Chemistry, Massachusetts Institute of Technology, Cambridge, Massachusetts*

CEDRIC LONG (12), *Viral Oncology Pro-*

gram, Frederick Cancer Research Center, Frederick, Maryland

LEWIS N. LUKENS (24), Department of Biology, Wesleyan University, Middletown, Connecticut

RICHARD P. MCPARTLAND (13), Department of Medicine, Medical College of Ohio, Toledo, Ohio

GIULIO MAGNI (37), Laboratory of Applied Biochemistry, University of Camerino, Camerino, Italy

V. G. MALATHI (53), Department of Medicine, New York University Medical Center, New York, New York

GLADYS F. MALEY (54), Division of Laboratories and Research, New York State Department of Health, Albany, New York

PATRICIA MANESS (41), Department of Biochemistry, The University of Texas System Cancer Center, M. D. Anderson Hospital and Tumor Institute, Houston, Texas

REYAD MAY (70), Department of Microbiology, University of Pittsburgh, School of Medicine, Pittsburgh, Pennsylvania

ALTON MEISTER (3), Department of Biochemistry, Cornell University Medical College, New York, New York

R. P. MIECH (64), Division of Biological and Medical Science, Brown University, Providence, Rhode Island

R. W. MILLER (9, 24), Chemistry and Biology Research Institute, Research Branch, Canada Agriculture, Ottawa, Ontario, Canada

GREGORY MILMAN (73, 74), Department of Biochemistry, School of Hygiene and Public Health, The Johns Hopkins University, Baltimore, Maryland

MASATAKA MORI (17), Department of Biochemistry, Chiba University, School of Medicine, Inohana, Chiba, Japan

ALLAN J. MORRIS (35), Department of Biochemistry, Michigan State University, East Lansing, Michigan

KATHERINE M. MUIRHEAD (27), Department of Medicine, University of Rochester School of Medicine, Rochester, New York

GENE R. NATHANS (66), Department of Biochemistry, The University of Chicago, Chicago, Illinois

JAN NEUHARD (55), Enzyme Division, University Institute of Biological Chemistry B, Copenhagen K, Denmark

DONALD P. NIERLICH (23), Department of Bacteriology and Molecular Biology Institute, University of California, Los Angeles, California

PER NYGAARD (68), Enzyme Division, University Institute of Biological Chemistry, Copenhagen K, Denmark

MAX P. OESCHGER (63), Department of Microbiology, Georgetown University School of Medicine and Dentistry, Washington, D. C.

SHIRO OHNOKI (25), University of St. Louis School of Medicine, St. Louis, Missouri

ANNE S. OLSEN (74), Laboratory of Biochemistry, National Cancer Institute, National Institutes of Health, Bethesda, Maryland

ANTONIO ORENGO (38, 41), The University of Texas System Cancer Center, M. D. Anderson Hospital and Tumor Institute, Houston, Texas

R. E. PARKS, JR. (49, 64, 67, 72, 79), Division of Biological and Medical Science, Brown University, Providence, Rhode Island

LAWRENCE M. PINKUS (3), Division of Gastroenterology, Nassau County Medical Center, East Meadow, New York

TAPAS K. PRADHAN (61), Department of Oncology, Howard University, Cancer Research Center, Washington, D. C.

LANSING M. PRESCOTT (6), Department of Biology, Augustana College, Sioux Falls, South Dakota

ALAN R. PRICE (36), Department of Biological Chemistry, The University of Michigan, Ann Arbor, Michigan

WILLIAM H. PRUSOFF (45), Department of Pharmacology, Yale University, New Haven, Connecticut

B. C. ROBERTSON (70), Department of Microbiology, University of Pittsburgh School of Medicine, Pittsburgh, Pennsylvania

BONNIE ROBISON (49), Division of Biological and Medical Science, Brown University, Providence, Rhode Island

DANIEL G. ROTH (2), Department of Medicine, The University of Chicago, Chicago, Illinois

IVAN K. ROTHMAN (53), Department of Medicine, New York University Medical Center, New York, New York

NAOTO SAKAMOTO (28), Office for Life Science Promotion, The Institute of Physical and Chemical Research, Honkomagome, Tokyo, Japan

C. RICHARD SAVAGE, JR. (13), Department of Biochemistry, Temple University School of Medicine, Philadelphia, Pennsylvania

H. K. SCHACHMAN (5), Department of Molecular Biology and The Virus Laboratory, Wendell M. Stanley Hall, University of California, Berkeley, California

VERN L. SCHRAMM (33), Biochemistry Department, Temple University School of Medicine, Philadelphia, Pennsylvania

MARIANNE SCHWARTZ (59), University Institute of Biological Chemistry B, Copenhagen K, Denmark

J. J. SCOCCA (57), Department of Biochemistry, The Johns Hopkins University, School of Hygiene and Public Health, Baltimore, Maryland

ROBERT SILBER (53), Department of Medicine, New York University Medical Center, New York, New York

ROBERT F. SILVA (20), Laboratory of Tumor Virus Genetics, National Cancer Institute, National Institutes of Health, Bethesda, Maryland

BRITT-MARIE SJÖBERG (30), Medical Nobel Institute, Department of Biochemistry I, Karolinska Institute, Stockholm, Sweden

THOMAS SPECTOR (29), Wellcome Research Laboratories, Burroughs Wellcome Company, Research Triangle Park, North Carolina

GEORGE R. STARK (18), Department of Biochemistry, Stanford University School of Medicine, Stanford, California

J. D. STOECKLER (72), Division of Biological and Medical Science, Brown University, Providence, Rhode Island

D. PARKER SUTTLE (18), Department of Biochemistry, Stanford University School of Medicine, Stanford, California

ROBERT L. SWITZER (1), Department of Biochemistry, University of Illinois, Urbana, Illinois

MASAMITI TATIBANA (17), Department of Biochemistry, Chiba University, School of Medicine, Inohana, Chiba, Japan

LARS THELANDER (30), Medical Nobel Institute, Department of Biochemistry I, Karolinska Institute, Stockholm, Sweden

K. TOMITA (10, 11), Department of Health Chemistry, Faculty of Pharmaceutical Sciences, Kyoto University, Kyoto, Japan

OSCAR TOUSTER (34), Department of Molecular Biology, Vanderbilt University, Nashville, Tennessee

THOMAS W. TRAUT (21), Department of Biochemistry, University of Southern California, School of Medicine, Los Angeles, California

PAUL P. TROTTA (3), Memorial Sloan Kettering Cancer Center, Department 5801, New York, New York

KENNETH K. TSUBOI (62), Department of Pediatrics, Stanford Medical Center, Stanford, California

K. UMEZU (11), Mitsubishi Chemical Industries, Ltd., Kawasaki, Japan

POUL VALENTIN-HANSEN (39), Enzyme Division, University Institute of Biological Chemistry B, Copenhagen, Denmark

SUE-MAY WANG (43), Department of Biochemistry, The Ohio State University, Columbus, Ohio

HERBERT WEINFELD (13), Department of Medicine C, Roswell Park Memorial Institute, Buffalo, New York

DAVID F. WENTWORTH (52), Department of Biochemistry, University of North Carolina, Chapel Hill, North Carolina

CHRISTINE WHITE (2), The Pritzker School of Medicine, The University of Chicago, Chicago, Illinois

JOHN M. WHITELEY (15), Department of Biochemistry, Scripps Clinic and Research Foundation, Keeney Park, La Jolla, California

JAMES R. WILD (19), *Genetics Section, Texas A & M University, College Station, Texas*

LARRY G. WILLIAMS (16), *Division of Biology, Kansas State University, Manhattan, Kansas*

RICHARD WOLFENDEN (52), *Department of Biochemistry, University of North Carolina, Chapel Hill, North Carolina*

DOW O. WOODWARD (26), *Department of Biological Sciences, Stanford University, Stanford, California*

E. W. YAMADA (56), *Department of Biochemistry, University of Manitoba, Winnipeg, Canada*

YING R. YANG (5), *Virus Laboratory, University of California, Berkeley, California*

A. YOSHIMOTO (10, 11), *Department of Biochemistry, Niigata College of Pharmacy, Niigata, Japan*

LEONA G. YOUNG (78), *Department of Biochemistry, Emory University, Atlanta, Georgia*

Preface

Over the last decade there has been a tremendous increase in the basic information concerning the enzymology of purine and pyrimidine metabolism and in the development of new procedures for the study of this field. This volume attempts to cover the enzymes involved in the biosynthetic, the degradative, and the salvage pathways of purine and pyrimidine nucleotides. Both the purification methods and the properties of pertinent enzymes are presented. When possible, enzymes from eukaryotic as well as prokaryotic sources are included.

We wish to thank the numerous authors for their valuable contributions to this volume and for their excellent cooperation. We would also like to thank Mrs. Barbara Baum, Mrs. Varian Hagglund, Mrs. Ruth Lightfoot, and Miss Betty Rooney for their valuable assistance in dealing with the correspondence involved with this volume, and the staff of Academic Press for their courtesy and efforts during its production.

PATRICIA A. HOFFEE
MARY ELLEN JONES

METHODS IN ENZYMOLOGY

EDITED BY

Sidney P. Colowick and Nathan O. Kaplan

VANDERBILT UNIVERSITY
SCHOOL OF MEDICINE
NASHVILLE, TENNESSEE

DEPARTMENT OF CHEMISTRY
UNIVERSITY OF CALIFORNIA
AT SAN DIEGO
LA JOLLA, CALIFORNIA

METHODS IN ENZYMOLOGY

EDITORS-IN-CHIEF

Sidney P. Colowick   Nathan O. Kaplan

VOLUME XXXIV. Affinity Techniques (Enzyme Purification: Part B)
Edited by WILLIAM B. JAKOBY AND MEIR WILCHEK

VOLUME XXXV. Lipids (Part B)
Edited by JOHN M. LOWENSTEIN

VOLUME XXXVI. Hormone Action (Part A: Steroid Hormones)
Edited by BERT W. O'MALLEY AND JOEL G. HARDMAN

VOLUME XXXVII. Hormone Action (Part B: Peptide Hormones)
Edited by BERT W. O'MALLEY AND JOEL G. HARDMAN

VOLUME XXXVIII. Hormone Action (Part C: Cyclic Nucleotides)
Edited by JOEL G. HARDMAN AND BERT W. O'MALLEY

VOLUME XXXIX. Hormone Action (Part D: Isolated Cells, Tissues, and Organ Systems)
Edited by JOEL G. HARDMAN AND BERT W. O'MALLEY

VOLUME XL. Hormone Action (Part E: Nuclear Structure and Function)
Edited by BERT W. O'MALLEY AND JOEL G. HARDMAN

VOLUME XLI. Carbohydrate Metabolism (Part B)
Edited by W. A. WOOD

VOLUME XLII. Carbohydrate Metabolism (Part C)
Edited by W. A. WOOD

VOLUME XLIII. Antibiotics
Edited by JOHN H. HASH

VOLUME XLIV. Immobilized Enzymes
Edited by KLAUS MOSBACH

VOLUME XLV. Proteolytic Enzymes (Part B)
Edited by LASZLO LORAND

VOLUME XLVI . Affinity Labeling
Edited by WILLIAM B. JAKOBY AND MEIR WILCHEK

VOLUME XLVII. Enzyme Structure (Part E)
Edited by C. H. W. HIRS AND SERGE N. TIMASHEFF

Methods in Enzymology

Volume LI
PURINE AND PYRIMIDINE
NUCLEOTIDE METABOLISM

Section I

Activation of Ribose Phosphate

[1] Phosphoribosylpyrophosphate Synthetase (Ribose-5-phosphate Pyrophosphokinase) from *Salmonella typhimurium*

By ROBERT L. SWITZER and KATHARINE J. GIBSON

ATP + D-ribose-5-phosphate → 5-phosphoribosyl-α-1-pyrophosphate + AMP

Assay Method[1,2]

Principle. Phosphoribosylpyrophosphate (PRPP) synthetase can be conveniently assayed in most tissues by measuring transfer of radioactivity from $[\gamma\text{-}^{32}\text{P}]$ATP to PRPP (and the products of acid hydrolysis of PRPP) after removal of unreacted ATP by charcoal adsorption.[3] A small amount of unadsorbed radioactivity is corrected for by control tubes from which enzyme is omitted. The procedure can be applied to crude cell extracts containing "ATPase" activities if the apparent activity of a control lacking ribose-5-phosphate is subtracted from each experimental measurement. Extracts containing ribose-5-phosphate kinase,[4] ribulose-5-phosphate kinase,[5] or ribokinase[6] activities may give erroneous results and should be assayed by one of the alternative methods listed.

Reagents

Triethanolamine, 0.1 *M*, potassium phosphate, 0.1 *M*, EDTA, 0.75 m*M*, buffer, pH 8.0 at 37°: mix 50 ml 1.0 *M* triethanolamine (free base), 30 ml 1.0 *M* K$_2$HPO$_4$, 20 ml 1.0 *M* KH$_2$PO$_4$, 3.75 ml 0.1 *M* NaEDTA, pH 7, and H$_2$O to a final volume of 500 ml (pH at 4° is 8.6)
Ribose-5-phosphate, 0.1 *M*, neutralized to pH 7.5 with NaOH
ATP 0.1 *M*, neutralized to pH 7.5 with NaOH
MgCl$_2$, 0.1 *M*
$[\gamma\text{-}^{32}\text{P}]$ATP,[7] 0.1–5 mCi/μmole
NaF, 0.5 *M*

[1] R. L. Switzer, *J. Biol. Chem.* **244**, 2854 (1969).
[2] R. L. Switzer, *J. Biol. Chem.* **246**, 2447 (1971).
[3] A. Kornberg, I. Lieberman, and E. S. Simms, *J. Biol. Chem.* **215**, 389 (1955).
[4] J. B. Alpers, Vol. 42, p. 120.
[5] B. A. Hart and J. Gibson, Vol. 42, p. 115.
[6] B. W. Agranoff and R. O. Brady, *J. Biol. Chem.* **219**, 221 (1956); A. Ginsburg, *ibid.* **234**, 481 (1959).
[7] Available commercially or synthesized by the method of I. M. Glynn and J. B. Chappell, *Biochem. J.* **90**, 147 (1964).

$HClO_4$, 5% (v/v)

Norit suspension, 20% (v/v), acid washed according to the procedure of Zimmerman[8]

"Carrier": 5 mg/ml bovine serum albumin, dissolved in 50 mM sodium pyrophosphate, which has been neutralized to pH 7 with H_3PO_4

Procedure

The relation components are mixed to yield the following concentrations in 0.50 ml final volume: 50 mM triethanolamine–50 mM potassium phosphate–0.37 mM EDTA buffer, 5 mM ribose-5-phosphate, 2 mM ATP, 100,000–300,000 cpm [γ-^{32}P]ATP, 25 mM NaF (when added), and 5 or 10 mM MgCl$_2$ (added last to avoid formation of a precipitate). It is usually convenient to prepare a mixture of these components (except MgCl$_2$) of sufficient volume so that an aliquot of the mixture can be pipetted into a series of 12 × 75 mm tubes. Water is then added so that the volume of liquid in each tube after MgCl$_2$ and enzyme addition is 0.50 ml. Enzyme solutions should be diluted with standard assay buffer, containing 1 mg/ml bovine serum albumin. After bringing the reaction mixtures to 37°, the reaction is initiated by addition of enzyme and mixing. After incubation at 37° for a suitable time, usually 10 min, the reaction is terminated by addition of 0.5 ml cold 5% $HClO_4$. The tubes are kept on ice for at least 10 min, then 0.3 ml of a uniform suspension (magnetic stirrer) of 20% acid-washed Norit is added with vigorous mixing to each tube except the control tubes which will be used to determine total radioactivity; 0.3 ml H_2O is added to these latter tubes. The tubes are again kept on ice for at least 10 min, then 0.2 ml of "carrier" solution is added to each tube with mixing. The carrier serves to ensure complete removal of PRPP and its acid breakdown products from the charcoal and to facilitate removal of finely divided charcoal during subsequent centrifugation, which is for 3–5 min in a clinical centrifuge. A sample (usually 300 μl) of the clear supernatant liquid from each tube is accurately pipetted into a scintillation vial. A convenient scintillant is 10 ml/vial of 0.4% diphenyloxazole in 7:8 (v/v) methanol:toluene.

The radioactivity in each vial is corrected for that found in an identical control from which ribose-5-phosphate was omitted. After the first four steps of the purification procedure, ATPase activity is removed and blanks lacking ribose-5-phosphate become equal to those lacking enzyme, so that two or three blank tubes suffice for an entire series of assays. After these purification steps it is also no longer necessary to

---

[8] S. B. Zimmerman, Vol. 6, p. 258.

include NaF in the reaction mixture. Activity in μmoles PRPP formed per minute per milliliter under the above conditions, which defines a unit of activity, is calculated as follows:

$$\text{units/ml} = \frac{\text{corrected cpm}}{\text{total cpm}} \times \mu\text{moles ATP in assay} \times \frac{1}{\text{min of assay}} \times \text{dilution factor}$$

In determining the μmoles of ATP in the assay it may be necessary to include the μmoles of ATP contributed by the [γ-^{32}P]ATP, depending on its specific activity.

Comments and Alternative Assays

The assay procedure can be used with a variety of crude cell extracts, so long as NaF is included and appropriate blanks are subtracted.[9,10] When the effects of nucleotides or other charcoal-adsorbable compounds are to be investigated, it is necessary to include equivalent concentrations of these in the blank mixtures.[11] EDTA is included in the assay because PRPP synthetase is inhibited by low levels of certain divalent cations, even in the presence of excess $MgCl_2$.[11] This assay can be adapted to any other kinase reaction in which the phosphorylated product does not adsorb to charcoal.

In those systems in which excess ATPase or other pentose transforming plus kinase activities prevent use of the ^{32}P transfer assay, coupling of PRPP synthetase to a phosphoribosyltransferase reaction provides a suitable assay. Reaction of accumulated PRPP with radioactive adenine using adenine phosphoribosyltransferase is most commonly used.[12,13] The radioactive nucleotide is separated from adenine by electrophoresis[12] or chromatography.[13] A similar assay using hypoxanthine-guanine phosphoribosyltransferase has also proven useful.[14,15] The reaction of PRPP with orotate to form orotidylic acid and uridylic acid forms the basis of a spectrophotometric assay for PRPP synthetase,[3,16] but this assay is not sufficiently sensitive for kinetic work. An alternative form of this assay in which release of $^{14}CO_2$ from [7-^{14}C]orotate is measured has also been described.[3,17]

[9] M. N. White, J. Olszowy, and R. L. Switzer, *J. Bacteriol.* **108**, 122 (1972).
[10] M. G. Johnson, S. Rosenzweig, R. L. Switzer, M. A. Becker, and J. E. Seegmiller, *Biochem. Med.* **10**, 266 (1974).
[11] R. L. Switzer and D. C. Sogin, *J. Biol. Chem.* **248**, 1063 (1973).
[12] I. H. Fox and W. N. Kelley, *J. Biol. Chem.* **246**, 5739 (1971).
[13] M. A. Becker, L. J. Meyer, and J. E. Seegmiller, *Am. J. Med.* **55**, 232 (1973).
[14] D. G. Roth, E. Shelton, and T. F. Deuel, *J. Biol. Chem.* **249**, 291, 297 (1974).
[15] A. Hershko, A. Razin, and J. Mager, *Biochim. Biophys. Acta* **184**, 64 (1969).
[16] J. G. Flaks, Vol. 6, p. 473.
[17] G. H. Reem, *Science* **190**, 1098 (1975).

With partially purified PRPP synthetase the formation of radioactive AMP from [^{14}C]- or [^{3}H]ATP can be followed directly by chromatographic or electrophoretic separation.[12] For studies of the steady-state kinetics of PRPP synthetase the reverse reaction can be assayed spectrophotometrically by coupling PRPP-dependent ATP formation from AMP to the combined action of hexokinase and glucose-6-phosphate dehydrogenase.[2,11] These reactions have also been used to form an activity stain for PRPP synthetase.[10]

Protein Determination

Protein concentrations are determined by the colorimetric procedure of Lowry *et al.*[18] with crystalline bovine serum albumin as a standard. With the purified enzyme a color value at 660 nm of 7.3 cm^{-1}/mg of enzyme was determined, based on dry weight of the enzyme.[19]

Purification Procedure[1,19,20]

Growth of Cells

Salmonella typhimurium LT-2 or strain Su 422, a methionine auxotroph, is cultured at 37° on the E medium of Vogel and Bonner[21] containing 0.5% glucose and, in the case of strain Su 422, 50 mg methionine per liter. The auxotrophic strain is used to provide a convenient check for contaminating bacteria, but it happens to contain PRPP synthetase at about twice the specific activity of the wild type. Large quantities of cells are grown in a 200-liter fermenter, harvested at the end of exponential growth, and can be stored frozen at −20°C indefinitely. The yield of cells is about 1500 g wet cell paste per 200 liters culture fluid.

Cell Extraction. Five hundred grams of frozen cell paste are thawed in 2500 ml cold 50 m*M* potassium phosphate, pH 7.5, and homogenized by exposure to 30-sec agitation periods in a Waring Blendor maintained at 0°. The blender is filled to the top to minimize foaming. The uniform suspension is passed through a Manton–Gaulin mill (precooled with ice water, followed by cold 50 m*M* potassium phosphate, pH 7.5, to 2–4°) at 8000 psi. The effluent should be maintained at 15–18° (not in excess of 22°) and collected in an ice-cooled flask. When the effluent has cooled to

[18] O. H. Lowry, N. J. Rosebrough, A. L. Farr, and R. J. Randall, *J. Biol. Chem.* **193**, 265 (1951).
[19] K. R. Schubert, R. L. Switzer, and E. Shelton, *J. Biol. Chem.* **250**, 7492 (1975).
[20] K. R. Schubert, Ph.D. Thesis, University of Illinois, Urbana (1975).
[21] H. J. Vogel and D. M. Bonner, *J. Biol. Chem.* **218**, 97 (1956).

5–8°, it is passed through the cold mill once more as above. The twice-treated solution is cooled to 5° in iced stainless-steel centrifuge cans and centrifuged at 23,000 g for 60 min at 0–4°. The cloudy yellow supernatant fluid is collected and the precipitate is discarded.

Streptomycin Sulfate–Heat Treatment. One-tenth volume of 10% (w/v) streptomycin sulfate (Sigma) in H_2O is added to the crude extract with stirring. The milky suspension is then heated to 54° within 5 min by suspension of 600–800 ml batches in a 6-liter flask in a boiling-water bath with continuous swirling of the liquid in the flask. When the temperature of the solution reaches 54°, the flask is transferred to a 55° bath and maintained at 55° for exactly 5 min. The solution is then quickly cooled to 5° by transfer to ice-cooled stainless-steel centrifuge cans. When all of the streptomycin-treated extract has been heated and cooled, the fluid is centrifuged at 23,000 g for 60 min at 4°. The clear yellow supernatant liquid is collected in an iced 4-liter beaker, and the gray precipitate is discarded.

First Ammonium Sulfate Precipitation. Solid $(NH_4)_2SO_4$ (Mann, enzyme grade) (209 g/liter) is added very slowly with stirring, and the slightly cloudy solution (35% saturated) is allowed to stand in ice for 30 min without stirring. The precipitate is collected by centrifuging in stainless-steel centrifuge cans at 23,000 g for 60 min at 4°, and the clear yellow supernatant fluid is discarded. The very small volume of somewhat sticky precipitate is dissolved in sufficient 50 mM potassium phosphate buffer, pH 7.5, containing 144 g $(NH_4)_2SO_4$ per liter (25% saturated) to yield a solution containing 1 to 2 mg protein per ml (usually about 125 ml). The resultant uniformly milky suspension is frozen in 10 to 30 ml samples in polypropylene tubes in a Dry Ice–acetone bath and stored at −20° overnight or longer. This freezing step usually results in an increase in total enzyme activity and improved yield of the subsequent purification steps.

First Acid Precipitation. The frozen solution from the previous step is quickly thawed, diluted to 1–2 mg protein per ml, if necessary, and adjusted to pH 4.6 by drop by drop addition of 1 N acetic acid at 0°. The solution is immediately centrifuged at 17,000 g for 10 min, the supernatant fluid is discarded, and the white precipitate is resuspended in sufficient 50 mM potassium phosphate, pH 7.5, to yield a solution of 1–2 mg protein per ml (usually about 100 ml). This solution is quick-frozen as in the previous step.

Second Ammonium Sulfate Fractionation. The solution from the previous step is thawed and centrifuged for 10 min at 17,000 g at 4° to remove any precipitate. One-half volume of 50 mM potassium phosphate, pH 7.5, saturated with $(NH_4)_2SO_4$ is added to the solution, which stands in ice for 10 min and is then centrifuged for 15 min at 17,000 g at 4°. The clear or slightly cloudy supernatant fluid is discarded. The precipitate usually consists of a white soft layer above a smaller volume of a more dense light or dark tan layer, which often contains black specks. Best purification requires that the upper white layer be physically separated from the darker layer with a small stainless-steel spatula. The white material is suspended in sufficient 50 mM potassium phosphate buffer, pH 7.5, containing 100 g $(NH_4)_2SO_4$ per liter (18% saturated) to yield a final protein concentration of 1–2 mg/ml (usually about 100 ml). The precipitate is thoroughly extracted with this solution by trituration with a glass rod at 4° for 15–20 min, yielding a uniform milky solution. The solution is centrifuged at 17,000 g for 20 min at 4°, the supernatant fluid is decanted, and the precipitate is dissolved in sufficient 50 mM potassium phosphate, pH 7.5, to yield a solution that is about 1 mg protein per ml (about 80 ml). The cloudy solution is quick-frozen as in previous steps.

Second Acid Precipitation. The solution from the previous step is quickly thawed, centrifuged for 10 min at 17,000 g at 4°, and the precipitate, if any, is discarded. The pH of the supernatant fluid is quickly brought to 4.6 with 1 N acetic acid, as in the first acid precipitation, and centrifuged at once for 10 min at 17,000 g at 4°. The white precipitate is redissolved in sufficient 50 mM potassium phosphate, pH 7.5, to yield 1–2 mg protein per ml (about 50 ml). This slightly cloudy solution is quick-frozen and stored as above.

Further Ammonium Sulfate Fractionation. The purity of the enzyme preparation is routinely assessed at this stage by electrophoresis in polyacrylamide gels containing SDS.[22] If the preparations are not better than 95% pure, the second ammonium sulfate precipitation procedure is repeated one or more times until no further improvement in purity can be obtained. This process often leads to substantial losses in total activity and increases in specific activity. Although specific activity is not a very reliable indicator of purity because of variability in the amount of activation that occurs upon freezing, highly purified preparations usually have specific activities of 100 to 150 units per milligram protein.

Affinity Chromatography.[23] We have recently developed a useful final step to be used after or instead of repeated ammonium sulfate fractionation for purifying PRPP synthetase, namely affinity chromatog-

---

[22] K. Weber, J. R. Pringle, and M. Osborn, Vol. 26, p. 3.
[23] K. J. Gibson, M. F. Roberts, and R. L. Switzer, unpublished experiments.

raphy either on a volume of Sepharose-bound N^6-(6-aminohexyl)-ATP (AG-ATP, type II, P. L. Laboratories), which is mixed with 3 volumes of Sepharose 4B, or on Sepharose-bound Blue dextran. The enzyme is adsorbed at 4° to columns of either adsorbent in standard triethanolamine-phosphate assay buffer plus 2 mM MgCl$_2$ at a flow rate of about two column volumes per hour. The column is washed with about five column volumes of 50 mM potassium phosphate, pH 7.5, to remove triethanolamine, which interferes with colorimetric protein determination. The enzyme is eluted with 50 mM potassium phosphate, pH 7.5, containing 25 mM each of MgCl$_2$ and ATP. Recoveries of enzyme activity are typically about 90%. The degree of purification depends on the specific activity of the starting material; the procedure is particularly useful in "cleaning up" preparations that have low specific activity or substantial impurities after ammonium sulfate fractionation. In all cases so far examined affinity chromatography has yielded enzyme with specific activity of 90 units/mg or greater that appears to be >95% pure on SDS polyacrylamide gel electrophoresis. Traces of faster migrating impurities can be detected on heavily loaded gels. The capacity of Sepharose-bound Blue dextran is lower than that of Sepharose-bound ATP, but both resins yield similar recoveries and degrees of purification.

Third Acid Precipitation.[23] Another recently developed precedure has been found to remove the major contaminant remaining after repeated ammonium sulfate fractionation gives no further purification. This procedure has been used in place of affinity chromatography and yields enzyme of equal purity, as judged by SDS-gel electrophoresis. Enzyme from the preceding ammonium sulfate fractionation is dissolved in 50 mM potassium phosphate at 1.5 to 2 mg protein/ml, and the pH is lowered to 5.8 with 1 N acetic acid. The solution is centrifuged at once for 10 min at 17,000 g at 4°; the supernatant liquid is decanted and immediately adjusted to pH 7.5 with 1 N NaOH. This supernatant fraction contains >85% of the enzyme activity, while the major contaminant is quantitatively removed.

A summary of the purification procedure is shown in the table.

Properties[1,2,11,19,20]

Stability and Storage

Purified PRPP synthetase is quite stable when stored in 50 mM potassium phosphate, pH 7.5, at −20° after rapid freezing in Dry Ice–acetone. Repeated freezing and thawing slowly denature the enzyme and should be avoided. The enzyme is generally stable to manipulations at 4–20° in the same buffer for 24 to 48 hr. It is irreversibly denatured by

PURIFICATION OF PHOSPHORIBOSYLPYROPHOSPHATE SYNTHETASE

Fraction	Volume (ml)	Protein (mg/ml)	Total activity (μmoles product/min)	Specific activity (μmoles product/min mg^{-1})	Recovery (%)
Crude extract (from 500 g cell paste)	2560	20	7680	0.15	100
Streptomycin–heat	2680	10	8840	0.33	115
1st (NH$_4$)$_2$SO$_4$ fractionation	152	3.8	8120	14	106
1st acid precipitation	102	3.2	7640	23	99
2nd (NH$_4$)$_2$SO$_4$ fractionation	70	1.3	6980	77	91
2nd acid precipitation	39	2.0	7640	98	99
3rd (NH$_4$)$_2$SO$_4$ fractionation	30	1.8	5400	103	70
Affinity adsorption on ATP-Sepharose[a]	—	—	—	130	—

[a] This final step was performed on only a portion of the material from the preceding step. Similar results were obtained with Blue dextran Sepharose.

reduction of the inorganic phosphate concentration to less than 5 mM and requires 25 mM P$_i$ or higher for complete stability; Mg^{2+} ions and ATP stabilize the enzyme, but will not replace P$_i$.

ATP binding studies with the method of Hummel and Dreyer[24] have shown that the purified enzyme contains an impurity, probably a denatured form of PRPP synthetase, that binds ATP and may interfere with other physical studies.[20]

Substrates, Activators, Inhibitors

Salmonella PRPP synthetase is specific for ATP; GTP, ITP, CTP, and UTP have less than 3% of the activity of ATP. The K_m for ATP under standard assay conditions is 46 μM. The α,β-methylene analogue of ATP is a linear competitive inhibitor with respect to ATP with a K_i of 30 μM.[20] The specificity of the enzyme for ribose-5-phosphate is also very high; no other sugar phosphate substrates are known. The analogue DL-1,4-anhydroribitol-5-phosphate is a competitive inhibitor with respect to ribose-5-phosphate (K_i = 10 mM), but other sugar phosphates do not inhibit.[25] The K_m for ribose-5-phosphate is 0.16 mM. In the presence of ADP, ribose-5-phosphate displays substrate inhibition, but not in its absence. The precise mechanism of this inhibition is not

[24] J. P. Hummel and W. J. Dreyer, *Biochim. Biophys. Acta* **63**, 530 (1962).
[25] R. L. Switzer and P. D. Simcox, *J. Biol. Chem.* **249**, 5304 (1974).

known, but kinetic evidence suggests that it is the consequence of binding of ADP at an allosteric site.[11]

The Michaelis constants for AMP and PRPP in the reverse reaction are 0.32 mM and 0.29 mM, respectively. The maximal velocity of the reverse reaction is less than 10% of the forward reaction. The equilibrium constant for the reaction in the direction of PRPP synthetase has been estimated to be 28 at pH 7.5 at 37°.

PRPP synthetase requires two activators: P_i and a divalent cation, Mg^{2+} or Mn^{2+}. The requirement for P_i is quite specific and kinetically complex; 25 to 50 mM P_i is required for optimal PRPP synthesis, but the optimum for the reverse reaction is 5 mM. It is likely that P_i binds at a specific activator site and, at higher concentrations, at the substrate sites as well. The requirement for Mg^{2+} or Mn^{2+} is 2-fold: to form the MgATP complex, which is the true substrate, and as the free cation. The K_a for Mg^{2+} from kinetic experiments is 0.6 mM.

The pH optimum of the PRPP synthesis reaction is between 8.0 and 8.6.

A number of ribonucleotides are inhibitors of PRPP synthetase. Purine nucleoside di- and triphosphates are most effective. All nucleotides, except ADP, are competitive with respect to ATP. ADP, which is a much more effective inhibitor than any other, displays very complex kinetics of inhibition. A detailed analysis of nucleotide inhibition is published elsewhere.[11]

Ca^{2+} ions are also effective inhibitors of PRPP synthetase, even in the presence of excess Mg^{2+} ions. The enzyme is not completely inhibited, but rather has about 40% of normal activity at maximal inhibition. Inhibition is half maximal at 50 μM Ca^{2+} (10 mM Mg^{2+}, pH 8). Ca^{2+} also desensitizes PRPP synthetase to ADP inhibition and alters the substrate saturation curve for ribose-5-phosphate. Other inhibitory cations are Co^{2+}, Cu^{2+}, Cd^{2+}, and Ni^{2+}. Ba^{2+}, Fe^{2+}, Mn^{2+}, and Zn^{2+} do not inhibit at concentrations below 1 mM.

Molecular Weight, States of Aggregation[19]

The molecular weight of the subunit of *Salmonella* PRPP synthetase is 31,000 ± 3000. Native enzyme exists in several states of aggregation. Under assay conditions the predominant form has a molecular weight of about 160,000. Physical and electron microscopic studies indicate that this fundamental unit is an asymmetric assembly of five identical subunits. At 4°, pH 7.5, and higher protein concentrations, a form that appears to be a dimer of the fundamental unit predominates. Aggregates of higher molecular weight also occur, which accounts for the poor solubility of the enzyme and turbidity of concentrated solutions.

[2] Ribosephosphate Pyrophosphokinase (Rat Liver)[1]

By Daniel G. Roth, Christine White, and Thomas F. Deuel

$$\text{Ribose-5-phosphate} + \text{ATP} \xrightarrow[\text{P}_i]{\text{Mg}^{2+}} \text{5-phosphoribosylpyrophosphate} + \text{AMP}$$

Rat liver ribosephosphate pyrophosphokinase (phosphoribosylpyrophosphate synthetase, EC 2.7.6.1) catalyzes an unusual pyrophosphoryl transfer reaction to form phosphoribosylpyrophosphate (PRPP), an important precursor of nucleotides for both the *de novo* and salvage pathways of purine synthesis, as well as for the synthesis of pyrimidine and pyridine nucleotides.

Assay Method A

Principle. For kinetic studies of the purified enzyme, the charcoal binding assay of Switzer is used.[2] This method measures the transfer of radioactivity from $[\gamma\text{-}^{32}\text{P}]$ATP to the product, PRPP. The labeled product is hydrolyzed in acid to inorganic phosphate. Unreacted ATP is removed by adsorption to charcoal and centrifugation of the adsorbed complex. This assay gives highly satisfactory results in kinetic studies with purified enzyme. It is not suitable for use in crude extracts.

Reagents
Potassium phosphate buffer, 1 *M*, pH 8.0
Magnesium chloride, 1 *M*
EDTA, disodium, 10 m*M*
Ribose-5-phosphate, sodium, 50 m*M*
ATP, sodium, 100 m*M*, pH 7.0
Bovine serum albumin, 1 mg/ml
$[\gamma\text{-}^{32}\text{P}]$ATP, prepared from ^{32}P-phosphoric acid by the method of Glynn and Chappell[3]
Acid-washed charcoal, 30% (v/v), prepared by washing 160 g charcoal (Norit-A) in 1 liter of water, followed by washing in 2

---

[1] This work was supported by contract EY-26-C-02-0069 awarded to the Franklin McLean Memorial Research Institute (operated by the University of Chicago for the United States Energy Research and Development Administration) and by Grant CA-13980 from the National Institutes of Health. Thomas F. Deuel holds Faculty Research Award No. 133 from the American Cancer Society.

[2] R. L. Switzer, *J. Biol. Chem.* **244**, 2854 (1969).
[3] I. M. Glynn and J. B. Chappell, *Biochem. J.* **90**, 147 (1964).

liters of 1 N HCl, by placing the mixture under gentle vacuum for 1–2 hr, and by washing with water until the pH is above 3.0 and adjusting the settled volume to 30% (v/v)

Perchloric acid, 5% (v/v)

Carrier solution, 5 mg/ml bovine albumin in 50 mM sodium pyrophosphate, adjusted to pH 7.0 with phosphoric acid

Procedure. The samples to be assayed contain: 50 μl potassium phosphate buffer, 5 μl magnesium chloride, 25 μl EDTA, 20 μl ribose-5-P, 25 μl albumin, 8 μl ATP, [γ-^{32}P]ATP sufficient to provide a final specific radioactivity of 100–200 cpm/nmole ATP, enzyme, and water to a final volume of 0.5 ml. Samples are incubated at 37° for 30 min. The reaction is stopped by the addition of 0.5 ml 5% perchloric acid. The samples are chilled on ice for 10 min, and 0.2 ml of the charcoal suspension is added and mixed vigorously. After 10 min, 0.2 ml carrier solution is added and mixed by agitation. The samples are centrifuged at 2000 g for 10 min, and 0.6 ml of the clear supernatant is removed for counting. After addition of 5 ml scintillation fluid (Aquasol, New England Nuclear), samples are counted in a liquid scintillation counter.

Assay Method B

Principle. The assay is used to monitor enzyme activity with crude or partially purified extracts. It measures the reaction of PRPP with hypoxanthine in the presence of hypoxanthine-guanine phosphoribosyltransferase (H-G PRTase, EC 2.4.2.8) to form inosine monophosphate.

Hypoxanthine + phosphoribosylpyrophosphate → inosine monophosphate + pyrophosphate

Unreacted ^{14}C-hypoxanthine is removed by filtration through DEAE paper discs; IMP, the product of the reaction, is retained by the DEAE disc.

Reagents

Potassium phosphate buffer, 0.7 M, pH 7.6

Magnesium sulfate, 70 mM

EDTA, 10 mM, disodium

Ribose-5-phosphate, sodium, 50 mM

ATP, 50 mM, disodium

[8-^{14}C]hypoxanthine, made by diluting [8-^{14}C]hypoxanthine (60 mCi/mmole, Schwarz-Mann) 1:20 with hypoxanthine, 4 mM, and filtering through DEAE paper (Whatman DE-81)

H-G PRTase, prepared from human erythrocytes as Fraction IV by

the method of Krenitsky *et al.*,[4] and concentrated to give
approximately 1 μmole IMP formed/min/ml
Inosine monophosphate, 0.1 M

Procedure. The incubation mixture contains: 2 μl potassium phosphate buffer, 5 μl magnesium sulfate, 1 μl EDTA, 2 μl ribose-5-phos-
phate, 2 μl ATP, 10 μl ^{14}C-hypoxanthine, 2 μl erythrocyte H-G PRTase,
the desired amount of extract to be assayed, and water to a final volume
of 70 μl. A blank sample is included which lacks ribose-5-phosphate.
Samples are incubated at 37° for 30 min, and the reaction is stopped by
placing the samples in a boiling water bath for 30 sec. After cooling,
10 μl of the 0.1 M IMP is added as carrier. Aliquots of 50 μl are spotted
on 2.4-cm DEAE paper discs which have been wetted with 4 mM
hypoxanthine. The discs are washed by pouring 50 ml water through the
disc in 10-ml aliquots under gentle vacuum, followed by 10 ml ethanol,
dried, and counted in 5 ml scintillation fluid.

Definition of Unit. For Method A, one unit is defined as the amount
of enzyme catalyzing the conversion of 1 μmole of ATP per minute at
37° into 5'-AMP and PRPP under the conditions specified. For Method
B, one unit is defined as the amount of enzyme which results in the
formation of 1 μmole of IMP per minute at 37° under the conditions
defined above. *Specific activity* is defined as units per milligram protein.

Purification

All purification steps are carried out at 0–4° unless otherwise
specified. The acid, streptomycin, and ammonium sulfate precipitates
are stirred for 20 min prior to centrifugation. Centrifugations are
performed at 10,000 g for 10 min.

Acetone Powder. Livers are removed from adult Sprague-Dawley
rats, and acetone powder is prepared by the method of Tabor.[5] Acetone
powder prepared in this manner is stable at −20° for up to at least 6
months. Eighty grams of rat liver acetone powder are suspended in 1.2
liters of Enzyme Buffer, which consists of 50 mM potassium phosphate,
pH 7.6; 2.5 mM 2-mercaptoethanol; 0.1 mM EDTA; 6 mM magnesium
chloride; and 0.3 mM ATP. The sample is stirred for 2 hr and centrifuged to remove insoluble material.

Acid Precipitation. Over a 20-min period the sample is brought to pH
5.8 by the slow addition of 1.2 M acetic acid. After stirring for 5 min the
precipitate is removed by centrifugation and discarded. The supernatant

[4] T. A. Krenitsky, R. Papaioannou, and G. B. Elion, *J. Biol. Chem.* **244**, 1263 (1969).
[5] H. Tabor, this series, Vol. 1, p. 609.

is brought to pH 5.3, stirred for 5 min and centrifuged as above. The precipitate is resuspended in 200 ml Enzyme Buffer, using a tissue grinder to disperse the precipitate into very small particles. The sample is stirred overnight, centrifuged, and the supernatant retained.

Streptomycin Precipitation. Streptomycin sulfate, 10% (w/v), was prepared freshly for each purification in water and chilled on ice. The enzyme sample is brought to pH 6.0 with 1.2 M acetic acid, and 0.2 volume of the streptomycin solution is quickly added. The sample is stirred for 15 min, centrifuged, and the supernatant discarded. The precipitate is dissolved in 150 ml Enzyme Buffer and stirred for a minimum of 4 hr. The use of smaller volumes of buffer is accompanied by incomplete recovery of activity.

Heating to 55°. The enzyme must be concentrated prior to the heat step. This may be done by ultrafiltration, or more quickly by ammonium sulfate precipitation. Solid ammonium sulfate is added to 25% saturation. The sample is stirred, centrifuged, and the precipitate discarded. The supernatant is brought to 80% saturation, stirred, and centrifuged. The precipitate dissolves readily in 25 ml Enzyme Buffer. By use of a glass centrifuge tube, the sample is placed in a 55° water bath and stirred continuously for 10 min. After chilling on ice, the denatured protein is removed by centrifugation.

Agarose Column Chromatography. A column of agarose A 1.5 m (BioRad) (0.9 × 60 cm) is poured and washed with three bed volumes of Enzyme Buffer. The enzyme solution was concentrated to 5–10 ml in an ultrafiltration cell (Amicon) using a PM 30 membrane. The enzyme is applied to the column and eluted with Enzyme Buffer, collecting 4-ml fractions. Eluted protein is detected spectrophotometrically by monitoring the optical density at 280 nm. Selected fractions are assayed for activity, which begins to appear at the void volume of the column. The active fractions are pooled and concentrated by ultrafiltration.

PURIFICATION OF RAT LIVER RIBOSEPHOSPHATE PYROPHOSPHOKINASE[a]

Fraction	Volume (ml)	Total protein (ml)	Total activity (units)	Yield (%)	Specific activity (units/mg)
Acetone powder	1185	23,450	14,100	100	0.0006
Acid precipitate	198	1,150	8,500	60	0.0074
Streptomycin precipitate	151	198	9,640	68	0.0487
55° heat	24	22	4,450	32	0.202
Agarose chromatography	7.4	0.9	947	6.7	1.05

[a] Adapted from *J. Biol. Chem.* **249,** 292 (1974), with permission of the publishers. See Ref. 6.

The results of a typical purification are presented in the table. The procedure has been used on numerous occasions and has proven to be highly reproducible. Purification of 1500- to 3000-fold are regularly achieved, depending on the specific activity of the acetone powder fraction.

Properties

Stability. The purified enzyme retains about 50% of the original activity after 2 months' storage in Enzyme Buffer at 4°. More rapid loss of activity occurs during storage at −20°. At low protein concentrations the enzyme is inactivated during the assay, resulting in a nonlinear relationship between the reaction velocity and the enzyme concentration.[6] High enzyme concentrations, albumin (50 μg/ml), EDTA (1 mM), or dithiothreitol (1 mM) stabilize the enzyme during assay. The instability of the enzyme has been found in all the enzyme preparations tested and is felt to be due to a contaminant reacting with a labile sulfhydryl group on the enzyme. Careful attention must be given to the instability of the enzyme at low enzyme concentrations during assay in inhibition studies. The results of inhibition studies obtained when the dilute enzyme is not stabilized are quite different than those obtained when albumin and EDTA are present.

Requirements. Inorganic phosphate is required for activity. Magnesium ion is required to provide Mg-ATP^{-2}, the preferred substrate for the enzyme. In addition, free Mg^{2+} stimulates activity, while free ATP^{-4} is inhibitory. The pH optimum for the purified enzyme is 8.0–8.8.

Molecular Weight. The purified enzyme is in an aggregated state and appears in electron micrographs as long, linear stacks of individual enzyme molecules.[7] The aggregated enzyme is excluded from Agarose 1.5 m during gel filtration experiments, but in 1 M NaCl it is reversibly disaggregated to a lower-molecular-weight form. The subunit molecular weight estimated from polyacrylamide gel electrophoresis in sodium dodecyl sulfate and mercaptoethanol is approximately 40,000.

Specificity. Strict specificity is demonstrated for the pyrophosphoryl acceptor, ribose-5-P. No other related compounds were effective acceptors under the conditions of study. Nucleoside triphosphates other than ATP and dATP were minimally effective or without effect. Arsenate and sulfate can partially substitute for the requirement for phosphate.

[6] D. G. Roth and T. F. Deuel, *J. Biol. Chem.* **249,** 297 (1974).
[7] D. G. Roth, E. Shelton, and T. F. Deuel, *J. Biol. Chem.* **249,** 291 (1974).

Kinetic Properties. The apparent K_m for ribose-5-P is 0.29 mM, and substrate inhibition occurs at ribose-5-P concentration above 1.5 mM. Free Mg^{2+} ions convert the Mg-ATP saturation curve from a sigmoidal function to a hyperbolic function, resulting in markedly increased activity at low Mg-ATP concentrations. In the presence of optimal magnesium concentrations, the apparent K_m for Mg-ATP is 0.22 mM.

Inhibitors. As mentioned above, the results of inhibitor studies depend on the conditions of the assay. When albumin and EDTA are not present to stabilize the dilute enzyme, apparent inhibition of activity is seen with a variety of dissimilar compounds which do not inhibit the stabilized enzyme. Using the full assay system, inhibition occurs with ADP and dADP. Inhibition by both compounds is competitive with respect to ATP. The apparent K_i for ADP is 0.15 mM. Inhibition by AMP is noncompetitive with respect to both substrates. A wide variety of purine pyrimidine bases, nucleosides, and nucleotides result in no significant inhibition of activity in the standard assay.

Section II
De Novo Pyrimidine Biosynthesis

A. Single Enzymes
Articles 3 through 15

B. Enzyme Complexes
Articles 16 through 21

[3] Glutamine-dependent Carbamyl-phosphate Synthetase (*Escherichia coli*); Preparation of Subunits

By PAUL P. TROTTA, MICHAEL E. BURT, LAWRENCE M. PINKUS, LEONARD F. ESTIS, RUDY H. HASCHEMEYER, and ALTON MEISTER

Glutamine-dependent carbamyl-phosphate synthetase is one of thirteen presently known enzymes (glutamine amidotransferases) that catalyze the utilization of the amide nitrogen atom of glutamine in various biosynthetic reactions.[1,2] Glutamine-dependent carbamyl-phosphate synthetase catalyzes the following reaction[3]:

$$\text{L-Glutamine} + \text{HCO}_3^- + 2\,\text{ATP} \qquad\qquad\qquad (1)$$

$$+ \text{H}_2\text{O} \xrightarrow{\text{Mg}^{2+},\text{K}^+} \text{NH}_2\text{CO}_2\text{PO}_3^{2-} + 2\,\text{ADP} + \text{P}_i + \text{L-glutamate}$$

The enzyme also catalyzes carbamyl-phosphate synthesis when glutamine is replaced by ammonia. The properties of this enzyme differ substantially from those of the carbamyl-phosphate synthetase found in liver mitochondria of ureotelic vertebrates which utilizes only ammonia and exhibits a requirement for *N*-acetyl-L-glutamate.[4-6] The latter enzyme catalyzes the formation of carbamyl phosphate used primarily for the synthesis of arginine in the urea cycle. Mammalian tissues also contain a glutamine-dependent carbamyl-phosphate synthetase, which is similar to that found in *E. coli*, and which catalyzes the synthesis of carbamyl phosphate utilized for the biosynthesis of pyrimidines.[7,8]

The glutamine-dependent carbamyl-phosphate synthetase of *E. coli* produces carbamyl phosphate which is used for both the synthesis of arginine and for pyrimidine biosynthesis.[9] The synthesis of this enzyme is repressed by arginine and by uracil. The activity of the enzyme is inhibited by UMP and to a lesser extent by UDP and UTP,[10] and it is

[1] A. Meister, *PAABS Rev.* **4**, 273–299 (1975).

[2] J. M. Buchanan, *Adv. Enzymol.* **39**, 91–184 (1973).

[3] P. M. Anderson and A. Meister, *Biochemistry* **4**, 2803–2809 (1965).

[4] P. P. Cohen, in "The Enzymes" (P. D. Boyer, H. Lardy, and K. Myrbäck, eds.), 2nd ed., Vol. 6, pp. 477–494. Academic Press, New York, 1962.

[5] M. E. Jones, *Annu. Rev. Biochem.* **34**, 381 (1965).

[6] M. E. Jones, in "Methods in Enzymology" (S. P. Colowick and N. O. Kaplan, eds.), Vol. 5, pp. 903–925. Academic Press, New York, 1962.

[7] S. E. Hager and M. E. Jones, *J. Biol. Chem.* **242**, 5674–5680 (1967).

[8] M. E. Jones, *Adv. Enzyme Regul.* **9**, 19–49 (1971).

[9] A. Pierard and J. M. Wiame, *Biochem. Biophys. Res. Commun.* **15**, 76–80 (1964).

[10] P. M. Anderson and A. Meister, *Biochemistry* **5**, 3164–3169 (1966).

METHODS IN ENZYMOLOGY, VOL. LI

activated by IMP,[10] ornithine,[11-15] and ammonia.[13-15] The enzyme catalyzes several partial reactions[15,16] which appear to reflect individual steps in the overall synthesis reaction; these include the bicarbonate-dependent hydrolysis of ATP:

$$ATP + H_2O \xrightarrow{\text{HCO}_3{}^-,\text{Mg}^{2+},\text{K}^+} ADP + P_i \qquad (2)$$

This reaction reflects the activation of bicarbonate by ATP. Pulse-labeling experiments with $H[^{14}C]O_3{}^-$ and $[^{32}P]ATP$ indicated intermediate formation of an activated form of CO_2.[3] A pulse-labeling experiment with $H[^{14}C]O_3{}^-$ and the frog liver ammonia-utilizing carbamyl-phosphate synthetase gave a similar result.[17] Earlier studies on the latter enzyme showed that ^{18}O is transferred from $HC[^{18}O_3]^-$ to inorganic phosphate in the overall reaction.[18] The activated CO_2 intermediate formed in the reaction catalyzed by the *E. coli* glutamine-dependent carbamyl-phosphate synthetase has been identified as carbonic-phosphoric anhydride ("carboxy phosphate") by isolation from the enzyme as the corresponding trimethyl derivative.[19]

The enzyme also catalyzes the synthesis of ATP from carbamyl phosphate and ADP:

$$ADP + \text{carbamyl phosphate} + H_2O \xrightarrow{\text{Mg}^{2+},\text{K}^+} NH_4{}^+ + HCO_3{}^- + ATP \qquad (3)$$

This reaction may reflect reversal of the phosphorylation of enzyme-bound carbamate.[16] In addition, the enzyme exhibits glutaminase activity[20]:

$$\text{L-Glutamine} + H_2O \rightarrow \text{L-glutamate} + NH_4{}^+ \qquad (4)$$

Studies on the hydrodynamic properties of the enzyme show that it can exist in two different monomer conformations which exhibit, respectively, sedimentation coefficients $(s_{20,w}^0)$ of 7.3 S and 8.7 S.[14] Both monomers can undergo rapid reversible self-association; allosteric acti-

[11] A. Pierard, *Science* **154**, 1572 (1966).
[12] P. M. Anderson and S. V. Marvin, *Biochem. Biophys. Res. Commun.* **32**, 928–934 (1968).
[13] P. P. Trotta, R. H. Haschemeyer, and A. Meister, *Fed. Proc., Fed. Am. Soc. Exp. Biol.* **30**, 31 (1971).
[14] P. P. Trotta, L. F. Estis, A. Meister, and R. H. Haschemeyer, *J. Biol. Chem.* **249**, 482–491 (1974).
[15] P. P. Trotta, L. M. Pinkus, R. H. Haschemeyer, and A. Meister, *J. Biol. Chem.* **249**, 492–499 (1974).
[16] P. M. Anderson and A. Meister, *Biochemistry* **5**, 3147–3163 (1966).
[17] V. Rubio and S. Grisolia, *Biochemistry* **16**, 321–329 (1977).
[18] M. E. Jones and L. Spector, *J. Biol. Chem.* **235**, 2897–2901 (1960).
[19] S. G. Powers and A. Meister, *Proc. Natl. Acad. Sci. U.S.A.* **73**, 3020–3024 (1976).
[20] V. P. Wellner and A. Meister, *J. Biol. Chem.* **250**, 3261–3266 (1975).

vators promote oligomer formation while allosteric inhibitors promote dissociation. Sedimentation equilibrium studies show that the average molecular weight of the monomer is 163,000 ± 4000.[14] The monomer can be dissociated into two nonidentical polypeptide chains by high concentrations of urea or sodium dodecylsulfate, and also by treatment with succinic or maleic anhydrides.[21,22] The molecular weights of the heavy and light subunits obtained in sodium dodecylsulfate are, respectively, about 130,000 and about 42,000. The subunits obtained by treatment of the enzyme with sodium dodecylsulfate and other agents mentioned above were found to be, as expected, enzymically inactive; furthermore, under these conditions dissociation is irreversible.

Considerable insight into the structure–function relationships in the enzyme was obtained after it was discovered that a relatively mild solvent perturbation can promote reversible dissociation of the enzyme into nonidentical subunits with retention of catalytic activities.[15,23] Thus, treatment of the enzyme with 1 M potassium thiocyanate in the presence of 0.1 M potassium phosphate (pH 7.6) leads to dissociation into light and heavy subunits; upon removal of the thiocyanate by dialysis, substantial reassociation can be demonstrated. Furthermore, as discussed below, the separated heavy and light subunits are catalytically active and the separated heavy subunit is responsive to allosteric effectors and undergoes an apparent self-association in potassium phosphate.

Dissociation of Carbamyl-Phosphate Synthetase Monomer into Catalytically Active Heavy and Light Subunits

The isolation of highly purified glutamine-dependent carbamyl-phosphate synthetase from *E. coli* was described in detail in an earlier volume of this series.[24] The heavy and light subunits of the enzyme may be separated by gel filtration as follows.[15] A column (60 × 1.6 cm) of Sephadex G-200 is equilibrated with 0.1 M potassium phosphate buffer (pH 7.6) containing 1 M potassium thiocyanate and 5 mM EDTA at 4°. It is important that the buffer be deoxygenated during chromatography by

[21] P. P. Trotta, L. M. Pinkus, V. P. Wellner, L. Estis, R. H. Haschemeyer, and A. Meister, *in* "The Enzymes of Glutamine Metabolism" (S. Prusiner and E. R. Stadtman, eds.), pp. 431–482. Academic Press, New York, 1973.

[22] S. L. Mathews and P. M. Anderson, *Biochemistry* **11,** 1176 (1972).

[23] P. P. Trotta, M. E. Burt, R. H. Haschemeyer, and A. Meister, *Proc. Natl. Acad. Sci. U.S.A.* **68,** 2599–2603 (1971).

[24] P. M. Anderson, V. P. Wellner, G. A. Rosenthal, and A. Meister, *in* "Methods in Enzymology" (H. Tabor and C. W. Tabor, eds.), Vol. 17, Part A, pp. 235–243. Academic Press, New York, 1970.

bubbling nitrogen through the buffer reservoir. The enzyme (1–2 mg in a volume of 1 ml) is applied to the column, and elution is carried out with the same buffer used for equilibration. Blue dextran 2000 and N-2,4-dinitrophenyl-methionine are used as convenient markers of the void volume and the total available column volume, respectively. A representative gel filtration preparation is given in Fig. 1. In the separation shown here, a relatively large amount of enzyme (9 mg) was added to the column. Under these conditions, a component appears (in the void volume) prior to elution of the heavy and light subunits; this is an aggregate of the heavy subunit. Such aggregate formation may be avoided by using initial protein loading concentrations of 1–2 mg/ml. The fractions containing the heavy subunit are treated with dithiothreitol to give a final concentration of 2 mM. These fractions are then pooled and dialyzed against 0.15 M potassium phosphate buffer (pH 7.8) containing

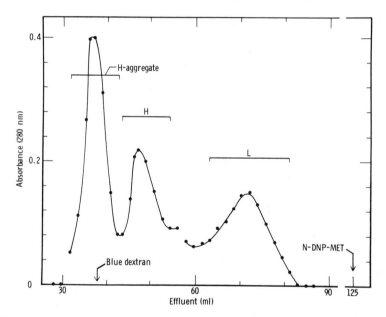

FIG. 1. Separation of the subunits of carbamyl-phosphate synthetase by gel filtration in potassium thiocyanate. A column of Sephadex G-200 (60 × 0.9 cm) was equilibrated with a solution containing 1.0 M potassium thiocyanate, 100 mM potassium phosphate (pH 7.6), and 5 mM EDTA at 4°. The enzyme [9 mg in 1 ml of 150 mM potassium phosphate buffer (pH 7.8) containing 0.5 mM EDTA] was applied; the column was eluted at a flow rate of about 4 ml/hr. Fractions of 0.9 ml were collected. The peaks containing the heavy (H) and light (L) subunits were separately pooled and concentrated. The purity of each subunit was established by sodium dodecyl sulfate–polyacrylamide gel electrophoresis. Blue dextran 2000 and N-2,4-dinitrophenyl methionine (N-DNP-MET) were employed as markers of the void volume and the total available column volume, respectively.

0.5 mM EDTA and 5 mM dithiothreitol to remove potassium thiocyanate. After dialysis, the solution is concentrated by ultrafiltration to yield a solution containing about 0.5 mg/ml of the heavy subunit. The fractions containing the light subunit are pooled and concentrated by ultrafiltration; the concentrated solution of the separated light subunit (about 0.5 mg/ml) is then dialyzed against 0.15 M potassium phosphate buffer (pH 7.8) containing 0.5 mM EDTA. The subunits obtained in this manner retain their initial enzymic activities for several days when stored at 0°.

The procedure described above differs in certain respects from that initially described[23] for the separation of the heavy and light subunits in that dithiothreitol was omitted from the elution buffer. Dithiothreitol inhibits the glutaminase activity of the light subunit under certain conditions.[25] The enzymic activities exhibited by the heavy subunit are inhibited in solutions exposed to air and for this reason deaerated buffers prepared by flushing with nitrogen are used.

Physical Properties of the Separated Heavy and Light Subunits and of the Isolated Enzyme

Sedimentation velocity studies of the enzyme in 1 M potassium thiocyanate show two boundaries which exhibit sedimentation coefficients of 2.9 S and 4.8 S, respectively.[14,23] Removal of the potassium thiocyanate by dialysis leads to reformation of a boundary that corresponds in sedimentation coefficient to that of the original enzyme monomer; under these conditions, most of the initial carbamyl-phosphate synthetase activity remains (see below). The subunits obtained by gel filtration on a column of Sephadex G-200 equilibrated with 1 M potassium thiocyanate as described above exhibit sedimentation coefficients of 4.62 ± 0.06 S and 2.96 ± 0.10, respectively, in a Tris·HCl–sodium chloride buffer.[14] On the other hand, a sedimentation coefficient of 7.6 S is found in potassium phosphate buffer.[15] The direction of this change is similar to that found in studies on the isolated enzyme. In distinction, the separated light subunit exhibits substantially the same sedimentation behavior in Tris and in potassium phosphate buffers. The separated heavy and light subunits exhibit mobility on sodium dodecyl-sulfate–polyacrylamide gel electrophoresis which is identical to that found for the respective subunits observed when the isolated enzyme is subjected to the same gel electrophoresis procedure.

The amino acid compositions of the isolated enzyme and of the separated heavy and light subunits have been determined.[15] The sums of

[25] P. P. Trotta, L. M. Pinkus, and A. Meister, *J. Biol. Chem.* **249**, 1915–1921 (1974).

the values obtained for the amino acid compositions of the individual subunits are in good agreement with those found for the isolated enzyme.

Enzymic Activities Exhibited by the Separated Heavy and Light Subunits and by the Reconstituted Enzyme

Neither the heavy nor the light subunit catalyzes glutamine-dependent carbamyl-phosphate synthesis (see the table). However, the reconstituted enzyme (prepared by adding an excess of light subunit to heavy subunit) catalyzes glutamine-dependent carbamyl-phosphate synthesis at a rate close to that exhibited by the isolated enzyme. The separated heavy subunit catalyzes ammonia-dependent synthesis of carbamyl phosphate, substantial bicarbonate-dependent ATPase activity, and synthesis of ATP from carbamyl phosphate and ADP. It is notable,

ENZYMIC ACTIVITIES OF SEPARATED HEAVY AND LIGHT SUBUNITS AND OF
RECONSTITUTED CARBAMYL-PHOSPHATE SYNTHETASE (CPS)

Enzymic activity[a]	Heavy subunit[b]	Light subunit[c,d]	Reconstituted enzyme (heavy + light subunits)[d,e]	Isolated enzyme
Glutamine-dependent CPS[f]	0	0	30[b]	29[b]
NH$_3$-dependent CPS[f]	18.4	0	23[b]	23[b]
ATPase (HCO$_3{}^-$-dependent)[g]	3.2	0	4.0[b]	4.4[b]
ATP synthesis (ADP + CP)[h]	1.3	0	2.5[b]	3.4[b]
Glutaminase[i]	0	0.060[c]	0.27[c]	0.58[c]
γ-Glutamylhydroxamatase[i,j]	0	0.044[c,k]	30[b]	31[f]

[a] The catalytic activities are expressed in terms of micromoles per nmole of enzyme protein per hour.

[b] Expressed per nmole of heavy subunit.

[c] Expressed per nmole of light subunit.

[d] Activities were determined after incubation of the protein with dithiothreitol (150 mM) under N$_2$ in potassium phosphate buffer (150 mM; pH 7.8) at 25° for 5 hr. (Under these conditions, the glutaminase activity of the light subunit is activated.)

[e] Prepared by adding a 4- to 5-fold excess of light subunit to heavy subunit, except for the glutaminase activity, which was determined in the presence of equimolar concentrations of the heavy and light subunits.

[f] Formation of carbamyl phosphate.

[g] Formation of ADP.

[h] Formation of ATP.

[i] Formation of glutamate.

[j] Determined in presence of ATP and HCO$_3{}^-$.

[k] Unaffected by addition of ATP and HCO$_3{}^-$.

however, that the partial activities are somewhat lower than the respective activities catalyzed by the isolated enzyme and the reconstituted enzyme (see below). The heavy subunit does not catalyze the hydrolysis of glutamine. On the other hand, the only catalytic activity exhibited by the light subunit is glutaminase activity. When the glutaminase activity of the light subunit is examined under the conditions used for determination of this enzyme activity using 20 mM L-glutamine, only about 20% of the activity exhibited by the isolated or reconstituted enzymes is found. This is explained by the finding that the affinity of the separated light subunit for glutamine is substantially lower than that of the isolated or reconstituted enzyme. Thus, when the light subunit is assayed for glutaminase activity in the presence of 180 mM L-glutamine, it exhibits 80–90% of the activity found with the reconstituted enzyme.

It is necessary to add about a 4-fold molar excess of light subunit to the heavy subunit to achieve maximal glutamine-dependent carbamyl-phosphate synthetase in the reconstituted enzyme. The required excess of light subunit may reflect the presence of some damaged light subunits; thus, thiocyanate may inactivate the light subunit, perhaps by thiocarba-mylation of the protein.

Allosteric Properties of the Separated Heavy and Light Subunits

Studies on the effects of various allosteric effectors on the synthesis of ATP from carbamyl phosphate and ADP catalyzed by the isolated heavy subunit demonstrated effects that closely resemble those seen with the isolated enzyme.[15] L-Ornithine, IMP, and NH_4^+ produce marked increases in ATP synthesis activity under these conditions, whereas UMP produces substantial inhibition. The order of effectiveness of the allosteric activators (L-ornithine > IMP > NH_4^+) is the same as that observed for the glutamine-dependent activity of the native enzyme. It is notable that the degree of activation by ornithine, IMP, and NH_4^+ diminishes significantly at high concentrations of ADP,[15] indicating that these allosteric activators increase the affinity of the enzyme for ADP in a manner apparently similar to the way in which these compounds increase the affinity of the isolated enzyme for ATP.[10]

Subunit Functions and Interactions

The dissociation of carbamyl-phosphate synthetase into heavy and light subunits which retain catalytic activities makes it possible to determine the catalytic functions of these component parts of the enzyme and thus to consider the manner in which two polypeptide chains interact during the normal catalytic reaction. The heavy subunit exhibits all of the catalytic activities of the isolated enzyme except those that involve glutamine. The separated heavy subunit has the binding sites for the allosteric effectors, and in addition those structural features of the enzyme that are responsible for its ability to undergo reversible self-association. On the other hand, the only catalytic function associated with the separated light subunit of the enzyme is its ability to bind and cleave glutamine and glutamine analogs.

The data provide a number of clues to the manner in which the subunits interact. For example, the binding of ATP and bicarbonate (in the presence of magnesium and potassium ions) to the heavy subunit enhances the rate of glutamine cleavage on the light subunit; thus, the isolated enzyme catalyzes a very low rate of glutaminase activity, but cleavage of glutamine is greatly increased in the presence of the other substrates. In addition, the presence of catalytic amounts of ATP, Mg^{2+}, and bicarbonate stimulates the hydrolysis of γ-glutamylhydroxamate about 200-fold. Interactions of the heavy and light subunits is required for such stimulation. The findings suggest that the formation of carboxy phosphate on the heavy subunit induces a conformational change that affects the rate of cleavage of glutamine on the light subunit. Other studies have shown the treatment of the enzyme with L-2-amino-4-oxo-5-chloropentanoate increases the bicarbonate-dependent ATPase activity substantially[26] and that it decreases the apparent K_m value for ammonia in the ammonia-dependent carbamyl-phosphate synthetase reaction.[27] The chloroketone binds to a specific sulfhydryl group on the light subunit, and such binding increases the ability of the heavy subunit to cleave ATP and to bind or use ammonia. These observations suggest that the binding of glutamine to the isolated enzyme in the normal catalytic reaction must facilitate its own cleavage by increasing the formation or utilization of carboxy phosphate. In addition, the binding of glutamine seems to increase the affinity of the heavy subunit for ammonia transferred to it from the light subunit.

Other evidences of interaction between the two subunits of the enzyme are evident from the data given in the table. For example, the low glutaminase activity of the separated light subunit is greatly increased upon addition of the heavy subunit. As noted above, this effect appears to be due to an increase in the affinity of the light subunit for glutamine. The apparent K_m value for glutamine is about 150 mM for the separated light subunit, while the K_m value for the native enzyme (and the reconstituted enzyme) is close to 1 mM. Such an increase in affinity

[26] E. Khedouri, P. M. Anderson, and A. Meister, *Biochemistry* **5**, 3442–3557 (1966).
[27] L. M. Pinkus and A. Meister, *J. Biol. Chem.* **247**, 6119–6127 (1972).

for glutamine might be explained by a conformational change in the light subunit associated with its interaction with the heavy subunit; alternatively, a portion of the glutamine molecule may bind to a site on the heavy subunit. It seems notable that the ammonia-dependent carbamyl-phosphate synthetase activity, bicarbonate-dependent ATPase activity, and the ATP synthesis activity of the heavy subunit are substantially increased upon addition of the light subunit (see the table). Thus, the light subunit seems to contribute in some manner to the catalytic activity of the heavy subunit, reflecting an important interaction between the subunits. Other evidences of interactions between the heavy and light subunits have emerged from studies on the pH-dependence of the glutaminase activities exhibited by the native enzyme, isolated light subunit, and chemically modified native enzyme.[28] The glutaminase activity of the native enzyme exhibits two pH optima (at about pH 4.2 and 9.4), while the separated light subunit exhibits a single pH optimum at about pH 6.7. When the heavy and light subunits are recombined, the two pH optima characteristic of the native enzyme return.

The procedure used for separation of the nonidentical subunits of *E. coli* carbamyl-phosphate synthetase is also applicable to the separation of the corresponding subunits of this enzyme in *Aerobacter aerogenes*[29] and *Salmonella typhimurium*.[30]

[28] P. P. Trotta, V. P. Wellner, L. M. Pinkus, and A. Meister, *Proc. Natl. Acad. Sci. U.S.A.* **70**, 2717–2721 (1973).

[29] P. P. Trotta, K. E. B. Platzer, R. H. Haschemeyer, and A. Meister, *Proc. Natl. Acad. Sci. U.S.A.* **71**, 4607–4611 (1974).

[30] A. T. H. Abdelal and J. L. Ingraham, *J. Biol. Chem.* **250**, 4410–4417 (1975).

[4] Carbamoyl-phosphate Synthetase (Glutamine): *Salmonella*[1]

By JOHN L. INGRAHAM and AHMED T. H. ABDELAL

$$HCO_3^- + 2\ ATP + glutamine \xrightarrow[Mg^{2+}]{K} carbamoyl\ phosphate$$
$$+\ glutamate + 2\ ADP + phosphate$$

In *Salmonella typhimurium,* a single species of carbamoyl-phosphate synthetase catalyzes the synthesis of all carbamoyl phosphate which serves two metabolic functions: it is an intermediate in the synthesis of

[1] See Volume 17A for earlier articles on glutamine-dependent carbamoyl-phosphate synthetase from *Escherichia coli* (22) and from *Agaricus bisporus* (23).

arginine and in the synthesis of pyrimidines. The enzyme is subject to cumulative repression by arginine and a cytosine compound as well as feedback control.[2] The enzyme from *Escherichia coli* has also been studied in some detail,[3,4] and it is quite similar to the one from *Salmonella typhimurium* with respect to its physical, catalytic, and regulatory properties.

Assay Method

Principle. A sensitive radiochemical assay is based on the principle that carbamoyl phosphate formed by the reaction can be readily converted to hydroxyurea by a subsequent nonenzymic reaction with hydroxylamine[5]; ^{14}C derived from [^{14}C] bicarbonate is thus converted to a nonvolatile product which remains in the reaction mixture following acidification. Because NH_4^+ can replace glutamine as a nitrogen donor in the reaction catalyzed by the large (α) subunit, assays are sometimes done with NH_4^+ rather than glutamine. If so, the concentration of nitrogen donor must be increased approximately 10-fold to compensate for the different K_m values.

Reagents

    1 M triethanolamine buffer (pH 8.0)
    1 M KCl
    120 mM ATP
    160 mM MgCl$_2$
    100 mM [^{14}C]NaHCO$_3$ (20,000–50,000 cpm/μmole)
    100mM glutamine
    1 M NH$_4$Cl
    1.2 M hydroxylamine-HCl
    60% (w/v) trichloroacetic acid
    Liquid scintillation counting mixture[6]: Two volumes of toluene [containing 0.4% PPO (2,5-diphenyloxazole) and 0.01% POPOP (1,4-bis-2(5 phenyloxazolyl)-benzene)] and one volume of Triton X-100.

Procedure. Small test tubes containing 50 μl each of triethanolamine buffer, KCl, ATP, MgCl$_2$, [^{14}C]NaHCO$_3$, glutamine (or NH$_4$Cl) and 150 μl of water are held in a constant temperature bath at 37° for 2 min, and

[2] A. Abdelal and J. L. Ingraham, *J. Biol. Chem.* **244**, 4033 (1969).
[3] S. L. Mathews and P. M. Anderson, *Biochemistry* **11**, 1176 (1972).
[4] P. O. Trotta, L. M. Pinkus, R. H. Haschemeyer, and A. Meister, *J. Biol. Chem.* **249**, 492 (1974).
[5] R. L. Levine and N. Kretchmer, *Anal. Biochem.* **42**, 324 (1971).
[6] M. S. Patterson and R. C. Greene, *Anal. Chem.* **37**, 854 (1965).

then the reaction is initiated by the addition of 50 μl of enzyme. After 10 min the reaction is terminated by the addition of 50 μl of hydroxylamine-HCl and immediately placing the mixture in boiling water. After 10 min the mixture is cooled and 100 μl of trichloroacetic acid are added. The mixture is shaken at room temperature to drive off excess $[^{14}C]O_2$, a 500 μl sample is taken to a vial containing 7 ml of counting fluid,[6] and the mixture is counted in a liquid scintillation spectrometer.

Comments. Crude extracts are dialyzed by application on Sephadex G-25 columns (0.9×6 cm) equilibriated with $0.1\ M$ potassium phosphate buffer, pH 7.6, containing 0.5 mM EDTA. Dialyzed extracts exhibit higher activities than nondialyzed extracts.

Definition of Unit and Specific Activity. One unit of carbamoyl-phosphate synthetase is defined as the amount of enzyme that catalyzes the formation of 1 nmole of carbamoyl phosphate per minute under the conditions of assay. The specific activity of a preparation is defined as units of carbamoyl phosphate activity present per milligram of protein.

Purification Procedure

A procedure involving conventional methods of purification has been published.[2] The enzyme can also be purified by successive sucrose density gradient centrifugations,[7] a method which takes advantage of the differing oligomeric forms of the enzyme in the presence of ornithine and in the presence of UMP.[7] The following procedure is based on affinity chromatography employing immobilized glutamine.[7]

Reagents
Culture medium: basal salts medium (007)[8] containing 4% glucose
Buffer A: 0.1 M potassium phosphate (pH 7.6) containing 0.5 mM EDTA
Buffer B: 0.02 M potassium phosphate (pH 7.6) containing 0.5 mM EDTA
Buffer C: 0.02 M potassium phosphate (pH 7.0) containing 0.5 mM EDTA
Buffer D: 0.5 M potassium phosphate (pH 7.0) containing 0.5 mM EDTA
Buffer E: 0.04 M potassium phosphate (pH 7.0) containing 0.5 mM EDTA 1% protamine sulfate (Sigma) adjusted to pH 5.0 with 1 N KOH
6-Amino-hexanoyl-glutamine-Sepharose: 20 ml containing 800 mg of

[7] A. T. H. Abdelal and J. L. Ingraham, *J. Biol. Chem.* **250**, 4410 (1975).
[8] D. J. Clark and O. Maaløe, *J. Mol. Biol.* **23**, 99 (1967).

6-amino-hexanoyl-Sepharose (P-L Biochemicals) are mixed with 12 ml of 0.2 M glutamine (pH 4.7) and 125 mg of ethyl-3(3-demethylaminopropyl) carbodiimide-HCl and the mixture is shaken at room temperature for 24 hr. The gel is then washed on a column successively with several volumes each of: 1 M NaCl, 1 M NaCl in 0.1 M Tris (pH 8.0), 1 M NaCl in 0.05 M sodium formate (pH 3.0), 1 M NaCl in 0.1 M Tris (pH 8.0), and finally 0.04 M potassium phosphate (pH 7.6).

Growth of Bacteria. Starvation for pyrimidine derepresses the synthesis of carbamoyl-phosphate synthetase; three successive starvations increases the activity over 7-fold. Thus a pyrimidine auxotroph of *Salmonella typhimurium* LT2 (JL1018, *pyrF146*) is used as a source of the enzyme. Two 80-liter fermentors containing 62 liters of culture medium are inoculated with 8 liters of starter cultures grown in 2-liter batches of culture medium in 4-liter Erlenmeyer flasks on a rotary shaker. Both starter cultures and fermentors are maintained at 37° and initially contain 5 μg of uracil/ml. The fermentor is aerated at 70 liters/min. Increments of 5 μg uracil/ml are added after cessation of growth as indicated by measurements of turbidity; pH is maintained above 6.5 by the periodic addition of solid KOH; the culture is harvested in a refrigerated Sharples supercentrifuge when the absorbance at 420 nm reaches about 10.

Preparation of Crude Extract. Cells are harvested, washed once in water, suspended at 0.4 g (wet weight)/ml Buffer A, and disrupted by two passages through a Manton-Gaulin high-pressure homogenizer at 9000 lb/in², maintaining the temperature below 15°. The homogenate is centrifuged for 30 min at 27,000 g. The clarified supernatant is the crude extract.

Removal of Nucleic Acids. Protamine sulfate (9 ml) is added with stirring at 0° for 10 min, and then the suspension is clarified by centrifugation at 18,000 g for 30 min.

Ammonium Sulfate Fractionation. EDTA (dipotassium salt; pH 7.6) is added at a final concentration of 0.5 mM. The enzyme solution was fractionated with solid ammonium sulfate at 0°. The fraction precipitating between 40 and 55% saturation which contained 90% of carbamoyl-phosphate synthetase activity was dissolved in buffer A and dialyzed against buffer B.

DEAE-Cellulose Chromatography. The dialyzed solution is applied to a 5 × 40 cm column of DEAE-cellulose (Whatman DE-52) previously equilibrated with buffer B. The column is developed at a rate of 8 ml/min

PURIFICATION OF CARBAMOYL-PHOSPHATE SYNTHETASE FROM *Salmonella typhimurium*,
STRAIN JL1018

Fraction	Volume (ml)	Protein (mg)	Total units × 10³	Specific activity (units/mg protein)
Crude extract	720	16,200	226.6	13
Protamine sulfate supernatant	1,150	11,600	346.6	30
Ammonium sulfate fraction	120	4,270	291.6	68
DEAE-cellulose eluate	110	275	145.8	529
Hexanoyl-glutamine Sepharose	90[a]	31.5	131.3	4166

[a] The total eluate is a result of two separate applications.

with a linear gradient of phosphate buffer (equal volumes of buffers C and D, total volume 800 ml). Eluate containing 0.15 to 0.24 M phosphate (peak of activity is 0.2 M) is collected, pooled, and concentrated by precipitation at 60% ammonium sulfate.

Affinity Chromatography. The ammonium sulfate precipitate, dissolved in a minimum volume of buffer A and dialyzed against buffer E, is pumped onto a 0.9 × 21 cm column of 6-amino-hexanoyl-glutamine-Sepharose at a flow rate of 20 ml/hr at 4°. Following washing of the column with buffer E until no further protein emerges, the column is developed by application of a linear gradient of 0 to 0.2 M KCl in buffer E (total volume, 200 ml) at a rate of 60 ml/hr. Enzyme elutes between 0.06 and 0.10 M KCl with a peak of activity at 0.07 M KCl. This preparation is homogeneous as judged by disc polyacylamide gel electrophoresis. The table summarizes the steps of purification.

Physical Properties

Size and Subunit Structure. The holoenzyme has a molecular weight of 150,000. It is composed of two subunits. The larger (α) has a molecular weight of 110,000, and the smaller (β) has a molecular weight of 45,000.

Catalytic Properties. The pH optimum for glutamine activity is broad with an optimum between 7.6 and 8.5. Ammonia activity rises steadily up to pH 9.0. The holoenzyme carbamoyl-phosphate synthetase has three substrates (HCO_3^-, glutamine, and ATP) and four effectors: UMP which inhibits the activity, and ornithine, IMP, and phosphoribosyl-l-pyrophosphate, which stimulate it. The ATP saturation curves with glutamine and ammonium chloride are sigmoidal with a Hill coefficient of 1.5. Although the maximum velocity of carbamoyl-phosphate synthetase with ammonia is only 60% of that with glutamine, the affinity for ATP is much higher with ammonia. Values for $s_{0.5}$ for ATP, determined from the Hill plots, are 0.2 mM at 100 mM ammonium chloride and 10 mM HCO_3^- and 1.4 mM at 10 mM glutamine and 10 mM HCO_3^-. The rate of synthesis of carbamoyl phosphate is a hyperbolic function of glutamine concentration: K_m is 0.25 mM at 12 mM ATP and 10 mM HCO_3^-. All effectors act by altering the affinity of the enzyme for ATP. UMP decreases the affinity while ornithine, IMP, and phosphoribosyl-l-pyrophosphate increase it.

The physiological significance of UMP as a negative effector and ornithine as a positive effector of activity is apparent. Increased concentrations of UMP, an end-product of one of the pathways (pyrimidine) to which carbamoyl phosphate contributes, indicate sufficiency of carbamoyl phosphate for this pathway. Increased concentrations of ornithine, which is the second reactant of the reaction utilizing carbamoyl phosphate in the arginine pathway, indicates deficiency of carbamoyl phosphate for arginine biosynthesis. Thus the combined effects of UMP and ornithine on the activity of carbamoyl-phosphate synthetase serve to maintain the intracellular concentration of carbamoyl phosphate at an optimal level over a wide range of growth conditions. The physiological significance of the effects of IMP and phosphoribosyl-l-pyrophosphate is less clear.

Reversible Dissociation of Subunits of Carbamoyl-phosphate Synthetase[7]

Reagents

Buffer F: 0.1 M potassium phosphate, pH 7.6, containing 1 M potassium thiocyanate, 3 mM ATP, 2 mM glutamine, 1 mM ornithine, 0.2 mM UMP, 4 mM $MgCl_2$, 10 mM NH_4Cl, and 1 mM EDTA

Enzyme (6 mg in 2 ml) is applied to a 108 × 1.6 cm column of Sephadex G-200 equilibrated against buffer F. The column is developed at a flow rate of 8 ml/hr. After 70 ml (void volume) have passed through column, a 45-ml fraction is collected which contains the large subunit. The next 4 ml are discarded and then the next 21 ml, which contain the small subunit, are collected. Before assay, the separate eluates are dialyzed against buffer A.

Subunits separated as described above retain full catalytic and regulatory properties. However, unlike the holoenzyme, the separate subunits are not stable in buffer A at 4°. The addition of 2 mM dithiothreitol, 2 mM ATP, 4 mM $MgCl_2$, 10 mM NH_4Cl, 1 mM ornithine, and 0.2 mM UMP stabilized the large subunit for 4 weeks at 4°. Combination of dialyzed subunits results in their immediate reassociation into holoenzyme.

Enzymic Properties of Subunits. The α subunit catalyzes the synthesis of carbamoyl phosphate from ammonia but not glutamine. The rate of carbamoyl-phosphate synthesis by the separated α subunit is sigmoidal, and the Hill plot yielded an interaction coefficient of 1.5. The $s_{0.5}$ value for ATP was 0.3 mM which is close to that obtained for the ammonia-dependent activity of the native enzyme. The β subunit exhibits glutaminase activity. Ornithine, IMP, and phosphoribosyl-pyrophosphate clearly stimulate the ammonia activity of the large subunit. This activity, however, was only slightly inhibited by high concentrations of UMP. The addition of the small subunit restores the ability to utilize glutamine as well as normal sensitivity to UMP.

[5] Aspartate Transcarbamoylase (*Escherichia coli*): Preparation of Subunits

By YING R. YANG, MARC W. KIRSCHNER, and H. K. SCHACHMAN

Carbamoyl phosphate + L-aspartate → carbamoyl aspartate

Aspartate transcarbamoylase (ATCase, EC 2.1.3.2: carbamoyl phosphate: L-aspartate carbamoyltransferase) catalyzes the first reaction unique to pyrimidine biosynthesis; in *E. coli* regulation is achieved in part by feedback inhibition of the enzyme by CTP, an end-product of the pathway.[1,2] The purified enzyme when treated with the mercurial, *p*-mercuribenzoate, is known to dissociate into two types of subunits, one responsible for catalysis, termed the catalytic subunit, and the other which binds both the inhibitor, CTP, and the activator, ATP.[3] After separation of the two types of subunits and removal of the mercurial, the enzyme can be reconstituted by mixing the catalytic and regulatory subunits.

The procedure for preparation of the discrete subunits after dissociation of the enzyme with *p*-mercuribenzoate was described in detail by Gerhart and Holoubek.[4] A modified procedure devised by Kirschner[5]

[1] R. A. Yates and A. B. Pardee, *J. Biol. Chem.* **221**, 757 (1956).
[2] J. C. Gerhart and A. B. Pardee, *J. Biol. Chem.* **237**, 891 (1962).
[3] J. C. Gerhart and H. K. Schachman, *Biochemistry* **4**, 1054 (1965).
[4] J. C. Gerhart and H. Holoubek, *J. Biol. Chem.* **242**, 2886 (1967).
[5] M. W. Kirschner, Ph.D. Thesis, University of California, Berkeley (1971).

was found to give higher yields of regulatory subunit and to reduce the time required for the separation. An alternative procedure involving heat treatment of the enzyme has also been used for the preparation of catalytic subunit.[6] The method described here is slightly modified from Kirschner's method and provides high yields of purified catalytic and regulatory subunits. It has been used routinely in our laboratory with ATCase from both *E. coli* and *Salmonella typhimurium*.

Reagents

100 mg purified ATCase

6.3 mg 1-(3-chloromercuri-2-methoxypropyl)urea, known as neohydrin or chloromecodrin (ICN, K & K Laboratories, Inc.)

20 g DEAE-cellulose (Schleicher & Schuell Co.)

0.01 M Tris-Cl, 0.1 M KCl, pH 8.7

0.01 M Tris-Cl, 0.5 M KCl, pH 8.7

20 mM K_2HPO_4, pH 8.7

3.6 M ammonium sulfate containing 5 mM 2-mercaptoethanol (the pH of the solution is adjusted with KOH to a value of 7 as measured with indicator paper)

25 mM Tris-Cl containing 2 mM 2-mercaptoethanol and 0.2 mM zinc acetate, pH 8.0

Dissociation of Enzyme into Subunits

A sample of 100 mg of ATCase at 15–20 mg/ml is dialyzed overnight at 4° against two changes of 500 ml of 0.01 M Tris-Cl containing 0.1 M KCl at pH 8.7 in order to reduce the concentration of 2-mercaptoethanol and EDTA in the stock enzyme solution. One milliliter of neohydrin solution, containing 6.3 mg of neohydrin in 0.01 M Tris-Cl at pH 8.7, is added to the enzyme solution which had been equilibrated at room temperature. The addition is performed rapidly, and the solution is mixed by gently inverting the tube several times. After 15 min of incubation at room temperature the reaction mixture is examined by zone electrophoresis on cellulose polyacetate strips in order to test whether the enzyme is completely dissociated.

Electrophoresis

Zone electrophoresis is used routinely before the mercurial-treated enzyme is loaded on the DEAE-cellulose column. Samples are applied in 20 mM potassium phosphate buffer at pH 8.7 onto a cellulose polyace-

[6] J. P. Rosenbusch and K. Weber, *J. Biol. Chem.* **246,** 1644 (1971).

tate strip (Gelman Sepraphore III), and electrophoresis is performed for 10 min at 250 V in a Microzone Electrophoresis Cell (Model R101, Beckman Instruments, Inc., Spinco Division). The protein is fixed and stained by immersion of the membrane in a solution of Ponceau-S (Beckman Spinco) for 5 min, and then the membrane is rinsed well in 5% acetic acid and dried at room temperature for storage.

Typical patterns for the native enzyme and the mixture of separated catalytic and regulatory subunits are shown in the top two strips in Fig. 1. The catalytic subunit yields a narrow band which migrates slightly faster than that corresponding to ATCase. In contrast, the regulatory subunit in the unfractionated mixture exhibits a very broad zone and migrates much more slowly than ATCase. At this stage the regulatory subunit has probably lost some of its zinc ions which may have been replaced in part by mercury.[7] As a consequence, there is a greater tendency for the regulatory dimers to dissociate into monomers,[8] and this reversible equilibrium leads to a broader zone than that observed when zinc ions are present in the preparation of the regulatory subunit (see bottom pattern in Fig. 1).

Separation of Catalytic and Regulatory Subunits by DEAE-Cellulose Chromatography

A column (0.9 cm × 25 cm) is packed with DEAE-cellulose which had been pre-equilibrated with 0.01 M Tris-Cl containing 0.1 M KCl at pH 8.7. The packed column is washed with 100 ml of the same buffer. Fibrous-form resin gives a flow rate between 20–30 ml/hr whereas preswollen microgranular form (Whatman DE 52) usually yields a much slower flow rate. Both produce satisfactory separations of the subunits.

The sample of mercurial-treated ATCase is applied to the column and then is washed with 45 ml of the same buffer (0.01 M Tris-Cl containing 0.1 M KCl at pH 8.7). This procedure elutes the regulatory subunit. The column is then washed with 100 ml of 0.01 M Tris-Cl containing 0.5 M KCl at pH 8.7 in order to elute the catalytic subunit. Fractions of 3 ml volume are collected and examined spectrophotometrically at 280 nm in order to locate the two subunits. A typical elution pattern is shown in Fig. 2.

The yield of regulatory subunit is greater than 90% of the theoretical value and that for the catalytic subunit is generally 85–90% of the theoretical yield. Concentrations of the catalytic and regulatory subunits

[7] M. E. Nelbach, V. P. Pigiet, Jr., J. C. Gerhart, and H. K. Schachman, *Biochemistry* **11**, 315 (1972).

[8] J. A. Cohlberg, V. P. Pigiet, Jr., and H. K. Schachman, *Biochemistry* **11**, 3396 (1972).

Origin

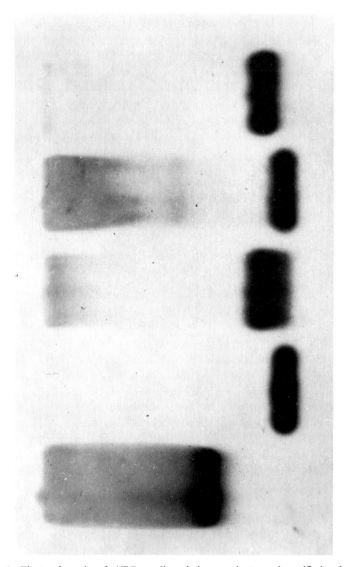

FIG. 1. Electrophoresis of ATCase dissociation products and purified subunits on cellulose polyacetate strip. Patterns from top to bottom are: (1) native ATCase; (2) ATCase treated with neohydrin to cause complete dissociation into catalytic and regulatory subunits; (3) partially dissociated ATCase with the broad band at the right corresponding to undissociated enzyme, ATCase-like molecules lacking one regulatory subunit, and free catalytic subunit; (4) purified catalytic subunit; and (5) zinc acetate treated regulatory subunit.

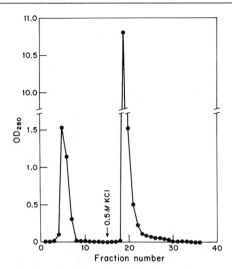

FIG. 2. Chromatographic separation of subunits on DEAE-cellulose. A sample of 100 mg of neohydrin-dissociated ATCase was loaded on a 0.9 × 20 cm column containing DEAE-cellulose equilibrated with 0.01 M Tris-Cl containing 0.1 M KCl at pH 8.7. The elution of the regulatory subunit with this buffer is seen by the peak in optical density at 280 nm in the fractions 4 to 7. After fraction 15 was collected the KCl concentration was increased to 0.5 M which led to the elution of the more highly charged catalytic subunit in fractions 19 to 25. The flow rate was 20–30 ml/hr, and each fraction was 3 ml.

are determined spectrophotometrically based on extinction coefficients (cm² mg⁻¹) of 0.72 and 0.30, respectively, at 280 nm.

Characterization of the Purified Subunits

The fractions (4 to 7) containing regulatory subunit are pooled and 2-mercaptoethanol and zinc acetate are added to give a final concentration of 10 mM and 2 mM, respectively. Addition of zinc acetate prior to the thiol causes the formation of a white precipitate which disappears when the reducing agent is added; only a slight loss of regulatory subunit occurs in this process. If the 2-mercaptoethanol is added before the zinc acetate there is no precipitation. The protein solution is then dialyzed against 500 ml of 25 mM Tris-Cl containing 2 mM 2-mercaptoethanol and 0.2 mM zinc acetate at pH 8.0. For storage of the regulatory subunit the protein is precipitated by dialysis against 3.6 M ammonium sulfate containing 5 mM 2-mercaptoethanol at pH 7.0.

The fractions (19 to 25) containing catalytic subunit are pooled and 2-mercaptoethanol is added to give a concentration of 10 mM. Precipitation of the protein is performed by dialyzing it overnight at 4° against 3.6 M ammonium sulfate containing 5 mM 2-mercaptoethanol at pH 7.0.

The precipitated protein is collected by centrifugation and resuspended in any desired buffer.

Both the catalytic and regulatory subunits are characterized in terms of purity and extent of aggregation by sedimentation velocity experiments and by electrophoresis in polyacrylamide gels. The catalytic subunit gives a single symmetrical boundary in the ultracentrifuge with a sedimentation coefficient about 5.8 S. The boundary for the regulatory subunit in the sedimentation velocity experiment migrates slowly (sedimentation coefficient is 2.8 S), and it is significantly broader than that for the catalytic subunit. This broadening is to be expected since the regulatory subunit has a much lower molecular weight than the catalytic subunit.[8]

The catalytic subunit exhibits a single band in polyacrylamide gel electrophoresis which migrates with a much larger mobility than the intact enzyme. Similarly a single band is observed for the purified regulatory subunit which has a slightly greater mobility than the intact enzyme. The band for the regulatory subunit is usually much broader than that observed for the freshly prepared sample. On prolonged storage the regulatory subunit exhibits a broad smear on electrophoresis but a much sharper band can be obtained by increasing the *N,N*-methylene-bis-acrylamide in the gel or by adding a small amount (0.6 m*M*) of zinc acetate in the gel and the lower buffer (L. Davis, unpublished). The broadening is probably attributable to the association-dissociation equilibrium exhibited by the regulatory subunit which is observed in the absence of zinc ions.[8]

Discussion

The procedure presented here is analogous to that of Gerhart and Holoubek[4] except that neohydrin is used in place of *p*-mercuribenzoate for the dissociation of ATCase and DEAE-cellulose is used for the chromatographic separation of the subunits instead of DEAE-Sephadex. With these changes the required column size and the time of separation are reduced. Within a few hours the purified subunits may be separated in high yield from 100 mg of ATCase. Moreover, both subunits are obtained in relatively concentrated solutions directly from the column.

The only difficulty experienced with this procedure results from storage of the neohydrin for long periods at room temperature. With such preparations the dissociation of ATCase proceeds slowly, and the prolonged reaction leads to the precipitation of the catalytic subunit. Incomplete dissociation of the enzyme leads to the formation of an

ATCase-like species lacking one regulatory subunit[9-11] in addition to the two types of subunits and some undissociated ATCase. Such a mixture is shown in the third pattern in Fig. 1. This difficulty can be avoided by purifying the neohydrin according to the following procedure. One gram of neohydrin is dissolved in 100 ml of 0.05 M KOH (pH about 11–12) and filtered through Whatman No. 50 filter paper. The filtrate is titrated drop by drop with concentrated HCl to pH 2.0 at 4°. A precipitate forms in several minutes and is collected on Whatman No. 50 filter paper. The extraction is then repeated, and the white precipitate is recovered and dried in a desiccator under vacuum. The dried purified neohydrin is stored in a freezer and used when needed for the dissociation of the enzyme.

Acknowledgments

This work was supported by NIH research grant GM 12159 from the National Institute of General Medical Sciences, and by grant PCM-76-23308 from the National Science Foundation.

[9] Y. R. Yang, J. M. Syvanen, G. M. Nagel, and H. K. Schachman, *Proc. Natl. Acad. Sci. U.S.A.* **71**, 918 (1974).
[10] M. Bothwell and H. K. Schachman, *Proc. Natl. Acad. Sci. U.S.A.* **71**, 3221 (1974).
[11] D. R. Evans, S. C. Pastra-Landis, and W. N. Lipscomb, *Proc. Natl. Acad. Sci. U.S.A.* **71**, 1351 (1974).

[6] Aspartate Carbamyltransferase (*Streptococcus faecalis*)

By TA-YUAN CHANG, LANSING M. PRESCOTT and MARY ELLEN JONES

Aspartate + carbamyl phosphate ⇌ carbamyl aspartate + P_i + H^+

Assay Methods

Method 1. A colorimetric measurement of carbamyl aspartate production developed by Prescott and Jones[1] was used in enzyme purification analysis and specific activity determination. This assay in our hands was reproducible and linear from 0.01 to 0.2 μmole of carbamyl aspartate with optical density values ranging from 0.04 to 0.65 at 466 nm.

[1] L. M. Prescott and M. E. Jones, *Anal. Biochem.* **32**, 408 (1969).

Reagents. The standard assay mixture contained 0.1 M Tris·HCl (pH 8.5), 10 mM carbamyl phosphate, 50 mM L-aspartate (pH 8.5), and enzyme to yield a final volume of 1 ml.

Procedure. The reaction was started by adding either carbamyl phosphate or enzyme after preincubation of the reaction mixture for 1 min at 25°. The incubation time for the assay was usually 10 min. The reaction was terminated by the addition of 1 ml of 1 M perchloric acid, and part of the total reaction mixture was taken for the colorimetric carbamyl aspartate assay. A blank containing the same components but omitting enzyme source was used as the control.

Definition of Unit and Specific Activity. A unit of enzyme activity is defined as that amount of enzyme which catalyzes the formation of 1 μmole of carbamyl aspartate or phosphate per minute under standard assay conditions. The protein concentration was determined using the method of Oyama and Eagle.[2]

Method 2. The [^{14}C]carbamyl phosphate assay of Bethell *et al.*[3] was used when more rigorous kinetic analysis was performed.

Assay Reagents and Procedure. The 0.5- or 1-ml reaction mixture contained 5 mM Tris·HCl with varying levels of substrates, inhibitors, and aspartate transcarbamylase. The final pH of the reaction mixture was 8.5. Reaction was performed at 25°. Incubation time ranged from 10 to 20 min. The reaction was stopped by addition of 0.5 or 1 ml of 1 M perchloric acid. After acidification the test tubes were processed according to the procedure of Bethell *et al.*[3] Controls without enzyme were routinely run and gave reproducible background values. The specific activity of [^{14}C]carbamyl phosphate used was selected so that the counts per minute in each experimental sample were greater than 500. The initial velocity was linear up to about 40% substrate consumption when carbamyl phosphate was limiting, or to about 20% substrate consumption when aspartate was the limiting substrate. All experiments were performed within these ranges.

[2] V. I. Oyama and H. Eagle, *Proc. Soc. Exp. Biol. Med.* **91,** 305 (1956).
[3] M. R. Bethell, K. E. Smith, J. S. White, and M. E. Jones, *Proc. Natl. Acad. Sci. U.S.A.* **60,** 1442 (1968).

Growth of Cells for Enzyme Source

Cells of *Streptococcus faecalis* R (ATCC #8043) were grown as arginine-adapted culture and harvested as described by Jones.[4]

Purification Procedure

Step 1. Disruption of Cells and Extraction of Enzyme. Cells were disrupted with a Braun Model MSK Mechanical Cell homogenizer (Brownwill Scientific) using 0.17–0.18 mm glass beads, according to the method of Bleiweis *et al.*[5] Frozen *S. faecalis* cells (0.52 kg) were thawed, suspended in 1.56 liter of ice-cold, 20 mM potassium phosphate buffer at pH 6.6, and held at 4°. Tri-n-butyl phosphate (14 ml) was added as antifoaming agent. For each single operation, 30 ml of diluted cell paste and 30 g of glass beads were mixed in a 75-ml glass flask. The stoppered flask was shaken for 3 min at 4000 oscillations/min in a stream of liquid CO_2 delivered at a rate sufficient to prevent heating of the chamber. After the disruption step was complete, the entire homogenate was centrifuged with a Sorvall GSA rotor at 10,000 rpm at 4° for 30 min; the supernatant was collected and dialyzed against three changes of 16 liters of 20 mM potassium phosphate buffer (pH 6.6) at 4°. The dialyzed homogenate had a total volume of 1950 ml.

Step 2. Streptomycin Sulfate Precipitation. Streptomycin sulfate solution (324 ml of a 5% solution in H_2O) was added drop by drop to 1950 ml of homogenate which had a protein concentration of 25.8 mg/ml. After the addition of streptomycin, the entire solution was kept overnight at 4° without disturbance to allow for the complete precipitation of nucleic acids. The precipitated protein was collected as in step 1 and discarded.

Step 3. pH 4.8 Precipitation. Approximately 30 ml of 0.5 M acetic acid were required to bring the pH of 1 liter of the streptomycin sulfate supernatant to 4.8. The acetic acid must be added as quickly as thorough mixing permits. The time usually required was 10–20 min. As soon as the acid addition was complete, the precipitate was centrifuged down as in step I. The precipitate was then resuspended in 0.1 M potassium phosphate buffer (pH 6.6) using a tissue homogenizer with a Teflon pestle. Normally a volume of buffer about 5% the volume of streptomycin sulfate supernatant was used. The resuspended pH 4.8 precipitate

[4] M. E. Jones, *in* "Methods in Enzymology" (S. P. Colowick and N. O. Kaplan, eds.), Vol. 5, p. 903. Academic Press, New York.
[5] A. S. Bleiweis, W. W. Karakawa, and R. M. Krause, *J. Bacteriol.* **88**, 1198 (1964).

was then dialyzed overnight at 4° against two 3-liter changes of 0.05 M phosphate buffer (pH 6.6). The pH 4.8 precipitate did not immediately dissolve completely when suspended in the 0.1 M buffer, but it completely dissolved during the dialysis. The dialyzed pH 4.8 precipitate solution was usually stored frozen at $-20°$ until it could be further purified. The enzyme is very stable in this condition.

Step 4. Hydroxylapatite Chromatography. A column, 16 cm in diameter, made in a Büchner funnel with fritted disc (porosity C, 3000-ml capacity), was packed with hydroxylapatite gel and equilibrated with 1 mM potassium phosphate buffer (pH 6.6) at 4°. The column height can vary from 5 to 9.5 cm without affecting the resolution of the column. Flow rate of the column was kept at about 200 ml/hr. The protein sample obtained from step 3, dialyzed against 1 mM potassium phosphate buffer (pH 6.6) at 4°, was applied to the column. The amount of protein applied was 4–5 mg of protein/ml of bed volume. The column was eluted with about two column volumes of 40 mM potassium phosphate buffer (pH 6.6), which removed large amounts of proteins without ATCase activity. The column was then eluted with 50 mM potassium phosphate buffer (pH 6.6) to elute ATCase activity. Usually 3–4 column volumes of 50 mM potassium phosphate buffer were sufficient to elute the enzyme, but sometimes a severe "tailing effect" was seen and as many as 6 column volumes of 50 mM phosphate buffer were required in order to elute all of the ATCase activity from the column. The fractions with ATCase activity were pooled and concentrated using an Amicon ultrafiltration cell fitted with a PM-10 membrane to yield a solution containing 6–8 mg of protein/ml. The concentrated protein solution had an appreciable amount of insoluble material which was centrifuged down and discarded since it contained no ATCase activity. The clear supernatant solution was dialyzed against 3 changes of 50 mM potassium phosphate buffer (pH 6.6) at 4°, and was stored at 4°. The hydroxylapatite column could be regenerated by washing with 4–5 column volumes of 0.4 M potassium phosphate buffer (pH 6.6) and used again several times.

Step 5. Ammonium Sulfate Fractionation. The 43 ml of enzyme solution (6 mg of protein/ml) from step 4 were brought to 43% saturation with saturated ammonium sulfate solution. This was done by very slow addition of the saturated solution at a rate of 2 ml/5 min with thorough mixing. The suspension was then kept at 4° for 1 hr undisturbed, and after it was centrifuged at 1700 rpm with a Sorvall SS-34 rotor for 10 min, the precipitate was discarded. The supernatant was then kept at 4° undisturbed for 1 day, after which it was centrifuged again; the precipitate was collected and dissolved in 4 ml of cold 50 mM potassium phosphate buffer (pH 6.6). This solution, which contained most of the

ATCase activity, was dialyzed against two changes of 1 liter of 10 mM potassium phosphate buffer (pH 6.6) containing 0.28 mM KCl and 2 mM mercaptoethanol.

Step 6. DEAE-Sephadex Column Chromatography. A 1.5 × 15 cm DEAE-Sephadex A-50 column (Pharmacia) was packed at 4° and was equilibrated with 10 mM potassium phosphate buffer (pH 6.6) containing 0.28 M KCl and 2 mM mercaptoethanol; the pressure drop over the bed was kept near 17–18 cm throughout the eluting process, and the flow rate was about 26 ml/hr. A 5-ml sample containing 13 mg of protein/ml (from step 5) was carefully layered on top of the column, after which a 300-ml linear KCl elution gradient, from 0.28 to 0.47 M, was used. Under the conditions described above, two peaks with ATCase activity were invariably obtained regardless of the amount of protein sample applied on the column (10–67 mg). This phenomenon could also be demonstrated with a DEAE-cellulose column (Bio-Rad Laboratories). Rechromatography of peak I (the peak that is eluted earlier) and peak II (the second peak) established that peak I was an "artificial" peak.[6] Therefore, only fractions from peak II were collected for further purification. Later it was found that the appearance of peak I could be totally prevented by running the DEAE-Sephadex or DEAE-cellulose column at a much faster flow rate, so the enzyme did not remain in the column too long (≤ 2 hr for the DEAE-cellulose column, or ≤4 hr for the DEAE-Sephadex column). Inclusion of 2 mM mercaptoethanol in the eluting buffers seemed to aid against production of peak I.[6]

Step 7. DEAE-Cellulose Column Chromatography. A 22 × 0.9 cm DEAE-cellulose column was packed at room temperature and then placed in a 4° cold room. It was equilibrated with 10 mM potassium phosphate buffer (pH 6.6) containing 0.28 M KCl and 2 mM mercapto-ethanol. The pressure over the bed should be high enough to give a flow rate of at least 50 ml/hr; the sample from step 6 (5.5 ml, dialyzed once against 500 ml of equilibrating buffer at 4°) was carefully layered on top of the column. Equilibrating buffer (6 ml) was used to rinse the protein from the glass above the column bed, and then the column was eluted with a 220-ml linear KCl elution gradient, from 0.28 to 0.47 M. Fractions containing ATCase activity were collected, pooled, and concentrated in the usual manner.

Step 8. (DEAE-Sephadex Column Rechromatography. An 11 × 0.9 cm DEAE-Sephadex A-50 column was packed as described for step 6. It was equilibrated with 10 mM potassium phosphate buffer containing 0.34 M KCl and 2 mM mercaptoethanol. The flow rate was about 16 ml/hr. A

---

[6] T. Y. Chang and M. E. Jones, *Biochemistry* **13**, 629 (1974).

sample from step 7 (12 ml, dialyzed once in 1 liter of equilibrating buffer at 4°) was layered on top of the column, and a 200-ml linear KCl gradient, from 0.34 to 0.47 M, was used as eluant. Of each 2-ml fraction, 1 μl was assayed for ATCase activity. In this case, the enzyme reaction was terminated directly by addition of the antipyrine–acetyl monoxime solution, and the terminated total reaction mixture was used for the colorimetric assay[1] to measure content of carbamyl aspartate formed during the reaction. Those fractions which had a ratio of absorbance for carbamyl aspartate at 466 nm and absorbance for protein at 280 nm of 3.2–3.6 were pooled and concentrated. After being dialyzed against three changes of 1 liter of 5 mM sodium phosphate buffer (pH 6.6), this purest enzyme fraction was stored in glass tubes and kept at −20°. The summary of a representative purification sequence is presented in Table I. This procedure has been successfully reproduced many times over a period of 2 years.

Criteria of Purity. The homogeneity of the purest enzyme fraction (specific activity = 350) was assessed by analytical gel electrophoresis at pH 9.5 according to the procedure of David.[7] The stained gel was scanned with a Gilford spectrophotometer equipped with an automatic gel scanner. The enzyme was estimated as 90–95% pure. The electrophoretic pattern obtained after disc gel electrophoresis in the presence of sodium dodecyl sulfate according to the procedure of Weber and Osborn[8] gave essentially a single band, confirming that this enzyme was at least 90–95% pure.

Molecular Weight of the Native Enzyme and Subunit Structure. The molecular weight of the native enzyme is 128,000 ± 6000, determined by short column sedimentation equilibrium according to the procedure of Van Holde and Baldwin.[9] Using sucrose gradient centrifugation according to the method of Martin and Ames,[10] the molecular weight of ATCase after purification procedure step 6 is estimated to be 125,000 ± 1000.

The molecular weight of the subunits of this ATCase determined according to the method of Weber and Osborn[8] is 32,500 ± 500. Thus, this ATCase contains four subunits of the same or nearly equal size.

General Properties. The pure enzyme is very stable in phosphate buffer at pH 6.6. It is stable at 0.1–1 mg/ml concentration at −20° for at least 2 years, or at 4° for at least 2 months without changing its physical

[7] B. J. David, *Ann. N. Y. Acad. Sci.* **121**, 404 (1964).
[8] K. Weber and M. Osborn, *J. Biol. Chem.* **244**, 4406 (1969).
[9] K. E. Van Holde and R. L. Baldwin, *J. Phys. Chem.* **62**, 734 (1958).
[10] R. G. Martin and B. N. Ames, *J. Biol. Chem.* **236**, 1372 (1961).

TABLE I

PURIFICATION OF ATCase FROM *Streptococcus faecalis*

Purification state	Vol (ml)	Protein (mg/ml)	Total protein (mg)	Total activity (units)	Specific activity (units/mg)	Stage recovery (%)
I. Homogenate	1950	25.8	50,200	14,600	0.29	(100)
II. Streptomycin supernatant	2250	16	36,000	13,700	0.39	94
III. pH 4.8 precipitate	358	45	16,100	12,200	0.76	90
IV. Hydroxylapatite column (eluate)	1840	0.2	368	9,579	26	70
Amicon filtration cell concentration, and centrifugation	43	6.0	258			
V. Ammonium sulfate fraction (43–54% cut)	5.0	13.4[a]	67	6,720	100	70
VI. DEAE-Sephadex column eluate	5.5	3.5[a]	19.3	5,400[a]	280	80
VII. DEAE-cellulose column eluate	12	1.04[a]	12.5	4,000	320	74
VIII. DEAE-Sephadex column rechromatography eluate	7.0	1.28	9.0	3,140	350	78

[a] The protein concentrations were determined by measuring the absorbance at 280 nm and assuming that the ratio of protein concentration (mg/ml)/OD_{280nm} = 1.2, since this ratio holds true for the purest enzyme fraction.

or kinetic properties. It is also stable in Tris·HCl buffer at pH 8.5 at 4° for at least 3 weeks. No effect of freezing and thawing has been detected. The ratio of absorbance of this ATCase was $A_{280}/A_{260} = 1.80 \pm 0.05$. Enzyme activity is linear within a broad range of protein concentrations (0.022–33.6 g/ml). The antigenic properties of this ATCase are distinctly different from those of *Escherichia coli* ATCase, for there is no cross-reaction between the enzyme from *S. faecalis* and antibodies[11] formed to the catalytic subunit of *E. coli* ATCase in an immunodiffusion analysis.

Effect of Sulfhydryl Inhibitors[12]

The enzyme is inhibited by $HgCl_2$, *p*-mercuribenzoate, and mersalyl with 50% inhibition being observed at 10^{-6} *M* for $HgCl_2$ and at 10^{-5} *M* for the organic mercurials. The substrates, either carbamyl phosphate or aspartate, and the products, either carbamyl aspartate or orthophosphate, can protect fully against the mercurial inhibition if they are present singly before the mercurial is added. In addition, a nonsubstrate anion, inorganic fluoride, also protected fully. The effect of *p*-mercuribenzoate was unusual, for when the highly purified enzyme was initially preincubated with 3×10^{-5} *M* *p*-mercuribenzoate, only 20% of the control activity remained (i.e., no mercurial was added in the preincubation); but if it was preincubated with 3×10^{-4} *M* *p*-mercuribenzoate, 60% of the control activity was observed. This unusual effect has not been investigated further.

Kinetic Properties and Nature of an Allosteric Activator Site

The kinetic properties of this enzyme have been investigated.[6] At low levels of carbamyl phosphate (0.1 m*M*), aspartate (1 m*M*), and buffer [5 m*M* Tris·HCl (pH 8.5)], many inorganic or organic anions greatly activated the reaction, but each anion activated to a different extent (Table II). In fact, under properly chosen conditions, every type of anion tested in our laboratory, including ATP and CTP, could stimulate the reaction to a certain extent. Urea and glycerol, two unionized compounds, had a minimal effect on the enzyme. Figure 1 shows that the effect of acetate, one of the better activator molecules tested, is competitive in nature against L-aspartate as the variable substrate. This suggests that at high aspartate levels addition of an activator would have no effect on the enzyme reaction; in other words, acetate and aspartate

[11] M. R. Bethell, R. von Fellenberg, M. E. Jones, and L. Levine, *Biochemistry* **7**, 4315 (1968).
[12] L. M. Prescott and M. E. Jones, *Biochemistry* **9**, 3783 (1970).

TABLE II

PERCENTAGE ACTIVATION OF CONTROL BY VARIOUS ORGANIC ACIDS AT 25 mM[a]

Salt used	% Activation
Formate	5600
Propionate	5000
Butyrate	5700
Isobutyrate	5800
L-α-Aminobutyrate	1400
γ-Aminobutyrate	400
D-Aspartate	5700
L-Alanine	800
β-Alanine	600
Carbamyl-β-alanine	5000

[a] Sodium was used as countercation.

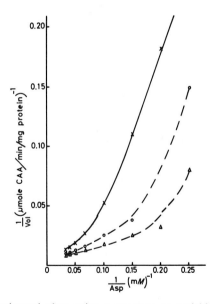

FIG. 1. Double reciprocal plots using L-aspartate as variable substrate at different concentrations of sodium acetate: (×) 0 mM; (○) 10 mM (△) 20 mM acetate. One milliliter of reaction mixture contained 5 mM Tris·HCl (pH 8.5), 0.1 mM [^{14}C]carbamyl phosphate (0.764 × 10^6 cpm/μmol), 0.037 μg of ATCase, and different concentrations of sodium acetate. The reaction time at 25° was 10 min (at 0 mM sodium acetate) or 5 min (at 10 and 20 mM sodium acetate). The [^{14}C]carbamyl phosphate assay was used. CAA = carbamyl aspartate. From Chang and Jones.[6] Reprinted with permission from *Biochemistry* **13**, 629 (1974). Copyright by The American Chemical Society.

appear to compete for the same activator site. The activation effects of chloride ion varied with the nature of the countercation used in the experiment. Nitrogen salts always gave better activation than simple inorganic salts[6] (e.g., NH_4Cl produced about twice the activation as was observed with an equal amount of NaCl).

In the absence of added salt, nonlinear kinetics (carbamyl phosphate and aspartate saturation curves deviate from normal Michaelis–Menten type curves) are observed. At high salt concentrations, 1/vol vs. 1/(carbamyl phosphate) and 1/vol vs. 1/(aspartate) plots could be brought back to linear curves, as the apparent K_m of carbamyl phosphate was reduced dramatically from 13 to 0.2 mM.

The data are consistent with the existence of an allosteric activator site, different from the substrate sites, which is sensitive to all anions that have been tested, including substrates and products of the reaction. We suspect that L-aspartate may be the "preferred" activator molecule for this enzyme, since it has two anionic carboxylic groups and an unhindered NH^+_3 group in the same molecule. The data in Fig. 1 suggest that the major function of the activator molecule is to influence (facilitate) the binding of substrate(s) rather than to facilitate the chemical transformation step(s) during the reaction.

Mechanism of Action

Steady-state kinetic analysis[13,14] on both forward (toward carbamyl aspartate) and reverse direction (toward carbamyl phosphate formation), as well as enzyme–substrate binding analyses were studied with this ATCase. An asymmetrical random mechanism has been constructed as a tentative scheme for the kinetic mechanism of this enzyme.

Acknowledgment

This work was supported by grants from the National Institutes of Health (Grant HD-02148 and 06538) and from the National Science Foundation (Grant GB-5354 and 31537).

[13] T. Y. Chang and M. E. Jones, *Biochemistry* **13**, 638 (1974).
[14] T. Y. Chang and M. E. Jones, *Biochemistry* **13**, 646 (1974).

[7] Aspartate Carbamyltransferase (*Pseudomonas fluorescens*)

By LINDA B. ADAIR and MARY ELLEN JONES

Aspartate + carbamyl phosphate $\rightleftharpoons$ carbamyl aspartate + P_i − H^+

In preliminary studies[1] aspartate carbamyltransferase from *Pseudomonas fluorescens* exhibited kinetic and feedback inhibition characteristics different from the ATCase of other bacteria. The enzyme from this bacteria was thus isolated and its characteristics studied.[2]

The procedures to be described are derived from an earlier fractionation procedure.[2]

Assay Methods

Method I

A colorimetric method of the determination of ATCase[3] was used for following the purification. For this method the enzyme assay solution contained (unless otherwise stated) 100 mM Tris-Cl, pH 8.5; 10 mM aspartate, pH 8.5; 1 mM lithium carbamyl phosphate; and enzyme in a final volume of 1 ml. The reaction mixture, minus enzyme, was incubated for 1 min. The enzyme was added, and the reaction incubated for 10 min. The reaction was stopped by the addition of 1 ml of color mixture. The color mixture was prepared immediately before use and consisted of 2 parts of the antipyrine/H_2SO_4 reagent and 1 part of the oxime reagent below.

1. Antipyrine/H_2SO_4 reagent: 5 g/liter of antipyrine in 50% (v/v) sulfuric acid.

2. Oxime reagent: 0.80 g of diacetylmonoxime in 100 ml of 5% (v/v) acetic acid.

This reagent should be stored at 4° in a dark bottle covered in aluminum foil.

After the color mixture was added to the enzyme reaction, the tubes were capped with marbles and placed in a 60° water bath for 120 min. The absorbancy at 466 nm was determined and compared with the color produced by standard amounts of carbamyl L-aspartate.

[1] J. Neumann and M. E. Jones, *Arch. Biochem. Biophys.* **104**, 438-477 (1964).
[2] L. B. Adair and M. E. Jones, *J. Biol. Chem.* **247**, 2308-2315 (1972).
[3] L. M. Prescott and M. E. Jones, *Anal. Biochem.* **32**, 408-419 (1969).

METHODS IN ENZYMOLOGY, VOL. LI

Method II

When low concentrations (below 20 μM) of CAP carbamyl phosphate were used in the kinetic studies, the radioactive assay of Bethell *et al.*[4] must be used. The [^{14}C]CAP used for this assay was obtained from New England Nuclear (1.97 mCi/mmole). Before use it was recrystallized from cold ethanol since 10% of the total radioactivity was a contaminant. Nine volumes of ice-cold absolute ethanol were added to a 0.1 to 0.2 M [^{14}C]CAP solution. The material remained at 0° for 15 min. The precipitated [^{14}C]CAP was removed by vacuum filtration, washed once with cold 95% ethanol and washed once with cold ether. The CAP was dried rapidly by placing in a 25° vacuum desiccator with P_2O_5. After 30 min, the P_2O_5 was stirred to give a new surface. This procedure was repeated until ether vapors could not be detected and the P_2O_5 remained dry. The CAP was dissolved in cold water to yield a 0.1 to 0.2 M solution. The solution was divided into 10 μl aliquots that were frozen in separate tubes.

The reaction mixture contained 100 mM Tris-Cl, pH 8.5, 10 mM aspartate, pH 8.5, and the desired concentration of [^{14}C]CAP in a final volume of 1 ml. The mixture, minus enzyme, was preincubated for 1 min at 30°, the enzyme added, and the reaction allowed to proceed for 10 min. The reaction was stopped by the addition of 0.1 ml of 70% perchloric acid, and the tube was capped and heated for 3 min in a 100° water bath. After cooling the tube, CO_2 gas was bubbled through the solution for 30 min. The entire sample was transferred to a scintillation vial containing a napthalene–dioxane mixture and counted.

Protein concentration was determined by the method of Oyama and Eagle.[5] A unit of enzyme activity is defined as that amount of enzyme that catalyzes the formation of 1 μmole of carbamyl aspartate per minute. The specific activity of the enzyme is the number of units of activity per mg of protein.

The activity of the enzyme varies greatly with pH. The optimum activity of the enzyme is observed at pH 8.5[2].

Purification

All purification steps were carried out at 4° and at no time after the frozen bacterial cells were initially disrupted was the enzyme frozen.

Step 1. Crude Extract. Seventy-five grams of frozen cell paste of *Pseudomonas fluorescens* (General Biochemicals, cat no. 150300) were

[4] M. R. Bethell, K. E. Smith, J. S. White, and M. E. Jones, *Proc. Natl. Acad. Sci. U.S.A.* **60**, 1442-1449 (1968).
[5] V. I. Oyama and H. Eagle, *Proc. Soc. Exp. Biol. Med.* **91**, 305-307 (1956).

thawed and suspended in 50 ml of homogenizing buffer (10 mM potassium phosphate, pH 7.5; 0.5 mM mercaptoethanol; and 20 μM EDTA). The cells were disrupted by passing the suspended cells twice through a French press under a pressure of 16,000 psi. The resulting homogenate was then centrifuged for 1 hr at 15,000 g.

Step 2. Streptomycin Sulfate Addition. To the supernatant was slowly added with vigorous stirring 50 ml of a solution containing 10% streptomycin sulfate and 10 mM potassium phosphate, pH 7.5. The resulting precipitate was allowed to settle for 1 hr and then removed by centrifuging for 30 min at 15,000 g. The supernatant was dialyzed overnight against two changes of 4 liters each of homogenizing buffer. If further precipitation occurred, it was removed by centrifugation.

Step 3. Ammonium Sulfate Fractionation. To dialyzed enzyme from step 2, 24.6 g of solid ammonium sulfate were added slowly, resulting in 30% saturation. After settling for 1 hr the precipitate was removed by centrifugation and the supernatant was brought to 43% saturation by the further addition of 11.2 g ammonium sulfate. The suspension was allowed to settle for 3 hr, and the precipitate containing the enzyme was collected by centrifugation. The pellet was dissolved in 6 ml of homogenizing buffer and dialyzed overnight against two changes of the homogenizing buffer.

Step 4. Sephadex G-200 Chromatography. About 4 ml of the dialyzed enzyme from step 3 containing 40 mg protein per ml were applied to the bottom of a Pharmacia column, 100 × 2.5 cm containing Sephadex G-200. The Sephadex G-200 slurry was prepared[6] by adding 25 g of Sephadex G-200 slowly into a 2-liter beaker containing 1.3 liters of rapidly boiling buffer (10 mM potassium phosphate, pH 7.5, and 50 mM KCl). The heat source was removed after half of the Sephadex had been added, and stirring was continued throughout addition until no lumps were visible. The suspension was cooled, the fines decanted, and the resulting slurry used to pack the column.

After the enzyme was applied, the column was regulated to a flow rate of 18 ml/hr and fractions of 3.5 ml were collected and concentrated in an Amicon ultrafiltration cell fitted with a PM-10 membrane to a concentration of about 2.5 mg of protein per ml. At this point the enzyme was stored until several batches of enzyme could be brought to this stage of purity. The enzyme is very stable in this condition.

Step 5. Hydroxylapatite Chromatography. Hydroxylapatite was obtained commercially or prepared using a modification of the methods of

[6] M. R. Bethell and M. E. Jones, *Arch. Biochem. Biophys.* **134,** 352-365 (1969).

Siegelman *et al.*[7] and Sizer and Jenkins,[8] as described by Prescott.[9] The hydroxylapatite for column use was prepared simply by suspending it in 5 mM potassium phosphate buffer at pH 7.5 and pouring directly into a column. The column with a fritted glass lower support was 8 cm in diameter and was fitted with Whatman No. 1 filter paper. The hydroxy-lapatite was poured to a height of 2.5 cm and filter paper placed on top of that to prevent disturbing the bed. If the column was poured to a greater height, the flow rate was extremely slow. With a large-diameter column, a large quantity of the enzyme could be purified at one time.

The enzyme from step 4 was dialyzed overnight against a 5 mM potassium phosphate buffer, pH 7.5. About 40 ml (2.5 mg protein per ml) of the enzyme were layered on the column. The column was then washed with 700 ml of 5 mM phosphate buffer. Then 15 mM potassium phosphate buffer, pH 7.5, were then added and 10 ml fractions collected. The column maintained a flow rate of 60 ml/hr under a constant pressure of 5 cm buffer. Fractions with ATCase activity were collected and concentrated in an Amicon filtration cell to a protein concentration of 1 mg/ml.

The column could be generated again by washing with 4 liters of 0.4 M potassium phosphate and then re-equilibrating with 5 mM potassium phosphate. With a regenerated column, the ATCase came off in the 5 mM phosphate wash, but appeared to have the same purity (as judged by electrophoresis) and specific activity as that coming off in the 15 mM phosphate wash with a freshly prepared column.

Step 6. Preparative Polyacrylamide Gel Electrophoresis. The method used for the preparative gel electrophoresis was that described in the instruction manual of the "Poly-Prep" electrophoresis apparatus by Buchler Instruments, Inc. The procedures are fully described in the Tris-glycine system of Jovin, Chrambach, and Naughton.[10]

The height of the resulting gel was 1.1 cm. This was the shortest gel that would adhere to the column support.

The enzyme from step 5 (4 mg) was diluted to 20 ml in 50 mM Tris buffer, pH 8.9. To this were added 0.6 g of sucrose and 0.1 ml of 0.01% (v/v) Bromphenol blue. Using a current of 50 mA, the enzyme was eluted from the gel 17.5 hr after elution of the Bromphenol blue. The

[7] H. W. Siegelman, G. A. Weiczorek, and B. C. Turner, *Anal. Biochem.* **13**, 402-404 (1965).

[8] I. W. Sizer, and W. T. Jenkins, *in* "Methods in Enzymology" (S. P. Colowick and N. O. Kaplan, eds.), Vol. 5, pp. 677-684. Academic Press, New York, 1962.

[9] L. M. Prescott, Ph.D. Dissertation, Brandeis University, Waltham, Massachusetts (1969).

[10] T. Jovin, A. Chrambach, and M. A. Naughton, *Anal. Biochem.* **9**, 351-369 (1964).

tubes with enzyme activity were combined and again concentrated in an Amicon ultrafiltration cell.

The purity of the ATCase was determined by two criteria: sedimentation equilibrium studies and analytical disc gel electrophoresis. Sedimentation to equilibrium performed according to the method of Yphantis[12] and described in Adair and Jones[2] showed only one protein species to be present. Analytical disc gel electrophoresis carried out by the procedure of Davis[13] showed one protein band, and this band coincided directly with enzymic activity.

Physical Properties

Molecular Weight

The molecular weight was determined by sedimentation to equilibrium[12] and by sucrose gradient centrifugation.[14] The sedimentation to equilibrium showed the enzyme to have a molecular weight of 360,000. The molecular weight by sucrose gradient centrifugation was found to be 365,000.

Subunit Structure

When the enzyme from step 6 was reduced with sodium dodecyl sulfate (SDS) by the method of Weber and Osborn[15] and then subjected to electrophoresis, a single band was obtained with a molecular weight of 180,000. The experiment was repeated after incubating the enzyme in 1% SDS and 1% mercaptoethanol for 4 hr followed by a 12-hr dialysis in 0.1% SDS and 0.1% mercaptoethanol; the same subunit weight was obtained.

Renaturation of ATCase

The enzyme (fraction 5) denatured with SDS as described above had no ATCase activity.[2] The activity was measured by the radioactive assay since SDS interferes with the color development of the normal assay. The denatured enzyme was then dialyzed against 5 liters of 10 mM sodium phosphate, pH 7.0, at room temperature. The buffer was changed every hour for 4 hr. The enzyme was then dialyzed against 5 liters of cold buffer for 36 hr. After the dialysis 1% of the activity

[11] S. E. Hager and M. E. Jones, *J. Biol. Chem.* **240**, 5667-5673 (1967). See Table I, this chapter.
[12] D. A. Yphantis, *Biochemistry* **3**, 297-317 (1964).
[13] B. J. Davis, *Ann. N. Y. Acad. Sci.* **121**, 404-427 (1964).
[14] R. G. Martin and B. N. Ames, *J. Biol. Chem.* **236**, 1372-1379 (1961).
[15] K. Weber and M. Osborn, *J. Biol. Chem.* **244**, 4406-4412 (1969).

TABLE I

PURIFICATION OF APCASE FROM P. FLUORESCENS[a]

Purification step	Protein (mg/ml)	ATCase activity[b] (units/ml)	Specific activity (units/mg)	Fold enrichment	Recovery (%)	CPSase[c] (units/mg)
I. Homogenate	76.9	2.64	0.034		100	12,660
II. Streptomycin sulfate supernatant	43.7	1.75	0.040	1.2	92	702
III. Ammonium sulfate fraction 30 to 43% saturation	66.8	13.66	0.204	6	68	162
IV. Sephadex G-200 eluate	2.65	3.12	1.18	35	48	0
V. Hydroxylapatite eluate	0.74	10.65	14.4	425	44	
VI. Preparative gel electrophoresis			3.16[d]			

[a] From Adair and Jones.[2]

[b] A unit is defined as that amount of enzyme that catalyzes the formation of 1 μmole of carbamyl aspartate per min under the standard assay conditions (30° incubation temperature; 10^{-3} M aspartate; 10^{-2} M CAP; and 0.10 M Tris-Cl buffer, pH 8.5).

[c] Relative units similar to those of Hager and Jones,[11] expressed as cpm/μmole of citrulline per mg of protein. See page 55 for ref. *11*.

[d] It is estimated that the homogeneous protein might have had a specific activity of about 144. The cause of the loss of enzymic activity with the pure enzyme is probably the extreme dilution at which the enzyme comes off the electrophoresis column. The enzyme loses 50% activity in 3 hr at a concentration of less than 10 μg/ml.

returned, and sucrose gradient centrifugation showed the enzyme to have a molecular weight of 360,000.

Carbamyl-phosphate Synthetase Activity

Yeast[16] and Neurospora,[17] are known to possess an ATCase–carbamyl-phosphate synthetase complex. To see if *P. fluorescens* had a similar complex, CPSase activity was measured at each stage of the purification process. Since the CPSase activity declines (Table I) at each step and cannot be detected in the purified enzyme, it is assumed that no such complex exists in the pseudomonads.

Kinetics

Substrate saturation curves of ATCase from some strains of the pseudomonads show hyperbolic curves with limiting CAP, but become

TABLE II
NUCLEOTIDE INHIBITION CONSTANTS[a]

Inhibitor	$K_i{}^b$ when CAP is limiting (μM)	$K_i{}^c$ when aspartate is limiting (mM)
CTP	7.0	7.5
UTP	8.5	9.0
ATP	8.5	9.0
GTP	—[d]	7.0
CDP	100	6.0
PP$_i$	0.4	No inhibition[e]
P$_i$	100	No inhibition[f]
Ribose-5-P	1500	No inhibition[f]
CMP	No inhibition[f]	No inhibition[f]
Cytosine	No inhibition[f]	No inhibition[f]
Cytidine	No inhibition[f]	No inhibition[f]

[a] From Adair and Jones.[2]

[b] Competitive inhibition with CAP limiting, i.e., 7.5, 15, and 30 μM with aspartate = 10 mM.

[c] Noncompetitive inhibition with aspartate limiting, i.e., 1, 2, or 4 mM with CAP = 10 mM.

[d] K_i for GTP could not be determined because all sources of GTP tested had PP$_i$ as a contaminant. Since the K_i for PP$_i$ is so low the results with GTP were invalid.

[e] PP$_i$ does not inhibition below 5 mM PP$_i$ when the CAP concentration is 10 mM. Above 5 mM PP$_i$, it begins to compete with CAP.

[f] Tested as high as 10 mM with no inhibition.

[16] P. F. Lue and J. G. Kaplan, *Can. J. Biochem.* **48**, 155-159 (1970).

[17] L. G. Williams, S. Bernhardt, and R. H. Davis, *Biochemistry* **9**, 4329-4335 (1970).

sigmoidal in the presence of the inhibitor UTP.[1] The enzyme from the bacteria supplied by General Biochemicals is also inhibited by UTP and other nucleotides, but the saturation curve for CAP in the presence of UTP is hyperbolic and not sigmoidal. From Lineweaver-Burk double reciprocal plots, the K_m for CAP was 14 μM and the apparent K_m for aspartate was 2.75 mM^2.

Inhibitor studies done when either CAP or aspartate were limiting gave the results seen in Table II. When CAP was the limiting substrate the enzyme showed a similar inhibition with CTP, UTP, and ATP. However, PP_i was a powerful inhibitor. From this and other data[2] it appears the PP_i is the most important factor in the inhibition with limiting CAP, and the structure of the base itself is relatively unimportant. The inhibition of the enzyme by the nucleotides with limiting CAP is competitive.[2]

By contrast, with limiting aspartate and saturating CAP, the nucleotides inhibit noncompetitively. Again, all the nucleotides inhibit to the same extent, but PP_i, which inhibits strongly with limiting CAP, no longer inhibits.

Acknowledgment

This work was supported by National Science Foundation Grant GB7929, and by Grants HD-02148 and 5 F02HD40300 from the National Institutes of Health.

[8] Dihydroorotate Dehydrogenase (*Escherichia coli*)[1]

By DORIS KARIBIAN

Dihydroorotate → orotate + 2 H$^+$

Assay Methods

The enzyme is membrane-bound and linked with the electron transport system of the cell[2]. When the system is intact the enzyme can

[1] R. A. Yates and A. B. Pardee, *Biochim. Biophys. Acta* **221**, 743–756 (1956).
[2] W. H. Taylor and M. L. Taylor, *J. Bacteriol.* **88**, 105–110 (1964).

therefore be assayed either as an oxidase[1,3] or as a dehydrogenase. When the system is not intact the enzyme can be assayed as a dehydrogenase using such electron acceptors as ferricyanide, various quinones, or 2,6-dichlorophenolindophenol (DCIP).[2,4] The oxidase and DCIP-reducing methods are described here. They have both been used with extracts from B and K12 strains grown on various minimal salts and rich media. With particulate preparations the specific oxidase activity at pH 8.3 is usually about 40% higher than the specific DCIP-reducing activity at pH 7.0. If both assay methods are used in parallel it is convenient to use the same buffer, 100 mM Tris·HCl, pH 7.6, instead of the different ones indicated below. Anaerobically grown cells have very low activity.

Dihydroorotate Oxidase

Principle. Enzyme activity is determined spectrophotometrically by following the rate of absorbance increase at 290 nm which is associated with the appearance of orotate.

Reagents
  Tris·HCl, 1 M, pH 8.3, 0.3 ml
  Enzyme: crude extract or membrane fraction, 20 mg protein/ml, 1–50 μl
  Sodium dihydroorotate, 0.01 M, 0.3 ml; store cold

Procedure. To each of two 3-ml quartz cuvettes of 1-cm light path add, in order, buffer, distilled water, and enzyme. Let equilibrate at 25° in the spectrophotometer, adjust the absorbance of the control to zero at 290 nm, and then add substrate to the test cuvette with good mixing. Follow the rate of absorbance increase in the test cuvette.

Specific activity is defined as nmoles of orotate formed per minute per milligram protein (assayed by the method of Lowry *et al.*[5]), taking the molar extinction of orotate at 290 nm as 6.2 × 10³.[1]

Dihydroorotate Dehydrogenase

Principle. Under conditions in which the bacterial oxidase system is inactive or absent, the enzyme transfers reducing groups from the substrate to the blue dye DCIP which on reduction no longer absorbs light at 600 nm.

[3] J. R. Beckwith, A. B. Pardee, R. Austrian, and F. Jacob, *J. Mol. Biol.* **5**, 618–634 (1962).
[4] C. T. Kerr and R. W. Miller, *J. Biol. Chem.* **243**, 2963–2968 (1968).
[5] O. H. Lowry, N. J. Rosebrough, A. L. Farr, and R. J. Randall, *J. Biol. Chem.* **193**, 265–275 (1951).

Reagents
Sodium phosphate buffer, $1 M$, pH 7.0, 0.3 ml
KCN, 0.1 M, 0.15 ml
Triton X-100, 1.0%, 0.3 ml; store cold
DCIP, 0.001 M, 0.12 ml
Sodium dihydroorotate, 0.01 M, 0.3 ml

Procedure. To three numbered 3-ml cuvettes of 1-cm light path add buffer, KCN, Triton X-100, water, and enzyme. Add DCIP to cuvettes 2 and 3, mix well, let equilibrate to 25°, adjust absorbance of cuvette 1 to zero at 600 nm, add substrate to cuvettes 1 and 3, readjust control absorbance to zero, and follow rate of absorbance decrease in the DCIP-containing cuvettes. Subtract the decrease observed in cuvette 2 from that in cuvette 3. With crude enzyme preparations the correction may be appreciable, especially during the first 2–3 min before stabilizing at an acceptably low level. With purified enzyme there is a negligible decrease in the absorbance of vessel 2.

Use sufficient enzyme to give a net absorbance change of between 0.020 and 0.045/5 min during the first 10–15 min. Only initial rates are used. One unit of activity reduces 1 nmole DCIP per minute.

Specific activity is expressed as units/mg protein taking the $\epsilon 600nm$ for DCIP as 20×10^3.

Purification Procedure

Although this procedure was developed with a pyrE⁻ derivative of K12 strain AT1243[6] other derepressible or leaky pyr⁻ strains (except pyrD⁻) should do.

Growth Conditions. Cells are grown aerobically at 37° on medium containing 50 mM sodium-potassium phosphate (pH 7.0), 0.2% ammonium sulfate, 0.01% $MgSO_4 \cdot 7 H_2O$, 0.001% $CaCl_2$, 0.2% Difco Casamino acids (technical), 0.2% glucose (autoclaved and added separately), and 8 mg uracil per liter. When the uracil is depleted growth stops (or is reduced), and culture is incubated 2 more hours with aeration before the cells are harvested. Derepression of the pyrimidine pathway increases cellular dihydroorotate dehydrogenase activity 5–10-fold. Freezing the cells does not impair activity.

Unless otherwise stated all steps are carried out at 0–4°.

Membrane Particle Preparation. The cells (20–200 g wet weight) are washed in 40 mM Tris·HCl, pH 7.6, suspended in the same buffer (2g/

[6] D. Karibian and P. Couchoud, *Biochim. Biophys. Acta* **364**, 218–232 (1974).

10ml), and broken in a French pressure cell at 4 tons/in². The extract is treated at room temperature with deoxyribonuclease and ribonuclease (1 μg each/ml) until the solution is no longer viscous, and it is then centrifuged at 100,000 g for 1 hr. The pellet is washed twice with one-fifth the original volume of the Tris buffer, and it is then resuspended in 4 mM phosphate (pH 7.0)–5 mM MgCl₂ to yield a protein concentration of 3 mg/ml.

Solubilization. Add 1 ml of 10% Triton X-100 per 100 ml and centrifuge at 150,000 g for 1 hr; discard the pellet. The supernatant is concentrated in a Diaflo filter with an XM50 membrane to a protein concentration of 5 mg/ml.

Ammonium Sulfate Fractionation. Solid salt is added slowly with stirring to a concentration of 40% saturation. After 1 hr of stirring the preparation is centrifuged 30 min at 10,000 g. The supernatant is then brought to 50% saturation in the same way and recentrifuged. The pellet is resuspended in 10 mM Tris·HCl (pH 8.4)–0.01% Triton X-100 and dialyzed against the same buffer with ammonium sulfate added to 0.6 M. Since the enzyme is stable for months at this stage, and less so after the next two steps, 10–15 mg portions are taken for further purification.

Agarose Filtration. The dialysate above is centrifuged at 150,000 g for 1 hr and the protein concentration adjusted to 3 mg/ml. The preparation is passed over a Biogel A-1.5, 200–400 mesh column (2.6 × 50 cm for a 3–5 ml sample) equilibrated with 10 mM Tris·HCl (pH 8.4)–0.01% Triton X-100–0.6 M ammonium sulfate solution. Activity comes off at 1.8 void volumes. The best fractions are collected and concentrated on a Diaflo XM50 filter to 1–2 ml.

PURIFICATION PROCEDURE FOR DIHYDROOROTATE DEHYDROGENASE (*E. coli*)

Preparation	Total activity units (nmoles product per min)	Protein (mg)	Specific activity (units/mg)
Crude extract	141.5	18,900	7.5
Washed particles	83.5	1,326	63
Solubilized enzyme 150,000 g supernatant	102.0	246	414
Precipitate 40–50% saturation (NH₄)₂SO₄	38.2	37.4	1020
Agarose column peak	18.1	6.6	2730
DEAE-cellulose column	6.8	1.8	3630

DEAE-Cellulose Chromatography. The above concentrate is loaded on a DEAE-cellulose column (1.6 × 5 cm) equilibrated with 50 mM sodium-potassium phosphate (pH 6.8)–0.1 M ammonium sulfate and then washed with the column buffer. Fractions of 0.5 ml are taken in tubes containing 50 μl 1% Triton X-100. Under these conditions the enzyme does not adhere to the column. Peak activity comes off at 2 column volumes.

The specific activity attained varies from 400 to 800 times that of the crude extract. Since polyacrylamide gel electrophoresis shows very little contaminating protein in all cases, the variability is probably due to enzyme inactivation.

Properties

Stability. All the above preparations through the ammonium sulfate step retain activity indefinitely at −18°. The particulate enzyme is also heat stable at 60° when mM orotate or dihydroorotate is present. After solubilization of the enzyme Triton X-100 is indispensable.

Stimulators and Inhibitors. Nonionic detergents such as Triton X-100, Brij 35 (0.02%), and Nonidet P42 (0.4%), ammonium salts (0.4 M acetate, phosphate, and sulfate), and bovine serum albumin stimulate DCIP-reducing activity increasingly with purification. Triton X-100 and ammonium salts together stimulate synergistically. Phospholipids, especially diphosphatidylglycerol, stimulate solubilized activity.[7]

Deoxycholate, dodecyl sulfate, and free fatty acids inhibit.

Orotate is a noncompetitive inhibitor in the absence, and a competitive inhibitor in the presence, of Triton X-100.

pCMB treatment has no effect on activity.

Molecular Weight. Molecular weight is estimated to be about 67,000 by agarose filtration.

Kinetics. $K_{m(app)}$ for dihydroorotate at pH 7.6 in the presence of Triton X-100 is $1 \times 10^{-5} M$.

Thermodynamics. Activation energies are 9.9 kcal and 16.8 kcal/mole, respectively, above and below the phase change temperature which is 19° in lipid-rich (> 0.15 μmoles lipid phosphate/mg protein) preparations and 15–16° in lipid-poor preparations with Triton X-100.

Other Properties. This enzyme does not contain flavins. There is evidence that the physiological electron acceptor is ubiquinone under

[7] D. Karibian, *Biochim. Biophys. Acta* **302**, 205–215 (1973).

aerobic conditions and menaquinone under anaerobic conditions.[8] The enzyme is associated with the inner side of the bacterial cytoplasmic membrane. It can be detached from it by the action of phospholipase A_2 (*Naja naja* or pig pancreatic enzyme) followed by adjustment of the pH to 8.4. Enzyme solubilized in this way has, however, a strong tendency to form inactive aggregates.

[8] N. A. Newton, G. B. Cox, and F. Gibson, *Biochim. Biophys. Acta* **244**, 155–166 (1971).

[9] Dihydroorotate Dehydrogenase (*Neurospora*)

By R. W. MILLER

Dihydroorotate + quinone → orotate + hydroquinone

Enzymes catalyzing the oxidation of dihydroorotate were originally identified in anaerobic bacteria[1] and mammalian cells[2] establishing the reaction as an obligatory step in the *de novo* synthesis of the pyrimidine heterocyclic ring. Constitutive enzymes responsible for catalyzing this step have been extensively investigated in prokaryotes[3] and fungi.[4] Although the molecular and catalytic properties of the enzyme differ depending on the source, biosynthetic dihydroorotate dehydrogenase (E.C. 1.3.3.1[5]) is a mitochondrial enzyme in eukaryotic cells[6,7] The oxidation of the substrate may be linked to reduction of components of the mitochondrial respiratory chain through quinones which are the primary electron acceptors for the enzyme purified as described in this article. The enzyme is a lipoprotein which can be isolated in catalytically active form from the mitochondrial membrane with nonionic detergents. Special treatment is required to prevent reaggregation or complete inactivation of the enzyme. Only dihydroorotate (2.5–4 mM) is capable of full protection of the isolated enzyme during purification.[6] Pyridine nucleotides cannot serve as electron acceptors for the fungal enzyme *in situ* nor can molecular oxygen serve as a primary electron acceptor for the isolated enzyme.

[1] I. Lieberman and A. Kornberg, *Biochim. Biophys. Acta* **12**, 223 (1953).
[2] R. Wu and D. W. Wilson, *J. Biol. Chem.* **223**, 195 (1956).
[3] W. H. Taylor, M. L. Taylor, and D. F. Eames, *J. Bacteriol.* **91**, 2251 (1966).
[4] R. W. Miller, *Arch. Biochem. Biophys.* **146**, 256 (1971).
[5] See properties of enzyme for consideration of reactivity with oxygen.
[6] R. W. Miller, *Can. J. Biochem.* **53**, 1288 (1975).
[7] J. J. Chen and M. E. Jones, *Arch. Biochem. Biophys.* **176**, 82 (1976).

Assay of Enzyme Activity

For routine assays of dihydroorotate dehydrogenase activity the reaction mixture provides a primary electron acceptor (quinone) which functions at all stages of purification and a secondary, readily observable acceptor which is reduced by the primary acceptor. Detergent-dispersed ubiquinone-30 (Sigma) is utilized as primary electron acceptor because it provides maximal activity as compared with simpler quinones.[4] Dichlorophenol-indophenol serves as secondary acceptor for routine assays. Menadione linked to reduction of cytochrome c has been used for assay of a mammalian dihydroorotate dehydrogenase.[7] In crude cellular extracts, electron transport to oxygen via cytochrome oxidase must be blocked.

The assay mixture is prepared by adding 3.3 μmole of ubiquinone-30 to 2 ml of 0.2 M Tris·HCl buffer, pH 8, containing 1% Triton X-100. The mixture is warmed at 37°C until turbidity is no longer visible. Sodium dihydroorotate, 5.5 μmole, and dichlorophenol-indophenol, 1.3 μmole, are added and the final volume is adjusted to 30 ml with distilled water. KCN, 1.7 mM is included for assay of the membrane-bound enzyme. The reaction mixture is dispensed (3.0 ml aliquots) into spectrophotometer cuvettes. Reference and control cuvettes are prepared with 3.0 ml of the same mixture lacking dye and dihydroorotate, respectively.

Enzyme samples, 5–20 μl, are added to the cuvettes thermostated at 37°C. For particulate samples, the enzyme is also added to the reference cuvette lacking dye. Absorbance at 610 nm is recorded as a function of time with a double-beam spectrophotometer, and the maximum rate of decrease in absorbance is determined after an initial lag period of approximately 1 min has elapsed. Correction for nonspecific reduction of the dye in crude preparations is made by determining the rate in the control cuvette lacking dihydroorotate and subtracting it from the rate obtained with the substrate.

A unit of enzyme activity is defined as that amount which causes the reduction of 1 μmole of dye per minute. Dichlorophenol-indophenol reduction is stoichiometrically equivalent to oxidation of dihydroorotate.

Growth of Mycelium

Wild-type *Neurospora crassa* (Commonwealth Mycological Institute No. EM 5297A) is grown on a defined medium containing (per liter): ammonium tartrate, 5 g; sucrose, 20 g; biotin, 5 μg; KH_2PO_4, 1.0 g; $MgSO_4$·7 H_2O, 0.5 g; NaCl, 0.1 g; $Na_2B_2O_7$·10 H_2O, 88 μg; $NH_4MO_7O_{24}$·4 H_2O, 64 μg; $ZnSO_4$·7 H_2O, 8.8 mg; $CaCl_2$, 0.27 mg; $MnCl_2$·4 H_2O, 0.72

mg; $FeCl_3 \cdot 6 \, H_2O$, 0.96 mg. The medium is dispensed in 1-liter amounts into 2.8-liter Fernbach flasks and inoculated with conidia and hyphae from slants. The flasks are aerated by shaking on a New Brunswick gyrotory shaker for 48 hr. The mycelium is harvested on a 450 mm Büchner funnel, washed with distilled water, and sucked dry.

The matted mycelium is mixed with twice the weight of glass homogenizing beads (3M Co., Superbrite, size 100), and 0.05 M Tris·HCl buffer, pH 7.5, containing 0.01 M disodium EDTA, is added. A volume of buffer equal to 1.5 times the cell weight is sufficient to form a thick slurry. The mixture is blended for 15 min in a 1 quart Mason jar. The temperature is maintained below 10°C with an ice bath while 90–100 V are applied with an autotransformer to a Sorvall Omimixer. Mannitol (0.5 M) may be included in the buffer if relatively intact mitochondrial membranes are desired. However, maximum yields of enzyme activity are recovered if the cells are disrupted in hypotonic media.

Purification

Membrane-bound dihydroorotate dehydrogenase is sedimented between 1000 and 100,000 g in the hypotonic buffer or between 800 and 12,000 g in isotonic media. The pelleted membrane preparation is lyophilized and then extracted sequentially with precooled (-10°C) n-butanol, acetone, and anhydrous ether to remove bulk lipids. The cake is allowed to warm to 25°C to vaporize ether, and then stored dry at -20°C.

The lyophilized powder (20–40 g) is homogenized (glass–Teflon homogenizer) with 300 ml of 0.05 M Tris·HCl, pH 8.8, containing 1% (v/v) Triton X-100 (Rohm and Haas), 10 mM EDTA, and 10 mM sodium dihydroorotate. The extract is heated at 50°C for 7 min in a water bath and then cooled in ice to 2°C. All subsequent steps are carried out at 2°C. n-Butanol, cooled to -20°C, is added drop by drop with stirring to give an 8.5% (v/v) mixture. The two-phase mixture is centrifuged at 105,000 g for 60 min. The resulting pellet may be washed with the extraction medium and resedimented. The combined supernatant fluids are brought to 85% saturation with solid $(NH_4)_2SO_4$, and the resulting precipitate is collected at 15,000 g for 10 min. A floating pellet is combined with a pellet packed at the bottom of the centrifuge tubes and mixed with 30 ml of 0.003 M Tris·HCl buffer, pH 8.2, containing 0.02% (v/v) Triton X-100, 1 mM EDTA, and 4 mM dihydroorotate. The suspension is centrifuged at 15,000 rpm in a Sorvall SS-1 rotor, and all pellets are reextracted 3–4 times with aliquots of the same buffer. The enzyme is recovered completely in the combined supernatant fluids which are clarified by brief centrifugation at 105,000 g. To remove ammonium sulfate, the enzyme solution is applied to a column (9 cm $\times$ 16 cm) of Sephadex G-15 (Pharmacia) and eluted with the buffer specified above under gravity-induced flow. The colored protein band is collected and the resulting fractions pooled. The fractions are concentrated by ultrafiltration maintaining equilibration with the dilute buffer which is the starting buffer for chromatographic removal of impurities.

Ion-exchange chromatography is carried out at 4°C with a 5.5 cm $\times$ 35 cm column of DEAE-cellulose (Whatman DE-52) which has been preequilibrated with the starting buffer. After the enzyme has been bound, the column is washed with 100 ml of the same buffer and then eluted with a gradient which is formed in a beaker positioned above the column. The mixing beaker containing 600 ml of the dilute buffer is connected to the column and to a second beaker through siphons. The second 2-liter beaker is filled with 0.04 M Tris·HCl buffer, pH 7.8, containing dihydroorotate, 4 mM, NaCl, 0.12 M, and 10 mM EDTA. Peak enzyme activity is eluted after 1 liter of eluant has passed through the column. Catalytically active fractions are pooled and concentrated with an Amicon Corp. Model 52 ultrafiltration cell having a 43-mm Diaflo membrane, type XM-50. The enzyme may be stored frozen at -20°C for several days without loss of activity.

Phospholipase Treatment

A phospholipase A_2 (E.C. 3.1.1.4) solution is prepared[8] from *Naja* venom (Sigma Chemical Co.). Aliquots of the concentrated enzyme solution are treated with this preparation (0.5 mg lyophilized venom per 10 mg protein) in a reaction mixture containing 1 mM CaCl$_2$, 10 mM dihydroorotate, and 0.4 M Tris·HCl buffer, pH 7.6. After incubation for 2 hr at 30°C, the mixture is diluted with 3 volumes of 0.16 M Tris·HCl buffer, pH 8.5, containing 4 mM EDTA and 0.2% Triton X-100. Prior to gel filtration chromatography, the solution is reconcentrated by ultrafiltration and sucrose is added to give a 10% (w/v) solution.

Gel Filtration Chromatography

Columns of Sepharose 6B (Pharmacia, 1.6 cm $\times$ 7.5 cm) are equilibrated with 0.05 M sodium phosphate buffer, pH 7.0, 0.1% (v/v) Triton X-100, and 2.5 mM dihydroorotate. Enzyme samples are applied to the column in 1.0 ml of 10% sucrose by layering under the buffer. Elution is carried out with the same buffer under gravity-induced flow

---

[8] D. Karibian, *Biochim. Biophys. Acta* **302**, 205 (1973).

while eluant volume and protein concentration are monitored. Fractions having maximal specific activity are pooled and concentrated to 1–2 ml. Active enzyme having a minimal observed molecular weight[6] is obtained on gel filtration chromatography after phospholipase treatment as summarized in the table. The detergent prevents nonspecific reaggregation with inactive lipoproteins.

Preparative Electrophoresis

Dihydroorotate dehydrogenase may be obtained in a form which migrates as a single band of protein or analytical gel electrophoresis by a preparative electrophoretic procedure.[6] However, in view of a noted decline in specific activity of the enzyme during this procedure, it is of questionable value. Reaggregation of lipoproteins may occur during stacking of the protein on the acrylamide gel.

Properties of Purified Dihydroorotate Dehydrogenase

The enzyme is a lipoprotein which contains solvent-extractable (25°C) fatty acids, phospholipids, and Triton X-100 even after extensive purification. It undergoes nonspecific aggregation with catalytically inactive proteins in the absence of Triton X-100 added above the critical micelle concentration. The minimum observed molecular weight[6] is 210,000 g protein per mole after phospholipase treatment and gel filtration. Enzyme protein accounts for approximately 25% of total protein which is present at this stage of purification. Four catalytically inactive bands are observed on analytical gel electrophoresis. Although

SUMMARY OF PURIFICATION OF DIHYDROOROTATE DEHYDROGENASE FROM *Neurospora*

Fraction	Total activity (EU)[a]	Specific activity (EU/mg)[a]	Overall recovery (%)
Lyophilized powder extract	16,600	0.05	[100]
Sephadex G-25 eluate	15,000	0.75	90
Active fractions from DEAE chromatography	11,000	7.0	63
Phospholipase-treated enzyme	10,800	20	65
Sepharose 6B eluate	9,130	25	55
Preparative electrophoresis concentrate	4,980	15	30

[a] Enzyme unit (EU) defined as 1 μmole dihydroorotate oxidized per minute, in 3.0 ml reaction mixture.

the electrophoretically purified protein migrates as a single band on analytical gel electrophoresis, treatment with sodium dodecyl sulfate yields 4–5 bands, indicating the possibility of nonidentical subunits or catalytically inactive contaminants.

Cofactors

The Sepharose-purified enzyme contains 1 mole of acid-releasable (pH < 3) FMN and 1.4 mole of ferric iron per 120,000 g of catalytically active protein. Acid-labile sulfide has not been detected by extraction of the purified enzyme with zinc acetate followed by formation of methelene blue.[9] Dihydroorotate stabilizes the enzyme against inactivation during dialysis, treatment with mercurials, and heat. Orotate is not as effective in this respect.[6] In the absence of demonstrable oxidase activity, the substrate must reduce the FMN and presumably maintains the enzyme in a stable conformation. Kinetic data[4] indicate a ping-pong type reaction mechanism with soluble quinones acting as the second substrate. Inhibitor studies are compatible with the existence of a hydrophilic pyrimidine binding site and a hydrophobic ubiquinone binding site.[6]

Inhibitors

Orotate is a competitive inhibitor ($K_i = 1.5 \times 10^{-5} M$) of dihydroorotate binding ($K_m = 3.5 \times 10^{-4} M$)[10] although no reduction of orotate has been demonstrated in the presence of reduced quinones. Fatty acids (18-carbon chains) inhibit noncompetitively with respect to ubiquinone at concentrations between 50 and 100 μg/ml. Removal of fatty acids during the initial purification results in a 40% increase in dehydrogenase activity in extracts of the lyophilized powder. Thenoyltrifluoroacetone (1 mM) inhibited the purified enzyme 50%, indicating a possible electron-transport role for iron.

Electron Transport

Triton X-100 [> 0.12% (v/v)] may alter the conformation of the solubilized enzyme producing catalytic properties differing from those of the enzyme *in vivo*. Detergent-induced release of dihydroorotate dehy-

[9] R. W. Miller, *Anal. Biochem.* **34**, 181 (1970).
[10] K_m determined at a saturating concentration of benzoquinone. Due to the ping-pong reaction mechanism, apparent K_m for dihydroorotate will depend on acceptor concentration in the soluble enzyme assay system.

drogenase from mitochondrial membranes eliminates all directly measureable electron transport from the substrate to oxygen. Of compounds
effective in inhibiting dihydroorotate oxidation in conjunction with the
mitochondrial electron transport chain,[6,11] only thenoyltrifluoroacetone
retains inhibitory activity with the solubilized enzyme from *Neurospora*.
The lack of catalytic oxidase activity may be due to inhibition by
intermediate products of oxygen reduction such as superoxide anion or
to inaccessibility of a reduced FMN cofactor to oxygen. Due to the
necessity for lipid removal for complete solubilization of the enzyme it is
possible that lipophilic cofactors are lost during purification.

Oxidase activity may be reconstituted in the purified enzyme by
adding an autoxidizable electron carrier such as phenazine methosulfate
(which is reduced to a free radical) plus dihydroorotate. Ubiquinone is
an absolute requirement for this oxidase activity.[6] Superoxide anion,
which is generated by autoxidation of the phenazine, can either reduce
dyes or dismute to oxygen and hydrogen peroxide. Hence, in the
presence of superoxide dismutase and phenazine methosulfate, dichlorophenol-indophenol reduction is completely suppressed because all electrons are utilized to produce hydrogen peroxide. Reactivity with molecular oxygen is strikingly stimulated pointing again to superoxide inhibition
of internal electron transport within the enzyme catalytic centers.

[11] H. J. Forman and J. Kennedy, *J. Biol. Chem.* **250**, 4322 (1975).

[10] Orotate Phosphoribosyltransferase (Yeast)

By A. YOSHIMOTO, T. AMAYA, K. KOBAYASHI, and K. TOMITA

$$\text{Orotate} + \text{5-phosphoribosyl-pyrophosphate} \rightleftarrows \text{orotidylate} + \text{PP}_i$$

Assay Method [1-3]

Principle. The orotate phosphoribosyltransferase reaction can be
assayed by coupling with orotidylate decarboxylase and following the
conversion of orotidylate formed to UMP at 295 nm.

[1] K. Umezu, T. Amaya, A. Yoshimoto, and K. Tomita, *J. Biochem. (Tokyo)* **70**, 249
(1971); A. Yoshimoto, K. Umezu, K. Kobayashi, and K. Tomita, see this volume [11].
[2] I. Lieberman, A. Kornberg, and E. S. Simms, *J. Biol. Chem.* **215**, 403 (1955).
[3] J. G. Flaks, see Vol. 6 [69].

Reagents

Tris·HCl buffer, 0.1 M, pH 8.0

MgCl$_2$, 10 mM

2-Mercaptoethanol, 0.1 M

Sodium orotate, 5 mM

PRPP,[4] 1 mM

Orotidylate decarboxylase, purified to the DEAE-Sephadex A-50 stage.[1] The enzyme preparation can be stored safely for several weeks in a 10 mM phosphate buffer (pH 5.8–7.5) containing 5 mM 2-mercaptoethanol at 4°, or in the same buffer containing 50% (v/v) glycerol and 5 mM 2-mercaptoethanol at −20°. A partially inactivated preparation obtained after longer storage may be restored to its original activity by incubating with 5 mM 2-mercaptoethanol at 4°.

Procedure. Into a quartz cuvette with a 1-cm light path are placed 0.4 ml of Tris·HCl buffer, 0.2 ml of MgCl$_2$, 0.01 ml of 2-mercaptoethanol, 0.2 unit of orotidylate decarboxylase, 0.05 ml of PRPP, 0.02 unit or less of orotate phosphoribosyltransferase, and 0.04 ml of sodium orotate to a total volume of 1 ml. The reaction is initiated by the addition of either the enzyme or orotate. The linear initial rate of decrease in absorbance at 295 nm is measured at 25°, using the molar extinction coefficient of 3.95×10^6 cm^{-1} M^{-1} for the conversion of orotate to UMP.

Definition of Unit and Specific Activity. One unit of enzyme catalyzes the conversion of 1 μmole of orotate to oritidylate per minute. Specific activity is expressed as units per milligram of protein. Protein is determined by the method of Lowry *et al.*[5] with bovine serum albumin as standard, or spectrophotometrically.[6]

Purification Procedure

All operations are carried out at 0–4° unless otherwise indicated.

Step 1. Crude Extract. Pressed baker's yeast (10 kg) is suspended in a mixture of 6 liters of 0.3 M potassium phosphate buffer (pH 8.0) and 1 liter of toluene and gently stirred for 4 hours at 30°, occasionally adjusting the pH to 8.0 with 5 M KOH. After cooling to 4°, the mixture

[4] PRPP: 5-phosphoribosyl-pyrophosphate, prepared by incubating ribose 5-phosphate and ATP with ribose 5-phosphate pyrophosphokinase as described by Flaks,[3] or purchased from commercial sources.

[5] O. H. Lowry, N. J. Rosebrough, A. L. Farr, and R. J. Randall, *J. Biol. Chem.* **193**, 265 (1951).

[6] E. Layne, see Vol. 3 [73].

is centrifuged at 15,000 g for 20 min, and the supernatant fluids are filtered through a few layers of gauze to remove fluffy materials.

Step 2. Ammonium Sulfate Fractionation. After the crude extract is adjusted to pH 5.0 with 8 M acetic acid while stirring in the presence of octanol, solid ammonium sulfate is added to obtain 50% saturation.[7] After stirring for 30 min, the mixture is filtered by gravity overnight, and the precipitate collected on the filter paper is dissolved in a minimum volume of 25 mM Tris·HCl buffer (pH 8.0), and adjusted to pH 8.0 with 1 M KOH. The enzyme solution is dialyzed exhaustively against 10 mM Tris·HCl buffer (pH 8.0) overnight, and then centrifuged to remove insoluble materials.

Step 3. Ethanol Fractionation. The enzyme solution from the previous step is made 50 mM in MnCl$_2$ by adding 1 M MnCl$_2$, stirred for 30 min, and then centrifuged to remove nucleic acids.[8] To the supernatant fluid, 2 M acetate buffer, pH 6.0 (1/7.5 volume), and 10 mM orotate (1/8 volume) are added, and the mixture is cooled to $-2°$; orotate is added as a protecting agent. To this mixture, 99% ethanol chilled to $-25°$ is slowly added, while stirring, to a concentration of 15% (v/v). After stirring for 5 min at $-15°$, the precipitate is removed by centrifugation at 15,000 g for 10 min at $-15°$. The supernatant fluid is treated with the additional portion of chilled ethanol to increase the concentration to 50%. The precipitate collected by centrifugation at $-15°$ is dissolved in 25 mM Tris·HCl (pH 8.0), stirred for 30 min, and centrifuged. The precipitate is extracted once more with the same buffer, and the combined extracts are dialyzed exhaustively against 25 mM Tris·HCl buffer (pH 8.0) overnight.

Step 4. Heat Treatment. The dialyzed fraction is made 2 mM in MgCl$_2$ and 1 mM in orotate. Aliquots (250–300 ml) are rapidly heated to 53°, maintained at this temperature for 5 min, and then quickly cooled to 2°. The precipitate is removed by centrifugation.

Step 5. C$_\gamma$ Gel Adsorption. The enzyme solution from the previous step is adjusted to pH 7.0 with 1 M acetic acid and stirred gently with 0.4 volume of aluminum hydroxide C$_\gamma$ gel[9] (46 mg dry weight/ml) for 10 min. The precipitate collected by centrifugation at 8000 g for 10 min is immediately extracted with the original volume of 0.2 M potassium phosphate buffer (pH 8.0) containing 1 mM orotate for 10 min, and the precipitate is removed by centrifugation at 12,000 g for 10 min.

[7] E. A. Noltman, C. J. Gubler, and S. A. Kuby, J. Biol. Chem. 236, 1225 (1961).
[8] S. Korkes, A. D. Campillo, I. C. Gunsalus, and S. Ochoa, J. Biol. Chem. 193, 721 (1951).
[9] S. P. Colowick, see Vol. 1 [11].

Step 6. Sephadex G-100 Gel Filtration. The C_γ gel fraction from the previous step is adjusted to pH 7.0 with 1 M acetic acid and brought to 65% saturation by the addition of solid ammonium sulfate. The precipitate is collected by centrifugation at 15,000 g for 20 min and dissolved in 25 mM Tris·HCl buffer (pH 8.0), adjusted to pH 8.0, and then dialyzed against the same buffer. The dialyzed solution is divided into portions (70–80 ml) and applied to a column (6 × 84 cm) of Sephadex G-100 equilibrated with 10 mM Tris·HCl buffer (pH 8.0) containing 1 mM orotate and 40 mM NaCl. The enzyme is eluted with the same buffer at a flow rate of 80–100 ml/hr. Fractions of 15 ml are collected, and those having specific activity of higher than 0.8 are pooled.

Step 7. DEAE-Cellulose Chromatography. The Sephadex G-100 fraction is dialyzed against 10 mM potassium phosphate buffer (pH 8.0) containing 1 mM orotate, and then applied to a column (3 × 32 cm) of DEAE-cellulose equilibrated with the same buffer. The enzyme is eluted with a (1500-ml) linear gradient of potassium phosphate, 10 mM to 200 mM (pH 8.0), containing 1 mM orotate. Fractions of 8 ml are collected at a flow rate of 80–100 ml/hr. The enzyme is eluted in two peaks, Fractions I (minor) and II (major), with 45–55 mM and 60–80 mM potassium phosphate, respectively.[1,10] The main portions of Fraction II with specific activity higher than 4.5 are pooled for the next purification step.

Steps 8 and 9. Hydroxylapatite Chromatography. Fraction II is dialyzed against 10 mM potassium phosphate buffer (pH 8.0) containing 1 mM orotate and then applied to a hydroxylapatite column (2.5 × 37 cm) equilibrated with the same buffer. Elution is performed with a (1-liter) linear gradient of potassium phosphate, 10 mM to 200 mM (pH 8.0), containing 1 mM orotate. The pooled active fraction is dialyzed against the same buffer as above, and rechromatographed on a second hydroxylapatite column (3 × 34 cm) under similar conditions. A 1800-fold purification with 10% recovery is achieved as summarized in the table.

The orotate phosphoribosyltransferase from various animal sources, including calf thymus,[11] cow brain,[12] rat liver,[13] Ehrlich ascites carci-

---

[10] Crude sonic extract or autolysate of baker's yeast is also separated into two fractions when directly chromatographed on a DEAE-cellulose column; on rechromatography on the same column, each fraction once separated is eluted as a single peak without any interconversion.

[11] D. K. Kasbekar, A. Nagabhushanam, and D. M. Greenberg, *J. Biol. Chem.* **239**, 4245 (1964).

[12] S. H. Appel, *J. Biol. Chem.* **243**, 3924 (1968).

[13] G. K. Brown, R. M. Fox, and W. J. O'Sullivan, *Biochem. Pharmacol.* **21**, 2469 (1972).

TABLE I

PURIFICATION OF OROTATE PHOSPHORIBOSYLTRANSFERASE FROM BAKER'S YEAST (10 KG)

Step and fraction	Volume (ml)	Total units	Total protein (mg)	Specific activity (units/mg)	Recovery (%)
1. Crude extract (autolysate)	10,000	13,700	309,000	0.044	100
2. Ammonium sulfate (0–50%)	2,860	11,500	154,000	0.074	84
3. Ethanol (15–50%)	1,780	9,580	50,400	0.191	70
4. Heat treatment (53°, 5 min)	1,830	8,620	30,300	0.284	63
5. C_γ gel	1,680	5,690	12,000	0.475	42
6. Sephadex G-100	492	4,740	1,580	3.0	34
7. DEAE-cellulose	118	2,450	186	13.1	18
8. First hydroxylapatite	100	2,190	73	30.0	16
9. Second hydroxylapatite	84	1,380	17	81.6	10

noma,[14] murine leukemia,[15] and human erythrocytes,[16] has been known to copurify with orotidylate decarboxylase. However, the two enzymes from baker's yeast, like the enzymes from brewer's yeast, are easily separated from each other; steps 1–3 especially are very effective in separating the two enzymes.

Properties

Stability and Purity. The enzyme is stable in the pH range of 7.5 to 9.5 for at least 6 months at $-20°$, but slowly loses activity when stored at $4°$. Orotate (1 mM) is effective in stabilizing the enzyme stored at $4°$. Polyacrylamide gel electrophoresis of the purified enzyme from hydroxylapatite shows that it contains one major and several minor bands.

Molecular Weight and Sedimentation Coefficient. Fractions I and II have molecular weights of 32,000 and 39,000, respectively, as determined by gel filtration on a calibrated Sephadex G-100 column. The sedimentation coefficient of Fraction II is 3.4 S as determined by sucrose density gradient centrifugation.

pH Optimum. Both Fractions I and II showed a similar pH optimum at 8.5–9.0.

[14] W. T. Shoaf and M. E. Jones, *Biochemistry* **12**, 4039 (1973).
[15] P. Reyes and M. E. Guganig, *J. Biol. Chem.* **250**, 5097 (1975).
[16] G. K. Brown, R. M. Fox, and W. J. O'Sullivan, *J. Biol. Chem.* **250**, 7352 (1975).

Michaelis Constants. Apparent K_m values for orotate and PRPP are 33 μM (at 100 μM PRPP) and 62 μM (at 200 μM orotate), respectively, for both Fractions I and II. Apparent K_m values for orotidylate and PP_i in the reverse reaction (pyrophosphorolysis of orotidylate) with Fraction II are 8.3 μM (at 2.5 mM PP_i) and 220 μM (at 2.5 mM orotidylate), respectively.

Activators and Inhibitors. Magnesium ion is required for optimal activity. The enzyme is not markedly inhibited by various nucleotides and sulfhydryl reagents tested, but it is completely inhibited by 5 mM EDTA, probably due to the removal of Mg^{2+}. Orotidylate inhibits the reaction competing with either orotate or PRPP. K_i values for orotidylate are 8.3 μM with orotate as the variable substrate, and 6.3 μM with PRPP as the variable substrate, respectively.

[11] Orotidylate Decarboxylase (Yeast)

By A. YOSHIMOTO, K. UMEZU, K. KOBAYASHI, and K. TOMITA

$$\text{Orotidylate} \rightleftarrows \text{UMP} + CO_2$$

Assay Method[1,2]

Principle. A spectrophotometric assay, based on the formation of UMP from orotidylate, is employed. The reaction is followed by the decrease in absorbance at 285 nm.

Reagents

Potassium phosphate buffer, 0.1 M, pH 6.0
2-Mercaptoethanol, 0.1 M
Orotidylate,[3,4] 1 mM
Enzyme. The enzyme is diluted in 10 mM potassium phosphate buffer (pH 7.5), containing 5 mM 2-mercaptoethanol, to give a solution with an activity not greater than 10 units/ml.

[1] K. Umezu, T. Amaya, A. Yoshimoto, and K. Tomita, *J. Biochem. (Tokyo) 70,* 249 (1971); A. Yoshimoto, T. Amaya, K. Kobayashi, and K. Tomita, see this volume [10].

[2] I. Lieberman, A. Kornberg, and E. S. Simms, *J. Biol. Chem.* **215,** 403 (1955).

[3] K. Mitsugi, A. Kamimura, E. Nakazawa, and S. Okumura, *Agric. Biol. Chem.* **28,** 828 (1964).

[4] Orotidylate is prepared from orotidine and *p*-nitrophenylphosphate by the phosphate transfer reaction using lyophilized *Serratia marcescens* cells[3] as described previously.[1]

Procedure. The method of Lieberman *et al.*[2] is modified by including 2-mercaptoethanol to fully activate the enzyme in the assay mixture. In a quartz cuvette with a 1-cm light path are placed 0.5 ml of the buffer, 0.01 ml of 2-mercaptoethanol, 0.005 to 0.1 unit of enzyme, and 0.05 ml of orotidylate to a total volume of 1 ml. The reaction is initiated by the addition of orotidylate, and the decrease in absorbance at 285 nm is measured at 25°, using the molar extinction coefficient of 1.65×10^6 $cm^{-1} M^{-1}$ for the conversion of orotidylate to UMP.

Definition of Unit of Enzyme Activity. One enzyme unit is defined as the amount of enzyme which catalyzes the decarboxylation of 1 μmole of orotidylate to UMP per minute. Protein is measured by the method of Lowry *et al.*,[5] or spectrophotometrically.[6]

Purification Procedure

All operations are carried out at 4° unless otherwise stated.

Step 1. Crude Extract (Autolysate). Pressed baker's yeast (5 kg) is crumbled, suspended in a mixture of 0.3 M K₂HPO₄ (5 liters) and toluene (500 ml), and gently stirred at 40° for 7.5 hr; pH of the mixture drops to about 6.3. After cooling to 5°, the autolysate mixture is centrifuged at 15,000 g for 10 min, and the supernatant fluid is filtered through a few layers of gauze.

Step 2. Ammonium Sulfate Fractionation. The autolysate is adjusted to pH 7.0 with 1 M NaOH. Solid ammonium sulfate is slowly added to the autolysate with constant stirring until 45% saturation is reached.[7] After 30 min of additional stirring, the precipitate is removed by filtration by gravity using a fluted filter paper overnight. The filtrate is adjusted to pH 7.0 with 1 M NaOH and brought to 80% saturation with additional solid ammonium sulfate. The precipitate is collected by filtration as above, dissolved in 10 mM phosphate buffer (pH 7.5) containing 5 mM 2-mercaptoethanol, and dialyzed against two changes of the same buffer. Since the enzyme is inactivated fairly rapidly in the medium containing no SH compounds, 5 mM 2-mercaptoethanol is added to the buffer (10 mM potassium phosphate, pH 7.5) used in the subsequent purification steps.

Step 3. Ethanol Fractionation. The dialyzed active fraction of step 2

[5] O. H. Lowry, N. J. Rosebrough, A. L. Farr, and R. J. Randall, *J. Biol. Chem.* **193,** 265 (1951).

[6] E. Layne, see Vol. 3 [73]

[7] E. A. Noltmann, C. J. Gubler, and S. A. Kuby, *J. Biol. Chem.* **236,** 1225 (1961).

is adjusted to pH 6.7 with 1 M acetate buffer (pH 4.5) and cooled to $-2°$. Ethanol (99%) chilled to $-45°$ is slowly added to the chilled aliquots (300 ml) of the enzyme with constant stirring at $-10°$ until the ethanol concentration of 43% (v/v) is reached. (The yield of enzyme is markedly reduced if the mixture is cooled to below $-10°$ and frozen.) After standing for 10 min at $-10°$, the precipitate is removed by centrifugation at 10,000 g for 10 min at $-10°$. The supernatant fluid is then treated with additional chilled ethanol to increase the ethanol concentration to 80% (v/v). The precipitate is collected by centrifugation as above, suspended in the buffer, and dialyzed against the same buffer; the precipitate is removed by centrifugation.

Step 4. First Hydroxylapatite Chromatography. The ethanol fraction is adsorbed to a hydroxylapatite column (4 × 41 cm) equilibrated with the buffer. Elution is performed with a (2600-ml) linear gradient of potassium phosphate, 10 mM to 200 mM, containing 5 mM 2-mercapto-ethanol. Fractions of 17 ml are collected at a flow rate of 40 ml/hr. Fractions with specific activity higher than 0.04, which are eluted with 60–90 mM buffer, are pooled and dialyzed against the 10 mM buffer.

Step 5. DEAE-Sephadex A-50 Chromatography. The active fraction of step 4 is applied to a DEAE-Sephadex A-50 column (4 × 35 cm) equilibrated with the 10 mM buffer. The enzyme is eluted with a (3000-ml) linear gradient of NaCl, 0–200 mM, in the same buffer. Fractions of 18 ml are collected at a flow rate of 90 ml/hr. The active fractions eluted with 60–110 mM NaCl are pooled.

Step 6. Sephadex G-100 Gel Filtration. The active fraction of step 5 is concentrated (to about 20 ml) with a DiaFlo Apparatus using a PM-10 Ultrafilter and then applied to a Sephadex G-100 column (4 × 75 cm) equilibrated with 50 mM potassium phosphate buffer (pH 7.5) containing 5 mM 2-mercaptoethanol. Fractions of 9 ml are collected at a flow rate of 90 ml/hr, and those with specific activity higher than 3.3 are combined and dialyzed against the buffer.

Step 7. Second Hydroxylapatite Chromatography. The dialyzed solution from step 6 is adsorbed on a second hydroxyapatite column (2 × 10 cm) equilibrated with the buffer as in step 4. Elution is performed with a (300-ml) linear gradient of phosphate, 10–200 mM, as before. Fractions of 3 ml are collected, and the active fractions are pooled.

As summarized in the table, the final specific activity of 40 represents a 6000-fold purification over the crude extract, with a yield of 22%. In contrast to the orotidylate decarboxylase from various mammalian

TABLE I

PURIFICATION OF OROTIDYLATE DECARBOXYLASE FROM BAKER'S YEAST[a]

Step and fraction	Volume (ml)	Total units	Total protein (mg)	Specific activity (units/mg)	Recovery (%)	Total units of orotate phosphoribosyl transferase[b]
1. Crude extract (autolysate)	6100	1100	174,600	0.0063	100	3010
2. Ammonium sulfate (45–80%)	890	675	46,100	0.0147	61	492
3. Ethanol (43–80%)	650	445	14,800	0.0301	40	138
4. First hydroxylapatite	500	393	1,800	0.218	36	120
5. DEAE-Sephadex A-50	340	343	156	2.20	31	2
6. Sephadex G-100	82	320	61	5.22	29	0
7. Second hydroxylapatite	32	240	6	40.0	22	0

[a] Started with 5 kg of pressed baker's yeast.
[b] Assayed as described previously.[1]

sources[8–13] which copurify with the orotate phosphoribosyltransferase, the enzyme from baker's yeast becomes practically free from the orotate phosphoribosyltransferase after the first few steps of purification procedures.

Properties

Stability and Purity. The enzyme is stable in 10 mM potassium phosphate buffer (pH 5.8–7.5) containing 5 mM 2-mercaptoethanol for several weeks at 0–4°, but is unstable to repeated freezing and thawing. The enzyme is conveniently stored in 10 mM potassium phosphate buffer (pH 7.5) containing 5 mM 2-mercaptoethanol and 50% (v/v) glycerol at −20°. An enzyme preparation which is slightly inactivated by prolonged storage can be reactivated to its original level by incubating with 5 mM 2-mercaptoethanol.

The purified enzyme from hydroxylapatite is almost homogeneous on polyacrylamide gel electrophoresis, but still contains a minute amount (<5%) of a faster moving band[14]; the main protein band coincides with that of the enzyme activity.

Molecular Weight and Sedimentation Coefficient. The molecular weight of the enzyme is estimated to be 51,000 by the Sephadex G-100 gel filtration method, and its sedimentation coefficient is estimated to be 4.0 S by the sucrose density gradient centrifugation.

Subunit. When the denatured and reduced enzyme is subjected to polyacrylamide gel electrophoresis in the presence of sodium dodecyl sulfate, two bands appear; the molecular weight of the main band is estimated to be 26,000 and that of the faint minor band (probably of the undissociated protein) is estimated to be 50,000.[14] Apparently the enzyme consists of two identical subunits.

pH Optimum. The activity of yeast orotidylate decarboxylase is optimal at pH 5.5.

Michaelis Constant. The apparent K_m value for orotidylate is 5 μM,

[8] D. K. Kasbekar, A. Nagabhushanam, and D. M. Greenberg, *J. Biol. Chem.* **239**, 4245 (1964).
[9] S. H. Appel, *J. Biol. Chem.* **243**, 3924 (1968).
[10] G. K. Brown, R. M. Fox, and W. J. O'Sullivan, *Biochem. Pharmacol.* **21**, 2469 (1972).
[11] W. T. Shoaf and M. E. Jones, *Biochemistry* **12**, 4039 (1973).
[12] P. Reyes and M. E. Guganig, *J. Biol. Chem.* **250**, 5097 (1975).
[13] G. K. Brown, R. M. Fox, and W. J. O'Sullivan, *J. Biol. Chem.* **250**, 7352 (1975).
[14] K. Kobayashi, A. Yoshimoto, and K. Tomita, unpublished data.

which is close to the value (7–8 μM) reported for the enzyme from brewer's yeast.[15]

Inhibitors. The enzyme is 70–100% inhibited by *p*-chloromercuribenzoate, 5,5'-dithio*bis*(2-nitrobenzoate), or Hg^{2+} at 10 μM, but is not significantly inhibited by iodoacetate and iodoacetamide at 1 mM. GMP, GDP, CMP, and CDP are not much inhibitory at 0.5–1 mM. (Strong inhibitions observed previously[1] must have been due to some technical errors.) Uridine nucleotides and EDTA show no inhibition in the standard assay.

[15] R. E. Handschumacher, *J. Biol. Chem.* **235**, 2917 (1960).

[12] Cytidine Triphosphate Synthetase

By Cedric Long and Daniel E. Koshland, Jr.

Cytidine triphosphate synthetase (CTPS) was initially characterized from *Escherichia coli* by Lieberman[1] and shown to catalyze the formation of CTP from UTP, ATP, magnesium, and ammonia. A similar reaction was later demonstrated in mammalian tissues[2] and bacteria[3] which used L-glutamine as an amino donor and required a guanosine nucleotide for maximum activities. Following these observations the enzyme was extensively purified from *E. coli*, and a kinetic analysis revealed a high degree of cooperativity in substrate binding to the enzyme.[4] Thus, both UTP and ATP showed strong site–site interactions, and GTP was found to be an allosteric effector for the glutamine reaction. Part of the high positive cooperativity was shown to proceed through conformational changes involving substrate-induced dimer–tetramer aggregation.[5] Such features have made CTPS an intriguing model for regulatory proteins.

Assay Method   *recorder moves .2"/min.   read @ 289 nm 0–1 scale*

Principle. The enzyme assay measures the amination of UTP spectrophotometrically at 291 nm where UTP has an extinction coefficient

2 ml total vol. in cuvette (includes enzyme)

[1] I. Lieberman, *J. Biol. Chem.* **222**, 675 (1956).
[2] H. O. Kammen and R. B. Hurlbert, *Cancer Res.* **19**, 654 (1959).
[3] K. B. Chakraborty and R. B. Hurlbert, *Biochim. Biophys. Acta* **47**, 607 (1961).
[4] C. W. Long and A. B. Pardee, *J. Biol. Chem.* **242**, 4715 (1967).
[5] C. W. Long, A. Levitzki, and D. E. Koshland, Jr., *J. Biol. Chem.* **245**, 80 (1970).

METHODS IN ENZYMOLOGY, VOL. LI

of 182 and CTP 1520 at pH 7.0. The assay can be carried out with either L-glutamine or ammonia serving as the amino group donor.

Reagents
    L-glutamine reaction mixture (total of 1.0 ml) contains:
    Tris-acetate, 20 mM, pH 7.2
    ATP, 0.5 mM
    MgCl$_2$, 10 mM
    UTP, 0.5 mM
    L-glutamine, 2 mM
    GTP, 0.1 mM
    Ammonia reaction mixture was similar except:
    (NH$_4$)$_2$ SO$_4$, 10 mM, replaced L-glutamine and GTP, and the assay was carried out at pH 8.15 with Tris-acetate, 20 mM.

Procedure. The substrates were preincubated for 3 min in cuvettes at 38°C. The enzyme preparation was added and contents mixed. The linear increase in absorbance at 291 nm was measured as a function of time on a Gilford recording spectrophotometer. Blank rates were obtained by omitting UTP from the reaction mixture. One unit of enzyme activity is defined as the amount of enzyme catalyzing the formation of 1 μmole of CTP per minute in 1.0 ml of reaction mixture at 38°C.

Purification Procedure

All operations are performed at 0–4°C. The original purification procedure[4] was modified to give a homogeneous protein.[5-7]

Crude Extract. To 850 g of *E. coli* B were added 1000 ml of 20 mM sodium phosphate buffer (pH 7.0) which contained 1 mM sodium EDTA and 2 mM L-glutamine. The cells were disrupted for 10 min in a 10-kc Raytheon sonic oscillator in 75-ml quantities. The cell debris was removed by centrifugation at 13,000 g for 45 min. The temperature was maintained below 6°C throughout subsequent steps.

Streptomycin Sulfate Treatment. To 1650 ml of the above extract 800 ml of 10% streptomycin sulfate were added with stirring over a period of 2 hr. The resulting suspension was centrifuged at 13,000 g for 30 min to remove the precipitate.

First Ammonium Sulfate Treatment. To 2300 ml of the above supernatant, 575 g of ammonium sulfate (enzyme grade) were added with stirring over a 60-min period. The resultant precipitate was re-

[6] A. Levitzki and D. E. Koshland, Jr., *Biochemistry* **10**, 3365 (1971).
[7] A. Levitzki and D. E. Koshland, Jr., *Biochim. Biophys. Acta* **206**, 473 (1970).

moved by centrifugation at 13,000 g for 30 min and dissolved in 1000 ml of the original buffer.

Second Ammonium Sulfate Treatment. To 1050 ml of above solution, 200 g of ammonium sulfate were added, with stirring over a period of 45 min. The precipitate was removed by centrifugation at 13,000 g for 30 min. To the remaining supernatant, an additional 60 g of ammonium sulfate were added with stirring over a 20-min period. The precipitate was dissolved in 150 ml of the original buffer, and the 175 ml which resulted were eluted in five batches from a column of Sephadex G50 (6 × 40 cm) that had been previously equilibrated with 20 mM, sodium phosphate buffer, pH 7.4, 1 mM EDTA, 4 mM L-glutamine, and 70 mM β-mercaptoethanol. This buffer was used in all subsequent steps.

DEAE-Sephadex Chromatography. The desalted enzyme solution was applied to a DEAE-Sephadex A-50 column (6 × 30 cm) equilibrated with the above buffer, washed with 200 ml of the same buffer, and eluted with a linear gradient from 0–0.20 M $(NH_4)_2SO_4$. The total gradient volume was 2400 ml, and 9-ml fractions were collected at a flow rate of 1 ml/min. The enzyme activity eluted midway through the gradient at 0.08–0.09 M $(NH_4)_2SO_4$. The peak enzyme-containing fractions were pooled, precipitated with 55% saturated ammonium sulfate, and desalted by passage through a Sephadex G100 column (4 × 75 cm) at a flow rate of 60 ml/hr.

Gel Filtration. The fractions from the Sephadex column with maximum activity were pooled and 3.0 ml were applied to a Biogel A 0.5 M column (2 × 85 cm) that had been equilibrated with buffer containing 0.75 M ATP, 0.75 M UTP, and 0.01 M $MgCl_2$. Under these conditions it was found[5,7] that the enzyme elutes as a tetramer of 210,000 molecular weight. The fractions containing maximum activity are pooled, precipitated with 55% ammonium sulfate, and reapplied to the Biogel A 0.5 M column in the absence of substrates. Under these conditions the activity elutes from the column at a point corresponding to a molecular weight of 105,000. A final purification is obtained by precipitating the pooled enzyme-containing fractions with 60% ammonium sulfate and applying 2.0 ml of the enzyme to a column of Sephadex G200 (2.5 × 90 cm) equilibrated with buffer without nucleotides. The enzyme eluting from the column is pure as judged from polyacrylamide gel electrophoresis.

Comments on the Purification Procedure. The ammonium sulfate, DEAE-Sephadex A50, and gel filtration steps result in good purification with minimal loss of activity and good yields. The aggregation of the enzyme in the presence of substrate effectors provides a unique way to separate CTP synthetase from other proteins which copurify with it and

PURIFICATION OF CYTIDINE TRIPHOSPHATE SYNTHETASE

Step	Activity (units)	Specific activity (μmoles CTP/min/mg)	Yield of pooled fractions (%)
Streptomycin sulfate	390	—	100
Third $(NH_4)_2SO_4$	260	0.05	67
DEAE-Sephadex	160	0.30	41
Sephadex G-100	150	1.30	38
Biogel A (0.5 M) + ATP + UTP + MgCl$_2$	150	2.00	38
Biogel A (0.5 M), no nucleotides	112	4.60	29
Sephadex G-200	100	5.80–6.10	26

otherwise would be difficult to remove. The overall purification from the crude extract is about 1000-fold and routinely resulted in a final yield of about 25%. A summary of the purification procedure is given in the table.

Properties of the Purified Enzyme

Requirements. The amination of UTP by the glutamine assay shows strict dependence on ATP, Mg^{2+}, and GTP with a pH optimum of 7.0 in Tris-acetate buffer. The ammonia assay does not require the allosteric effector GTP and has a pH optimum of 8.2. The pure enzyme when stored at a concentration of 1–4 mg/ml at 4°C in 20 mM sodium phosphate buffer, pH 7.2, 2 mM L-glutamine, 1 mM EDTA, 70 mM β-mercaptoethanol, and 20% glycerol is stable for 3 months. Enzyme solutions that have lost activity may be reactivated by the addition of β-mercaptoethanol.

Structure. CTP synthetase exists as a dimer in the absence of substrates with a molecular weight of 105,000 ± 2000 as measured by equilibrium centrifugation or gel filtration on Sephadex G200.[5] The enzyme can be split under denaturing conditions in guanidine hydrochloride or urea to two identical subunits with molecular weight of 52,000.[5] Polyacrylamide gel electrophoresis of the purified protein at pH 8.0 and 9.5 in the presence of 8 M urea gave a single band.[8] In the presence of UTP, Mg^{2+}, and ATP at concentrations used in the assay, the enzyme aggregates to a tetrameric form with molecular weight of 210,000 at

[8] A. Levitzki, W. B. Stallcup, and D. E. Koshland, Jr., *Biochemistry* **10**, 3371 (1971).

protein concentrations as low as 5 μg/ml.[5,9] The enzyme cannot be denatured to its monomeric form at pH values near its isoelectric point, 5.5.[8] CTP synthetase from mammalian liver has also been found to aggregate in the presence of ATP and UTP.[10]

Kinetics and Mechanism. Purified CTP synthetase exhibits positive cooperativity for the substrates ATP and UTP[4,11] and negative cooperativity for an effector GTP[12,13] and the substrate L-glutamine.[5,8,11] The initial kinetic analysis of the reaction indicated strong site–site interactions for ATP and UTP with Hill coefficients of 3.8 and 3.4, respectively. The high degree of cooperativity was in part explained by the substrate-induced dimer–tetramer aggregation. Both UTP and ATP were able to bring about this transition separately but were most efficient in combination.

Initial studies[4] on the role of GTP indicated that this compound had no effect on the ammonia reaction but greatly accelerated the glutamine reaction. A detailed kinetic study[12] provided evidence that saturating levels of GTP increased the V_{max} of the glutamine reaction 10-fold and decreased the K_m for glutamine by a factor of six. The CTP synthetase from mammalian liver has not been found to exhibit negative cooperativity for GTP binding.[13]

The effect of GTP on the formation of the enzyme–glutamine complex was verified by measuring the reaction of enzyme with the glutamine analog 6-diazo-5-oxonorleucine (DON). It was found that GTP accelerates the inactivation of CTP synthetase by enhancing the reaction of the affinity label DON with essential SH groups.[8,12] The interaction of CTP synthetase with DON initially presented an enigma since only two molecules were found to react per tetramer and to destroy activity.[5] This result was subsequently shown to be an extreme case of negative cooperativity where the DON-induced conformational changes are transmitted to remaining subunits and prevent binding of unfilled sites.[8] This phenomena was termed "half of the sites reactivity" and was found to occur in other proteins as well.

[9] A. Levitzki and D. E. Koshland, Jr., *Biochemistry* **11,** 247 (1972).
[10] R. P. McPartland and H. Weinfeld, *J. Biol. Chem.* **251,** 4372 (1976).
[11] A. Levitzki and D. E. Koshland, Jr., *Proc. Natl. Acad. Sci. U.S.A.* **62,** 1121 (1969).
[12] A. Levitzki and D. E. Koshland, Jr., *Biochemistry* **11,** 241 (1972).
[13] C. R. Savage, Jr. and H. Weinfeld, *J. Biol. Chem.* **245,** 2529 (1970).

[13] CTP Synthetase of Bovine Calf Liver[1]

By HERBERT WEINFELD, C. RICHARD SAVAGE, JR., and RICHARD P. MCPARTLAND

$$\text{UTP + ATP + L-glutamine} \xrightarrow[\text{GTP}]{\text{Mg}^{2+}} \text{CTP + ADP + L-glutamate + P}_i$$

Hurlbert and Kammen[2,3] were the first to find that either UMP, UDP, or UTP could be converted to a ribonucleotide of cytosine by an enzyme present in the cytosol of rat liver and Novikoff hepatoma. The enzyme purified from rat liver[4] and bovine liver[4,5] utilizes UTP, and the product is CTP.[4] The stoichiometry of the reaction[5] is the same as that of the reaction catalyzed by the *Escherichia coli* B enzyme,[6,7] and in both cases GTP is a strong activator.[3-7] The bacterial enzyme has been designated EC 6.3.4.2, UTP: ammonia ligase (ADP-forming),[8] but both enzymes are usually called CTP synthetase. Despite the identical stoichiometries and similar polymerizing properties,[5,7] the enzymes have markedly different kinetic properties.[4,6,7,9] There is evidence that a CTP synthetase is present in other mammalian cells.[10-13]

Assay Methods

The rate of formation of CTP can be determined by two methods.[4] In the first method CTP formation is followed by an increase in absorbance

[1] Supported in part by grant AM-1337 and CA-5016, National Institutes of Health, United States Public Health Service.

[2] H. O. Kammen and R. B. Hurlbert, *Cancer Res.* **19**, 654 (1959).
[3] R. B. Hurlbert and H. O. Kammen, *J. Biol. Chem.* **235**, 443 (1960).
[4] C. R. Savage and H. Weinfeld, *J. Biol. Chem.* **245**, 2529 (1970).
[5] R. P. McPartland and H. Weinfeld, *J. Biol. Chem.* **251**, 4372 (1976).
[6] C. W. Long and A. B. Pardee, *J. Biol. Chem.* **242**, 4715 (1967).
[7] D. E. Koshland, Jr. and A. Levitzki, *in* "The Enzymes" (P. D. Boyer, ed.), 3rd ed., Vol. 10, p. 539. Academic Press, New York, 1974.
[8] "Enzyme Nomenclature," Recommendations (1972) of the Commission on Biochemical Nomenclature, p. 325. Am. Elsevier, New York, 1972.
[9] R. P. McPartland and H. Weinfeld, *Fed. Proc., Fed. Am. Soc. Exp. Biol.,* **34**, 549 (1975).
[10] R. P. McPartland, M. C. Wang, A. Bloch, and H. Weinfeld, *Cancer Res.* **34**, 3107 (1974).
[11] R. W. Brockman, S. C. Shaddix, M. Williams, J. A. Nelson, L. M. Rose, and F. M. Schabel, Jr., *Ann. N.Y. Acad. Sci.* **255**, 501 (1975).
[12] D. D. Genchev, *Experientia* **29**, 789 (1973).
[13] R. C. Jackson, J. C. Williams, and G. Weber, *Cancer Treat. Rep.* **60**, 835 (1976).

at 290 nm since uridine nucleotides are almost completely transparent at this wavelength. In the second method, the formation of labeled CTP from [2–^{14}C]UTP is measured.

Reagents for Assays (Final Concentrations)
  ATP, 8 mM
  UTP or [^{14}C]UTP, 0.2 mM. The specific activity in the radioassay
    is 0.5 μCi/μmole.
  2-Mercaptoethanol, 50 mM
  L-glutamine, 55 mM
  MgCl$_2$, 18 mM
  Tris-Cl, 35 mM. The final pH of the assay mixture is 7.2.

Procedure

Method 1. After incubating 1 ml of the mixture containing the enzyme at 37°, the reaction is stopped by the addition of 0.1 ml of 4 N HClO$_4$ and protein is removed by centrifugation at 1000 g for 10 min at room temperature. The optical density at 290 nm is read against a blank where UTP is added to a complete reaction mixture after the addition of HClO$_4$. Under these conditions the molar extinction coefficient of CTP was observed[4] to be 5.4 × 10^3, and this value is used to calculate the amount of CTP formed.

Method 2. The reaction in 1 ml containing [^{14}C]UTP is stopped after an appropriate time of incubation at 37° by the addition of 0.1 ml of 3 N HCl. CTP and UTP are hydrolyzed to CMP and UMP by heating the mixture in a boiling water bath for 1 hr. One μmole of carrier CMP is added, and the solution is diluted 10-fold with water and applied to a 1 × 4 cm column of Dowex 50-8XH$^+$, 200–400 mesh, at room temperature. UMP is washed from the column with three 3-ml portions of 10^{-3} M HCl. CMP is then eluted with 24 ml of 1 N LiOH. The eluate is neutralized with 1 ml of glacial acetic acid, and the radioactivity is determined by scintillation spectrometry of a 2-ml aliquot of the elutate added to 10 ml of a scintillation phosphor.[14] Blank [^{14}C]UTP values are obtained by adding [^{14}C]UTP to a complete reaction mixture just after the addition of 3 N HCl. In most cases less than 2% of the radioactivity retained by the column can be attributed to [^{14}C]UTP. At 50% counting efficiency, 100% conversion to [^{14}C]CTP yields 8800 cpm in the 2-ml aliquot.

The two assays agree within 10% and the formation of CTP is directly proportional to the amount of enzyme protein incubated and to

[14] W. J. Steele, N. Okamura, and H. Busch, *J. Biol. Chem.* **240,** 1742 (1965).

the time of incubation at 37° up to at least 20% conversion of the UTP to CTP.[4] The formation of 1 μmole of CTP per hour represents 1 unit of enzyme activity. Protein is determined utilizing the absorbance ratio 280/260 nm[15] with bovine serum albumin or ovalbumin as the standard.

Purification

Initial attempts to purify CTP synthetase from liver cytosol were unsuccessful due to marked enzyme instability. It was later found that the activity could be preserved by maintaining the enzyme in buffers containing both L-glutamine and a sulfhydryl-protecting compound. A 350-fold purification of the stabilized enzyme was subsequently achieved using classical protein precipitation and adsorption techniques.[4] The enzyme was further purified[5] 1000-fold and more from liver cytosol using differential ammonium sulfate precipitation and extraction followed by sucrose density gradient centrifugation under conditions known to polymerize the bacterial enzyme.

Unless otherwise stated, all procedures are carried out at 2–4° and all centrifugations are for 10 min at 15,000 g.

Step 1. Fresh bovine calf liver is minced with scissors or passed through a meat grinder, and a 25% homogenate (w/v) is prepared in 0.01 M Tris-Cl buffer, pH 8.0, containing 0.25 M sucrose and 0.1 M 2-mercaptoethanol. A batch of 400 ml is blended in a Waring Blendor at low speed for 5 sec at 80 V and 30 sec at 60 V. The homogenate is centrifuged for 1 hr at 105,000 g. The supernatant fluid, Fraction I, is carefully removed by aspiration from the pellet which apparently contains an inhibitor of the enzyme.

Step 2. To 1000 ml of Fraction I are added 282 ml of aqueous ammonium sulfate solution saturated at 4°. After standing for 20 min, 191 ml of the ammonium sulfate solution are added with slow mechanical stirring. Twenty minutes after the second addition the precipitate is collected by centrifugation and dissolved in 250 ml of GMT buffer which contains 0.1 M L-glutamine, 0.1 M mercaptoethanol, and 0.035 M Tris-Cl, final pH 7.4; the volume is adjusted with GMT buffer to provide a protein concentration of 6.5–7.5 mg/ml. This solution is designated Fraction II.

Step 3. Fraction II is combined with an equal volume of calcium phosphate gel [16] in water (1 mg solids per ml). The ratio of gel solids to

[15] O. Warburg and W. Christian, *Biochem. Z.* **310**, 384 (1942).
[16] L. A. Heppel, *in* "Methods in Enzymology" (S. P. Colowick and N. O. Kaplan, eds.), Vol. 2, p. 572. Academic Press, New York, 1955.

protein is about 1:7. The mixture is intermittently swirled for 20 min, and the gel is collected by centrifugation at 1000 g for 10 min. The pellet is washed once with 12 ml of cold GMT buffer before the gel is dissolved in 36 ml of GMT buffer containing 0.05 M EDTA. The solution is dialyzed overnight against 1 liter of GMT buffer with one change after 10 hr. The solution, Fraction III, is clarified by centrifugation and adjusted to a protein concentration of 4.0–4.5 mg/ml with GMT buffer.

Step 4. A 1% solution of streptomycin sulfate is added to the Fraction III solution with stirring until the final concentration of streptomycin sulfate is 0.45%. After standing for 20 min the precipitate is removed by centrifugation and the supernatant solution is designated Fraction IV.

Step 5. CTP synthetase is concentrated from Fraction IV by adding a saturated solution $(NH_4)_2SO_4$, preadjusted to pH 7.2 with concentrated NH_3 solution, to a final concentration of 35% of saturation. After 20 min the precipitate, Fraction V, is collected by centrifugation.

The enzyme can be stockpiled conveniently as Fraction V in the lyophilized form by dissolving the precipitate in 0.035 M Tris-Cl buffer, pH 7.4, containing 0.01 M L-cysteine before freeze-drying. In solution in GMT buffer Fraction V has a half-life of 4 days at 4°. When lyophilized in Tris buffer containing L-cysteine and stored at $-10°$, the enzyme is relatively stable for several months with a 30–40% loss in activity.[4]

Step 6. The Jakoby procedure[17] is carried out with fresh or lyophilized Fraction V using neutralized saturated $(NH_4)_2SO_4$. Fraction V protein, 20–25 mg, is dissolved in 8 ml of ice-cold GMT buffer and precipitated at 2° by adding cold saturated $(NH_4)_2SO_4$ to 35% of saturation. After standing for 20 min, the precipitate is collected by centrifugation in the cold for 20 min. The supernatant is discarded. The pellet is mixed at 2° with 5 ml of GMT buffer containing $(NH_4)_2SO_4$ at 30% of saturation, and after standing for 10 min the precipitate is collected by centrifugation, discarding the supernatant. The pellet is then extracted with 5 ml of cold GMT buffer containing $(NH_4)_2SO_4$ at 25% of saturation. After standing for 10 min at 2° the mixture is centrifuged and the supernatant is saved and designated the S-25 supernatant. The pellet is extracted twice with 5 ml of cold GMT buffer containing $(NH_4)_2SO_4$ at 20% of saturation, centrifuging between extractions. The supernatants from the two extractions are combined and designated S-20. The pellet is discarded. S-25 and S-20 are allowed to stand for 1 hr at room temperature. Precipitates will form. The two

[17] W. B. Jakoby, *Anal. Biochem.* **26**, 295 (1968).

preparations are centrifuged at 15,000 g for 10 min at room temperature. The supernatant from S-20 is used to dissolve the pellet from S-25 at room temperature. The solution is chilled to 2° and brought to 35% of saturation with saturated $(NH_4)_2SO_4$. The precipitate, Fraction VI, formed after 20 min is collected by centrifugation, dissolved in 1 ml of GMT buffer, and assayed for protein. Fraction VI can be stored in the lyophilized state as described above and dissolved in GMT buffer for further purifications. The recovery of Fraction VI from Fraction V varies from 30% to 50%.

Step 7. Fraction VI is dissolved in GMT buffer to provide a protein concentration of 6–18 mg/ml, and the solution is dialyzed against GMT buffer until sulfate-free. Aliquots of 0.1 or 0.2 ml are layered on 5%–20% linear sucrose gradients prepared in GMT buffer, and the samples are centrifuged at 41,000 rpm in a Beckman-Spinco SW41 rotor for 18.3 hr at 5°. Fractions of 0.5 ml each are collected from the pierced bottoms of six gradient tubes. Corresponding fractions are combined, and each solution is assayed for enzyme activity. The active fractions are pooled and precipitated by the addition of neutral saturated ammonium sulfate to achieve 35% of saturation. After 20 min, the precipitate is collected by centrifugation, dissolved in 1 ml of GMT buffer, and dialyzed against GMT buffer until free of sulfate. Alternatively, it can be dissolved in the L-cysteine-Tris buffer (see step 5) and lyophilized and stockpiled.

Step 8. Polymerization to Fraction VIII. Fraction VII, 0.5–2 mg of protein in 0.1 ml of GMT buffer, is layered over 5–20% linear sucrose gradients prepared in GMT buffer containing kinetically saturating concentrations of ATP, UTP, GTP, and $MgCl_2$ (see the list of reagents). Centrifugation and fractionation of the gradients followed by assay are performed as in step 7. The active fractions are pooled, and the enzyme is precipitated with neutralized ammonium sulfate at 35% of saturation and dissolved in 1 ml of GMT buffer. The solution is dialyzed against GMT buffer until it is free of sulfate and nucleotides. The resulting solution is Fraction VIII. For protein determination, aliquots are precipitated with trichloroacetic acid, the precipitates dissolved in 0.1 M NaOH, and the protein determined[18] using bovine serum albumin as standard. Enzyme assays are performed on separate aliquots.

Five separate preparations of Fraction VIII had an average specific activity of 0.59 units/mg protein. Since the usual cytosol specific activity is about 0.50×10^{-3}, the entire procedure yields about a 1000-fold purification. Table I details the results of the purification. Although

---

[18] O. H. Lowry, N. J. Rosebrough, A. L. Farr, and R. J. Randall, *J. Biol. Chem.* **193**, 265 (1951).

TABLE I
PURIFICATION OF CTP SYNTHETASE FROM BOVINE LIVER[4,5]

Fraction	Units	Protein (mg)	Specific activity[a] (units/mg protein)	Yield (%)
I. 105,000 g supernatant	15.1	32,600	4.63×10^{-4}	100
II. First ammonium sulfate	15.1	3,140	4.80×10^{-3}	100
III. Ca$_3$(PO$_4$)$_2$ adsorption	11.6	351	3.30×10^{-2}	77
IV. Streptomycin	7.6	100	7.60×10^{-2}	50
V. Second ammonium sulfate	6.1	35	1.74×10^{-1}	40
V. Second ammonium sulfate, pooled and lyophilized	18.0	133	1.35×10^{-1}	40[b]
VI. Differential precipitation	6.1	38	1.61×10^{-1}	14
VII. Sucrose gradient centrifugation	2.7	11	2.45×10^{-1}	6
VII. Sucrose gradient centrifugation, pooled and lyophilized	2.3	7.7	2.99×10^{-1}	10[b]
VIII. Polymerized	0.8	0.88	9.10×10^{-1}	3.4

[a] The radioassay was used throughout.
[b] Based on the cytosols from which these pools were obtained.

slight purification resulted from step 6, its retention produced a purer enzyme after polymerization when homogeneity was examined. Occasionally a 2000-fold purification results.

The homogeneity of Fraction VIII was analyzed using polyacrylamide gel electrophoresis.[5] Replicate gels were assayed for enzymic activity, stained for protein, and scanned. Two protein bands of equal intensity of staining were observed, but only one had enzymic activity. Fraction VIII is at least 50% pure.

Properties of the Enzyme

Some physical properties are given in Table II. Under most conditions liver CTP synthetase exhibits classical Michaelis–Menten kinetics[4] in sharp contrast to the cooperative behavior of the bacterial enzyme. However, the liver enzyme does exhibit sigmoidal kinetics as a function of the concentration of UTP when 0.2 mM CTP is present and ATP is saturating.[9] A strong inhibitor of the enzyme is 3-deazauridine 5'-triphosphate (3-deazaUTP) which competes with UTP.[10]

The molecular weights and sedimentation coefficients of the enzyme have been determined in the absence and presence of kinetically

TABLE II
CHARACTERISTICS OF BOVINE LIVER CTP SYNTHETASE

Kinetics[4,9]	K_m ($M \times 10^5$)	Hill coefficient
Compound varied[a]		
UTP	7	1.0
UTP; ATP subsaturating	25	1.0
UTP; CTP 0.2 mM	—	2.5
ATP	91	1.0
GTP	7	1.1
L-glutamine	21	1.0
Molecular weights[5]		
Monomer	128,000[b]	
Dimer	263,000[b]	
Sedimentation coefficients[5]		
Monomer	6.8 S[c]	
Dimer	10.1 S[c]	

[a] ATP, GTP, UTP, and L-glutamine were present at saturating concentrations unless otherwise stated. CTP was present only where indicated.
[b] Determined by thin-layer gel chromatography.
[c] Determined by sucrose density-gradient centrifugation.

saturating levels of ATP, UTP, GTP, L-glutamine, and MgCl$_2$. The results[5] indicate that the liver enzyme can undergo substrate-dependent dimerization as does the bacterial enzyme.[7] ATP and UTP present simultaneously are required for maximal dimerization.[5]

[14] TMP Synthetase from *Lactobacillus casei*[1]

By R. BRUCE DUNLAP

Assay Methods

Principle. The methylenetetrahydrofolate-dependent reductive methylation of dUMP is followed spectrophotometrically at 340 nm by

[1] Supported in part by NIH grant CA12842 and a Faculty Research Award (FRA-144) to the author from the American Cancer Society.

observing the formation of 7,8-dihydrofolate.[2,3] Alternatively, enzyme activity is measured by monitoring the progress of the reaction via the loss of tritium to the aqueous solvent from [5-^{3}H]dUMP[3,4] or by use of a filter assay which traps the 5-fluorodUMP-methylenetetrahydrofolate-thymidylate synthetase ternary complex.[5]

Procedures. SPECTROPHOTOMETRIC ASSAY.[2,3] To a 1-cm light path cuvette add 0.1 ml of $2 \times 10^{-3} M$ (±)-L-5,10-methylenetetrahydrofolate (see below), 0.2 ml of 0.5 M potassium phosphate buffer, pH 6.8, enzyme, and water to a volume of 0.9 ml. The dUMP-independent change in absorbance at 340 nm is followed for 1–3 min in a recording spectrophotometer whose sample cell compartment is thermostatically controlled at 25°. The thymidylate synthetase reaction is initiated by the addition of 0.1 ml of $10^{-3} M$ dUMP to the cuvette, and the resulting absorbance change, corrected for the dUMP-independent blank rate, is used together with the differential extinction coefficient of $6.4 \times 10^3 M^{-1}$ cm^{-1} at 340 nm for 7,8-dihydrofolate[2] to calculate the enzymic rate. One unit of enzyme activity is defined as the amount of enzyme required to synthesize 1 μmole of thymidylate per minute under these conditions. A stock solution of $(2 \times 10^{-3} M)$ (±)-L-methylenetetrahydrofolate is prepared by dissolving 6 mg of (±)-L-tetrahydrofolate[6,7] in 5 ml of a solution containing 0.05 M NaHCO$_3$, 0.07 M formaldehyde, and 0.25 M 2-mercaptoethanol. This solution is prepared fresh daily and is kept in an ice bath under a gentle stream of argon when being used for assays.

TRITIUM RELEASE ASSAY.[3,4] To a 10×100 mm test tube add 0.05 ml of $2 \times 10^{-3} M$ (±)-L-methylenetetrahydrofolate (see above), 2.5 μmoles of potassium fluoride, 50 μmoles of potassium phosphate buffer, pH 6.8, enzyme, and water to make the total volume 0.5 ml. The assay is initiated by the addition of 0.05 ml of $10^{-3} M$ [5-^{3}H]dUMP (3.6 μCi/ μmole). Following an incubation period of 10 min at 25° under argon, the reaction is terminated by the addition of 0.1 ml of 1 M HCl. Washed Norit A (40 mg), suspended in 0.5 ml of water, is added to the tube, the contents are mixed, and filtration through a Millipore membrane is performed to remove the charcoal which adsorbs the unreacted [5-^{3}H]dUMP. A 0.2-ml aliquot of the filtrate is added to 10 ml of scintillant (5 g of butyl-PBD, 100 ml of Bio-Solv BBS2, and 900 ml of toluene) and

[2] A. J. Wahba and M. Friedkin, *J. Biol. Chem.* **236,** PC11 (1961).
[3] R. B. Dunlap, N. G. L. Harding, and F. M. Huennekens, *Biochemistry* **10,** 88 (1971).
[4] M. I. S. Lomax and G. R. Greenberg, *J. Biol. Chem.* **242,** 109 (1967).
[5] D. V. Santi, C. S. McHenry, and E. R. Perriard, *Biochemistry* **13,** 467 (1974).
[6] Y. Hatefi, P. T. Talbert, M. J. Osborn, and F. M. Huennekens, *Biochem. Prep.* **7,** 89 (1960).
[7] W. E. Caldwell, J. A. Lyon, and R. B. Dunlap, *Prep. Biochem.* **3,** 323 (1973).

assayed for tritium by scintillation counting. Corrections must be made for blanks in which enzyme is omitted or inactivated by HCl prior to addition of the substrates.

Purification Procedure

An amethopterin-resistant strain of *Lactobacillus casei* was obtained as described by Dunlap *et al.*[3] by culturing *Lactobacillus casei* var. *rhamnosus* (ATCC 7469) in the presence of increasing levels of amethopterin, until the bacteria grew readily in media containing a drug concentration of 1×10^{-5} M. The drug-resistant organism is maintained on slants containing 1×10^{-5} M amethopterin or is stored conveniently in the lyophilized form at -70°.

Growth of Bacterial Cells. Amethopterin-resistant *L. casei* was grown on the 400-liter scale essentially according to the procedure previously described,[3] except that the separate autoclaving and charcoal treatment of the enzyme-hydrolyzed casein was found to be unnecessary and, thus, was omitted. The complete medium, containing enzyme-hydrolyzed casein, amino acids, bases, salts, glucose, and vitamins but no amethopterin, was inoculated with 1.5 liters of a log-phase culture of the amethopterin-resistant organism containing 1×10^{-5} M amethopterin. Growth at 37° was monitored by turbidity and pH changes, and after 8–10 hr the cells were harvested in late-log phase by centrifugation. The wet, unwashed cell paste was stored at -70°. The yield of wet cells is 4–5 g/liter of medium.

General. Unless otherwise stated, all operations were conducted at 0–5°. A summary of data for a typical purification run is provided in the table.

Step 1. Sonic Extract. Frozen cells (150 g) of amethopterin-resistant *L. casei* were thawed overnight in 7 volumes of 50 mM Tris·HCl buffer, pH 7.3, containing 50 mM KCl, 10 mM 2-mercaptoethanol, and 1 mM EDTA. On the following morning the pH of the suspension (~pH 5) was adjusted to 7.3 with 6 M NaOH. Suitable aliquots of this suspension were placed in a 400-ml rosette flask immersed in an ice-water bath, and the cells were disrupted with a Heat Systems–Ultrasonics Sonifier Cell Disruptor Model W185 at a setting of 9 for five 5-min periods. Temperatures during sonification were maintained at or below 15°. The pH of the sonicate (1200 ml) was adjusted to 7.3 with 6 M NaOH. Centrifugation at 21,000 g for 40 min yielded a pale yellow cell-free extract.

Step 2. Ammonium Sulfate Fractionation. Solid ammonium sulfate

PURIFICATION OF THYMIDYLATE SYNTHETASE FROM AMETHOPTERIN-RESISTANT *Lactobacillus casei*

Step and fraction	Volume (ml)	Protein (mg/ml)	Activity (units/ml)	Specific activity (units/mg)	Total units	Recovery (%)
1. Cell-free extract	1100	12	0.4	0.033	440	100
2. Ammonium sulfate dialysate	300	14	1.2	0.086	360	82
3. CM-Sephadex pool	160	2.4	2.0	0.85	320	73
4. Hydroxyapatite pool	150	0.72	1.87	2.5	280	63
4'. DEAE-Sephadex pool	50	2.0	5.0	2.5	250	57

was added over a period of 10 min to the stirred cell-free extract until 35% saturation (19.6 g/100 ml) was attained. The resulting suspension was stirred an additional 15 min, centrifuged at 21,000 g for 40 min, and the precipitate was discarded. The supernatant was treated with sufficient ammonium sulfate (18.6 g/100 ml) to attain 65% saturation. After an additional 15 min of stirring, the suspension underwent centrifugation at 21,000 g for 40 min to yield about 100 g of an intensely yellow precipitate. This pellet was dissolved in a minimal volume of 50 mM Tris·HCl buffer, pH 7.3, containing 50 mM KCl, 10 mM 2-mercaptoethanol, and 1 mM EDTA. The resulting protein solution was dialyzed overnight against 6 liters of the preceding buffer, followed by dialysis for two 5-hr periods each against 6 liters of 5 mM Tris·HCl, pH 6.5, containing 10 mM KCl, 10 mM 2-mercaptoethanol, and 1 mM EDTA. Steps 1 and 2 are conveniently completed over an 8-hr period.

Step 3. Carboxymethyl-Sephadex Chromatography. The dialyzed protein solution was loaded on a carboxymethyl-Sephadex column (5 × 90 cm) which had been equilibrated in the latter buffer. The column was washed with the same buffer (~1 liter); many of the contaminating materials were eluted directly as evidenced by the appearance of a series of opaque and then highly yellow fractions from the column. Both thymidylate synthetase and dihydrofolate reductase activities were strongly retained by the cation exchanger. When the absorbance at 280 nm of the effluent had decreased to 0.1, the column was eluted with a linear gradient consisting of 1 liter of 50 mM Tris·HCl buffer, pH 6.5, 50 mM KCl, 10 mM 2-mercaptoethanol, and 1 mM EDTA in the mixing chamber and 1 liter of 50 mM Tris·HCl, pH 7.7, 700 mM KCl, 10 mM 2-mercaptoethanol, and 1 mM EDTA in the reservoir chamber. If necessary, once the gradient was exhausted, washing of the column was continued with reservoir buffer. Elution profiles[8] for the carboxymethyl-Sephaex column illustrate the fact that thymidylate synthetase elutes toward the end of the gradient and is completely separated from the dihydrofolate reductase activity which immediately precedes it.[9]

Step 4. Hydroxyapatite Chromatography. The pool of thymidylate synthetase activity obtained from the carboxymethyl-Sephadex step was dialyzed against 6 liters of 50 mM potassium phosphate, pH 6.8, containing 10 mM 2-mercaptoethanol for 12 hr. The dialyzate was centrifuged for 15 min at 12,000 g to remove a small amount of white precipitate prior to loading on a hydroxyapatite column (2.5 × 25 cm)

[8] J. A. Lyon, A. L. Pollard, R. B. Loeble, and R. B. Dunlap, *Cancer Biochem. Biophys.* **1**, 121 (1975).

[9] J. K. Liu and R. B. Dunlap, *Biochemistry* **13**, 1807 (1974).

equilibrated in the same buffer. The column was then washed over a 12-hr period with about 300 ml of 110 mM phosphate buffer, pH 6.8, containing 10 mM 2-mercaptoethanol. Thymidylate synthetase activity was eluted from the column in a broad peak by washing with 130 mM phosphate buffer, pH 6.8, containing 10 mM 2-mercaptoethanol.

Step 5. DEAE-Sephadex Chromatography. As an alternative to the hydroxyapatite step, chromatography on DEAE-Sephadex can be employed to purify thymidylate synthetase to homogeneity. Thus, the carboxymethyl-Sephadex pool was dialyzed against two 6-liter volumes of 0.1 M Tris·HCl buffer, pH 7.3, containing 10 mM MgCl$_2$, 10 mM 2-mercaptoethanol, and 1 mM EDTA prior to loading on a DEAE-Sephadex column (2.5 × 35 cm) previously equilibrated in the same buffer. The column was then eluted with a linear gradient constructed with 250 ml of 0.1 M Tris·HCl buffer, pH 7.3, 10 mM MgCl$_2$, 10 mM 2-mercaptoethanol, and 1 mM EDTA in the mixing chamber and 250 ml of a similar buffer containing 200 mM MgCl$_2$ in the reservoir. Thymidylate synthetase eluted from this column in a single, sharp protein peak.

The entire purification scheme is accomplished over a period of 10 to 14 days. The purity of the enzyme following the hydroxyapatite step or the DEAE-Sephadex procedure was routinely analyzed by subjecting samples (75–100 μg protein) from fractions across the thymidylate synthetase peak to electrophoresis on 7.5% polyacrylamide gels. Both the purity of enzyme samples and their relative degree of activation were measured by employing gel electrophoresis to monitor the extent of the conversion of native enzyme, incubated with 5-fluorodUMP and methylenetetrahydrofolate, to its ternary complexes[10,11] (see below). As a final step in preparing thymidylate synthetase for investigations in this laboratory, homogenous samples of the enzyme are activated by dialysis for 12 hr against 2 liters of buffer containing 25 mM 2-mercaptoethanol. The specific activity of activated thymidylate synthetase ranges from 3.0–4.2 units/mg protein. In those studies requiring the absence of thiol, the 2-mercaptoethanol is conveniently removed from the activated enzyme by gel filtration on a Sephadex G-10 column (2 × 30 cm).

Properties

Physical Properties. Native thymidylate synthetase has a molecular weight of approximately 70,000 based on analytical polyacrylamide gel electrophoresis and gel filtration chromatography. Sedimentation veloc-

[10] J. L. Aull, J. A. Lyon, and R. B. Dunlap, *Microchem. J.* **19**, 210 (1974).
[11] J. L. Aull, J. A. Lyon, and R. B. Dunlap, *Arch. Biochem. Biophys.* **165**, 805 (1974).

ity studies in the presence of guanidine HCl and electrophoresis on sodium dodecyl sulfate–polyacrylamide gels indicate that the enzyme consists of two subunits, each having a molecular weight of 35,000. Results of end-group analyses coupled with cyanogen bromide cleavage studies suggest that the subunits share an identical primary structure.[12] The absorption spectrum of thymidylate synthetase exhibits a prominent shoulder at 292 nm and maxima at 282 nm and 278 nm ($E_{278} = 1.05 \times 10^5$ M^{-1} cm^{-1}). The magnitude of the molar extinction coefficient is largely accounted for by the presence of 12 tryptophan residues in the protein. The isoelectric point of the enzyme is 5.4.

Stability. The enzyme is stable over a period of months when stored at 0–5° in the presence of 10 mM 2-mercaptoethanol in capped vessels layered with argon. Although specific activities of enzyme samples stored in this manner decline gradually, activation of the stored enzyme by dialysis against thiol-containing buffer usually results in recovery of specific activities in the range of 2.5–3.5 units/mg. Thawing of frozen thymidylate synthetase solutions leads to irreversible precipitation of the protein and total loss of activity. We have not yet found conditions under which the enzyme can be successfully lyophilized.

pH Optimum and Kinetic Properties. In both phosphate and Tris-acetate buffers the enzyme displays a broad pH optimum in the range of 6.5–6.8 with dUMP as the substrate. In Tris-acetate buffers 10 mM Mg^{2+} enhances enzyme activity by some 30%. The K_m values for dUMP in Tris-acetate buffers are 5.1 × 10^{-6} M and 6.8 × 10^{-7} M, respectively, in the presence and absence of 10 mM Mg^{2+}. The K_m values for (±)-L-methylenetetrahydrofolate in Tris-acetate buffers are 3.2 × 10^{-5} M and 1.2 × 10^{-5} M, respectively, in the presence and absence of 10 mM Mg^{2+}. In Tris-acetate buffers the enzyme utilizes UMP as a substrate, the pH optimum is shifted to 5.9, and added Mg^{2+} has no effect on the rate of methylation. Phosphate buffers afford the enzyme considerable protection from heat inactivation as compared to Tris-acetate buffers.

Sulfhydryl Group Studies. Thymidylate synthetases are stabilized and their activities are stimulated to varying degrees by different thiols.[13] The enzyme is inhibited by *p*-chloromercuribenzoate, iodoacetamide, cyanide, H$_2$O$_2$, 2-hydroxyethyl disulfide, and 5,5′-dithiobis (2-nitrobenzoic acid). Amino acid analyses and chemical modification studies indicate the presence of 3.5 to 4 sulfhydryl groups per enzyme molecule. Recent studies with *p*-chloromercuribenzoate, *N*-ethylmaleimide, and

[12] R. B. Loeble and R. B. Dunlap, *Biochem. Biophys. Res. Commun.* **49,** 1671 (1972).
[13] M. Friedkin, *Adv. Enzymol.* **37,** 235 (1973).

iodoacetamide indicate total inactivation of the enzyme occurs on modification of 1.4 to 1.7 sulfhydryl groups per enzyme molecule.[14] Inactivation of the enzyme by these reagents is strictly correlated with a loss in ability of the enzyme to form covalent ternary complexes. The role of the catalytic sulfhydryl groups has been amplified by recent reports of the isolation of a peptide containing cysteine covalently attached to FdUMP[15,16] and by the observation that treatment of the ternary complexes with Raney nickel releases the covalently bound FdUMP.[17]

Ternary Complexes. The potent inhibitor, 5-fluorodUMP, forms covalent ternary complexes[18,19] with thymidylate synthetase (Form I) and the cofactor, (+)-methylenetetrahydrofolate, which migrate as two well-separated bands (Form II, Form III) when subjected to electrophoresis on polyacrylamide gels.[10] Treatment of the activated enzyme with large molar excesses of 5-fluorodUMP and cofactor yields a maximum conversion of Form I to 30% Form II and 70% Form III, as measured by gel-scanning techniques. Chromatography on Sephadex G-25 (1 × 50 cm) conveniently separates the mixture of ternary complexes from excess inhibitor and coenzyme. The mixture of ternary complexes, which exhibits absorption maxima at 375 nm, 322 nm, and 280 nm, can be fractionated on a preparative scale on carboxymethyl-Sephadex chromatography yielding pure Form II and Form III.[11] Labeling experiments indicate that the stoichiometry of 5-fluorodUMP to coenzyme to thymidylate synthetase is 1:1:1 in Form II and 2:2:1 in Form III. The absorption spectrum of pure Form II is characterized by absorption maxima at 280 nm ($1.14 \times 10^5 M^{-1} cm^{-1}$), 322 nm ($2.82 \times 10^4 M^{-1} cm^{-1}$), and 375 nm, yielding a 322/280 ratio of 0.25, while that for Form III features maxima at 280 nm ($1.38 \times 10^5 M^{-1} cm^{-1}$), 322 nm ($5.98 \times 10^4 M^{-1} cm^{-1}$), and 375 nm with a 322/280 ratio of 0.44. The extent of ternary complex formation can be readily monitored by absorption, circular dichroic, and fluorescence spectroscopy and by polyacrylamide gel electrophoresis.[20]

[14] P. C. Plese and R. B. Dunlap, *J. Biol. Chem.* **252**, 6139 (1977).
[15] R. L. Bellisario, G. F. Maley, J. H. Galivan, and F. Maley, *Proc. Natl. Acad. Sci. U.S.A.* **73**, 1848 (1976).
[16] A. L. Pogolotti, K. M. Ivanetich, H. Sommer, and D. V. Santi, *Biochem. Biophys. Rec. Commun.* **70**, 972 (1976)
[17] P. V. Danenberg and C. Heidelberger, *Biochemistry* **15**, 1331 (1976).
[18] D. V. Santi and C. S. McHenry, *Proc. Natl. Acad. Sci. U.S.A.* **69**, 1855 (1972).
[19] R. J. Langenbach, P. V. Danenberg, and C. Heidelberger, *Biochem. Biophys. Res. Commun.* **48**, 1565 (1972).
[20] H. Donato, Jr., J. L. Aull, J. A. Lyon, J. W. Reinsch, and R. B. Dunlap, *J. Biol Chem.* **251**, 1303 (1976).

[15] 5-Fluoro-2'-deoxyuridylate-agarose in the Affinity-Chromatographic Purification of Thymidylate Synthetase[1]

By John M. Whiteley

5-Fluoro-2'-deoxyuridine 5'-phosphate is a potent inhibitor of thymidylate synthetase[2,3] (K_i ~10^{-9} M) which binds covalently to the enzyme.[4] Replacement of the 5'-phosphate by a 5'-p-aminophenyl phosphate ester group reduces this interaction (K_i ~10^{-6} M),[5] and additionally provides a suitable substituent for linking the inhibitor to an agarose matrix. The synthesis and coupling of 5-fluoro-2'-deoxyuridine 5'-(p-aminophenyl phosphate) via a succinylaminohexyl linkage to agarose, and the use of this matrix for the efficient isolation of thymidylate synthetase from amethopterin-resistant *Lactobacillus casei*, are described.

Preparation of the 5-Fluoro-2'-deoxyuridine 5'-(p-aminophenyl phosphate)-agarose Column Material

5-Fluoro-2'-deoxyuridine 5'-(p-nitrophenylphosphate). Disodium p-nitrophenyl phosphate hexahydrate[6] (3.00 g) is converted to the acid form by passage through a 3 × 20 cm column of Dowex 50, and the aqueous effluent is lyophilized. The white solid is twice taken up in dry pyridine (20 ml) and evaporated to dryness.[7] 5-Fluoro-2'-deoxyuridine (2.00 g), after treatment with pyridine in the same manner, is then taken up in a 1:1 (v/v) mixture of dry pyridine and dimethylformamide (20 ml) and added to the p-nitrophenyl phosphate. The mixture is again evaporated to dryness. Dicyclohexyl-carbodiimide (2.50 g), dissolved in a

---

[1] Supported by Grants CA-11778 and CA-00106 (RCDA) from the National Cancer Institute, National Institutes of Health.

[2] S. S. Cohen, J. G. Flaks, H. D. Barner, M. R. Loeb, and J. Lichtenstein, *Proc. Natl. Acad. Sci. U.S.A.* **44**, 1004 (1958).

[3] L. Bosch, E. Harbers, and C. Heidelberger, *Cancer Res.* **18**, 335 (1958).

[4] D. V. Santi and C. S. McHenry, *Proc. Natl. Acad. Sci. U.S.A.* **69**, 1855 (1972).

[5] J. M. Whiteley, I. Jerkunica, and T. Deits, *Biochemistry* **13**, 2044 (1974).

[6] It has been observed that differing preparations of this ester greatly influence the yield of the desired product.

[7] (a) Solubility in this step is improved if a little dry dimethylformamide is added prior to evaporation. (b) Pronounced insolubility may indicate a lack of complete conversion of the sodium salt to the acid form. (c) Evaporation is achieved using a rotary evaporator with an external bath temperature of < 50° and a pressure of < 10 mm maintained with a vacuum pump.

further volume of the mixed pyridine-dimethylformamide solvent (40 ml), is then added to the residue, and the solution is stirred for 3 days at 37° in a flask protected from moisture by a $CaCl_2$ drying tube. Water (20 ml) is added, and the mixture is stirred for 2 hr. The precipitate which forms is removed by filtration, and the filtrate is evaporated to dryness. Water (20 ml) is again added to the residue, and insoluble material is removed by filtration. The filtrate is passed through the Dowex 50 column, and the aqueous eluate is adjusted to pH 7 with 0.1 M NH_4OH. The neutral solution is then chromatographed on a 4 × 50 cm column of DEAE-cellulose, previously equilibrated with 0.01 M ammonium acetate. The products are eluted with a 0.01–0.5 M linear gradient of ammonium acetate (1 liter in each reservoir) followed by a further 2 liters of the 0.5 M solution. Twenty-milliliter fractions are collected, and the effluent is monitored[8] for absorbance at 280 nm. Fractions containing material with maximum absorbance at 272 nm are combined and lyophilized. The residue is dissolved in water and passed, with the same solvent, through the column of Dowex 50. Fractions absorbing at 272 nm are again combined, adjusted to pH 7 with saturated barium hydroxide, and lyophilized. The product is dissolved in a minimum volume of water, and insoluble matter is removed by filtration. Addition of ethanol produces partial precipitation, and the mixture is allowed to stand at 4° for 12 hr. The precipitate is removed; further ethanol is added to the filtrate, and a second precipitate is isolated. This procedure is repeated until no further precipitation occurs; ether is then added to give a final precipitate. Each precipitate is collected and transferred rapidly to an evacuated desiccator. The absorbance spectrum of each sample is determined in 0.1 N HCl, and material (approximately 1.0 g) with a λ_{max} at 272 nm and $\epsilon > 12,000$ is considered sufficiently homogeneous for conversion to the p-amino derivative.

5-Fluoro-2'-deoxyuridine 5'-(p-aminophenyl phosphate). The previous product is reconverted to the acid form by passage through Dowex 50. The aqueous effluent is evaporated, and the gummy residue is taken up in methanol (100 ml); following the addition of 10% palladium charcoal (0.10 g), the mixture is hydrogenated in a Parr pressure hydrogenator at 35 psi for 1 hr. The catalyst is removed by filtration, and the filtrate, after concentration to 10 ml, is chromatographed on Dowex 50. Elution with water affords initially the unchanged starting material,

---

[8] Since in some preparations the resolution of this column is not ideal, aliquots from selected fractions should be diluted in 0.1 N HCl and carefully monitored for absorbance maximum. Unreacted 5-fluoro-2'-deoxyuridine (λ_{max} 268 nm) appears first, followed by the desired product (λ_{max} 272 nm), then 5-fluoro-2'-deoxyuridine 3,5-bis(p-nitrophenyl-phosphate) (λ_{max} 278 nm), and finally a mixture of products with $\lambda_{max} > 280$ nm.

followed by the product. Fractions containing the product are combined, neutralized with saturated barium hydroxide, then precipitated fractionally with ethanol as described above for the *p*-nitrophenyl ester.[9] Material with λ_{max} (0.1 *N* HCl) 268 nm and $\epsilon > 7500$ is considered suitable for the next step.

Coupling to Agarose. An aqueous solution (25 ml) of cyanogen bromide containing approximately 100 mg/ml is added to a stirred suspension of agarose (Sepharose 4B, wet volume 25 ml).[10–12] The mixture is adjusted to pH 10–11 with 4 *N* NaOH and is maintained at this pH for 10 min. The product is filtered rapidly, washed repeatedly with cold water and 0.1 *M* NaHCO$_3$, pH 9, then treated with 0.30 g of 1,6-diaminohexane in 5 ml of the same buffer. After gentle stirring at 4° for 24 hr, the product is collected by filtration, washed thoroughly with buffer and water, and then resuspended in cold water (30 ml). An aqueous suspension of succinic anhydride (3 g in 20 ml) is added to the stirred aminoalkyl-agarose mixture, and the reactants are maintained at 4° by an external ice–salt bath.[13] The pH is maintained at 6 for 2 hr by the addition of 4 *N* NaOH. After stirring for a further 12 hr at 4° the product is collected by filtration and washed repeatedly with water. The succinylaminohexyl-agarose is suspended in water (20 ml) and barium 5-fluoro-2'-deoxyuridine 5'-(*p*-aminophenyl phosphate) (0.04 g) and 1-ethyl-3(3'-dimethylaminopropyl) carbodiimide hydrochloride (0.60 g) are added. The pH is adjusted to 5.5 with 1 *N* HCl, and the suspension is stirred for 48 hr. The solid is washed thoroughly with 1.0 *M* phosphate buffer (pH 7) and water. The unbound nucleotide content is determined from the absorption of a known aliquot of the combined washings diluted into 0.1 *N* HCl.[5] Approximately 1.5 μmoles of ligand are usually bound per gram of wet agarose.

Blocking of Excess Carboxyl Groups. To minimize ion-exchange effects generated by uncoupled carboxyalkyl substituents on the agarose, the matrix from the previous step is suspended in a 1:1 (v/v) mixture of dimethyl-formamide and 0.1 *M* cacodylate buffer (pH 4.5) 50

---

[9] Formation of the barium salts of both the *p*-nitro- and *p*-aminophenyl esters leads to products of greater purity and stability than can be obtained with the free acid.

[10] This operation should be carried out in a well-ventilated hood.

[11] Stabilized, freeze-dried CNBr-activated Sepharose 4B is commercially available (Pharmacia Fine Chemicals, Inc.) and may be used in this step.

[12] Sepharose 4B is washed to remove preservatives and fine particles. The settled volume is determined by overnight sedimentation in a measuring cylinder, or by gentle centrifugation in a clinical centrifuge. The density of the sedimented material is approximately 1 g/ml at 20°.

[13] P. Cuatrecasas, *J. Biol. Chem.* **245,** 3059 (1970).

ml) containing glycinamide (1.10 g).[14] To this suspension is added 1-ethyl-3(3'-dimethylaminopropyl) carbodiimide hydrochloride (2.00 g) dissolved in the same mixture (10 ml), and the reactants are stirred at room temperature for 12 hr. The product is isolated by filtration and then washed sequentially with 0.1 N acetic acid, water, and 1.0 M potassium phosphate buffer (pH 7) prior to its use in a 2 × 10 cm column.

Assay Method

Thymidylate synthetase is assayed[15] by measuring the overall increase in absorbance at 340 nm caused by the conversion of 5,10-methylene-5,6,7,8-tetrahydrofolate to 7,8-dihydrofolate according to Eq. (1).

2'-Deoxyuridylate + 5,10-methylene-5,6,7,8-tetrahydrofolate →

5-methyl-2'-deoxyuridylate (thymidylate) + 7,8-dihydrofolate   (1)

The molar change in extinction at 340 nm for this reaction is 6.4 × 10³ M^{-1} cm^{-1}. Assays are performed at 30° in 0.1 M potassium phosphate buffer (pH 6.8) with 0.2 mM methylene-tetrahydrofolate and 0.1 mM deoxyuridylate. Protein is determined by the Lowry method.[16] One unit of enzyme activity is defined as the amount required to synthesize 1 μmole of thymidylate per minute under these conditions. Specific activity is expressed as units per milligram of protein.

Enzyme Purification Procedure

Cell-free extracts of amethopterin-resistant *L. casei* are prepared according to the procedure of Dunlap *et al.*[17] The column material is preequilibrated at 4° with 0.01 M phosphate (pH 7.0) containing 0.02 M mercaptoethanol. After dialysis against the same buffer, the cell-free extract is applied to the column. A typical purification sequence is illustrated in Fig. 1. The affinity material has a capacity to adsorb approximately 2 units of synthetase per gram wet weight of agarose derivative under the above conditions. Cell-free extracts of the resistant *L. casei* contain approximately 0.4 units/ml of synthetase activity with specific activities extending from 0.003 to 0.050. Step by step elution with a series of phosphate buffers ranging from 0.01 to 0.1 M (cf. legend

[14] H.-C. Chen, L. C. Craig, and E. Stauer, *Biochemistry* **11**, 3559 (1972).

[15] A. J. Wahba and M. Friedkin, *J. Biol. Chem.* **236**, PC11 (1961).

[16] O. H. Lowry, N. J. Rosebrough, A. L. Farr, and R. J. Randall, *J. Biol. Chem.* **193**, 265 (1951).

[17] R. B. Dunlap, N. G. L. Harding, and F. M. Huennekens, *Biochemistry* **10**, 88 (1971).

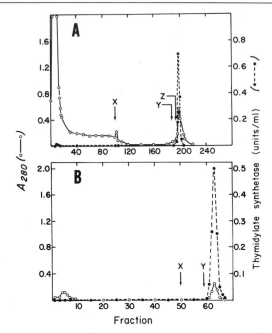

FIG. 1. (A) Affinity chromatography of a cell-free extract of amethopterin-resistant *L. casei*. The extract (37 ml; 0.36 unit/ml) was applied and elution begun with 0.01 *M* phosphate buffer (pH 7) containing 0.02 *M* mercaptoethanol. Step by step increments in buffer concentrations are indicated at points X (0.05 *M*), Y (0.1 *M*), and Z (0.2 *M*); 21-ml fractions were collected up to point X and thereafter the volume was reduced to 6 ml. (B) Rechromatography of peak fractions from (A) on the same column. Elution was initiated with 0.01 *M* phosphate buffer (pH 7.0) containing 0.02 *M* mercaptoethanol. Step by step increments were applied at X (0.1 *M*) and Y (0.3 M).

to Fig. 1), each containing 0.02 *M* mercaptoethanol, causes the elution of unwanted cellular constituents. The addition of 0.2 *M* phosphate containing 0.02 *M* mercaptoethanol causes the elution of a peak of enzymic activity. Fractions containing the enzyme activity are combined, concentrated (if necessary) to approximately 25 ml by ultrafiltration through a Diaflo pressure cell (Amicon PM 10 membrane at 40 psi, argon pressure), dialyzed against 0.01 *M* phosphate buffer, pH 7, containing 0.02 *M* mercaptoethanol, and reapplied to the affinity column material which has again been equilibrated with the phosphate–mercaptoethanol starting buffer. Repetition of the elution procedure gives enzyme, homogeneous by polyacrylamide electrophoresis, of specific activity[18] greater than 3.0.

---

[18] The specific activity varies with buffer ion (phosphate or Tris), and in the presence of Mg^{2+}.

PURIFICATION OF THYMIDYLATE SYNTHETASE FROM AMETHOPTERIN-RESISTANT *Lactobacillus casei*

Step	Protein (mg)	Volume (ml)	Total activity[a] (units)	Specific activity (units/mg protein)	Purification (-fold)	Recovery (%)
Cell-free extract	782	37	13.3	0.017	—	—
Affinity column						
(a) First passage	8.1	23	9.3	1.15	68	70
(b) Second passage	1.8	30	5.9	3.3	194	44

[a] One unit of enzyme activity is defined as the amount required to synthesize 1 μmole of thymidylate per minute.

Comments

The column effluent should be monitored for enzymic activity during the application of the 0.1 M buffer solution, and upon appearance of activity, the eluant should immediately be changed to the higher buffer strength. The sharpest band of enzyme and the highest specific activity are obtained if the strength of the final buffer in the second affinity column is increased to 0.3 M with respect to phosphate. The same column can be used for both preparative steps; however, for repeated use, it is convenient to prepare two identical columns and retain one for the primary application. The recovery of enzymic activity is approximately 70% with each column passage (see the table). If all emergent enzyme-containing fractions are combined, the recovery of activity increases to 85% with each passage; however, this gain is offset by a decrease in specific activity. Approximately 1–2% of the applied activity usually elutes with the initial peak of the first column. With some enzyme preparations a loss of activity occurs during the dialysis step prior to application of the partially purified enzyme to the second affinity column. To avoid this occurrence, the product from the first column may be applied directly to the second column, without dialysis. After each use, the column material is regenerated by washing with 0.5 M phosphate buffer, pH 7, containing 0.02 M mercaptoethanol, and then pre-equilibrating with the starting buffer. With this treatment, and continual maintenance of low temperature (4°), the affinity matrix is stable for at least 6 months.

Other Affinity-chromatographic Procedures for the Isolation of Thymidylate Synthestase

A discussion of the relative merits of column materials prepared by condensing 5-fluoro-2′-deoxyuridylate to agarose by alternate linkages is contained in an article by Whiteley *et al.*[19] In addition, Danenberg *et al.*[20] describe the preparation and use of 2′-deoxyuridine 5′-(6-*p*-aminobenzamido)hexyl phosphate bound to agarose, and Slavik *et al.*[21] report the use of a second substrate analog, tetrahydromethotrexate, bound to aminoethyl-agarose via its carboxylic acid groups, as an alternate matrix for the isolation of thymidylate synthetases from enzyme-poor sources.

[19] J. M. Whiteley, I. Jerkunica, and T. Deits, *in* "Immobilized Biochemicals and Affinity Chromatography" (R. B. Dunlap, ed.), p. 135. Plenum, New York, 1974.
[20] P. V. Danenberg, R. J. Langenbach, and C. Heidelberger, *Biochem. Biophys. Res. Commun.* **49**, 1029 (1972).
[21] K. Slavík, W. Rode, and V. Slavíková, *Biochemistry* **15**, 4222 (1976).

[16]Carbamyl-phosphate Synthetase (Glutamine): Aspartate Carbamyltransferase of *Neurospora*

By Larry G. Williams and Rowland H. Davis

$$2 \text{ ATP} + \text{L-glutamine} + CO_2 \xrightarrow{Mg^{2+}} \text{carbamyl phosphate} + \text{L-glutamate} + 2 \text{ ADP} + P_i$$

$$\text{Carbamyl phosphate} + \text{L-aspartate} \rightarrow \text{L-ureidosuccinate} + P_i$$

In *Neurospora* the initial enzymic activities of pyrimidine synthesis, carbamyl-phosphate synthetase (CPSase) and aspartate carbamyltransferase (ATCase), are determined by a single genetic locus and copurify as a single enzyme complex.[1] *Neurospora* also possesses a second CPSase which synthesizes carbamyl phosphate (carbamyl-P) specifically for use in arginine metabolism.[2] Although the arginine- and pyrimidine-specific CPSase have the same substrate requirements, they are under control of separate genetic loci and can clearly be distinguished on the basis of feedback inhibitors, derepression conditions, stabilizing factors, and molecular weight. The CPSase referred to in this article is the pyrimdine-specific CPSase (that is, its product is normally not available to the arginine pathway), which copurifies with ATCase. A peculiarity of the CPSase activity is its cold lability in the absence of its feedback effector, UTP.

Assay Method for Carbamyl-Phosphate Synthetase

Principle. CPSase activity is measured by the conversion of the labile product, carbamyl-P, to a stable compound which can be determined colorimetrically. The method currently used is to add NH_4Cl to the reaction mixture when it is completed, and to boil it. This converts carbamyl-P almost quantitatively to urea.[3,4] Alternatively, stable derivatives can be formed in the course of the reaction by trapping carbamyl-P enzymically. This requires addition of aspartate and ATCase or of ornithine and ornithine carbamoyltransferase to the CPSase reaction mixture to form ureidosuccinate or citrulline, respectively.[2,3] Urea, ureidosuccinate, and citrulline can all be measured by the same colorimetric procedure for determination of carbamyl groups.

[1] L. G. Williams, S. Bernhardt, and R. H. Davis, *Biochemistry* **9**, 4329 (1970).
[2] L. G. Williams and R. H. Davis, *J. Bacteriol.* **103**, 335 (1970).
[3] R. H. Davis, *Biochim. Biophys. Acta* **107**, 44 (1965).
[4] J. Yashphe and L. Gorini, *J. Biol. Chem.* **240**, 1681 (1965).

Reagents
    Tris-acetate buffer, 1 M, pH 7.5
    L-glutamine, 0.06 M
    $MgCl_2$, 0.12 M
    NH_4Cl, 1 M (freshly prepared)
    $KHCO_3$
    Disodium ATP

Procedure. The reaction mixture (0.50 ml) contains 0.05 ml L-glutamine, 0.05 ml $MgCl_2$, 0.25 ml H_2O, 15 μmoles $KHCO_3$, 6 μmoles ATP, and 0.10 ml enzyme preparation. Control mixtures lacking either ATP or $KHCO_3$ are also prepared. Where crude extracts with low activity are assayed, it may be desirable to increase the volume of extract and decrease the volume of H_2O added to the reaction mixture. Crude extracts must also be desalted by gel filtration. Following addition of enzyme, the reaction is incubated at 25° for 30 min. The reaction is terminated by addition of 0.30 ml of NH_4Cl, followed by 10 min in a boiling water bath. Precipitate, if any, is removed by low-speed centrifugation, and a portion of the supernatant is tested for the presence of urea by a modified Gerhart–Pardee colorimetric assay.[5] Modifications were a final 20-min incubation for color development at 25° and then placement of the reaction tubes in ice water to prevent loss of developed color and formation of spurious color. Color is read in a Klett–Summerson colorimeter using a No. 54 filter. This procedure will detect as little as 5 nmoles urea.

An enzyme unit is the quantity of enzyme that forms 1 μmole of carbamyl-P per minute under the conditions above.

Assay Method for Aspartate Carbamyltransferase

Principle. ATCase activity is measured by colorimetric determination of the product, L-ureidosuccinate.

Reagents
    Glycine·NaOH buffer, 1.0 M, pH 9.0
    L-aspartic acid, 0.10 M, pH 7.0
    Dilithium carbamyl-P, 0.02 M (dissolved just before use)
    $HClO_4$, 2.0 M

The reaction mixture (1.0 ml) contains 0.15 ml glycine buffer, 0.10 ml aspartate, 0.25 ml carbamyl-P, 0.30 ml H_2O, and 0.20 ml enzyme preparation. Control incubation mixtures lack aspartate. The reaction is

[5] J. C. Gerhart and A. B. Pardee, *J. Biol. Chem.* **237**, 891 (1962).

initiated by addition of enzyme. Following 15-min incubation at 25° the reaction is stopped by addition of 0.20 ml of $HClO_4$. The protein precipitate is removed by centrifugation, and portions of the supernatant are assayed for ureidosuccinate. The colorimetric assay employed is the same as that previously described for urea except that the incubation time for color development is 30 min. The assay can detect as little as 10 nmoles ureidosuccinate.

An enzyme unit is the quantity of enzyme that forms 1 μmole of ureidosuccinate per minute under the conditions stated above.

Purification Procedure[1]

Mycelium from an *arg-3, pyr-1* double-mutant strain is the source of the enzyme complex. The *arg-3* mutation eliminates arginine-specific CPSase activity, while the *pyr-1* mutation permits an approximately 5-fold derepression of synthesis of the CPSase–ATCase complex under conditions of pyrimidine starvation.[2] Mycelia of *arg-3, pyr-1* are grown from conidia (10^6 per milliliter) in standard Neurospora medium containing Vogel's salts, 15 g sucrose, 100 mg uridine, and 200 mg arginine per liter. Growth takes place on a reciprocating shaker at 25° in 700 ml medium in 2500-ml, low-form culture flasks. After 18–24 hr growth, mycelia are harvested without packing, rinsed in distilled water, and resuspended in the same volume of fresh medium lacking uridine. After 4 to 5 hr, derepression of pyrimidine enzymes has taken place. The standard buffer system used in enzyme extraction and purification is 0.05 M potassium phosphate, pH 7.3, containing $1 \times 10^{-3} M$ L-glutamine and $2 \times 10^{-4} M$ Cleland's reagent. Where noted, UTP, an inhibitor of cold inactivation of CPSase activity, is added to the standard buffer. All purification steps are conducted at room temperature, 22–25°, unless otherwise noted. Each purification step is devised to give maximum yield of the more labile CPSase activity while maintaining its sensitivity to feedback inhibition. See the accompanying purification table.

Step 1. Extraction. Mycelia are harvested on cheesecloth and pressed between paper towels to form a damp pad. Approximately 80 g of damp mycelial pad, an equal volume of standard buffer containing $7.5 \times 10^{-4} M$ UTP, and 16 g of fine sand are ground to a slurry with mortar and pestle at 0°. Sand and insoluble material are removed by 15-min centrifugation at 12,000 g and 5°. The crude supernatant (110 ml) is then centrifuged 30 min at 50,000 g at 5°. After removal of the heavy lipid pellicle (by pipet), the high-speed supernatant, containing all the recoverable CPSase and ATCase activity, is decanted from the residue and maintained at 0°.

PURIFICATION OF CPSASE AND ATCASE FROM *Neurospora*

Fraction	Volume (ml)	Protein (mg/ml)	Total units		Specific activity[a]		Recovery	
			CPSase	ATCase	CPSase	ATCase	CPSase	ATCase
1. Crude extract[b]	88.0	3.90	2.33	74.5	0.007	0.22	100	100
2. $(NH_4)SO_4$ I	22.0	5.95	2.32	80.1	0.018	0.61	100	107
3. Calcium phosphate gel	33.0	0.64	1.53	34.0	0.073	1.61	65	46
4. $(NH_4)_2SO_4$ II	1.3	13.30	1.53	27.7	0.077	1.60	57	37
5. Agarose 1.5 gel	8.4	0.90	0.92	19.3	0.122	2.55	39	26
6. DEAE	11.4	0.15	0.50	7.9	0.292	4.62	21	11

[a] Units/mg protein. Protein is determined via the Lowry method using a bovine albumin standard.
[b] Crude extract was prepared from a derepressed mycelium.

Step 2. First Ammonium Sulfate Fractionation. Solid $(NH_4)_2SO_4$ is immediately dissolved in the high-speed supernatant (2.5 g/10 ml). Following 5 min of stirring to dissolve the $(NH_4)_2SO_4$ and standing an additional 5 min at $0°$, the protein precipitate is collected by 25-min centrifugation at 12,000 g at $5°$. The supernatant is discarded. The precipitate is dissolved in 8 ml standard buffer (to about 15 mg of protein per milliliter) and desalted by passage of 4-ml portions through Sephadex G-25 gel filtration columns (3.3 cm diam. $\times$ 4.5 cm) equilibrated with the same buffer, at room temperature.

Step 3. Calcium Phosphate Gel Fractionation. Calcium phosphate gel (21 mg dry weight/ml) is prepared by a standard procedure.[6] The desalted protein solution from step 2 (about 20 ml) is stirred for 3–5 min with 4.0–4.5 ml of calcium phosphate gel at room temperature. The gel is sedimented by a brief centrifugation and discarded. An additional 13 to 14 ml of calcium phosphate gel are added to the supernatant. The second gel addition absorbs most of the CPSase and ATCase activity. The gel is collected by a brief low-speed centrifugation. The supernatant is discarded. Protein is eluted from the gel by successive washes (suspension of the gel in buffer with 5-min stirring, followed by centrifugation) with 10-ml portions of standard buffer (containing 1 $\times$ 10^{-4} M UTP) at pH 7.3, 7.8, and 8.1. The amount of calcium phosphate gel required to give optimum purification varies slightly from preparation to preparation. Therefore, a trial experiment must be carried out on a small portion of each preparation to determine the volumes of gel needed.

Step 4. Second $(NH_4)_2SO_4$ Fractionation. The three washes of the calcium phosphate gel are combined and cooled to $0°$, and protein is precipitated by the addition of solid $(NH_4)_2SO_4$ (3 g/10 ml). Following 5-min standing at $0°$ the precipitate is recovered by 15-min centrifugation at 50,000 g at $5°$.

Step 5. Agarose Gel Filtration. A Biogel agarose 1.5 m column (1.2 $\times$ 37.5 cm) is equilibrated at room temperature by the flow of 200 ml of standard buffer through the column. The precipitate of the second $(NH_4)_2SO_4$ fractionation is immediately redissolved in 1.25 ml of standard buffer and applied to the top of the gel column. A constant flow rate of buffer through the column is maintained by use of a Mariotte flask. The void volume of the column is 20 ml. About 85% of the recoverable enzyme activities are recovered in fractions eluting between 21 and 29 ml, which are then pooled.

---

[6] D. Keilin and E. F. Hartree, *Proc. R. Soc. London, Ser. B* **124**, 397 (1938).

Step 6. DEAE-Cellulose Fractionation. DEAE-cellulose (Biorad), washed with 0.5 N NaOH, 0.5 N HCl, and H_2O, is used to prepare a small column (1.1 × 4.0 cm). Following equilibration of the column at room temperature with 30 ml of standard buffer containing 1 × 10^{-4}, UTP, the pooled fractions from step 5 are immediately applied to the column. After the protein solution has entered the column bed the column is washed with 15 ml of standard buffer (1 × 10^{-4} M in UTP) and then eluted with a 100-ml exponential gradient of 0–0.5 M NaCl prepared in standard buffer (1 × 10^{-4} M in UTP). Fractions of 3.0 ml are collected, CPSase assays are run on a small portion of each fraction, and those fractions having appreciable activity are pooled. The pooled fractions are concentrated by $(NH_4)_2SO_4$ precipitation as described in step 4. The precipitate is redissolved in 1–2 ml of standard buffer containing 5 × 10^{-4} M UTP and stored at −80°.

Optimum recovery of CPSase activity is obtained if the purification is conducted over a 2-day period. The first day the preparation is taken through step 2 of the procedure. It is then frozen overnight at −80° in standard buffer 5 × 10^{-4} M in UTP, and the purification is completed the next day.

Properties

Stability and Purity. The CPSase activity of the enzyme complex is quite labile in the absence of stabilizers. In a buffered solution over 90% of this activity is irreversibly lost after only 45 min at 0°, while 50% of the activity is lost after 3 hr at 25°. Addition of glutamine to the buffer greatly slows loss of activity at room temperature (5% loss after 3 hr at 25%), while addition of UTP retards cold inactivation (25% loss after 3 hr at 0°).[2] The enzyme can be stored for several months if frozen at −80° in the presence of 5 × 10^{-4} M UTP and at a protein concentration about 0.5 mg/ml.

The final enzyme preparation appears to be about 50% pure based on sucrose gradient data where concomitant shifts in the major protein peak and the peak of enzyme activities occur in response to the presence or absence of UTP.[1]

CPSase–ATCase Enzyme Complex. The purification procedure is designed for maximum yield of CPSase; however, ATCase activity always copurifies with CPSase. The two activities cannot be separated by $(NH_4)_2SO_4$ fractionation or calcium phosphate gel adsorption. The distribution of the two activities from gel filtration columns, from ion-exchange columns, and from sucrose density gradients is identical. In no step of the procedure was it possible to isolate a second fraction which had only one of the two activities. Difference in yield between the two activities with some purification steps apparently is due to differential inactivation of the two active sites.

Molecular Weight.[1] On the basis of elution patterns from agarose gel filtration columns and sedimentation patterns in sucrose density gradients the enzyme complex is estimated to have a molecular weight of 650,000 ± 50,000. On sucrose density gradients the complex sediments at a value of 21 S in the absence of UTP and 15 S in the presence of UTP, suggesting a dissociation of enzyme subunits in the presence of the feedback inhibitor.

Other Properties. UTP, an end-product of the pyrimidine synthetic pathway, gave 50% inhibition of CPSase activity at a concentration of 1 $\times$ $10^{-4} M$ and complete inhibition at $1 \times 10^{-3} M$. UMP was about two-thirds as effective, but the cytidine nucleotides had no inhibiting effect.[2] None of the pyrimidine nucleotides had any effect on the V_{max} of ATCase activity. At pH 7.5 ATCase exhibits about 5% of the activity present at its pH optimum of 9.1.[7] The K_m of ATCase for carbamyl-P decreases from 83 μM at pH 9.1 to 32 μM at pH 7.5. The K_m of ATCase for carbamyl phosphate is doubled in the presence of $5 \times 10^{-4} M$ UTP.[8]

[7] R. H. Davis, *in* "Organizational Biosynthesis" (H. J. Vogel, J. O. Lampen, and V. Bryson, eds.), p. 302. Academic Press, New York, 1967.
[8] L. G. Williams, *Genetics* **77**, 270 (1974).

[17] A Multienzyme Complex of Carbamoyl-phosphate Synthase (Glutamine):Aspartate Carbamoyltransferase:Dihydoorotase (Rat Ascites Hepatoma Cells and Rat Liver)

By MASATAKA MORI and MASAMITI TATIBANA

L-Glutamine + 2 ATP + HCO^-_3 + H_2O →
carbamoyl phosphate + 2 ADP + P_i + L-glutamate
Carbamoyl phosphate + L-aspartate → carbamoyl-L-aspartate + P_i
Carbamoyl-L-aspartate ⇄ L-dihydroorotate + H_2O

Carbamoyl-P[1] synthase (glutamine) (EC 2.7.2.9) of higher animals, which was first discovered by Tatibana and Ito[2,3] in hematopoietic

[1] Abbreviations used are: Carbamoyl-P, carbamoyl phosphate; Me₂SO, dimethyl sulfoxide; Hepes, N-2-hydroxyethylpiperazine-N'-2-ethanesulfonic acid; PP-ribose-P, 5-phosphoribosyl 1-pyrophosphate.
[2] M. Tatibana and K. Ito, *Biochem. Biophys. Res. Commun.* **26**, 221 (1967).
[3] M. Tatibana and K. Ito, *J. Biol. Chem.* **244**, 5403 (1969).

mouse spleen and by Hager and Jones in Ehrlich ascites tumor cells[4] and fetal rat liver,[5] catalyzes the first step of *de novo* pyrimidine nucleotide biosynthesis and plays a key role in the control of this pathway.[6,7] A unique feature of the enzyme is that it exists as a multienzyme complex with aspartate carbamoyltransferase (EC 2.1.3.2) and dihydroorotase (EC 3.5.2.3), the second and third enzymes of the pathway.[8-12] The complex has recently been purified to homogeneity from rat ascites hepatoma cells[13] and from hamster cell mutants which overproduce these three enzymes.[14]

Assay for Carbamoyl-P Synthase Activity

Principle. The activity is measured with [^{14}C]bicarbonate as a substrate by following the formation of [^{14}C]citrulline in the presence of excess amounts of L-ornithine and ornithine carbamoyltransferase purified from bovine liver.[3,15]

Reagents
   Potassium Hepes buffer, 0.5 M, pH 7.0
   ATP, 0.2 M, adjusted to pH 7.0 with KOH
   $MgCl_2$, 0.6 M
   L-Glutamine, 0.2 M
   Dithiothreitol, 0.2 M
   L-Ornithine, 0.1 M, pH 7.0
   50%[16] Glycerol
   30% Me$_2$SO–5% glycerol; Me$_2$SO distilled at about 80° under a
      reduced pressure is used.
   KH[^{14}C]O$_3$ (2000 cpm/nmole), 0.5 M^3

[4] S. E. Hager and M. E. Jones, *J. Biol. Chem.* **242**, 5667 (1967).
[5] S. E. Hager and M. E. Jones, *J. Biol. Chem.* **242**, 5674 (1967).
[6] M. E. Jones, *Adv. Enzyme Regul.* **9**, 19 (1971).
[7] M. Tatibana and K. Shigesada, *Adv. Enzyme Regul.* **10**, 249 (1972).
[8] N. J. Hoogenraad, R. L. Levine, and N. Kretchmer, *Biochem. Biophys. Res. Commun.* **44**, 981 (1971).
[9] W. T. Shoaf and M. E. Jones, *Biochem. Biophys. Res. Commun.* **45**, 796 (1971).
[10] M. Mori and M. Tatibana, *Biochem. Biophys. Res. Commun.* **54**, 1525 (1973).
[11] K. Ito and H. Uchino, *J. Biol. Chem.* **248**, 389 (1973).
[12] R. J. Kent, R.-L. Lin, H. J. Sallach, and P. P. Cohen, *Proc. Natl. Acad. Sci. U.S.A.* **72**, 1712 (1975).
[13] M. Mori and M. Tatibana, *J. Biochem. (Tokyo)* **78**, 239 (1975).
[14] P. F. Coleman, D. P. Suttle, and G. R. Stark, this volume [18].
[15] M. Tatibana and K. Shigesada, *J. Biochem. (Tokyo)* **72**, 537 (1972).
[16] Concentrations of glycerol are expressed on a weight by volume basis and those of Me$_2$SO are on a volume by volume basis.

Ornithine carbamoyltransferase highly purified from bovine liver, 4000 units per ml.[17] Before use it is dialyzed against 20 mM Tris·HCl, pH 7.0, containing 0.5 mM ornithine to remove ammonia.

HCOOH, 3 M

Enzyme in a buffer containing 30% Me_2SO, 5% glycerol, and 1 mM dithiothreitol

Assay Cocktail. A cocktail solution, pH 7.0, containing 0.15 M potassium Hepes, 30 mM ATP, 45 mM $MgCl_2$, 1.5 mM L-ornithine, and 3.75% glycerol is prepared and remains stable for 6 months at $-20°$.

Procedure. The reaction mixture contains 100 μl of the cocktail, 5 μl of 0.2 M glutamine, 1.5 μl of 0.2 M dithiothreitol, 1 μl of ornithine carbamoyltransferase (4 units), 106.5 μl of water, 10 μl of $K[H^{14}C]O_3$, 30% Me_2SO–5% glycerol, and enzyme (the volume of Me_2SO–glycerol plus enzyme is 75 μl) in a volume of 300 μl. The mixture without enzyme is kept at 20° for 5 min, and the reaction is started by the addition of enzyme. After incubation for 15 min at 37°, the reaction is stopped by the addition of 50 μl of 3 M HCOOH. The mixture is quantitatively transferred to a scintillation vial and evaporated in a vacuum over solid KOH. The residual matter is dissolved in 0.5 ml of water, 10 ml of Bray's scintillation fluid are added, and the radioactivity is counted. This assay system, which contains 7.5% Me_2SO and 2.5% glycerol (final concentrations), is used to measure the maximal velocity. In kinetic studies such as those done to determine the effect of substrate concentration and of allosteric effectors, 10% glycerol is used instead of 7.5% Me_2SO and 2.5% glycerol and incubation is carried out for 10 min.

The method is not applicable to crude liver preparations of ureoteric animals; the enzyme has to be partially purified to remove the ammonia-dependent enzyme.[18] When extrahepatic tissue extracts are used as the enzyme source, the reaction product must be partially purified before counting.[3]

Assays for Aspartate Carbamoyltransferase and Dihydroorotase Activities

These assays are performed as described previously.[13,18]

[17] M. Marshall and P. P. Cohen, *J. Biol. Chem.* **247**, 1641 (1972). One unit is defined as the amount of enzyme that catalyzes the formation of 1.0 μmol of citrulline in 1.0 min at 37° in the standard system.

[18] M. Mori, H. Ishida, and M. Tatibana, *Biochemistry* **14**, 2622 (1975).

Purification Procedure of the Complex from Rat Ascites Hepatoma
Cells (AH 13)

Carbamoyl-P synthase in the complex is extremely unstable in usual
aqueous buffers, and purification procedures are carried out in the
presence of high concentrations of Me_2SO and glycerol as stabilizers.
Purification of the complex from rat ascites hepatoma cells is first
described, and then comments are made on the purification from rat
liver, which contains only one-tenth the amount of complex. All
manipulations are carried out at 0–4° unless otherwise indicated. Protein
is first precipitated with trichloroacetic acid[3] and determined by the
method of Lowry[19] with bovine serum albumin as a standard.

Buffer A: 50 mM potassium phosphate, 30% glycerol, 10 mM
$MgCl_2$, 1 mM EDTA, 3 mM L-glutamine, 1 mM dithiothreitol, 1
mM phenylmethyl sulfonylfluoride, pH 7.0. Buffers must be
adjusted to the desired pH after the addition of glycerol and/or
Me_2SO. Glutamine, dithiothreitol, and phenylmethyl sulfonylfluoride are added immediately before use.

Buffer B: 2 mM potassium phosphate, 30% Me_2SO, 5% glycerol, 0.5
mM EDTA, 1 mM dithiothreitol, pH 7.0. Dithiothreitol is added
immediately before use (same for Buffers C to G).

Buffer C: 30 mM potassium phosphate, 30% Me_2SO, 5% glycerol, 1
mM dithiothreitol, pH 7.8.

Buffer D: 150 mM potassium phosphate, 30% Me_2SO, 5% glycerol,
1 mM dithiothreitol, pH 7.8.

Buffer E: 210 mM potassium phosphate, 30% Me_2SO, 5% glycerol,
1 mM dithiothreitol, pH 7.8.

Buffer F: 20 mM potassium phosphate, 30% Me_2SO, 5% glycerol, 1
mM dithiothreitol, pH 7.2.

Buffer G: 30 mM potassium Hepes, 30% Me_2SO, 5% glycerol, 1mM
dithiothreitol, pH 7.2.

Saturated ammonium sulfate solution: Prepared at 25° by adding
ammonium sulfate to saturation to a solution containing 10%
glycerol, 50 mM potassium phosphate at pH 7.0, and 1 mM
EDTA, followed by neutralization with NH_4OH so that a 1:5
dilution has a pH of 7.0 at 4°. The solution is kept at 25°.

Hydroxylapatite: Calcium phosphate is prepared first by the
method of Siegelman *et al.*,[20] and is then treated with sodium
hydroxide as described by Levin.[21] The preparation thus ob-

[19] E. Layne, Vol. 4 [73].
[20] H. W. Siegelman, G. A. Wieczorek, and B. C. Turner, *Anal. Biochem.* **13**, 402 (1965).
[21] Ö. Levin, Vol. 5 [2].

tained has a sufficiently high flow rate. Good results are also obtained with spheroidal hydroxylapatite (BDH Chemicals Ltd., England).

Growth and Collection of AH 13 Cells. Yoshida ascites hepatoma cells, AH 13, (1 × 10⁶ cells per rat) are injected intraperitoneally into female Donryu strain rats (Nippon Rat Inc., Saitama, Japan), weighing 150 to 200 g. After 5 days, 30 ml of cold 0.9% NaCl are injected intraperitoneally, and the ascites fluid is collected and centrifuged at 4000 g for 5 min at 0°. The packed cells are washed once with cold 0.9% NaCl. About 3 g wet weight of ascites cells are obtained per rat.

Step 1. Extraction. The packed cells (about 100 g) are suspended in 4 volumes of buffer A and disrupted by sonic oscillation at 20 kHz for 40 sec in portions of 20 ml. The sonicated cells are centrifuged at 20,000 g for 15 min; the pellets are discarded.

Step 2. First Ammonium Sulfate Precipitation. While the supernatant solution is being gently stirred, 0.65 volume of cold saturated ammonium sulfate solution is added. The latter is cooled to 4° immediately before use. The enzyme solution thus treated is allowed to stand for 10 min and centrifuged at 20,000 g for 15 min. The precipitate is gently homogenized in buffer B (0.6 volume of the original cells), and the solution is centrifuged at 105,000 g for 1 hr. The supernatant is carefully removed so as not to disturb the loosely packed pellets.

Step 3. Second Ammonium Sulfate Precipitation. To the clear supernatant 0.65 volume of the saturated ammonium sulfate solution is added. After standing for 10 min the precipitate is collected by centrifugation at 20,000 g for 15 min and dissolved in buffer B (0.2 volume of the original cells). Two batches of this preparation are pooled at this step and used for the next purification step.

Step 4. First Hydroxylapatite Adsorption To the enzyme solution buffer E (0.5 volume of the original cells) is added, and the solution is mixed with hydroxylapatite (0.2 volume of the original cells), previously equilibrated with buffer D, and centrifuged at 1000 g for 5 min. The hydroxylapatite is washed twice with the same buffer (0.1 volume of the original cells), and the supernatant and washings are combined.

Step 5. Second Hydroxylapatite Adsorption. The enzyme solution is diluted with 1.5 volumes of buffer B and adjusted to pH 6.7 with 1 M acetic acid. The solution is then applied to a hydroxylapatite column (5.0 × 6.0 cm), previously equilibrated with buffer B. Elution is performed with a linear gradient of buffers C and E (4 column volumes, each) at a flow rate of 150 ml/hr. Active fractions are combined, diluted with the same volume of buffer B, adjusted to pH 6.8 with acetic acid, and applied to a smaller hydroxylapatite column (10 ml) previously equilibrated with buffer B. The enzymes are eluted with buffer E; the first 8 ml are discarded and the following 20 ml are pooled. Enzymes are then concentrated to about 4 ml by ultrafiltration at about 20° using an Amicon XM-300 membrane.

Step 6. Gel Filtration on Sepharose 6B. The enzymes are layered on a column (3.0 × 15.5 cm) of Sepharose 6B previously equilibrated with buffer F. Elution is carried out with the same buffer at a flow rate at 50 ml/hr. The three enzyme activities emerge in the void volume. Active fractions are pooled and concentrated by adsorption to a small hydroxylapatite column (1.0 ml) and are then eluted with buffer E.

Step 7. Sucrose Gradient Centrifugation. The enzyme solution (0.3–0.5 ml) is layered on a 5–20% sucrose gradient (4.5 ml) also containing 50 mM potassium Hepes, pH 7.2, 30% Me$_2$SO, 5% glycerol, 4 mM L-glutamine, and 1 mM dithiothreitol, and centrifuged at 65,000 rpm (302,000 g) for 6.5 hr at 18° in a Hitachi 65P ultracentrifuge using an RPS 65TA rotor. After centrifugation, the gradient is fractionated into 0.25-ml fractions. Those with the highest activities are pooled, concentrated, and dialyzed against buffer G in a collodion bag (Sartorius Membranfilter GmbH), and then stored in small portions at −80°.

The purification procedure may be stopped after steps 3, 5, or 6 by freezing the enzymes in a Dry Ice–acetone bath and storing at −80°. Table I summarizes the result of a typical purification. The purified enzyme appears homogeneous upon analytical ultracentrifugation, agarose-acrylamide composite gel electrophoresis, and polyacrylamide gel electrophoresis in sodium dodecyl sulfate.[13] Essentially the same procedure is applicable to the enzymes from rat liver,[18] hematopoietic mouse spleen,[3,8] and various ascites cells.[22]

Purification of the Complex from Rat Liver

The rat liver enzymes can be highly purified by much the same procedure as that used for the ascites hepatoma enzymes. Wistar strain rats, weighing 250–350 g, are anesthetized with ether and exsanguinated from the abdominal aorta. The livers are excised and cooled in 0.9% NaCl. These tissues are homogenized, in portions of 5 g, with 5 volumes of buffer A in a Potter–Elvehjem homogenizer with a Teflon pestle at

---

[22] K. Ito, S. Nakanishi, M. Terada, and M. Tatibana, *Biochim. Biophys. Acta* **220**, 477 (1970).

TABLE I

PURIFICATION OF CARBAMOYL-P SYNTHASE (GLUTAMINE):ASPARTATE CARBAMOYLTRANSFERASE: DIHYDROOROTASE FROM RAT ASCITES HEPATOMA CELLS (AH 13)[a]

Step	Total volume (ml)	Total activity (units)[b]			Total protein (mg)	Specific activity of synthase (munits/mg protein)	Synthase: carbamoyl-transferase: dihydroorotase ratio
		Synthase	Carbamoyl-transferase	Dihydro-orotase			
1. Crude extract	725	6.44	421	36.4	16,800	0.383	1:65:5.7
2. First ammonium sulfate	108	3.76	177	25.6	1,490	2.54	1:47:6.8
3. Second ammonium sulfate	40.7	3.44	169	21.2	730	4.71	1:49:6.2
4. First hydroxylapatite	390	2.66	125	17.3	268	10.0	1:47:6.5
5. Second hydroxylapatite	4.0	0.830	33.4	5.23	27.7	30.0	1:40:6.3
6. Sepharose 6B	1.2	0.574	21.2	3.59	4.41	130	1:37:6.2
7. Sucrose gradient	0.57	0.281	10.5	1.83	0.55	510	1:37:6.5

[a] The starting material was 177 g wet weight of AH 13 cells. Data taken from Mori and Tatibana.[13]

[b] One unit of enzyme activity is defined as that amount of enzyme which produces 1.0 μmole of product per minute at 37° in the respective standard assay systems.[13,18]

about 600 rpm for 40 sec and centrifuged at 20,000 g for 10 min. Steps 2 to 5 are carried out in the same manner, except that high-speed centrifugation at 105,000 g is performed for 1.5 hr. The second ammonium sulfate fraction still contains the activity of ammonia-dependent carbamoyl-P synthase I; the latter enzyme can be completely removed in steps 4 and 5. Through steps 1 to 5, the synthase is purified about 280-fold, and the purified preparation has a specific activity of 0.01 unit/mg of protein.[18] The enzyme can be further purified about 20-fold either by sucrose gradient centrifugation[18] or by gel filtration on Sepharose 6B.

Physical Properties of the Complex[13]

The complex of ascites hepatoma cells has a sedimentation coefficient ($s_{20,w}$) of 24.6 S at 10° in Hepes buffer, pH 7.2, containing 30% Me$_2$SO and 5% glycerol. Sedimentation equilibrium studies give a molecular weight of 870,000. The complex gives a single protein band corresponding to a molecular weight of 210,000 upon sodium dodecyl sulfate–polyacrylamide gel electrophoresis, suggesting that it is composed of four to five subunits of similar size. Available evidence also suggests that the subunits are composed of three distinct molecular species carrying the respective three enzyme activities. Electrofocusing of the complex in the presence of 30% Me$_2$SO and 5% glycerol gives a p*I* of 5.5.

Properties of Carbamoyl-P Synthase

No difference has been observed between the properties of the AH 13 enzyme and those of the rat liver enzyme.

Stability. The enzyme is extremely unstable in usual aqueous buffer solutions; the half-life is about 20 min at pH 7.0 and 0°. The enzyme is markedly stabilized by cryoprotectants such as Me$_2$SO, glycerol, and ethylene glycol. It is most stable in the presence of 30% Me$_2$SO, 5% glycerol, and 1 m*M* dithiothreitol; loss in the activity at 0° under these conditions is 11% per day. The same enzyme solution can be stored over 6 months at −80° without detectable loss in the activity. MgATP, MgUTP, and PP-ribose-P are also effective in stabilizing the enzyme,[3,4,18,23] but less so than the cryoprotectants. The enzyme is readily inactivated by sulfhydryl reagents such as hydroxymercuribenzoate (20 μM) and mersalyl (10 μM).

Kinetic Properties. Potassium ions are required for enzymic activity,[23] the apparent K_m for K$^+$ being 18 m*M*. The enzyme requires free

[23] R. L. Levine, N. J. Hoogenraad, and N. Kretchmer, *Biochemistry* **10**, 3694 (1971).

magnesium ion in addition to MgATP for its activity[15]; the optimal concentration of Mg^{2+} is about 5 mM but decreases to 2 mM by addition of 100 μM PP-ribose-P.[24] The enzyme utilizes ammonia in addition to glutamine as the nitrogen donor.Optimal activity in the presence of 10% glycerol with 3.3 mM glutamine as a substrate is observed at pH 6.8–7.2 but the optimum pH is around 7.8 with 20 mM NH_4Cl. The ratio of activity with 0.2 mM glutamine to that with 40 mM NH_4Cl at pH 7.0 is about 1.2. The apparent K_m values in 10% glycerol for L-glutamine and NH_4Cl are 21 μM and 15 mM, respectively, at pH 7.0; the apparent K_m values for MgATP and bicarbonate under the conditions are 1.7 and 10 mM, respectively. The cryoprotectants alter the apparent K_m for MgATP and bicarbonate; the K_m for bicarbonate is 20 mM in their absence and 1.0 mM in the presence of 30% Me_2SO.[25]

The enzyme activity is significantly altered by a number of nucleotides and related compounds.[2,18,23,26] Thus, UTP, UDP, UDP-glucose, CTP, dUDP, and dUTP inhibit the enzyme, while it is activated by PP-ribose-P. The most prominent are the inhibition by UTP and the activation by PP-ribose-P. These compounds exert their effects by altering the K_m for MgATP of the enzyme. Me_2SO and glycerol markedly alter the enzymic activity and modify the effects of allosteric effectors.[25] Effects of the cryoprotectants and allosteric effectors on kinetic parameters of the enzyme are shown in Table II.

ADP, a reaction product, inhibits the activity competitively with MgATP; the apparent K_i is 1.2 mM. Glycine inhibits carbamoyl-P synthesis from glutamine white it stimulates the synthesis from ammonia.[27] The inhibition is competitive with glutamine and the apparent K_i is 3.5 mM. Alanine is also inhibitory, but less so than glycine.

Polyamines inhibit the activity.[24] The degree of inhibition by 0.8 mM concentrations of spermine, spermidine, and putrescine is 70, 46, and 18%, respectively, when assayed with 1 mM ATP and 2 mM $MgCl_2$. Polyamines also modify the activation effect of PP-ribose-P.

The enzymes from mouse spleen and rat liver differ in the apparent K_m for MgATP (5.2 and 1.7 mM, respectively) as well as in quantitative sensitivities to allosteric effectors.[18] However, no other qualitative difference has been noted in the properties of the enzymes from various mammalian tissues.

Tissue Distribution. The enzyme is distributed widely in various mammalian normal and neoplastic tissues[3,4,22,28] as well as in avian liver[5]

[24] M. Mori and M. Tatibana, *Biochem. Biophys. Res. Commun.* **67**, 287 (1975).
[25] H. Ishida, M. Mori, and M. Tatibana, *Arch. Biochem. Biophys.* **182**, 258 (1977).
[26] M. Tatibana and K. Shigesada, *J. Biochem. (Tokyo)* **72**, 549, (1972).
[27] M. Mori and M. Tatibana, *Biochim. Biophys. Acta* **483**, 90 (1977).
[28] M. C. M. Yip and W. E. Knox, *J. Biol. Chem.* **245**, 2199 (1970).

TABLE II
EFFECTS OF CRYOPROTECTANTS AND ALLOSTERIC EFFECTORS ON KINETIC PARAMETERS
OF RAT LIVER CARBAMOYL-P SYNTHASE[a]

Cryoprotectant	Effector[b]	Maximal velocity[c] (munits/mg protein)	K_m for MgATP[d] (mM)	Hill coefficient[d]
None	None	160	7.0	1.1
	UTP	160	9.0	1.6
	PP-ribose-P	222	1.9	0.88
Me$_2$SO				
7.5%	None	222	1.6	1.1
	UTP	222	2.7	1.7
	PP-ribose-P	300	0.35	1.0
25%	None	62	0.10	1.1
	UTP	94	0.21	1.6
	PP-ribose-P	62	0.05	—
Glycerol				
10%	None	189	1.7	1.1
	UTP	189	5.5	1.4
	PP-ribose-P	222	0.5	1.3
30%	None	89	0.8	0.72
	UTP	89	1.6	1.3
	PP-ribose-P	108	0.2	—

[a] The enzyme preparation purified 2800-fold from the cytosol of rat liver was used. The activity was measured in a system containing 10% glycerol. Data taken from Ishida *et al.*[25]

[b] UTP at 0.2 mM and PP-ribose-P at 50 μM were used.

[c] The maximal velocity was obtained by extrapolation of double reciprocal plots to infinite substrate concentration.

[d] Apparent K_m values and Hill coefficients were obtained from a standard Hill plot.

and frog eggs.[12] The activity in mammalian tissues other than liver appears to be related to the growth rate of the tissue.

Properties of Aspartate Carbamoyltransferase and Dihydroorotase

See elsewhere.[29]

Dissociation of the Complex

The complex partially and irreversibly dissociated when the cryoprotectant concentrations are reduced.[18,29] The dissociation is prevented by

[29] W. T. Shoaf and M. E. Jones, *Biochemistry* **12**, 4039 (1973).

MgATP, MgUTP, and PP-ribose-P.[18] Limited proteolysis of the complex by pancreatic elastase dissociates the three component enzymes with retention of catalytic activities (Mori and Tatibana,[10] and unpublished observations); the synthase is recovered as a 197,000-dalton fragment, the carbamoyltransferase as a 145,000-dalton fragment, and dihydroorotase as 197,000- and 43,000-dalton fragments. The dissociated synthase has a K_m for MgATP of 7.6 mM which is 5 times higher than that of the enzyme within the complex, and their quantitative sensitivities to allosteric effectors are different.

Acknowledgment

Thanks are due to M. Ohara, Kyoto University, for assistance in the preparation of the manuscript.

[18] Purification of a Multifunctional Protein Bearing Carbamyl-phosphate Synthase, Aspartate Transcarbamylase, and Dihydroorotase Enzyme Activities from Mutant Hamster Cells

By PATRICK F. COLEMAN, D. PARKER SUTTLE, and GEORGE R. STARK

Carbamyl-P synthase (EC 2.7.2.9), aspartate transcarbamylase (EC 2.1.3.2), and dihydroorotase (EC 3.5.2.3), the first three enzymes of the pathway for *de novo* synthesis of pyrimidine nucleotides, have been partially purified in low yield from *Drosophila melanogaster,*[1] bullfrog eggs,[2] rat liver,[3,4] hematopoietic mouse spleen,[5] mouse Ehrlich ascites cells,[6] and human lymphocytes[7] and have been extensively purified from mouse ascites hepatoma cells.[8] In all cases, all three activities are associated with a single multienzyme complex of high molecular weight. We have purified to homogeneity large amounts of the complex from an SV40 (Simian virus 40) transformed baby hamster kidney cell line which

[1] B. Jarry, *FEBS Lett.* **70,** 71 (1976).
[2] R. J. Kent, R. Lin, H. J. Sallach, and P. P. Cohen, *Proc. Natl. Acad. Sci. U.S.A.* **72,** 1712 (1975).
[3] M. Mori and M. Tatibana, *Biochem. Biophys. Res. Commun.* **54,** 1525 (1973).
[4] M. Mori, H. Ishida, and M. Tatibana, *Biochemistry* **14,** 2622 (1975).
[5] N. J. Hoogenraad, R. L. Levine, and N. Kretchmer, *Biochem. Biophys. Res. Commun.* **44,** 981 (1971).
[6] W. T. Shoaf and M. E. Jones, *Biochemistry* **12,** 4039 (1973).
[7] K. Ito and H. Uchino, *J. Biol. Chem.* **248,** 389 (1972).
[8] M. Mori and M. Tatibana, *J. Biochem. (Tokyo)* **78,** 239 (1975).

overaccumulates this complex by nearly 100-fold. In many ways the complex from hamster cells is quite similar to those studied previously, but the availability of substantial amounts of pure protein has allowed more extensive characterization than was possible before. The three enzymes are linked covalently as a multifunctional protein, providing a simple explanation for the coordinate control of their synthesis in a series of mutant cell lines which overaccumulate the complex to varying degrees.[9]

Assay Methods

Carbamyl-P-Synthase (Glutamine)

Principle. The substrates glutamine, $H[^{14}C]O_3^-$, and ATP are converted by the enzyme to $[^{14}C]$carbamyl-P. In the presence of L-aspartate and aspartate transcarbamylase, the labeled carbamyl-P is converted to carbamyl aspartate, which is stable to acid and can be measured.

Reagents
Complete assay cocktail: sodium ATP, 25 mM; $MgCl_2$, 30 mM; potassium Hepes[10] buffer, 50 mM, pH 7.2 at 37°; KCl, 150 mM; L-glutamine, 3.3 mM; L-aspartic acid, 15 mM; $KHCO_3$, 16.7 mM; $Na_2[^{14}C]O_3$, 0.2 mM; dithiothreitol, 1 mM; Me_2SO, 7.5% (v/v); and glycerol, 2.5% (w/v)
50% Acetic acid
5% Hydrogen peroxide

Procedure. The buffer system used in the assays is essentially that described by Tatibana and Shigesada.[11] Aliquots of 0.4 ml of the complete assay cocktail are put into standard scintillation vials, and a total of 45 μl (enzyme sample and H_2O, added separately) is added to initiate each assay. Assays are incubated for 10 min at 37°. Endogenous aspartate transcarbamylase activity is sufficient to convert all carbamyl-P to the acid-stable carbamyl aspartate. The reaction is terminated by adding 0.2 ml of 50% acetic acid, and the contents of the vials are evaporated to dryness at 110°. Four drops of 5% hydrogen peroxide are added to each vial before adding scintillation fluid to eliminate chemiluminescence due to the presence of ATP and dithiothreitol.

Aspartate Transcarbamylase

Principle. $[^{14}C]$carbamyl-P is converted to acid-stable carbamyl aspartate in the presence of L-aspartate.

[9] T. D. Kempe, E. A. Swyryd, M. Bruist, and G. R. Stark, *Cell* **9**, 541 (1976).
[10] Hepes, *N*-2-hydroxyethyl piperazine-*N'*-2-ethane sulfonic acid.
[11] M. Tatibana and K. Shigesada, *J. Biochem. (Tokyo)* **72**, 537 (1972).

Reagents

Assay cocktail: 5 parts 0.2 M Tris·HCl, pH 8.5; 3 parts H_2O; 1 part 0.15 M L-aspartic acid, pH 8.5

6.5 mM [^{14}C]carbamyl-P

50% acetic acid

Procedure. The [^{14}C]carbamyl-P is prepared from KN[^{14}C]O (2.7 mCi/mmole, Amersham Searle) as described by Davies *et al.*[12] Aliquots of 0.9 ml assay cocktail are placed in standard scintillation vials and are kept at 0° after the addition of the enzyme. The vials are brought to 37° just before adding 50 µl of [^{14}C]carbamyl-P to initiate the reaction. Assays are incubated for 5 min at 37° and are terminated by adding 0.2 ml of 50% acetic acid. The samples are counted after evaporation to dryness at 110°.

Dihydroorotase

Principle. The enzyme reaction is run in the reverse direction (dihydro-L-orotate to carbamyl-L-aspartate). Carbamyl-L-aspartate is measured using a colorimetric assay.

Reagents

Assay buffer: 0.2 M Tris·HCl, pH 8.5

Antipyrene reagent: 5 g/liter antipyrene (1,5-dimethyl-2-phenyl-3-pyrazolone, Matheson, Coleman and Bell) in 50% (v/v) sulfuric acid

Monoxime reagent: 0.8 g 2,3-butanedione monoxime (Eastman) in 100 ml 5% (v/v) acetic acid

0.15 M dihydro-L-orotic acid (Sigma), pH 8.5

Procedure. Aliquots of 0.5 ml of the assay buffer are put into tubes at 0° followed by the enzyme. After a brief incubation at 37°, the reaction is initiated by adding 50 µl of 0.15 M dihydroorotic acid. The assays are incubated for 5 min at 37° and terminated by adding 0.6 ml of the antipyrine–monoxime solution (2 parts antipyrine:1 part monoxime). The carbamyl aspartate levels are determined using either of the colorimetric assays described by Prescott and Jones.[13]

Enzyme Purification

Carbamyl-P synthase, aspartate transcarbamylase, and dihydroorotase were copurified from the 165-23 cell line, a PALA (*N*-phosphonacetyl-L-aspartate)-resistant mutant selected from C13/SV28 cells (an SV40-

[12] G. E. Davies, T. C. Vanaman, and G. R. Stark, *J. Biol. Chem.* **245**, 1175 (1970).
[13] L. M. Prescott and M. E. Jones, *Anal. Biochem.* **32**, 408 (1969).

transformed diploid clone of BHK 21[9]). The 165-23 cells overaccumulate all three activities by about 100-fold. Cells were grown on plastic plates (Lux Scientific) in modified Eagle's medium (Microbiological Associates) containing 10% calf serum (Irvine Scientific), penicillin (100 mg/ml), and streptomycin sulfate (500 units/ml). The cells were incubated at 37° in an air atmosphere enriched to contain 7% CO_2. For each preparation, 20 100-mm confluent plates of 165-23 cells were grown with one passage from a frozen stock. The cells were removed with trypsin and transferred to 200 150-mm plates. When nearly confluent, the cells were scraped from each plate with a rubber policeman and collected by centrifugation for 5 min at 4000 rpm in a Beckman JA-10 rotor. The supernatant medium was decanted, and the cells were resuspended in 40 ml of fresh culture medium at 0° without calf serum but with the protease inhibitors benzamidine (2.5 mM) and soybean trypsin inhibitor (0.25 mg/ml). The cells were transferred to a 40-ml conical centrifuge tube and pelleted by spinning for 5 min at 1500 rpm; the supernatant solution was discarded, and the cells were washed a second time in the same medium. The supernatant solution was again discarded, and the packed cell pellet (12–16 ml) was resuspended in 2 volumes of a hypotonic buffer containing 10 mM triethanolamine, 5 mM $MgCl_2$, 2 mM $CaCl_2$, 1 mM dithiothreitol, 15 mM KCl, 6 mM L-glutamine, 6 mM L-aspartic acid, 10 mM benzamidine, and 1 mg/ml soybean trypsin inhibitor. The cells were allowed to swell for 5–10 min prior to lysis in a glass tissue homogenizer. The lysis procedure was designed to keep nuclei intact after cells are broken. Protease inhibitors were included in the wash and lysis steps in order to minimize a "nicking" reaction observed by SDS–polyacrylamide gel electrophoresis (see below). However, the inhibitors did not eliminate proteolysis completely, and nicking was minimized most effectively by reducing as much as possible the length of time between cell lysis and ammonium sulfate precipitation.

When 90–95% of the cells had been broken by homogenization (determined by vital staining with trypan blue), the lysate was made 30% in glycerol and 1 mM in dithiothreitol and centrifuged at 12,000 rpm for 15 min in a Beckman JA-20 rotor. The postmitochondrial supernatant solution was removed carefully, and the pellet was discarded. Nucleic acid was precipitated by adding 2.85 ml of a 5% solution of streptomycin sulfate for each 1000 A_{260} units (A_{260} times ml of solution). The mixture was stirred on ice for 15 min, and the precipitate was removed by centrifugation for 15 min at 18,000 rpm in a JA-20 rotor. RNase A was added to the supernatant solution to a final concentration of 10 μg/ml to degrade any remaining RNA from ribosomes or other ribonucleoproteins, and the solution was stirred gently for 20 min on ice. Immediately following this, 0.65 volume of a solution of potassium phosphate (50 mM, pH 7.0), 10% glycerol, and 1 mM EDTA, saturated at 0° with ammonium sulfate,[14] was added slowly, and the solution was stirred for 20 min on ice. The precipitate was collected by centrifugation at 18,000 rpm for 20 min in a JA-20 rotor. The pellet was redissolved in a minimal amount (1.5–2.5 ml) of buffer A [20 mM Tris·HCl, pH 7.4 (0°); 50 mM KCl; 4 mM L-glutamine; 4 mM L-aspartic acid; 0.1 mM EDTA; 1 mM dithiothreitol; 30% Me$_2$SO; and 5% glycerol]. The redissolved pellet was clarified by centrifugation at 10,000 rpm for 10 min in a JA-21 rotor prior to loading onto a 1.5 × 28 cm column of Bio-Gel A-5M (200–400 mesh) equilibrated with buffer A. Fractions were assayed for aspartate transcarbamylase activity. Typically, the activity peak was broad (Fig. 1), and trace amounts of smaller proteins were often present in the trailing half of the peak, as revealed by SDS gel electrophoresis. For this reason the two halves of the peak were pooled separately (for details, see legend to Fig. 1). Aliquots (1–2 ml) were transferred to small plastic vials and stored in liquid nitrogen. All three activities are stable for months under these conditions.

A summary of the purification is shown in Table I. The ratios of specific activities for all three enzymes are quite constant from the postmitochondrial supernatant through to the A-5M pools. The recovery

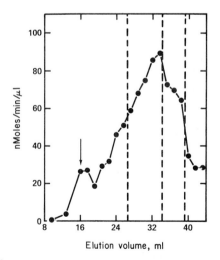

FIG. 1. Profile of aspartate transcarbamylase activity eluted from a Bio-Gel A-5M column. The ammonium sulfate precipitate, 84 mg protein in 1.65 ml of buffer A, was applied to a 1.5 × 28 cm column. Elution was at a rate of 5 ml/hr, and 1.6-ml fractions were collected. The void volume is indicated. The activity was divided into two pools: 27 to 33.5 ml (pool I) and 33.6 to 38.4 ml (pool II).

[14] M. Tatibana and K. Ito, *J. Biol. Chem.* **244**, 4504 (1969).

from the postmitochondrial supernatant solution of each activity is about 50%. Aspartate transcarbamylase and dihydroorotase were purified about 10-fold from the crude lysate to pool I, whereas carbamyl-P synthase was purified about 14-fold. This discrepancy may be due to the presence of low-molecular-weight inhibitors of carbamyl-P synthase in the crude lysate. The protocol is simplified by the fact that nearly 10% of the total cellular protein is associated with these three activities. The entire procedure can be completed in less than 24 hours with a yield of about 40 mg of pure protein from about 15 ml of packed cells.

Physical and Chemical Properties

SDS Gel Electrophoresis. SDS gels (4.5% polyacrylamide matrix, Weber and Osborn[15]) prepared from trichloroacetic acid precipitates made directly from the overproducing mutant (165-23) and its parent (SV28) show that a single band of high molecular weight is greatly intensified in the mutant (Fig. 2). SDS gels have been used to follow this polypeptide through the purification summarized in Table I. As shown in Fig. 3, the relative amount of the polypeptide does increase roughly in parallel with increases of the specific enzyme activities. In pool I from the A-5M column, there is only a single band associated with three enzyme activities (Fig. 3). The molecular weight of the polypeptide is 200,000 as determined by SDS gel electrophoresis.[16] It is apparent from Fig. 3 that limited proteolytic nicking of the 200,000-dalton polypeptide to a 190,000-dalton polypeptide is a problem. Only the 200,000-dalton species is apparent in precipitates made directly from unbroken cells (Fig. 2).

Titration of Aspartate Transcarbamylase Active Sites with PALA. Kempe *et al.*[9] showed that it is possible to determine the number of aspartate transcarbamylase active sites in a crude lysate by titration with [^{3}H]PALA (150 Ci/mole). Using this procedure, but with highly purified material from pool I, 4.3 ± 0.2 pmoles of [^{3}H]PALA were bound per microgram of enzyme complex in the peak fraction after separating protein-bound PALA from excess free PALA by gel filtration (average of four determinations in two separate experiments). One molecule of PALA is bound for each 230,000 daltons of protein, or each polypeptide of 200,000 daltons has 0.87 PALA binding sites.

Cross-linking Experiments. Covalent cross-linking of an oligomeric protein, followed by separation of the cross-linked species on SDS gels,

---

[15] K. Weber and M. Osborn, *J. Biol. Chem.* **244**, 4406 (1969).
[16] P. F. Coleman, D. P. Suttle, and G. R. Stark, *J. Biol. Chem.* **252**, 6379 (1977).

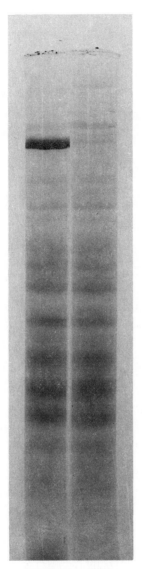

A   B

FIG. 2. SDS gel electrophoresis of the total cell protein from SV28 wild-type and 165-23 PALA-resistant lines. The cells from confluent plates of SV28 and 165-23 were removed by treatment with trypsin, collected by centrifugation, and washed once in Tris-saline buffer. The cells were resuspended in 0.5 ml of Tris-saline, and an equal volume of 10% trichloroacetic acid was added. Equal volumes of the resulting suspensions (15 μl) were further diluted in 5% trichloroacetic acid, and the precipitated protein was collected by centrifugation. The samples were denatured by heating at 60° for 10 min and run on SDS gels (4.5% acrylamide) as described by Weber and Osborn.[15] (A) 165-23; (B) SV28.

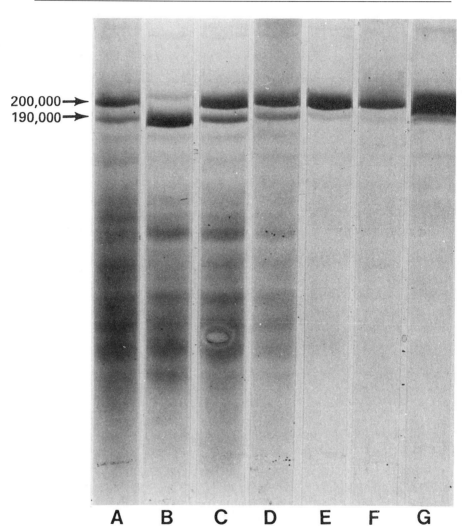

200,000→
190,000→

A   B   C   D   E   F   G

FIG. 3. SDS gel electrophoresis of samples from each step in the purification of the complex. Samples were precipitated in 5% trichloroacetic acid, and the protein was collected by centrifugation. The pellet was neutralized with 0.2 M NaOH and redissolved in denaturing buffer. Samples were denatured as in Fig. 2, and run on SDS gels (4.5% acrylamide). (A) Cell lysate, 107 μg; (B) postmitochondrial supernatant solution, 77 μg; (C) streptomycin sulfate supernatant solution, 65 μg; (D) RNase A treated supernatant solution, 60 μg; (E) ammonium sulfate pellet, 27 μg; (F) Bio-Gel A-5M pool I, 35 μg; (G) Bio-Gel A-5M pool II, 38 μg.

TABLE I

PURIFICATION OF CARBAMYL-P-SYNTHASE, ASPARTATE TRANSCARBAMYLASE, AND DIHYDROOROTASE FROM PALA-RESISTANT HAMSTER CELLS

Fraction	Protein concentration (mg/ml)	Total protein	Carbamyl-P synthase		Aspartate transcarbamylase		Dihydroorotase		Ratio of specific activities
			Specific activity[a]	Total units $\times 10^{-3}$	Specific activity[a]	Total units $\times 10^{-3}$	Specific activity[a]	Total units $(\times 10^{-3})$	
Crude lysate	21.5	1353	0.03	40.6	2.2	2990	0.32	433	1:74:11
Postmitochondrial supernatant	7.7	414	0.09	37.3	4.2	1751	0.63	261	1:47:7
Streptomycin sulfate supernatant	6.5	346	0.10	34.6	4.4	1514	0.75	259	1:43:7.5
RNAse supernatant	6.0	317	0.11	34.9	4.7	1480	0.80	254	1:42:7
Ammonium sulfate fraction	50.3	105.6	0.29	30.6	13.2	1392	2.13	225	1:45:7
A-5M column pool I	3.45	21.1	0.42	8.8	21.7	457	3.16	66.5	1:51:7.5
A-5M column pool II	3.78	23.5	0.40	9.4	19.3	452	2.66	62.4	1:48:7
Purification from crude lysate to pool I			14.0		9.8		9.9		
Recovery from postmitochondrial supernatant to A-5M pools			48%		52%		49%		

[a] Units (nanomoles of carbamyl-L-aspartate formed per minute at 37°) per microgram of protein.

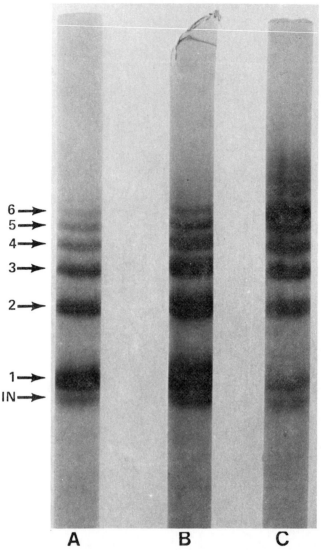

Fig. 4. Analysis of the covalently cross-linked species of the purified complex by SDS gel electrophoresis. The purified complex (pool I) was reacted at two concentrations of DTDI in buffer A made 10% in sucrose. The pH was adjusted to 8.5 by the addition of $\frac{1}{4}$ volume of 1.0 M triethanolamine hydrochloride, pH 9.0. The final protein concentration was 1.0 mg/ml in a total volume of 0.5 ml. Aliquots of the cross-linked sample were precipitated in 5% trichloroacetic acid and electrophoresed on polyacrylamide–agarose composite gels. The gels were prepared from a solution 2.75% in acrylamide, 0.075% in bisacrylamide, and 0.8% agarose heated to 100° for 20 min in a pressure cooker. Ammonium persulfate and TEMED were added to the hot solution and the gels poured immediately. (A) 0.25 mg/ml DTDI, 25 μg of protein on the gel; (B) 0.25 mg/ml DTDI, 50 μg of protein on the gel; (C) 5.0 mg/ml DTDI, 25 μg of protein on the gel. The band labeled 1N is the nicked monomer band. Bands 1 through 6 correspond to the cross-linked species with that number of monomer units.

usually indicates the number of subunits in each oligomer.[17] The enzyme complex (pool I, 1 mg/ml) was reacted with 0.25 mg/ml of DTDI,[18] and the cross-linked species were separated on 2.75% polyacrylamide–0.8% agarose SDS gels (Fig. 4). There are six major bands plus the 190,000-dalton protease cleavage product. At the DTDI concentration which gave maximum cross-linking (5 mg/ml), the pattern shifted toward species of higher molecular weight, predominantly trimers and hexamers. When the mobility of the bands relative to Bromphenol blue is plotted versus log molecular weight (using 200,000 for the monomer), a straight line is obtained,[16] showing that species larger than the monomer have molecular weights which are integral multiples of 200,000. The mobilities of the bands above the hexamer are appropriate for nonamer and dodecamer. In order to determine whether the larger species are formed by interoligomer cross-linking, the protein was diluted to 0.03 mg/ml 12 hr before addition of DTDI. Under these conditions trimer and hexamer predominate although larger species are still seen. Therefore, low concentrations of nonamer and dodecamer are probably present even at low concentrations of protein. The relative preponderance of trimers and hexamers shown in Fig. 4C indicates either that the material in pool I is a mixture of trimers and hexamers not in rapid equilibrium (since the basic pattern is retained at high dilution), or that the material is predominantly a hexamer, with more effective cross-linking within the trimeric substructures.

Electrophoresis in Nondenaturing Gels. Two major and two minor species of undenatured protein are resolved in gels run in a discontinuous Tris-glycine system (Fig. 5). In the experiment shown, the two major bands are about equal in intensity, although their relative amounts do vary somewhat in different samples. If we assume that the four bands represent trimers, hexamers, nonamers, and dodecamers of the 200,000-dalton subunit, and plot their mobilities against the logs of their molecular weights, all four points lie on a straight line.[16] This result would be expected for oligomers of a trimeric subunit. Since the charge-to-mass ratio of these oligomers would be the same, they would separate solely on the basis of their different molecular weights.

Sedimentation Coefficients and Frictional Ratios. A sedimentation velocity experiment was run on the purified protein at about 3 mg/ml in

---

[17] G. E. Davies and G. R. Stark, *Proc. Natl. Acad. Sci. U.S.A.* **66**, 651 (1970).

[18] DTDI is dimethyl-3,3'-(tetramethylenedioxy) dipropionimidate dihydrochloride, a bis-imidoester 12 atoms long, prepared according to R. F. Luduena, E. M. Shooter, and L. Wilson [*J. Biol. Chem.* **252**, 7006 (1977)], who found that it crosslinks tubulin more effectively than dimethylsuberimidate.

FIG. 5. Analysis by electrophoresis in a nondenaturing gel. The purified complex (26 μg) in buffer A was diluted to 50 μl and applied to 5% polyacrylamide–0.8% agarose gels. Composite gels were prepared as in Fig. 4. The following buffers were used: upper reservoir, 53 m*M* Tris-glycinate, pH 8.75; lower reservoir, 0.1 *M* Tris, pH 8.0; gel buffer, 1.5 *M* Tris, pH 8.9 diluted 1 to 4 in the gels; sample dilution buffer, 0.2 ml of 53 m*M* Tris-glycinate, pH 8.75, 0.2 ml 0.05% Bromophenol blue, 0.5 ml H_2O, 2 drops glycerol, and 50 μl 2-mercaptoethanol. Gels were run at 200 *V* for 1.75 hr at 6°.

53 mM Tris-glycine buffer, pH 8.7, at 6°.[16] The sedimentation pattern shows predominantly one peak with a sedimentation coefficient of 9.4 S ($s_{20,w}$ = 13.9) and a minor species (approximately 20% of total) sedimenting at 13.6 S ($s_{20,w}$ = 20.0). When the protein concentration is reduced to 2 mg/ml only the major slower sedimenting peak is observed. The observation of two peaks in the ultracentrifuge is consistent with the evidence provided by cross-linking and electrophoresis experiments which show the presence of both trimers and hexamers, not in rapid equilibrium.

A partial specific volume of 0.746 cm^3/g was calculated for the complex from the amino acid composition using the data of Cohn and Edsall.[19]

A molecular weight of 600,000 for the trimer and 1,200,000 for the hexamer, determined from SDS gels, was used along with the calculated S values and partial specific volume to compute a frictional ratio (f/f_0) of 1.67 for the major slower sedimenting peak and 1.84 for the minor peak. Most globular proteins have a frictional ratio between 1.0 and 1.5. A frictional ratio above 1.7 indicates a nonspherical shape quite unusual for an enzyme.

Amino Acid Analysis and N-terminal Analysis. The results of amino acid analysis are shown in Table II. Cysteine residues were carboxymethylated with iodoacetamide in 6 M guanidinium chloride, and tryptophan was not determined.

Analysis with [^{35}S]phenylisothiocyanate, using a sequenator, revealed no free N-terminus for the 200,000-dalton polypeptide. This result is not surprising since Brown and Roberts[20] have reported that 80% of the soluble proteins from Ehrlich ascites cells are N-α-acetylated. The sample contained less than 5% of protease-cleaved polypeptide by analysis on SDS gels.[16] The phenylisothiocyanate had a specific activity of 2200 cpm/nmole, and about 5 nmoles of protein were analyzed. If 5% of the N-termini were labeled, 0.25 nmole (550 cpm) of phenylthiohydantoin would have been formed. When a 24-hr autoradiogram is used, the limit of sensitivity is about 700 cpm of label; therefore the N-terminus of the 190,000-dalton cleavage product, if it has one, would not have been detected. A total of 6 cycles of the sequenator were run, and no phenylthiohydantoin was observed for any of the runs.[21]

[19] E. J. Cohn and J. T. Edsall, "Proteins, Amino Acids, and Peptides," p. 372. Van Nostrand-Reinhold, Princeton, New Jersey, 1943.

[20] J. L. Brown and W. K. Roberts, *J. Biol. Chem.* **251**, 1009 (1976).

[21] Prof. Ralph Bradshaw and Dr. Owen Bates of the Department of Biochemistry, Washington University, St. Louis, Mo. kindly performed the N-terminal analysis.

TABLE II

AMINO ACID COMPOSITION OF THE CARBAMYL-P-SYNTHASE, ASPARTATE
TRANSCARBAMYLASE, DIHYDROOROTASE COMPLEX[a]

Amino acid	Residues/200,000 daltons
Lysine	67
Histidine	38
Arginine	70
Aspartic acid	127
Threonine	112
Serine	148
Glutamic acid	183
Proline	127
Glycine	276
Alanine	246
Valine	173
Methionine	30
Isoleucine	80
Leucine	186
Tyrosine	31
Phenylalanine	50
Carboxymethyl cysteine	29

[a] The analysis was performed on a Beckman-Spinco Model 116 analyzer. The mole
fractions were normalized to alanine (8.75 mole %). Serine and threonine were
extrapolated to zero hydrolysis time and tryptophan was not determined.

Stability. All three enzyme activities are stable for months when
stored in buffer A at liquid nitrogen temperature. The removal of the
Me_2SO and glycerol results in dissociation of the oligomer and very
rapid loss of carbamyl-P synthase activity (half time about 12 hr at 0°).

Acknowledgment

We are grateful to Dr. R. L. Baldwin for many helpful discussions.

[19] Orotate Phosphoribosyltransferase: Orotidylate Decarboxylase (*Serratia*)

By W. L. Belser and James R. Wild

$$\text{orotate} + \text{PRPP} \rightarrow \text{orotidylate} + \text{PP}_i \rightarrow \text{uridylate} + \text{CO}_2$$

Based upon earlier work from our laboratory,[1,2] we suspected that the orotate phosphoribosyltransferase (OPRTase, EC 2.4.2.10):orotidylate decarboxylase (OMPdecase, EC 4.1.1.23) complex might be associated with an earlier pathway enzyme dihydroorotase (DHOase, EC 3.5.2.3). Thus we monitored this activity also during the purification and observed that separation of the OPRTase:OMPdecase complex from DHOase resulted in a loss of activity for the complex.[3] This could be restored by addition of partially purified DHOase. In monitoring purification of the complex, it is essential to complement the reaction with DHOase for accurate determination of activities of OPRTase and OMPdecase.

Assay

Principle. The standard assay used to monitor the purification of the OPRTase:OMPdecase complex is based upon conversion of orotate to orotidylate in the presence of orotate phosphoribosyltransferase (OPRTase) and upon the conversion of orotidylic acid (OMP) to uridylic acid (UMP) in the presence of orotidylate decarboxylase (OMPdecase).

Reagents. Composition of the assay mixtures is as follows:

OPRTase assay: The standard assay mixture contained 100 μmoles of Tris buffer (pH 8.8), 0.25 μmoles of orotate, 1.0 μmoles of MgCl$_2$, 0.6 μmoles 5'-phosphoribosyl-pyrophosphate (PRPP), bacterial extract, and water to a volume of 1.0 ml.

OMPdecase assay: The standard assay mixture contained 100 μmoles of Tris buffer (pH 8.8), 0.2 μmoles OMP, 0.5 μmoles MgCl$_2$, bacterial extract, and water to a volume of 1.0 ml.

Procedure. The assay mixture for the OPRTase assay was incubated for 3–5 min at 37° prior to initiation of the reaction by addition of

[1] W. S. Hayward and W. L. Belser, *Proc. Natl. Acad. Sci. U.S.A.* **53**, 1483 (1965).

[2] J. R. Wild and W. L. Belser, *Biochem. Genet.* **15**, 157 (1977).

[3] J. R. Wild and W. L. Belser, *Biochem. Genet.* **15**, 174 (1977).

METHODS IN ENZYMOLOGY, VOL. LI

orotate. The reaction was followed at 295 nm comparing the decrease in optical density due to disappearance of orotate with a blank containing no protein extract. A change of 3.67 OD units at 295 nm was equivalent to 1.0 μmole of substrate converted to product. This assay is based upon that described by Beckwith *et al.*,[4] but see notes under *Properties* for differences.

The assay mixture for OMPdecase assay[4] was incubated for 3–5 min at 37° before addition of OMP to initiate the reaction. This was compared to blanks lacking either substrate or enzyme extract. An optical density change of 1.38 OD units at 290 nm corresponds to 1.0 μmole of substrate converted to product.

Both assays were conducted in quartz cuvettes with a 1-cm light path using the Beckman DU2 spectrophotometer with a water-jacketed cell holder at 37°.

Units and Specific Activity. One unit of activity is defined as that amount of protein which catalyzes the conversion of 1.0 μmole of substrate to product per minute under the standard assay conditions. For calculating specific activities (units per milligram protein), protein concentrations were estimated by the Folin-Calcioteau method described by Lowry *et al.*[5] using desiccated bovine serum albumin as the protein standard.

Culture and Harvest of Cells

The organism used was *Serratia marcescens* strain HY.[6] Stocks were maintained on agar slants containing SME medium.[6] Inocula from slants were introduced into 1 liter SM medium[6] in 2.8-liter Fernbach flasks and incubated overnight on a rotary shaker at 30°. The entire contents of the flasks were delivered into Fermenter vessels containing 13 liters of SM medium and incubated at 30°. The vessels were sparged at approximately 3 liters per minute with line air and stirred at approximately 400 rpm in the New Brunswick Fermenter. Three vessels totaling 42 liters of cell culture were prepared each time and harvested in a Sharples Cream separator. The cell pellets were frozen at −20°, and a total of eight runs provided 1200 g of wet packed weight of cells.

[4] J. R. Beckwith, A. B. Pardee, R. Austrian, and F. Jacob, *J. Mol. Biol.* **5**, 618 (1962).
[5] O. H. Lowry, N. J. Rosebrough, A. L. Farr, and R. J. Randall, *J. Biol. Chem.* **193**, 265 (1951).
[6] E. L. Labrum and M. I. Bunting, *J. Bacteriol.* **65**, 394 (1953).

Purification

All operations are carried out at 2° unless otherwise noted.

Step 1. Preparation of Cell Extracts. First 1200 g of cells were thawed and suspended in 2 liters of 50 mM potassium phosphate buffer, pH 7.5, containing 0.1 mM EDTA and 0.1 mM DTT (PED buffer). The cells were broken by two passes through a Manton-Gauldin cell fractionator at 30,000 psi. The extract was centrifuged in a GSA rotor in the Sorvall RC2B centrifuge for 30 min, and the supernatant solution was clarified at 18,000 rpm in the SS34 rotor for 60 min.

Step 2. Streptomycin Sulfate Precipitation. Next 1200 ml of extract were adjusted to 1% streptomycin sulfate by addition of crystals and stirred for 2 hr. The precipitate was removed by centrifugation at 13,000 rpm for 30 min in the GSA rotor.

Step 3. Ammonium Sulfate Precipitation I. Orotate (2.5 μmoles/liter) was added to the extract and equilibrated in the cold for 15 min. The extract was divided into two batches and each was brought to 40% saturation by addition of 173 g of crystalline ammonium sulfate (Mann assay grade). After stirring gently for 4 hr, the precipitate was removed by centrifugation at 13,000 rpm and discarded. The 40% supernatant solution was adjusted to 70% saturation by addition of 250 g ammonium sulfate crystals/liter. After 4 hr of gentle stirring, the precipitate was collected by centrifugation and resuspended in 200 ml PED buffer at pH 7.5. The suspension was dialyzed overnight against a 50-fold excess of PED buffer at pH 7.5. An insoluble precipitate remained after dialysis, and it was removed by centrifugation.

Step 4. Sephadex G-200 Chromatography. Four 50-ml samples were passed over a Sephadex G-200 column equilibrated in PED buffer at pH 7.5. The column was 5.1 cm × 90 cm, and proteins were eluted with PED buffer at 40 ml/hr by upward flow. Fractions (10 ml) were collected, and OPRTase and OMPdecase were co-eluted in fractions 55–78, corresponding to a molecular weight of approximately 70,000. This operation was carried out at 6° in the cold room.

Step 5. DEAE-Cellulose Fractionation. The pooled fractions from G-200 were added to a column (5.0 × 40 cm) packed with DEAE-cellulose (Cellex D, 0.931 μeq/gm) that had been previously equilibrated with PED buffer, pH 7.5. After loading, the column was washed with 200 ml of PED buffer. Proteins were eluted by passing a linear KCl gradient over the column (0.05–0.5 M KCl in PED buffer, total volume 2.0 liters).

OPRTase and OMPdecase were recovered in fractions 65–85 (10-ml fractions). This operation was carried out at 6°. At this stage, DHOase was no longer associated with the complex.

Step 6. Ammonium Sulfate Precipitation II. Fractions from DEAE containing OPRTase and OMPdecase were pooled and brought to 45% saturation by addition of solid ammonium sulfate. After 4 hr with stirring, a small precipitate was removed by centrifugation, and the supernatant solution was adjusted to 65% saturation with solid ammonium sulfate. After 4 hr with stirring, the precipitate was recovered by centrifugation and resuspended in 5.0 ml PED buffer at pH 7.5. The suspension was dialyzed overnight against a 100-fold excess of PED buffer.

Step 7. DEAE-Cellulose Chromatography II. A DEAE-cellulose column (2.5 × 30 cm) was equilibrated with PED Buffer at pH 7.5 containing 0.05 M KCl. The dialyzed suspension was loaded onto the column and eluted with a linear (0.05–0.3 M) KCl gradient. OPRTase and OMPdecase co-eluted as a symmetrical peak with a slight trailing shoulder. This operation was carried out at 6°. A summary of the purification is shown in Table I.

At low protein concentrations (e.g., less than 100 μg/ml), the enzyme complex loses stability and both purification and yield are compromised. If low concentrations cannot be avoided, then stabilization by glycerol (5–20%) may be substituted.

Properties

Molecular Weight. Before the first DEAE fractionation the native OPRTase:OMPdecase complex has a molecular weight of approximately 70,000 as estimated by elution from a calibrated Sephadex G-200 column, and confirmed by an S value of 4.8 observed in a linear sucrose gradient. Following the first DEAE fractionation step, the molecular weight of the complex drops to 40,000 as estimated by elution from a calibrated Sephadex G-200 column, and confirmed by an S value of 3.0 in a linear sucrose gradient. The DEAE fractionation step separates DHOase from the complex. The ratio of OPRTase activity to OMPdecase activity is greater than 1.0 before the DEAE step, and drops below 1.0 following that step (Table I). Addition of partially purified DHOase to fractions containing the complex brings the ratio back to greater than 1.0. This has been interpreted to mean that the DHOase forms a weak aggregation complex with the OPRTase:OMPase complex to activate these latter enzymes, perhaps by inducing a conformational change.

TABLE I

PURIFICATION OF THE OPRTASE:OMPDECASE COMPLEX FROM *Serratia*

Fractionation step	Total volume (ml)	Total protein (mg)	Total activity (units)		Specific activity[a]		Ratio[b] OPRTase/ OMPdecase	Recovery (%)	
			OPRTase	OMPdecase	OPRTase	OMPdecase		OPRTase	OMPdecase
1. Crude	1200	18,000	33,300	25,020	1.85	1.39	1.33	100	100
2. Streptomycin SO₄	1200	7,770	25,786	20,435	3.19	2.63	1.42	74	81.6
3. (NH₄)₂ SO₄	220	4,170	32,359	22,018	7.76	5.28	1.45	97	88
4. G-200 pooled	305	388	8,904	7,806	22.95	20.12	1.16	27	31
5. DEAE 1 pooled	155	26	2,730	3,120	105.00	120.00	0.875[c] (1.18)[d]	8	12
6. (NH₄)₂ SO₄	5	25.5	2,779	3,544	109.00	139.00	0.790[c] (1.07)[d]	8	14
7. DEAE 2	35	16.5	4,372	5,362	265.00	325.00	0.814[c] (1.08)[d]	13	21

[a] Specific activity is in μmoles substrate converted to product per minute per milligram protein.

[b] This activity ratio represents a relative index of copurification of the two enzymes.

[c] Complex assayed in the absence of added DHOase.

[d] Numbers in parentheses are based upon specific activities obtained when partially purified DHOase was added to the complex prior to assay.

Kinetic Properties. OPRTase shows an absolute dependence on Mg^{2+} ions and possesses first-order saturation kinetics with $MgCl_2$ (apparent $K_m = 2 \times 10^{-6}$ *M*). Mg^{2+} is required by all OPRTase systems studied.[7-15] OPRTase from *Serratia* shows Michaelis–Menton kinetics with both PRPP and orotate. A linear reaction rate was observed when the enzyme was incubated with Mg^{2+} and PRPP and the reaction was initiated by addition of orotate. A nonlinear reaction rate was observed when the enzyme was incubated with Mg^{2+} and orotate and the reaction was initiated with PRPP. We infer from these results that there is sequential substrate binding to the enzyme in the presence of Mg^{2+}. Michaelis–Menton kinetics were observed for OPRTase with both PRPP and orotate concentration in the presence or absence of added DHOase. There was, however, an observed change in apparent K_m depending upon the presence or absence of added DHOase. The apparent K_m for PRPP in the presence of DHOase was 2.6×10^{-6} *M*, while in the absence of DHOase $K_m = 4.1 \times 10^{-6}$ *M*. The apparent K_m for orotate in the presence of DHOase was 2.3×10^{-6} *M* and in the absence of DHOase $K_m = 1.3 \times 10^{-6}$ *M*. The K_m's observed for *Serratia* OPRTase are comparable to values obtained with enzyme from other organisms.[9,11,16] The change in K_m accompanying separation of DHOase from the complex further supports our contention that DHOase activates the OPRTase:OMPdecase complex by aggregating and inducing some sort of conformational change.

OMPdecase shows Michaelis–Menton kinetics with increasing concentration of orotidylic acid, with an apparent $K_m = 1.0 \times 10^7$ *M*. Most K_m's reported for eukaryotic OMPdecase were significantly higher (2.0–8.4 $\times 10^{-6}$ *M*).[9,11-13] However, Kavipurapu and Jones[17] published a value of 8×10^{-7} *M* for mouse Ehrlich cites cells assayed in phosphate buffer, and the K_m falls to 4×10^{-7} *M* when assayed in Tris buffer.[18]

[7] I. Lieberman, A. Kornberg, and E. S. Simms, *J. Am. Chem. Soc.* **76**, 2844 (1954).

[8] I. Lieberman, A. Kornberg, and E. S. Simms, *J. Biol. Chem.* **215**, 403 (1955).

[9] K. Umezu, T. Amaya, A. Yoshimoto, and K. Tomita, *J. Biochem. (Tokyo)* **70**, 249 (1971).

[10] T. L. Pynadath and R. M. Fink, *Arch. Biochem. Biophys.* **118**, 185 (1967).

[11] D. K. Kasbekar, A. Nagabhushanam, and D. M. Greenberg, *J. Biol. Chem.* **239**, 4245 (1964).

[12] S. H. Appel, *J. Biol. Chem.* **243**, 3924 (1968).

[13] G. K. Brown, R. M. Fox, and W. J. O'Sullivan, *J. Biol. Chem.* **250**, 7352 (1975).

[14] W. T. Shoaf and M. E. Jones, *Biochemistry* **12**, 4039 (1973).

[15] W. Grobner and W. N. Kelley, *Biochem. Pharmacol.* **24**, 379 (1975).

[16] M. E. Jones, *Curr. Top. Cell. Regul.* **6**, 227 (1972).

[17] P. R. Kavipurapu and M. E. Jones, *J. Biol. Chem.* **251**, 5589 (1976).

[18] M. E. Jones, T. W. Traut, and P. R. Kavipurapu, *in* "Methods in Enzymology" (this volume).

These latter values are much more consistent with those we obtained for the enzyme from *Serratia*.

Separation of OPRTase and OMPdecase. Most attempts to separate the two activities resulted in total loss of both activities or only fractional recovery of OMPdecase activity. Hydroxylapatite cation-exchange chromatography, narrow pH range isoelectric focusing (pH 3.0–6.0), and preliminary studies with acrylamide gel electrophoresis resulted in destruction of both activities. In one experiment minimal recovery of OMPdecase was observed. OPRTase was consistently more labile than was OMPdecase. However, acrylamide gel electrophoresis in the presence of reduced glutathione permitted the separation and recovery of the two activities. Acrylamide gel electrophoresis was performed by the technique of Ornstein.[19] The stacking gels were 7.5%, pH 8.3, and the spacing gels were 7.5%, pH 9.5. Electrophoresis was performed at 5° at 2 mA per tube (0.6 × 5.0 cm). The upper Tris-glycine buffer (pH 9.8) contained 1 mM reduced glutathione and 0.1 mM EDTA. Gels were stained with naphthol blue-black for protein, and the enzymes were eluted from 0.25-cm slices of unstained companion gels. Protein stains showed three major protein bands and several minor bands. Two of the major bands corresponded to the positions of the two enzyme activities. Complete separation of OMPdecase and OPRTase was accomplished, although accompanied by considerable loss of activity. Approximately 80% of the OPRTase activity and 30% of the OMPdecase activity was lost.

Complementation Studies. OPRTase and OMPdecase recovered from the acrylamide gel separation were assayed by the standard assay, then mixed and incubated for 15 min and assayed under standard conditions (Table II). It seems clear from the striking loss of OPRTase activity, and the restoration of that activity when the complex is reformed, that the conformation of this enzyme is highly unstable. Complementation tests were also performed with the OPRTase:OMPdecase complex and DHOase recovered from the DEAE-cellulose fractionation, step 1 (Table III). The results of these two studies indicate that the accuracy of any assays designed to determine kinetic parameters or turnover numbers for OPRTase depend upon measurements made in the presence of both OMP decase and DHOase. Otherwise, the determinations will be spuriously lower than the corresponding physiological activity in the intact aggregate.

Notes on the OPRTase Assay. The assay as conducted by Beckwith *et al.*[4] called for incubation of enzyme with Mg^{2+} and orotate, and the

[19] L. Ornstein, "Disc Electrophoresis." Canal Industrial Corp., Bethesda, Maryland (1961).

TABLE II
COMPLEMENTATION BETWEEN ACRYLAMIDE GEL FRACTIONS CONTAINING OPRTASE OR
OMPDECASE

| | Enzyme activities | | | |
| | OPRTase assay | | OMPdecase assay | |
Gel fraction	Units[a]	%	Units[a]	%
Control[b]	0.01	1.6	0.01	0.7
OPRTase	0.65	100	0.02	1.4
OMPdecase	0.02	2.5	0.90	100
OPRTase + OMPdecase[c]	2.80	421	1.18	131

[a] Units are μmoles substrate converted per minute per milliliter.
[b] The control is the eluate from an area of the gel containing no protein band.
[c] 50 μliters of the eluate from each of the above fractions (e.g., OPRTase and OMPdecase) were mixed and incubated for 15 hours at 37° and activities were assayed.

TABLE III
COMPLEMENTATION OF OPRTASE:OMPDECASE COMPLEX BY DHOASE

| | Enzyme activities | | | |
| | Units[b] | | %[c] | |
Enzyme fraction and volume[a]	OPRTase	OMPdecase	OPRTase	OMPdecase
DHOase (35) 0.1 ml	0.47	0.40	—	—
OPRTase:OMPdecase (80) 0.05 ml	17.74	49.00	100	100
DHOase + complex (35) (80) 0.05 ml 0.05 ml	24.20	46.40	136	96
DHOase + complex (35) (80) 0.025 ml 0.05 ml	20.42	51.20	116	104
DHOase + complex (35)[d] (80) 0.05 ml 0.05 ml	17.80	—[e]	102	—

[a] Fraction (35) containing DHOase and fraction (80) containing the complex were *recovered* from DEAE-cellulose fractionation 1, step 5.
[b] Units are expressed as μmoles substrate converted per minute per milliliter extract.
[c] % is based upon using 100% for the uncomplemented activity of the complex.
[d] The DHOase fraction was heated prior to incubation with the complex.
[e] Not determined.

reaction was initiated by addition of PRPP. With the *Serratia* enzyme a perceptible lag in reaction rate occurs under these conditions (see section on *Kinetic Properties*). We modified this part of the procedure, and routinely initiate the reaction with orotate following incubation with Mg^{2+} and PRPP. In addition, the original assay calls for addition of excess OMPdecase to the assay system to remove OMP formed, since it product inhibits. Although OMP inhibits the *Serratia* enzyme we found it unnecessary to add OMPdecase, since the reaction in all cases was linear for 3–5 min without added OMPdecase.

Acknowledgment

This work was supported by grants from the National Science Foundation (NSF GB 5811) and the Office of Naval Research (Nonr 4413), and by funds from an Intramural Research Grant. One of us (JRW) was a National Science Foundation Graduate Fellow. We also acknowledge support from the National Institutes of Health, Grant 55289 to JRW.

[20] Orotate Phosphoribosyltransferase:Orotidylate Decarboxylase (Erythrocyte)

By ROBERT F. SILVA and DOLPH HATFIELD

Orotate phosphoribosyltransferase and orotidylate decarboxylase are purified together from mammalian sources and exist as a complex.[1-5] The assay and preparation of these enzymes from beef[1] and human[4,5] erythrocytes are described.

Assay Method

Principle. Orotate phosphoribosyltransferase activity may be measured spectrophotometrically by following the change in absorbance which results upon the conversion of a pyrimidine or purine base to the corresponding ribonucleotide. The assay is carried out with a pyrimidine or purine base in the presence of PP-ribose-P and Mg^{2+} as given in procedure A. The product of the reaction with a purine and PP-ribose-P as substrates is 3-*N*-ribosyl purine-5'-phosphate.[6] A second assay proce-

[1] D. Hatfield and J. B. Wyngaarden, *J. Biol. Chem.* **239**, 2580 (1964).
[2] S. H. Appel, *J. Biol. Chem.* **243**, 3924 (1968).
[3] P. R. Kavipurapu and M. E. Jones, *J. Biol. Chem.* **251**, 5589 (1976).
[4] G. K. Brown, R. M. Fox, and W. J. O'Sullivan, *J. Biol. Chem.* **250**, 7352 (1975).
[5] G. K. Brown and W. J. O'Sullivan, *Biochemistry* **16**, 3235 (1977).
[6] D. Hatfield and J. B. Wyngaarden, *J. Biol. Chem.* **239**, 2587 (1964).

dure (B) involves the use of ^{14}C-labeled bases. The products of the reaction mixture are separated by paper chromatography, and the radioactivity is measured in a liquid scintillation counter. Orotidylate decarboxylase activity may be determined spectrophotometrically as given in procedure C. Additionally, orotate phosphoribosyltransferase and orotidylate decarboxylase activities may be measured by the release of ^{14}C]O$_2$ from [carboxy-^{14}C]orotidylic acid; the transferase activity in this assay is determined in the presence of [carboxy-^{14}C]orotic acid, PP-ribose-P, and Mg^{2+} and depends upon the conversion of the corresponding product, orotidylic acid, to uridylic acid and ^{14}C]O$_2$ by the decarboxylase (procedure D).

Reagents

Orotic acid or [^{14}C]orotic acid
Uracil or [^{14}C]uracil
Xanthine or [^{14}C]xanthine
Uric acid or [^{14}C]uric acid
Orotidylic acid or [carboxy-^{14}C]orotidylic acid
PP-ribose-P
Tris·HCl
MgCl$_2$

Procedure A. The activity of erythrocyte orotate phosphoribosyl-transferase may be measured spectrophotometrically after the removal of hemoglobin.[1,7] Reactions are carried out at 37° in a 1-ml volume containing 3 mM MgCl$_2$, 50 mM Tris·HCl, 0.3 mM PP-ribose-P, pyrimidine or purine base, and 0.05 ml of enzyme. Reaction mixtures are incubated at 37° for 5 min before adding enzyme. The following purine and pyrimidine bases, their concentrations in assays, the pH at which assays are conducted, the wavelength for monitoring enzyme activity, and the change in absorbance which results upon the conversion of 0.1 μmole of base to the corresponding ribonucleotide are: orotic acid, 0.15 mM, pH 8.0, 295 nm, and $\Delta = -0.395$[7]; uracil, 0.15 mM, pH 9.5, 262 nm, and $\Delta = +0.25$; xanthine, 0.15 mM, pH 8.1, 265 nm, and $\Delta = +0.425$; and uric acid, 0.2 mM, pH 8.0, 305 nm, and $\Delta = +1.16$. The spectrophotometric assay of the transferase activity with orotic acid as the base substrate depends upon the conversion of the corresponding product, orotidylic acid, to uridylic acid by the decarboxylase.

Procedure B. Orotate phosphoribosyltransferase activity may be measured prior to the removal of hemoglobin[1] by procedure B. The reaction mixture is the same as described in procedure A, except that

[7] I. Lieberman, A. Kornberg, and E. S. Simms, *J. Biol. Chem.* **215**, 403 (1955).

0.4 μCi of the respective [14C]base is also added. ([Carboxy-14C]orotic acid can not be used in this assay.) Incubation times at 37° are: 5 min with orotic acid, 15 min with uracil, 10 min with xanthine, and 30 min with uric acid. At the end of the incubation, protein is precipitated by heating for 2 min in a boiling water bath and subsequently removed by centrifugation. A 0.2-ml aliquot of the supernatant is chromatographed on Whatman No. 1 filter paper in either 1-propanol-water (7:3; descending) or 4% sodium citrate (ascending). Controls containing the appropriate base and nucleotide are cochromatographed with aliquots from reaction mixtures so that bases and nucleotides may easily be identified on developed chromatograms. Ultraviolet-absorbing areas on developed chromatograms are detected with a UV lamp, cut out, and the radioactivity determined in a liquid scintillation counter.

Procedure C. Orotidylate decarboxylase activity may be measured spectrophotometrically[7] after the removal of hemoglobin. Reactions are carried out at 37° in a 1-ml volume containing 50 mM Tris·HCl, pH 7.4, 0.05 mM orotidylic acid, and 0.05 ml of enzyme. Reaction mixtures are incubated at 37° for 5 min before adding the enzyme. A decrease in absorbance of 0.165 at 285 nm corresponds to the conversion of 0.1 μmole of orotidylic acid to uridylic acid.

Procedure D. Orotidylate decarboxylase activity may be measured by the release of [14C]O_2 from [carboxy-14C]orotidylic acid.[4] This assay may be used to measure activity prior to the removal of hemoglobin as well as throughout the purification procedure. Orotate phosphoribosyltransferase activity may also be assayed by the release of [14C]O_2. The details of measuring the transferase and decarboxylase activities by this assay are described elsewhere in this volume in chapters dealing with these enzymes from mammalian tissues other than erythrocytes.

Definition of an Enzyme Unit and of Specific Activity. One unit of enzyme activity is defined as that amount which is required to transform 1 nmole of substrate in 1 min at 37°. The specific activity is defined as the units of enzyme activity per milligram of protein.

Purification Procedures

The transferase and decarboxylase have been purified from beef erythrocytes[1] as given in A. The enzymes have been purified from human erythrocytes[4,5] and the enzyme complex separated into monomer, dimer, and tetramer forms as given in B.

All purification steps are carried out at 0–4° unless indicated otherwise.

A. Beef Erythrocytes

Step 1. Preparation of Hemolysate. Heparin is added to freshly collected beef blood to a final concentration of 0.001%. The blood is then centrifuged at 7900 g for 15 min and the plasma and buffy coat removed. Erythrocytes are washed twice with 4 volumes of 0.9% NaCl, the packed cells lysed by sonic oscillation (Raytheon 10-KC sonic oscillator) for 3.5 min, and the stromal fraction removed by centrifugation at 16,300 g for 30 min. The transferase and decarboxylase are then purified about 3800-fold by steps 2 through 7 (Table I).

Step 2. Purification on DEAE-Cellulose. Immediately before use, DEAE-cellulose is washed with 1 volume of 1 M potassium phosphate, pH 7.4, followed by 4 volumes of distilled water. The DEAE-cellulose is resuspended in water to give 8 g of cellulose (dry weight) per 100 ml of water. Hemolysate (500 ml) is added to the DEAE-cellulose (500 ml). This mixture is stirred every 4 to 5 min for 45 min and poured onto a Büchner funnel with glass wool serving as a filter pad. Hemoglobin is removed by washing with 2–3 liters of 10 mM potassium phosphate, pH 7.4, in 200–300 ml portions. When the eluate becomes colorless, the DEAE-cellulose is washed with an additional 500 ml of buffer. The enzymes are eluted by placing the DEAE-cellulose into 500 ml of 500 mM potassium phosphate, pH 7.4. This mixture is stirred every 5 min for 30 min and then poured onto a Büchner funnel as described above. The eluate is collected and the elution step repeated a second time. The two eluate fractions are pooled and then poured onto solid $(NH_4)_2SO_4$ (46.2 g/100 ml of eluate). After the $(NH_4)_2SO_4$ dissolves, the resulting precipitate is collected by centrifugation at 16,300 g for 15 min. The precipitate is redissolved in a minimum of water, and this solution is dialyzed for 9 hr against 6 liters of 2 mM potassium phosphate, pH 7.5.

Step 3. First Ammonium Sulfate Fractionation. Saturated $(NH_4)_2SO_4$, pH 7.5, is slowly added with stirring to the dialyzed solution until the mixture becomes 40% saturated. The precipitate is removed by centrifugation, and the supernatant is brought to 60% saturation with $(NH_4)_2SO_4$, pH 7.5. The resulting precipitate is collected by centrifugation and dissolved in water. This solution is dialyzed for 3 hr against 3 liters of 2 mM potassium phosphate, pH 7.5.

Step 4. Heating and Second Ammonium Sulfate Fractionation. The dialyzed solution is incubated at 60° for 7 min and rapidly cooled in an ice bath. A second 40–60% $(NH_4)_2SO_4$ fraction is obtained and dialyzed as in step 3.

Step 5. Treatment with Alumina C_γ Gel. The dialyzed solution is stirred for 5 min with alumina C_γ gel (15 mg of solids per milliliter of

TABLE I

PURIFICATION OF OROTATE PHOSPHORIBOSYLTRANSFERASE AND OROTIDYLATE DECARBOXYLASE FROM BEEF ERYTHROCYTES[a]

Fraction	Volume (ml)	Protein (mg/ml)	Specific activity (orotic acid) (units/mg)	Purification (fold)	Yield of activity (%)	Relative specific activity (orotic acid = 1.0)			
						Uracil	Xanthine	Uric acid	Orotidylic acid
1. Hemolysate	500	366	0.185		100	0.24	0.50	0.012	2.68
2. DEAE-cellulose and 65% $(NH_4)_2SO_4$	196	7.0	9.8	53	40	0.60	0.79	0.028	2.10
3. 0.4–0.6 $(NH_4)_2SO_4$	36	10.6	27.2	147	31	0.43	0.64	0.027	2.21
4. 7 min at 60° and 0.4–0.6 $(NH_4)_2SO_4$	17	8.4	55	297	23	0.55	0.68	0.021	1.99
5. Alumina C_γ gel	10	2.3	172	930	12	0.47	0.65	0.026	2.31
6. DEAE-cellulose column	6	0.85	447	2416	6.7	0.60	0.72	0.035	2.13
7. $Ca_3(PO_4)_2$ gel	4	0.46	715	3865	3.9	0.52	0.69	0.027	

[a] The original publication of the data in this table was based on specific activity of the transferase with xanthine as the base substrate.[1] However, the data have been converted in this table to specific activity with orotic acid.

water) using 0.5 mg of gel per milligram of protein. The slurry is centrifuged and the gel discarded. Fresh gel is added to the supernatant using 1.0 mg of gel per milligram of the original protein concentration, and the whole is stirred for 15 min. The gel is collected and washed once in 5 ml of 10 mM potassium phosphate, pH 6.8. The enzymes are then eluted from the gel by two successive washings with 5 ml of 50 mM potassium phosphate, pH 6.8, for 15 min each.

Step 6. Chromatography on DEAE-Cellulose. The pooled supernatants from step 5 are dialyzed 2 hr against 3 liters of 5 mM potassium phosphate, pH 7.4. This solution is applied to a DEAE-cellulose column (1 × 8 cm), previously equilibrated with 10 mM potassium phosphate, pH 7.4. Enzymes are eluted by successive additions of 15-ml quantities of the following solutions containing 10 mM potassium phosphate, pH 7.4: the buffer solution, 50 mM NaCl, 100 mM NaCl, and 150 mM NaCl. Three-milliliter fractions are collected from the column, the protein concentration is determined in each fraction by measuring absorbance at 280 nm, and enzyme activities are detected by assaying each fraction as given in procedure A. The enzymes are eluted in 150 mM NaCl.

Step 7. Calcium Phosphate Gel Adsorption and Elution. Pooled fractions showing enzymic activity from step 6 are stirred for 10 min with a calcium phosphate gel solution (1 ml; 15–20 mg solids/ml). The gel is collected by centrifugation and washed once with 4 ml of 50 mM potassium phosphate, pH 8.0, for 5 min. The enzymes are eluted from the gel by stirring twice with 2-ml portions of 100 mM potassium phosphate, pH 8.0, for 10 min each. The eluate fractions are pooled. A summary of the purification procedure is given in Table I.

B. Human Erythrocytes

PROCEDURE I

This procedure[4] is a modification of that used for preparation of transferase and decarboxylase from beef erythrocytes. Although it is designed primarily for purification of human erythrocyte decarboxylase, it may also be used to prepare transferase.

Step 1. Preparation of Hemolysate. Human erythrocytes are obtained by venesection of polycythemic patients and are used within 3 days of collection. The cells are washed 3 times by suspension in 0.155 M KCl followed by centrifugation at 1200 g for 30 min. After each centrifugation, the cells to approximately 1 cm below the buffy coat are discarded to minimize contamination. The erythrocytes are lysed in 5 mM potassium phosphate, pH 7.4 (2 volumes of buffer per original

TABLE II
PURIFICATION OF OROTIDYLATE DECARBOXYLASE FROM HUMAN ERYTHROCYTES

Fraction	Volume (ml)	Protein (mg/ml)	Specific activity (units/mg)	Purification (fold)	Yield of activity (%)
1. Hemolysate	2800	95	0.0058		100
2. DEAE-cellulose-75% $(NH_4)_2SO_4$	154	52	0.09	16	47
3. $(NH_4)_2SO_4$ 0.4–0.6	32	22	0.68	117	31
4. Alumina C_γ gel	25	8.6	1.67	288	23

volume of blood) and the stromal fraction removed by centrifugation at 13,000 g. The decarboxylase is then purified about 300-fold by steps 2 through 4 (Table II).

Step 2. Purification on DEAE-Cellulose. The hemolysate is stirred with washed DEAE-cellulose (4 g dry wt/100 ml), previously equilibrated with 10 mM potassium phosphate, pH 7.4, in the proportion of 3:2 by volume. Unbound hemoglobin is removed by washing on a Büchner funnel, and the enzymes are eluted by washing twice in 0.5 M potassium phosphate, pH 7.4. The eluates are pooled and brought to 75% saturation with $(NH_4)_2SO_4$; the resulting precipitate is collected at 35,000 g, resuspended, and dialyzed against 10 mM potassium phosphate, pH 7.4.

Step 3. Ammonium Sulfate Fractionation. A 40–60% $(NH_4)_2SO_4$ fraction is obtained and dialyzed as in step 2.

Step 4. Treatment with Alumina C_γ Gel. The activity obtained from step 3 is adsorbed onto alumina C_γ gel (15 mg of gel/ml of water) using 0.5 mg of gel per milligram of protein, eluted in 500 mM potassium phosphate, pH 6.8, and stored at −15° in this buffer. A summary of this purification procedure is given in Table II.

PROCEDURE II

Step 1. Preparation of Hemolysate. The hemolysate is prepared as given in procedure I. The decarboxylase is then purified about 880-fold (Table III) by steps 2 and 3.[5]

Step 2. Purification on DEAE-Cellulose. The enzyme activities are adsorbed from the hemolysate onto DEAE-cellulose, and the unbound hemoglobin is removed as given in procedure I. The DEAE-cellulose is

TABLE III

PURIFICATION OF OROTATE PHOSPHORIBOSYLTRANSFERASE AND OROTIDYLATE DECARBOXYLASE FROM HUMAN ERYTHROCYTES

Fraction	Volume (ml)	Protein (mg/ml)	Specific activity		Purification		Yield	
			Transferase (units/mg)	Decarboxylase (units/mg)	Transferase (fold)	Decarboxylase (fold)	Transferase (%)	Decarboxylase (%)
1. Hemolysate	1066	105	0.0025	0.005			100	100
2. DEAE-cellulose: (NH$_4$)$_2$SO$_4$ ppt.	31	11	0.82	1.5	328	300	99.9	91
3. CM-Sephadex: (NH$_4$)$_2$SO$_4$ ppt.	6.4	3.4	0.98 (1.8[a])	4.4	392 (720[a])	880	7.6 (14[a])	17

[a] Assayed in the presence of 1 mM dithiothreitol and 10 μM EDTA.

then poured onto a column and further washed with 10 mM potassium phosphate, pH 7.4, until the optical density at 280 nm of the eluting buffer is less than 0.1. The enzymes are eluted in a linear gradient of potassium phosphate, pH 7.4, from 10–300 mM. The fractions containing activity are pooled and brought to 95% saturation with solid $(NH_4)_2SO_4$; the resulting precipitate is collected by centrifugation, resuspended, and dialyzed against 50 mM potassium phosphate, pH 7.4.

Step 3. Purification on CM-Sephadex. The enzymes obtained from step 2 are stirred with CM-Sephadex C-50, previously equilibrated with 10 mM potassium phosphate, pH 6.6, and the whole is poured onto a column (1.5 × 30 cm). Unbound protein is removed by washing the column in the same buffer, and the enzymes are eluted in a linear gradient of potassium phosphate, pH 6.8, from 10–250 mM. The fractions containing activity are pooled, and the activity is precipitated from solution, collected, and dialyzed as given in step 2 of this procedure. A relative loss in transferase activity is observed in step 3; however, this activity is effectively restored by assaying in the presence of dithiothreitol (see Table III).

SEPARATION OF MONOMER, DIMER, AND TETRAMER FORMS

The enzyme activities obtained by procedure I (Table II) may be separated into monomer, dimer, and tetramer forms on a G-200 Sephadex column (1.5 × 60 cm).[4] The column is equilibrated and the enzymes eluted in 10 mM potassium phosphate, pH 7.4, at a flow rate of 6 ml/hr.

The enzyme activities obtained by procedure II (Table III) are in the monomer form. However, the different molecular weight forms may be prepared after step 2 of this procedure by separation on a Sephadex G-150 column (1.5 × 80 cm).[5] The column is equilibrated and the enzymes eluted in 50 mM potassium phosphate, pH 7.4.

The monomer may be separated from the dimer and tetramer forms by its selective elution off DEAE-cellulose following the adsorption of enzyme activities from the hemolysate.[4] For example, 1 liter of hemolysate is stirred in a DEAE-cellulose suspension containing 4 g (dry weight) per 100 ml of 10 mM potassium phosphate, pH 7.4, in 200-ml aliquots. The cellulose is washed free of hemoglobin with 10 mM potassium phosphate, pH 7.4, between each application of hemolysate. This procedure is repeated until no further enzyme binds to the cellulose. The cellulose is poured onto a column (1.5 × 40 cm) and further washed with 10 mM potassium phosphate, pH 7.4, until the eluate is free of hemoglobin. The monomer form is eluted with a solution of 35 mM potassium phosphate, pH 7.4, containing 2 mM orotic acid.

The dimer and tetramer forms are eluted in 500 mM potassium phosphate, pH 7.4. Pooled fractions of the monomer and of the dimer and tetramer are brought to 75% saturation with $(NH_4)_2SO_4$. The resulting precipitates are collected and dissolved in 40 mM potassium phosphate, pH 7.4; these solutions are then dialyzed against the same buffer for 12 hr.

Properties

Substrate Specificity and K_m Values. Compounds which are substrates for the transferase, in addition to those reported above, are: thymine, 6-azathymine, 6-azauracil, 5-fluorouracil, 5-iodouracil, and 2,4-dihydroxypteridine.[1,6] The common feature of the substrates is a 2,4-diketo substitution on the pyrimidine portion of the ring structure. The following are average K_m values for several substrates of the transferase: orotic acid, 2.2×10^{-5} M; uracil, 1.4×10^{-4} M; xanthine, 2.1×10^{-4} M; uric acid, 1.8×10^{-3} M; PP-ribose-P with orotic acid as the base substrate (orotic acid = 0.15 mM), 1.5×10^{-5} M; and PP-ribose-P with xanthine as the base substrate (xanthine = 0.2 mM), 3.7×10^{-5} M. The K_m for orotidylic acid with the decarboxylase from beef erythrocytes was estimated to be lower than 10^{-5} M.[1] The K_m values for orotidylic acid with the monomer, dimer, and tetramer forms of the decarboxylase from human erythrocytes are $2.5(\pm0.5) \times 10^{-5}$ M, $3.3(\pm0.6) \times 10^{-6}$ M, and $6(\pm2.5) \times 10^{-7}$ M, respectively, and the relative V_{max} values are in a ratio of 50:11:2, respectively; i.e., the higher K_m results in weaker substrate binding and a higher V_{max}.[4] The K_m and V_{max} values for orotic acid with human phosphoribosyltransferase are approximately 4 times greater in the presence than in the absence of dithiothreitol; i.e., K_m = 36 and 10 μM and V_{max} = 144 and 32 nmole mg^{-1} hr^{-1}, respectively (range of orotic acid concentration = 5–200 μM; PP-ribose-P concentration = 250 μM; dithiothreitol concentration where used = 1 mM; and enzyme activity = 30% in the absence as compared to that observed in the presence of dithiothreitol).[5]

Mg^{2+} Requirement and pH Optima. Orotate phosphoribosyltransferase requires Mg^{2+} ions at a concentration of 2 mM or higher. The orotidylate decarboxylase does not require Mg^{2+} ions. The transferase exhibited a broad pH optimum with both orotic acid (pH 8.0–9.2) and xanthine (pH 7.6–8.6). However, the pH optimum is 9.5 with uracil and 6.0 with uric acid. The pH optimum for the decarboxylase is 7.2–7.8. In each case the pH optimum is near the pK value of the hydroxyl group. Therefore, the substrate is probably the monovalent anion in each instance.[1]

Inhibitors. PP_i at 10^{-3} M inhibits the transferase 50% with xanthine or uracil as the base substrate and 65% with orotic acid as the base substrate. Uridylic acid and 3-N-ribosylxanthine-5'-phosphate inhibit the transferase with xanthine as the substrate (K_i = 1.5 × 10^{-5} M for uridylic acid inhibition and 1.3 × 10^{-4} M for 3-N-ribosylxanthine-5'-phosphate inhibition). The decarboxylase from beef erythrocytes is inhibited 37% by uridylic acid at 10^{-3} M.[1] Phosphate is a competitive inhibitor of orotidylic acid, and the K_i is approximately 9 × 10^{-1} M for the monomer and dimer forms, and 2.2 × 10^{-4} M for the tetramer form of the human decarboxylase.[4] Other inhibitors of the monomer, dimer, and tetramer forms of the human decarboxylase and the corresponding K_i values are: 1-allopurinol-5'-phosphate, 70, 8.0, and 2.6; 1-oxipurinol-5'-phosphate, not determined for monomer, 0.004, and 0.0003; 7-oxipuri-nol-5'-phosphate, not determined for monomer, 0.8, and 0.06; 3-N-ribosylxanthine-5'-phosphate, not determined for monomer, 0.45, and 0.15; 9-N-ribosylxanthine-5'-phosphate, 90, 6.3, and 1.9; 6-azauridine-5'-phosphate, 0.6, 0.12, and 0.06; and uridylic acid, 2600, 88, and 11, respectively.[8] These data show that the inhibition constants are dependent upon the aggregation state of the human enzyme and that much tighter binding is obtained with the higher molecular weight forms. This trend is consistent with that observed for the K_m values of the substrate, orotidylic acid.

Stability, Molecular Weight, and Subunit Structure. The enzymes from beef erythrocytes were stable to heat under slightly acidic or basic conditions. For example, at pH 6.0 for 40 min at 50° the amount of activity lost with each substrate was: orotic acid, 22%; uracil, 15%; xanthine, 19%; uric acid, 19%; and orotidylic acid, 12%.[1] The tetramer form of the decarboxylase from human erythrocytes lost 10% activity at pH 7.4 after 10 min at 55°, while the monomer form lost about 80% activity under these conditions.[4]

The tetramer form of the human decarboxylase spontaneously dissociated into the monomer.[4] For example, after storage for 1 month at −15° in 50 mM potassium phosphate, approximately 60% of a sample of the tetramer was found to be in the monomer form. The tetramer may be stabilized under these conditions by addition of 0.15 mM oxypurinol. Additionally, oxypurinol induced formation of dimer and tetramer forms in a sample of the monomer.

Storage of the human enzyme complex for periods of greater than 1 month at −20° in 50 mM potassium phosphate, pH 7.4, had little effect on decarboxylase but transferase activity decreased as follows: (1)

[8] G. K. Brown and W. J. O'Sullivan, *Biochem. Pharmacol.* **26,** 1947 (1977).

Transferase activity measured with orotic acid and Mg-PP-ribose-P alone decreased with a half-life of approximately 6 weeks in which time almost all the activity was recoverable if dithiothreitol (1 mM) and EDTA (10 μM) were included in the assay; and (2) when the transferase activity measured with substrates alone fell to approximately 10% of the initial activity, the activity which was recoverable with dithiothreitol began to decline with a half-life of about 1 week.[5] The molecular weights, estimated by gel filtration on Sephadex G-200, of the monomer, dimer, and tetramer forms of the decarboxylase are 62,000, 115,000, and 250,000, respectively.[4,5]

The monomer may be further dissociated by prolonged storage at −20° or by treatment with guanidine hydrochloride.[5] These studies suggest that (1) decarboxylase activity is associated with two subunits of approximately 20,000 molecular weight each; (2) transferase activity is associated with two subunits of approximately 13,000 molecular weight each; and (3) the monomer is constituted of two decarboxylase subunits and attached to the two transferase subunits.

Stoichiometry, Equilibrium, and Pyrophosphorolysis. Equivalent amounts of 3-N-ribosylxanthine-5'-phosphate and PP$_i$ are formed when xanthine and PP-ribose-P are incubated with the transferase. An equilibrium constant for the uracil and xanthine reactions could not be determined since the corresponding products act as inhibitors of the reaction. A small degree of pyrophosphorolysis of 3-N-ribosylxanthine-5'-phosphate (approximately 4%) and of uridylic acid (less than 1%) is observed in the presence of transferase and PP$_i$. Pyrophosphorolysis of orotidylic acid could be readily demonstrated in transferase preparations which contained low decarboxylase activity.

Acknowledgments

The authors express their sincere appreciation to Dr. William O'Sullivan for making preprints of his work available to us and for reviewing the present manuscript.

[21] Orotate Phosphoribosyltransferase : Orotidylate Decarboxylase (Ehrlich Ascites Cell)

By MARY ELLEN JONES, PRABHAKAR RAO KAVIPURAPU, and THOMAS W. TRAUT

Complex U,[1] which catalyzes reaction 1, occurs in the mammalian organisms and has been purified from erythrocytes,[2] thymus,[3] and two cancer cells.[4,5]

$$\text{Orotate + phosphoribosyl pyrophosphate} \xrightarrow{Mg^{2+}} \text{UMP + CO}_2 \qquad (1)$$

This reaction occurs in intact cells[6] or in solution[7,8] with an insignificant accumulation of OMP, even though the complex contains two catalytic centers, an orotate (or pyrimidine) phosphoribosyltransferase, which catalyzes reaction 2, and an orotidylate decarboxylase, which catalyzes reaction 3. Reactions 2 and 3 sum to yield reaction 1.

$$\text{Orotate + P-rib-PP}^{10} \underset{}{\overset{Mg^{2+}}{\rightleftharpoons}} \text{orotidylate (OMP) + PP}_i \qquad (2)$$
$$\text{OMP} \longrightarrow \text{UMP + CO}_2 \qquad (3)$$

The decarboxylase (reaction 3) of the complex is sufficiently active so that the stoichiometric production of OMP and PP$_i$ by the phosphoribosyltransferase (reaction 2) can only be observed if reaction 3 is completely inhibited.[7] Reaction 3, however, can be followed when OMP is the sole substrate.

Assay Methods

Principles. Assay procedures reported include: (1) methods measuring the spectrophotometric changes[2] that occur when a pyrimidine base (reaction 1 or 2) is converted to a nucleotide or when OMP is converted to UMP[2] (reaction 3); (2) methods which separate the free base from the

---

[1] W. T. Shoaf and M. E. Jones, *Biochemistry* **12**, 4039 (1973).

[2] R. F. Silva and D. Hatfield, this volume.

[3] D. K. Kasebekar, A. Nagabhushanam, and D. M. Greenberg, *J. Biol. Chem.* **239**, 4245 (1964).

[4] P. R. Kavipurapu and M. E. Jones, *J. Biol. Chem.* **251**, 5589 (1976).

[5] P. Reyes and M. E. Guganig, *J. Biol. Chem.* **250**, 5097 (1975).

[6] S. E. Hager and M. E. Jones, *J. Biol. Chem.* **240**, 4556 (1965).

[7] T. W. Traut and M. E. Jones, *Biochem. Pharmacol.* **26**, 2291 (1977).

[8] K. Prabhakararao and M. E. Jones, *Anal. Biochem.* **69**, 451 (1975).

METHODS IN ENZYMOLOGY, VOL. LI

nucleotides[2] (reaction 1 or 2); and (3) methods measuring the release of $[^{14}C]O_2$ from $[^{14}C]$carboxyl-labeled orotate (reaction 1) or orotidylate (reaction 3). The release of $[^{14}C]O_2$ is useful for purification of the complex since orotate and OMP are not normal cellular constituents and do not dilute the added substrate. The separation of orotate, OMP, and UMP, or of the two nucleotides, OMP with UMP, from OA,[5] is most useful for the study of kinetic or equilibrium parameters of reaction 2 and for the simultaneous assignment of inhibitors to one or both of the two active centers of complex U.

A. Assay for Reaction 1 and/or Reaction 2: Release of $[^{14}C]O_2$ from $[7-^{14}C]$Orotic Acid

If no inhibitor is present for the decarboxylase (reaction 3), reaction 1 is limited in rate by reaction 2, and therefore the formation of $[^{14}C]O_2$ from $[7-^{14}C]$orotate is a measure of both reaction 1 and 2. If, however, an inhibitor is present for the decarboxylase, $[^{14}C]O_2$-release from $[7-^{14}C]$orotate will underestimate both reaction 1 and 2. The $[^{14}C]O_2$ produced enzymically is released from the reaction mixture by acidification of the sample (to denature complex U and to convert bicarbonate to CO_2), and is collected as carbonate with base placed in a center well. The base can be either ethanolamine in dioxane[1] or a NaOH-soaked paper strip.[8] The paper strip is a simpler and more economical absorbant. This assay is good for crude homogenates as well as pure enzyme provided an inhibitor of the decarboxylase is not present.

Reagents

    Tris·Cl buffer, pH 7.5, 0.2 M
    $MgCl_2$, 0.1 M
    Na P-rib-PP,[9] 3 mM
    $[7-^{14}C]$Orotate, 0.5 mM; 0.1 Ci/mole
    Dithiothreitol, 10 mM
    NaOH, 2 M
    Perchloric acid, 4 M

Procedure. Each assay vessel, a standard glass scintillation vial, contained 50 μl each of Tris, $MgCl_2$, $[^{14}C]$orotate, dithiothreitol, P-rib-PP, and sufficient water to give a volume of 0.5 ml after the enzyme is added. The enzyme should not produce more than 10 nmoles of product in the incubation interval which has been 10 or 20 min. The P-rib-PP is added to the other substrates and the water 1 min before the enzyme, at which time the vessel is placed at 37°. The enzyme is added at zero time,

---

[9] Some samples of P-rib-PP absorb ultraviolet light. These products contain an inhibitory nucleotide or base which can be removed by charcoal absorption (4).

and the vial is capped with a rubber stopper (serum type) containing a plastic center well (Kontes Glass Co., Vineland, New Jersey). The vial is incubated for 10 (or 20) min at 37°. The center well was filled beforehand with a 0.5 × 3 cm piece of Whatman 3 MM filter paper that had been dipped into the 2 M NaOH.[8] The excess NaOH was removed by blotting on paper or by pressing the strip against a solid support. After the strip was folded it was inserted into the center well. The reaction was terminated by injecting 0.2 ml of 4 M HClO$_4$ through the rubber cap with a syringe having a 22-gauge needle. The vials were then shaken for 90 min at 37°. At this time, the filter paper was removed from the center well and dried in a depression of a porcelain spot-plate using heat from 250 W incandescent bulbs. The paper was dry when it attained a pale yellow color. It was immediately transferred to 10 ml of toluene scintillation fluid containing 0.2 g of 1,4-bis-2-(5-phenyloxazolyl)benzene and 3 g of 2,5-diphenyloxazole per liter of toluene. The samples were counted in a scintillation counter along with suitable aliquots (ca. 5 μl) of the [^{14}C]orotate which was applied to the partially dried piece of the NaOH-soaked filter paper; these strips were counted in the same manner as the assay samples after they had been dried completely. The optical density of the [^{14}C]orotate solution (pH 7) used as substrate was measured at 279 nm, and this value was used to calculate the molarity of the [^{14}C]orotate, using a molar extinction coefficient of 7700.

After the samples have been counted the paper strips are removed with forceps. The vials are reused (sometimes for several years) until they have acquired 10 cpm above the counts obtained from a vial containing fresh scintillation fluid.

Determination of Units. The cpm observed for the experimental samples was divided by the cpm/nmole of the [^{14}C]orotate to determine the nmoles of [^{14}C]O$_2$ formed in the assay samples during the incubation period. This value was divided by the number of minutes the incubation continued at 37° to give the enzyme units which were nmoles/min.

B. Assay for Reaction 3: OMP Decarboxylase by [^{14}C]O$_2$ Release

The assay is similar to assay A above, except that [7-^{14}C]OMP is the sole substrate. This assay is appropriate for both crude or pure enzyme preparations.

Reagents

Tris·Cl buffer, pH 7.5, 0.2 M
[7-^{14}C]OMP,[10] 0.5 mM, 0.1 Ci/mole

[10] Abbreviations used are: P-rib-PP for phosphoribosyl pyrophosphate; OMP for orotidylic acid; Allo-PuMP for allopurinol ribonucleotide.

Dithiothreitol, 10 mM
NaOH, 2 M
Perchloric acid, 4 M

Procedure. The vessels used for the assay are the same as those used for procedure A above. Fifty microliters each of Tris buffer, OMP, and dithiothreitol and sufficient water to reach a final volume of 0.5 ml after the enzyme is added are preincubated 1 min at 37°. Enzyme (diluted so that no more than 20 nmoles of [^{14}C]O$_2$ are produced) is added at zero time, and the vessel is capped with the serum-type rubber stopper containing a plastic center well with a 0.5 × 3 cm strip soaked in the 2 M NaOH as described above. The vessel is incubated at 37° for 10 (or 20) min. The reaction was stopped, and the [^{14}C]O$_2$ was trapped and processed for scintillation counting as described in procedure A. The calculation of the units was based on standardization of the [^{14}C]OMP by measuring the absorbance at 267 nm (assuming a molar extinction coefficient of 9420) and by counting aliquots of the standard placed on semidry NaOH strips which were then dried and counted as the CO$_2$ strips.

Determination of Units. The cpm observed for the [^{14}C]O$_2$ released from [7-^{14}C]OMP was divided by the cpm/nmole of the [7-^{14}C]OMP to give nmoles of [^{14}C]O$_2$ formed during the incubation period, and this value was then divided by the time of incubation at 37° to yield the units of decarboxylase activity (nmoles/min).

C. Assay for Reaction 1, 2 (or 3)

In this procedure orotate, OMP, and UMP are separated from one another by thin-layer chromatography.[5,7] This system is advantageous whenever an inhibitor of reaction 3, the decarboxylase, is present, for under these conditions assay A above underestimates the transferase reaction (2). It is also useful when one wishes to estimate how effective inhibitors of the decarboxylase are when the OMP level is that maintained at a steady-state level by the complex itself, i.e., when orotate and P-Rib-PP are the substrates and only the OMP formed by the transferase is present, this assay is ideal.[7]

Reagents. The reagents for the incubations are the same as for assay A or B, except that [6-^{14}C]orotate or [6-^{14}C]OMP is used, and the 2 M NaOH and 4 M perchloric acid are unnecessary. Polyethyleneimine (PEI) cellulose thin-layer plates (Brinkmann Instruments, Los Angeles, CA) are necessary, along with jars for ascending chromatography and the solvent for the chromatography, 0.2 M LiCl, pH 5.5.[5]

Procedure. The standard reaction mixture had a total volume of 50 μl. For reactions 1 or 2, 5 μl each of [6-^{14}C]orotate, MgCl, Tris·Cl, dithiothreitol, and P-rib-PP were used with enzyme at a concentration such that no more than 20% of the orotate was used. After the 10 (or 20) min incubation at 37°, 5 μl of each reaction mixture were spotted directly onto a 20 × 20 cm PEI cellulose plate, and the spot was dried immediately with hot air. The time required for spotting and drying was 15–20 sec. Up to 15 samples can be spotted at 1.3-cm intervals along a line near one end of the plate. Before use, the thin layer is scored vertically with the end of a paper clip at 1.3-cm intervals so that each sample is discreet from the others even if the plate is not vertical in the solvent. To facilitate detection of the spots, every sample spot received 5 μl of a carrier solution containing orotate, OMP, and UMP, each at 5 mM. The plates are then placed in a 5 × 21 × 21 cm developing tank which has the LiCl solvent (250 ml) in the bottom. A metal carrier can be used so 5 plates can be developed simultaneously. The glass lid (preferably ground glass) is placed on the tank, and the solvent ascends the thin layer. After 2 to 3 hr, when the solvent is about 2 cm from the top of the plate, the thin-layer plate is removed and, after drying under a heat lamp for about 15 min, the spots for orotate ($R_f = 0.48$), UMP ($R_f = 0.21$), and OMP ($R_f = 0.03$) are detected under a UV lamp and cut out of the plastic-backed thin-layer plate. Such pieces (no larger than 2.2 × 1.3 cm) are placed in the toluene scintillation fluid described in procedure A. Only one piece of plastic sheet can be placed in each vial. These vials can be reused as long as the counts of the vial plus scintillation fluid are low. Normally, 90–95% of the sample cpm are recovered in the OMP, UMP, and orotate spots. Where total recovery of the sample cpm is desirable (e.g., due to the fact that the sample spot may not be exactly 5 μl but can vary somewhat), we have found it best to count most of the entire strip (from 0.5 cm before the origin to 1 or 2 cm beyond the leading edge of the orotate spot) so that the total cpm applied at the origin can be determined from the summation of the counts for the several pieces of plastic for a single sample. The orotate, UMP, and OMP areas are then expressed as percentages of this sum, and the average of the total cpm of all samples run in a given assay is taken as the "true" total cpm for a 5-μl sample. In those strips from samples with more or fewer total cpm/strip than the average, the cpm in the orotate, OMP, or UMP areas is normalized to the value expected if the strip had possessed the average cpm for a 5-μl sample. The nmoles of each product are determined as in methods A and B above, except that the [6-^{14}C]orotate or [6-^{14}C]OMP standard is spotted on a piece of the thin-layer plate for counting.[7]

Determination of Units. The units are nmoles/min at 37° and are determined as in methods A and B.

Dilution of Protein for Assay. Complex U is not stable in dilute solution, and therefore the assay is not linear with protein at low protein concentrations.[1,4] If the protein must be diluted for assay, it should be done immediately beforehand and in the presence of 1 mM dithiothreitol; Mg^{2+} and P-rib-PP also aid stabilization in dilute solutions.[4]

Protein Assays. If the enzyme does not contain dithiothreitol the protein can be assayed by the Lowry procedure.[11] Since the enzyme usually contains dithiothreitol (except in the initial purification steps) as a stabilizing agent, it is necessary to use a protein method that is not influenced by this reagent. The method of Ross and Schatz[12] which uses iodoacetic acid to react with the dithiothreitol before addition of the Lowry reagents is suitable.

Purification of Complex U.[4] All operations were carried out either in an ice bath or in a space at 4°, and all the centrifugations were at 30,000 g for 20 min. Enzyme solutions were stored at −20°. The exact quantities and volumes given below are for the purification reported in Table I; other fractionations were similar, but the detailed amounts or volumes varied somewhat.

Fraction 1: Whole Homogenate. Saline-washed Ehrlich ascites cells (136 g wet weight) which had been stored at −20° were thawed, and 4 volumes of deionized distilled water were added to allow the cells to swell at room temperature (21°). The swollen cells (30- to 40-ml portions) were homogenized (25 strokes) with the Potter–Elvehjem homogenizer. The total volume of the homogenate was 790 ml, and the pH was 7.2.

Fraction 2: 30,000 g Supernatant. The homogenate (790 ml) was centrifuged in a Sorvall refrigerated centrifuge at 30,000 g for 20 min. To the 662 ml of clear supernatant, 6.68 ml of 1 M potassium P_i buffer, pH 7.0, was added; the final P_i concentration was 10 mM, and the final pH was 6.7. The pH was readjusted to 7.0 with 0.5 N KOH if the enzyme was to be stored at this point.

Fraction 3: pH 5.7 Supernatant. The pH of fraction 2 (670 ml) was adjusted to 5.7 with 0.5 N acetic acid (12.5 ml). The enzyme solution was stirred for 15 min at this pH in ice, and was then centrifuged to remove the precipitated protein. The pH of the supernatant (664 ml) was readjusted to pH 7.0 with 8 ml of 0.5 N KOH.

Fraction 4: 40–65% Ammonium Sulfate Precipitate. Inactive protein of fraction 3 (670 ml) was precipitated at 40% ammonium sulfate

---

[11] E. Layne, *in* "Methods in Enzymology" (S. P. Colowick and N. O. Kaplan, eds.), Vol. 3, p. 447. Academic Press, New York, 1957.
[12] E. Ross and G. Schatz, *Anal. Biochem.* **54,** 304 (1973).

TABLE I

PURIFICATION OF COMPLEX U FROM EHRLICH ASCITES CARCINOMA[a]

Fraction	Volume (ml)	Protein (mg)	Orotic acid phosphoribosyltransferase			OMP decarboxylase			
			Specific activity[b]	Fold	Recovery (%)	Specific activity[b]	Fold	Recovery (%)	Ratio of specific activities
1. Whole homogenate	790	9350	0.77	1.0	100	1.4	1.0	100	1.8
2. 30,000 g supernatant	669	5490	1.3	1.7	99	2.4	1.7	99	1.8
3. pH 5.7 supernatant	670	2880	2.5	3.1	98	4.2	3.0	93	1.7
4. 40–65% (NH$_4$)$_2$SO$_4$ precipitate (dialyzed)	43.8	1205	5.8	7.5	97	9.9	7.1	92	1.7
5. A-50 batch eluate (dialyzed)	13.7	341	16	21	76	30	21	79	1.9
6. A-50 gradient eluate	1.6	75.3	59	77	62	87	62	50[c]	1.5
7. Calcium phosphate gel eluate	3.0	12.5	115	149	20	295	211	28	2.5

[a] The purification details and assay methods are described in the text. From fraction 4 onwards, 2 mM dithiothreitol was included in the buffer solutions, while in the final stages of fraction 6 and throughout step 7, 1 mM PP-Rib-P and dithiothreitol were present.

[b] Units (nmoles of [^{14}C]O$_2$ released from orotic acid or OMP/min at 37°) per milligram of protein.

[c] The lower recovery was due to the fact that part of the OMP decarboxylase had very little orotic acid phosphoribosyltransferase activity and therefore was not pooled with the fractions having both activities.

saturation by adding 162.8 g of special enzyme-grade $(NH_4)_2SO_4$ crystals (supplied by Schwartz/Mann, New York) and 0.5 N KOH as required to maintain a pH of 7.0. The solution was stirred constantly during, and for 15 min after, these additions. The precipitated protein was removed by centrifugation. To the supernatant, $(NH_4)_2SO_4$ (112.6 g) was slowly added, as described above, to raise the ammonium sulfate saturation to 65% and pH 7.0. The precipitated protein was collected by centrifugation. The 40–65% ammonium sulfate precipitate was dissolved in a minimum volume (e.g., 1/20 to 1/25 volume of fraction 3) of 10 mM potassium P_i buffer, pH 7.5, and the pH was adjusted to 7.5 with 0.5 N KOH.

The purification was carried through this step in 1 day to minimize loss of the two enzyme activities. The complex can be stored frozen at pH 7.0 or 7.5 at this stage for 2 months without significant loss of either enzyme activity; therefore, the complex is accumulated at this stage until a sufficient amount is available to maintain high protein concentrations throughout the subsequent steps.

Undialyzed fraction 4 is stable, but it becomes unstable after dialysis against phosphate buffer (pH 7.0 or 7.5). Since 2 mM dithiothreitol protects the enzyme activities, all solutions used for purification from fraction 4 onwards had dithiothreitol (2 mM). Fraction 4 (40 ml) was therefore dialyzed for 18 hr against 10 mM potassium P_i buffer, pH 7.5, containing 2 mM dithiothreitol. Precipitated protein was removed by centrifugation; the final volume was 43.8 ml.

Fraction 5: DEAE-Sephadex (A-50) Batch Eluate. Swollen DEAE-Sephadex (A-50; 52 ml) in 10 mM potassium P_i buffer with dithiothreitol, pH 7.5, was formed into a cake on a Whatman 50 filter-paper circle (11 cm) in a Büchner funnel. This cake was mixed in a beaker with dialyzed fraction 4 for about 30 min. The slurry was poured back onto a Büchner funnel, and the filtrate was removed under mild vacuum. After the Sephadex had been washed twice with 52 ml of 10 mM P_i buffer, 20 mM P_i buffer (60 ml) containing 75 mM KCl and dithiothreitol, pH 6.8, was mixed with the cake for 20 min, when the filtrate was removed as before. The process was repeated two times utilizing a total of 200 ml of buffer. Finally the complex was eluted by three washings of the cake, as above, with 200 ml of 20 mM P_i containing 400 mM KCl and dithiothreitol, pH 6.8; the protein released was collected in a clean suction flask. Complex U was immediately concentrated by precipitating it with ammonium sulfate (0–65% saturation). The precipitate was collected by centrifugation and was dissolved in a minimum volume of 20 mM P_i/dithiothreitol buffer, pH 7.5. The solution was dialyzed for 18 hr against two 500-ml portions of the same buffer. The final volume was 13.7 ml.

Fraction 6: DEAE-Sephadex (A-50) Gradient Eluate. A DEAE-Sephadex column (2.5 × 16.5 cm) was equilibrated with 20 mM potassium P_i buffer containing 75 mM KCl and dithiothreitol, pH 6.8. Solid KCl was added to fraction 5 to a concentration of 75 mM. For the fractionation shown in Table I, the complex was loaded at the bottom

TABLE II

Inhibitor	M	% Inhibition	
		Transferase[a]	Decarboxylase[a]
PP_i	10^{-4}	50	70
P_i	10^{-2}	75	50
Ribose-P	10^{-3}	0	50
6-Aza-UMP	10^{-7}	0	90
	10^{-3}	15	100
	10^{-2}	95	100
Allo-PuMP[10]	10^{-6}	0	40
UMP	10^{-4}	0	70
	10^{-3}	10	90
	10^{-2}	70	100
dUMP	10^{-3}	15	55
	5×10^{-3}	45	70
CMP	10^{-4}	0	50
TMP	10^{-2}	0	75
5-Br-UMP	10^{-4}	0	60
IMP	10^{-4}	15	65
	10^{-2}	65	80
AMP	5×10^{-5}	10	55
	10^{-2}	25	70
GMP	10^{-3}	35	35
UDP	10^{-3}	5	60
	5×10^{-3}	60	80
UTP	10^{-3}	10	30
	10^{-2}	100	100
ADP	10^{-3}	5	50
	10^{-2}	100	100
ATP	10^{-3}	0	10
	10^{-2}	100	100

[a] Both enzyme activities were assayed simultaneously by separating and measuring orotate, OMP, and UMP (assay C).[7] The reaction was started with 5 μM orotate, and the normal OMP concentration was about 50 nM in the uninhibited control.

of the column, and the upper plunger of the Pharmacia column was connected to an LKB fraction collector. Ascending development was carried out with a linear gradient formed with 100 ml of 75 mM KCl in 20 mM potassium P$_i$/dithiothreitol, pH 6.8, in the mixing chamber, and 100 ml of the same buffer, but 400 mM in KCl, in the reservoir; the flow rate was 20 ml/hr. The active fractions (eluted at KCl concentrations starting between 0.26 or 0.3 and ending at 0.35 M) were combined, concentrated by Amicon ultrafiltration cell (models 52 and 12) fitted with PM 10 membrane filter (Amicon Corporation, MA), and then dialyzed for 18 hr against two changes of 10 mM P$_i$/dithiothreitol buffer, pH 7.5, containing 1 mM PP-Rib-P.

Fraction 7: Calcium Phosphate Gel Column Fraction. Washed calcium phosphate gel, equilibrated with 10 mM potassium P$_i$/dithiothreitol buffer containing 1 mM PP-rib-P, pH 7.5, was placed in a column that had a sintered glass bottom. The outlet of the column was fitted to a vacuum flask connected to a water pump. The column (2.2 × 3 cm) was washed with 44 ml of the above buffer. Fraction 6 was adsorbed onto the gel and was washed with 7 ml of the same buffer. To elute complex U, 34 ml of 40 mM P$_i$/dithiothreitol buffer, pH 6.8, containing 1 mM PP-rib-P, was used. The eluate was adjusted to pH 7.5 and concentrated by Amicon ultrafiltration cell using a PM 10 membrane.

Properties

Equilibrium Constants for the Transferase[13] *(Reaction 2).* As illustrated above, the overall reaction (1) is not in equilibrium since the decarboxylase reaction is not reversible. Therefore, to measure the equilibrium constant for the transferase, the decarboxylase must be inhibited completely. This can be achieved by adding 0.1 mM azauridylic acid. The K_{eq} for the orotate phosphoribosyltransferase, measured in either the forward or reverse direction, is 0.07 {= (OMP)(PP$_i$)/[(P-rib-PP)(orotate)]}.

Molecular Weight. The molecular size of complex U depends on the solvent present.[1,4,13] Using sucrose[1,13] or Ficoll[4] gradients the apparent molecular weights[14] (±4000) are: 110,000 in 30% dimethylsulfoxide–5% glycerol;[1] 55,000 in 10 mM phosphate,[4] or in 50 mM Tris buffer,[1,13] or in 10 mM phosphate buffer with or without 0.2 mM P-rib-PP.[4] However, when the enzyme is in the assay mix for reaction 1 the apparent size is

---

[13] T. W. Traut and M. E. Jones, *J. Biol. Chem.* **252**, 8374 (1977).

[14] The standards used were bovine Hb (64,500); horse liver alcohol dehydrogenase (80,000); beef muscle lactic dehydrogenase (140,000); catalase (232,000).

68,000, and when 10 μM OMP is also present the apparent size is 79,000.[13] Regardless of what solvent was used,[1,4,13] the two active centers, transferase and decarboxylase, always sedimented together. One highly purified preparation which had lost the transferase activity[4] and retained only the decarboxylase activity sedimented at 55,000 in 10 mM phosphate buffer. Therefore, the denaturation of the transferase activity did not reduce the minimum molecular size of the complex. This result indicates that if the complex is composed of two distinct polypeptides (i.e., two subunits or multiples thereof), they bind very tightly to one another; a second alternative is that the two active centers are on a single polypeptide chain, i.e., they are part of a multifunctional protein.[15] The fact that the complex, in the presence of its substrates, has apparent weights intermediate between what seems to be a monomer (55,000) and what appears to be a dimer size (110,000) suggests that these might be conformer sizes, either dense monomers or an expanded dimer.

Reyes and Guganig,[5] studying a highly purified complex U from murine leukemia, also find the weight of the complex dependent on the solvent. They obtained a molecular weight of 69,000 in 50 mM phosphate, 1 mM P-rib-PP, and 1 mM dithiothreitol buffer. When 5 mM UMP is added the apparent weight became 98,000. Under both conditions the two activities of complex U sedimented together. If 100 μM OMP was added to the standard buffer, the decarboxylase activity sedimented with a major peak at 99,000 that had a shoulder at 65,000. The transferase activity did not follow the decarboxylase, for its major peak was at 65,000 with a shoulder at 99,000.[5] Unpublished experiments from this laboratory, using 50 mM Tris buffer with 40 μM OMP and dithiothreitol, give a major peak of 98,000 molecular weight with a shoulder whose peak is between 51,000–60,000; however, both the transferase and decarboxylase activities migrated together.

Grobner and Kelley[16] have observed that human erythrocyte complex U had three molecular forms in 10 mM P_i buffer which had Stokes radii of 37.4, 47.5, and 54.7 Å, using an agarose column as the molecular sieve. In sucrose gradient, these authors observed two forms that had $s_{20,w}$ values of 3.6 or 5.1. These authors estimate the molecular sizes of these three forms emerging from the agarose columns as 55,000, 80,000, and 113,000, respectively. Brown and O'Sullivan (in Silva and Hatfield[2]) have also observed three forms of human erythrocyte complex U in phosphate buffer by use of Sephadex sieves. However, they assign weights of 62,000, 115,000, and 250,000, respectively, to these species.

[15] K. Kirschner and H. Bisswanger, *Annu. Rev. Biochem.* **47**, 143 (1976).
[16] W. Grobner and W. N. Kelley, *Biochem. Pharmacol.* **24**, 379 (1975).

Both groups working with the erythrocyte complex find that nucleotides stabilize the largest[16] or larger forms.[2] Brown and O'Sullivan find that the monomeric, dimeric, and tetrameric erythrocyte forms have different K_m values (in Silva and Hatfield[2]) for OMP. The Ehrlich ascites cell complex does not have the larger forms when the buffer is 10 mM phosphate. Unpublished data from this laboratory with the Ehrlich ascites complex show that azauridylic acid (10^{-7} to 10^{-4} M in Tris buffer) stabilizes a conformer with an apparent size of 85,000 rather than the 113,000 size observed for the erythrocyte complex.[16] It is possible that the mouse ascites cell and human erythrocyte complexes differ in protein structure.

Kinetic Constants. The apparent K_m values[4] for the transferase and reaction 1 determined in 10 mM P_i buffer are: orotate, 2 μM; P-rib-PP, 16 μM; Mg^{2+}, 3 mM, when the nonvaried substrates were orotate, 40 μM; P-rib-PP, 1 mM; and Mg^{2+}, 5 mM. Under these conditions, OMP does not accumulate[8] but is maintained at a steady-state value between 0.05 to 0.1 μM.[7]

Inhibition of the transferase activity cannot be easily divorced from effects on the decarboxylase if CO_2 production is measured. If assay C, above, is used, or another assay which measures both OMP as well as UMP is used, proper values can be obtained.[7] Using method C,[7] it was found that pyrimidine and purine bases affected only reaction 1. Bases that produce 30–70% inhibition are: 5-F-orotate, at 0.05 mM; barbituric acid, at 0.5 mM; allopurinol or oxipurinol $\geq$ adenine $>$ 5-Cl-uracil $>$ uracil, at 1 mM; and azauracil, 5-bromouracil, or 5-fluorouracil, at 5 mM. Dihydroorotate and cytosine had to be 10 mM to produce this degree of inhibition. Nucleosides that produced between 30–70% inhibition had to be 10 mM, and such inhibition was observed with adenosine, orotidine, uridine, and azauridine. Nucleotide and phosphate compounds that inhibit are shown in Table II.

The decarboxylase activity of complex U has a K_m for OMP of 0.3 μM. This enzyme is not inhibited by free purine or pyrimidine bases but is inhibited by nucleotide and phosphate compounds, as detailed in Table II. The most effective inhibitor of the decarboxylase activity that has been reported is pyrazofurin-5'-phosphate, a 5-member ring analog of OMP that is 20 times as potent as 6-aza UMP.[17]

Nucleotides might be able to regulate this complex *in vivo* at the active center of the decarboxylase, and AMP, UMP, and perhaps IMP would seem to be the principal nucleotides to exert such control. The

[17] G. E. Gutowski, M. J. Sweeney, D. C. DeLong, R. L. Hamill, K. Gerzon, and R. W. Dyke, *Ann. N. Y. Acad. Sci.* **255**, 544 (1975).

complex activities are sensitive to mercurials and other sulfhydryl inhibitors. The pH optima for both activities is between 7.0 to 7.5. Tris is the best buffer when the substrates are not saturating; dilute orthophosphate buffers are as good when the substrates are saturating. Maleate, imidazole, Tes, and Mops buffers are to be avoided for assay and storage.

Gel Electrophoresis. The purest fraction of the complex from Ehrlich ascites cells is not pure,[4] as demonstrated by gel electrophoresis, even though it has similar, or slightly higher, specific activities than the complex isolated from thymus and murine leukemia cells. Gel electrophoresis must be carried out on gels that have had P-rib-PP and dithiothreitol pulled into them before the protein is subject to electrophoresis. When this is done, both the transferase and decarboxylase activities can be measured in gel slices;[4] if it is not done, one may be able to detect the decarboxylase activity of the complex but not the transferase activity.[3,4] As is true of the molecular weight determinations, the two activities of this complex travel together. The complex migrates in the gels more slowly when P-rib-PP and dithiothreitol are present.

Stability of the Enzymes. The enzymes are unstable to dilution, to pH values below 4 and above 9 for very short intervals, and to a narrower pH range for long intervals. The best stabilizing reagents are a mixture of dithiothreitol (2 mM), P-rib-PP (1 mM), and Mg^{2+} (5 mM), or high orthophosphate (0.3 M or more) or 6-aza-UMP (10^{-7} M). The latter two reagents, of course, inhibit one or both of the enzyme activities. The purified enzyme is more labile than the cruder preparations. The homogenate or the ammonium sulfate fraction has been stored at $-20°$ to $-60°$ for as long as 2 years with little loss of activity. Batches of step 6 enzyme that had 50 mg of protein per milliliter were quite stable, but batches with lower protein concentrations were more labile. Addition of foreign proteins, i.e., serum albumins, does not increase enzyme stability.

Section III
De Novo Purine Biosynthesis

[22] Amidophosphoribosyltransferase (Chicken Liver)

By Jerome M. Lewis and Standish C. Hartman

Amidophosphoribosyltransferase (EC 2.4.2.14) catalyzes the reaction

$$\text{Glutamine} + \text{5-phosphoribosyl pyrophosphate} + H_2O \xrightarrow{\text{Mg}^{2+}}$$
$$\text{5-phosphoribosylamine} + \text{glutamate} + \text{pyrophosphate}$$

This is the first reaction unique to purine nucleotide biosynthesis and a primary site of feedback regulation of the pathway. Various purine nucleotides have been found to act as allosteric inhibitors of amidophosphoribosyltransferase derived from avian,[1,2] mammalian,[3,4] and microbial sources.[5,6] The enzyme has been purified to a limited extent from certain of these sources, including human placenta,[7] but only the avian enzyme (from pigeon and chicken liver) has been obtained in a high degree of purity. We describe here two procedures for purification of the chicken liver enzyme. The first yields enzyme purified 2300-fold which is homogeneous as judged by disc gel electrophoresis and gel isoelectric focusing. The enzyme obtained by this procedure is insensitive to inhibition by purine nucleotides. In the second procedure the enzyme is obtained 55-fold purified in a form inhibitable by nucleotides.

Assay

There is no totally satisfactory assay method for amidophosphoribosyltransferase, for impure enzyme systems pose particular difficulties. The formation of labeled glutamate from glutamine is potentially a very sensitive method,[7] but one must be concerned with degradation products present in the glutamine used as substrate and with nonspecific hydrolysis of this substrate by glutaminases. Coupling glutamate formation to pyridine nucleotide reduction via glutamate dehydrogenase has been employed,[8] but we and other workers have found this method to show

[1] S. C. Hartman, *J. Biol. Chem.* **238**, 3024 (1963).
[2] P. B. Rowe and J. B. Wyngaarden, *J. Biol. Chem.* **243**, 6373 (1968).
[3] D. L. Hill and L. L. Bennett, *Biochemistry* **18**, 122 (1969).
[4] B. S. Tay, R. M. Lilley, A. W. Murray, and M. R. Atkinson, *Biochem. Pharmacol.* **18**, 936 (1969).
[5] D. P. Nierlich and B. J. Magasnik, *J. Biol. Chem.* **240**, 358 (1965).
[6] C. L. Turnbough, Jr. and R. L. Switzer, *J. Bacteriol.* **121**, 108 (1975).
[7] E. W. Holmes, J. A. McDonald, J. M. McCord, J. B. Wyngaarden, and N. Kelley, *J. Biol. Chem.* **248**, 144 (1973).
[8] J. B. Wyngaarden and D. M. Ashton, *J. Biol. Chem.* **234**, 1492 (1959).

an artifactual lag phase owing to the high K_m for glutamate. Glutaminases may also interfere. Formation of the specific product, phosphoribosylamine, has been followed by its enzymic conversion to phosphoribosylglycineamide in the presence of ATP, glycine, and phosphoribosylglycineamide synthetase.[9] The glycineamide derivative is then determined by a colorimetric procedure involving a complex of enzymic reactions. Phosphoribosylamine has also been assayed by conversion, in the presence of labeled cysteine, to an adduct of uncertain structure.[10] While this is a sensitive method, the possibility of interference by unknown compounds is a concern, particularly since the adduct being determined is uncharacterized.

The method we describe here, and which has been used in our work, depends upon measurement of the product pyrophosphate. Its drawbacks include its relative insensitivity, which makes it unreliable in impure preparations of low activity, and the possibility that pyrophosphate will be underestimated as a result of hydrolysis. However, with enzyme samples which are at least partially purified we find it to be reliable, reproducible, and linear both with respect to time and enzyme concentration. In brief, pyrophosphate is isolated as the manganous salt, hydrolyzed, and measured colorimetrically as orthophosphate, essentially according to Kornberg.[11] Fluoride is added to inhibit pyrophosphatases which may be present, and carrier pyrophosphate is added during isolation of the Mn salt to insure quantitative precipitation.

Reagents

18 mM sodium 5-phosphoribosyl pyrophosphate (PP-ribose-P) (Sigma)
72 mM L-glutamine
54 mM MgCl$_2$
96 mM KF
180 mM Tris-chloride, pH 8.0
1.0 M sodium acetate, pH 5.0
3.0 mM Na$_4$P$_2$O$_7$
0.1 M MnCl$_2$
0.01 M MnCl$_2$
10% acetone in water
1 N H$_2$SO$_4$
10 N H$_2$SO$_4$
2.5% ammonium molybdate tetrahydrate
1% Elon (Eastman Kodak) in 3% Na$_2$SO$_3$

[9] T. Nara, T. Komuro, M. Misawa, and S. Kinoshita, *Agric. Biol. Chem.* **33**, 739 (1969).
[10] G. L. King and E. W. Holmes, *Anal. Biochem.* **75**, 30 (1976).
[11] A. Kornberg, *J. Biol. Chem.* **182**, 779 (1950).

Procedure. The reaction mixtures, in 12-ml centrifuge tubes, consist of 0.10 ml each of Tris buffer, glutamine, $MgCl_2$, KF, PP-ribose-P, and enzyme (about 0.01 to 0.1 unit) in a total volume of 0.6 ml. The PP-ribose-P solution is added last to initiate the reaction. A control vessel is run in which glutamine is omitted from the complete system. When nucleotide inhibitors are employed an equimolar amount of $MgCl_2$ is included additionally in the assay system. After incubation at 25° for 10 min, 0.20 ml of 1.0 M sodium acetate, pH 5.0, is added to stop the reaction, followed by 0.10 ml of 3.0 mM $Na_4P_2O_7$ and 0.20 ml of 0.10 M $MnCl_2$. After 10 min the tubes are spun in a table-top centrifuge for 2 min. The supernatant solution is carefully decanted and discarded. The precipitate is then washed by suspending in 1.0 ml of 0.01 M $MnCl_2$ and 0.20 ml of 10% acetone. The suspension is centrifuged for 2 min and the supernatant liquid discarded.

To the precipitate is added 2.0 ml of 1.0 N H_2SO_4. The tubes are heated in a boiling water bath for 15 min, cooled, and centrifuged for 2 min. The following solutions are added to 20-ml test tubes with shaking after each addition: 2.8 ml of water, 0.20 ml of 10 N H_2SO_4, 0.80 ml of the hydrolysate, 0.40 ml of 2.5% ammonium molybdate, and 0.80 ml of the Elon solution. The absorbance of the solution at 700 nm is determined after 15 min using the control sample as blank. A standard curve is prepared by adding pyrophosphate in varying amounts to the assay mixture in the absence of enzyme. An absorbance of 1.0 corresponds to 1.62 μmoles of pyrophosphate.

A unit of enzyme is defined as that amount which gives a rate of pyrophosphate formation of 1.0 μmole/min at 25° under the conditions described.

Purification Procedure I

Extraction. Fresh chicken livers (4.54 kg) are homogenized in portions for 2 min in a blender with a total of 8 liters of 0.1 M ammonium citrate buffer, pH 5.0. The pH of the homogenate should be 5.3 ± 0.1. The homogenate is centrifuged at 25,000 g for 30 min after which the supernatant solution is carefully decanted through four layers of cheese cloth to remove floating fatty material.

First Heat Step. The pH of the extract is adjusted to 7.2 by the addition of 1 M K_3PO_4. The solution is heated with thorough stirring in a stainless-steel beaker held in a water bath at 70° until the temperature of the solution just reaches 60°, at which point cold water is circulated about the beaker to lower the temperature to 20°. The mixture is then stirred in an ice bath until the temperature reaches 4° when the precipitated protein is removed by centrifugation at 25,000 g for 45 min.

First Ammonium Sulfate Step. The pH of the enzyme solution is adjusted to 5.4 by drop by drop addition of glacial acetic acid, and solid ammonium sulfate is added to 45% saturation (278 g/liter). The precipitate is collected by centrifugation at 25,000 g for 30 min and dissolved in enough deionized water to give a total volume of 400 ml.

Precipitation at Low Ionic Strength. The protein solution is adjusted to pH 5.20 with glacial acetic acid and then dialyzed, in 5/8-inch dialysis tubing, against 24 liters of deionized water for a total of 6 hr, with the dialyzing water changed twice. After dialysis the sacs are opened and the pH of the solution is readjusted to 5.20 if necessary. The precipitate is collected by centrifugation for 45 min at 48,000 g, then thoroughly suspended in 50 ml of 0.05 M K_2HPO_4 containing 0.5% 2-mercaptoethanol. The pH of the suspension is adjusted to 7.4 with 1 M NaOH. After stirring the mixture at 0° for 1 hr, the soluble phase is separated by centrifugation at 48,000 g for 20 min. The precipitate is extracted again with 50 ml of deionized water containing 0.5% 2-mercaptoethanol for 30 min at 0°, followed by centrifugation as before. The combined extracts are then recentrifuged for 1 hr at 48,000 g.

Chromatography on DEAE-Cellulose. A suspension of 30 g of diethylaminoethylcellulose (Cellex D, Bio Rad Laboratories) in 4 liters of deionized water is adjusted to about pH 13 with 10 M NaOH and stirred for 15 min. The cellulose is washed repeatedly with deionized water by decantation until the pH of the water phase drops below 9. The suspension is then titrated with 1 M HCl to pH 7.4 with stirring. Fines are removed by suspension in water, settling, and decantation three times. The cellulose is then equilibrated with 0.05 M Tris-chloride, pH 7.4, 0.5% in 2-mercaptoethanol. A column of the DEAE-cellulose, 2.5 × 28 cm, is prepared by allowing the packing to settle while washing with equilibrating buffer under a 1-m pressure head. After settling the column is washed with 2 liters of the buffer. The protein solution is applied to the column, washed in with 200 ml of the equilibrating buffer, and eluted with a linear gradient, 0.05–0.30 M in Tris-chloride, pH 7.4. Three hundred milliliters of each limiting buffer are used, each 0.5% in mercaptoethanol. Fractions of 10 ml are collected every 15 min; the bulk of the enzyme activity is located at about 300–400 ml in the gradient. The brownish color of the enzyme affords a convenient marker during chromatography.

Second Ammonium Sulfate and Heat Steps. The enzyme solution from the column is fractionated with ammonium sulfate between 35 and 50% saturation. The solution is stirred in an ice bath while 20.9 g of solid ammonium sulfate per 100 ml of solution are slowly added. The suspension is allowed to stand in the ice bath for 30 min before it is separated by centrifugation at 48,000 g for 30 min. To the supernatant, 9.4 g of ammonium sulfate per 100 ml are added. After 30 min at 0° the precipitate is collected as before and dissolved in 3.5 ml of 0.02 M potassium phosphate buffer, pH 7.4, 0.5% in mercaptoethanol. The dark brown solution is then heated to 60° for 3.0 min in a water bath. The precipitated protein is removed by centrifugation at 48,000 g for 30 min.

Chromatography on Sephadex G-200 and Ultrafiltration. The protein solution is applied to a Sephadex G-200 column, 2.5 × 185 cm, previously equilibrated with 5 mM potassium phosphate buffer, pH 7.0, 0.5% in 2-mercaptoethanol. The same buffer is used for elution at a flow rate of 0.33 ml/min; 5.0 ml fractions are collected. The bulk of the enzyme appears in about 8 tubes. These fractions are immediately concentrated to a volume of 6 ml with an Amicon Ultrafiltration Cell, Model 52, using an XM-100 filter.

The purification procedure is summarized in Table I.

The highly purified enzyme is very unstable. The above procedure can be carried out in 4 days, and for best results it should be completed as quickly as possible. The pure enzyme loses about 50% of its activity in 2 days when stored at 4° in 0.5% mercaptoethanol. Evidently iron is lost and dissociation into subunits occurs. Activity is lost even more rapidly in the absence of 2-mercaptoethanol although dissociation into monomers does not occur under these conditions.

Purification Procedure II: Amidophosphoribosyltransferase Sensitive to Feedback Inhibition

The following procedure is modified from one described by Rowe and Wyngaarden for the enzyme from pigeon liver.[2] All steps are carried out at 4°.

Extraction. Fresh chicken livers (2.27 kg) are blended in portions with 4 liters of 0.01 M ammonium citrate buffer, pH 5.0, for 2 min. The suspension is centrifuged for 45 min at 25,000 g, and the supernatant phase is passed through four layers of cheesecloth to remove floating lipid.

Ammonium Sulfate Fractionation. The enzyme solution is brought to 35% saturation with solid ammonium sulfate (209 g/liter) and stirred overnight. The next day the supernatant phase is recovered by centrifugation for 45 min at 25,000 g and passed through four layers of cheesecloth. The solution is brought to 50% saturation with solid ammonium sulfate (94 g/liter), stirred for 1 hr, and centrifuged for 45 min at 25,000 g. The precipitate is dissolved in buffer A (0.05 M Trischloride, 1 mM MgCl$_2$, 0.5% mercaptoethanol, pH 8.0) to a final volume of 240 ml.

TABLE I
PURIFICATION OF AMIDOPHOSPHORIBOSYLTRANSFERASE BY PROCEDURE I

Step	Protein (mg)	Unit	Specific activity (units/mg)	Recovery (%)
Extraction	725,000	756	0.00104	100
Heat step 1	295,000	680	0.0023	90
Ammonium sulfate fractionation[a]	90,000	580	0.0064	77
Precipitation at pH 5.2	7,900	570	0.072	75
DEAE-cellulose	712	380	0.53	50
Heat step 2	255	321	1.26	42
Sephadex G-200	61	146	2.40	19

[a] Dialyzed to remove ammonium sulfate before analysis.

Ultracentrifugation. The solution is centrifuged for 2 hr at 160,000 g in a fixed-angle rotor of the ultracentrifuge, following which the supernatant is carefully decanted from the tubes through a layer of glass wool.

Chromatography on Sephadex G-75. The enzyme solution is layered on a Sephadex G-75 column, 9.0 × 85 cm, which has been extensively washed with buffer A. The same buffer is used for elution; fractions of 20 ml are collected at a flow rate of 1.5 ml/min. The enzyme emerges at about 1700 ml.

Chromatography on DEAE-Cellulose. The pooled fractions are centrifuged at 48,000 g for 30 min and then applied to a DEAE-cellulose column, 2.5 × 55 cm, prepared as described in the previous procedure and washed with buffer A. After washing the column with 400 ml of buffer A a linear gradient is applied, formed from 400 ml each of 0.05 M and 0.4 M Tris-chloride, pH 8.4, 1 mM in $MgCl_2$ and 0.5% in mercaptoethanol. Elution is carried out at 0.75 ml/min, and 15-ml fractions are collected. Fractions containing the bulk of the enzymic activity (about numbers 30 through 38) are pooled and immediately concentrated to 10 ml with an Amicon Ultrafiltration Cell using an XM-100 filter. The concentrated enzyme solution is then dialyzed against two portions of 1 liter of 2 mM Tris-chloride, pH 8.0, 1 mM in $MgCl_2$ and 0.5% in 2-mercaptoethanol.

This procedure is summarized in Table II.

The partially purified enzyme is most stable when stored (4°) at high protein concentration and in the presence of 2-mercaptoethanol. Dithiothreitol is not as effective in protecting the enzyme. Of the purification steps used in procedure I, the heat step, the precipitation at pH 5.2,

TABLE II
PURIFICATION OF AMIDOPHOSPHORIBOSYLTRANSFERASE BY PROCEDURE II

Step	Protein (mg)	Unit	Specific activity (units/mg)	Recovery (%)
Extraction	363,000	378	0.00104	100
Ammonium sulfate fractionation[a]	51,500	204	0.00395	54
Ultracentrifugation	43,000	185	0.0043	49
Sephadex G-75	8,700	93	0.011	25
DEAE-cellulose	4,800	29	0.060	7.8
Ultrafiltration	460	27	0.058	7.1

[a] Dialyzed to remove ammonium sulfate before analysis.

and the gel filtration on Sephadex G-200 cause extensive or complete desensitization toward inhibition by nucleotides, and therefore these steps cannot be used when a preparation retaining this property is required.

Properties

Physical. The enzyme obtained by procedure I appears homogeneous upon disc gel electrophoresis at pH 7.4 or pH 8.0, after staining with fast green or Coomassie blue. A single component is also observed by gel isoelectric focusing (pI = 5.2). Single sharp zones are seen after gel electrophoresis of the denatured protein in the presence of either sodium dodecyl sulfate or urea.[12]

A molecular weight of 210,000 is obtained from measurements of the sedimentation coefficient (9.3 S) and the diffusion coefficient (3.7×10^{-7} $cm^2 sec^{-1}$),[1] and a value of 202,000 daltons is obtained by gel filtration on Sephadex G-200.[12] Subunit molecular weight is estimated to be 53,000 daltons by SDS gel electrophoresis. Observation of a single component under these conditions, together with results from amino terminal sequencing and peptide mapping, shows the native enzyme to be a tetramer of identical subunits.[12] Under the conditions of isolation described, in the presence of either Tris or phosphate buffers, we have not observed dimeric species (MW = 1×10^5) of the type reported for the impure pigeon liver enzyme.[13] The tetramer contains 10–12 g-atoms of nonheme iron which gives it a characteristic brown color.[1,2] Contrary

[12] J. M. Lewis, unpublished results (1975–1977).
[13] R. Itoh, E. W. Holmes, and J. B. Wyngaarden, *J. Biol. Chem.* **251**, 2234 (1976).

to an earlier report from this laboratory indicating that the glutamine analogue, diazo-oxonorleucine (DON), binds covalently with a stoichiometry of one per tetramer,[14] values approaching 4 (3.6) are obtained with the freshly prepared, pure enzyme.[12]

Kinetic. The V_{max} with glutamine as substrate is essentially independent of pH between 6.0 and 9.0.[1] The specific activity of the enzyme corresponds to a turnover number of 127 moles of each product per mole active center per minute at 25°, assuming one catalytic site per monomer of 53,000 daltons. The apparent K_m for glutamine is $1.1 \times 10^{-3} M$.[1] That for PP-ribose-P has not been accurately measured with this enzyme because of the insensitivity of the assay system. Values in the general range of $10^{-4} M$ have been reported for this substrate with amidophosphoribosyltransferases from other biological sources, in some instances with nonhyperbolic kinetics being observed.[2,8] The catalytic mechanism is apparently ordered sequential, PP-ribose-P binding first.[1] PP-ribose-P also reacts enzymically with a number of amines and alcohols; e.g., with ammonia to give phosphoribosylamine, and with methanol to give methyl 5-phosphoriboside.[1]

Inhibition. Purine nucleotides inhibit the enzyme obtained by procedure II above. When magnesium levels in the assay system are adjusted to correct for the nonspecific chelating effect of nucleotides, the order of inhibitory effectiveness among the adenosine phosphates is ATP > ADP > AMP. Accurate K_i values cannot be determined with the pyrophosphate assay system, although they can be estimated to be on the order of 2–$4 \times 10^{-4} M$.[12]

Diazo-oxonorleucine and O-diazoacetylserine (azaserine) inhibit competitively with respect to glutamine, with K_i values of $1.9 \times 10^{-5} M$ and $4.2 \times 10^{-3} M$, respectively.[14] These agents also react as active site-directed irreversible inhibitors by alkylation of a critical cysteine SH group (carboxymethyl cysteine is recovered after reaction with [14C]azaserine and acid hydrolysis of the protein).[12] General sulfhydryl reagents such as hydroxymercuribenzoate, N-ethylmaleimide, and iodoacetate inhibit both the catalytic activity of the enzyme and the binding of labeled DON.[14] o-Phenanthroline causes inhibition of activity and precipitation of denatured enzyme through chelation of the ferrous iron.[1]

Note Added in Proof. Physical properties of the homogeneous amidotransferase from *Bacillus subtilis* are very similar to those described here, according to a recent report of J. Y. Wong, E. Meyer, and R. L. Switzer, *J. Biol. Chem.* **252**, 7424 (1977).

[14] S. C. Hartman, *J. Biol. Chem.* **238**, 3036 (1963).

[23] Phosphoribosylglycinamide Synthetase from *Aerobacter aerogenes*[1]

By DONALD P. NIERLICH

$$\text{5'-Phosphoribosylamine} + \text{glycine} + \text{ATP} \overset{Mg^{2+}}{\rightleftharpoons} \text{5'-phosphoribosylglycinamide} + \text{ADP} + \text{P}_i$$

PRG synthetase (5'-phosphoribosylamine:glycine ligase [ADP], EC 6.3.3.3) catalyzes the second step in the *de novo* path of purine biosynthesis in bacteria as well as higher organisms. Purification of the enzyme from pigeon liver has been described.[2] The preparation here is from *Aerobacter aerogenes* or, according to the current classification of the original strain, *Klebsiella pneumoniae*.

Assay Methods

The assay is complicated by the fact that the substrate, 5'-phorphoribosylamine, is very labile, hydrolysing with a half-life of less than 1 minute at pH 8.[3] In the assays described here substrate is formed in the assay mixture by the nonenzymic reaction between ribose 5-phosphate and NH_4^+. Alternatively, substrate can be synthesized chemically and added just prior to assay,[4] or generated enzymically in the reaction mixture from 5'-phosphoribosylpyrophosphate and glutamine.[5]

Two assays are described. The colorimetric assay is recommended for enzyme purification. It is fast, sensitive, and works well with crude enzyme preparations. The second, radioisotopic assay, is more readily quantitated. It does not work well with crude enzyme preparations.

Colorimetric Assay

Principle. The measurement is carried out in two steps: the formation of PRG and the PRG assay. The PRG assay is based on procedures first described by Warren and Buchanan and revised and described more fully by Flaks and Lukins.[2] Basically, enzymes contained in an extract of pigeon liver are used to carry out the formylation of PRG to 5'-phorphoribosylformylglycinamide. The formyl donor is IMP and the

[1] Supported by a grant from the U.S. Public Health Service.
[2] J. G. Flaks and L. N. Lukens, this series, Vol. VI [9].
[3] D. P. Nierlich and B. Magasanik, *J. Biol. Chem.* **240**, 366 (1965).
[4] L. Lukins and J. G. Flaks, Vol. VI [96].
[5] D. P. Nierlich and B. Magasanik, *J. Biol. Chem.* **240**, 358 (1965).

secondary product is 5′-phosphoribosyl-5-amino-4-imidazolecarboxamide, which in turn can be readily measured colorimetrically after diazotization by the Bratton and Marshall procedure.

Reagents for PRG Synthetase

Tris (hydroxymethyl)aminomethane·HCl (Tris·HCl; Sigma), 1.0 M, pH 8.0

Ribose 5-phosphate, sodium (Sigma), 0.2 M

NH_4Cl, 0.5 M

Glycine, 0.2 M

ATP, 0.025 M, adjusted to pH 6–8 and stored frozen

$MgCl_2$, 0.2 M

Reagents for Colorimetric Assay

IMP-EDTA solution, containing 25 mM IMP and 150 mM ethylenediaminetetraacetic acid, adjusted to neutrality with NaOH

Pigeon liver enzyme, lyophilized 15–30% ethanol fraction prepared as described by Flaks and Lukins[2] (chicken liver enzyme prepared in the same way can be used but is neither as potent nor stable), 5–20 mg/ml depending on potency, dissolved prior to use in cold potassium phosphate (K^+ is required) buffer, 0.03 M, pH 7.5, and centrifuged if necessary to clarify

TCA-HCl solution, containing 15% trichloroacetic acid in 1 N HCl

Sodium nitrite, 0.5%

Ammonium sulfamate, 0.5%

N-1-naphthyethylenediamine dihydrochloride, 0.1%

The latter three reagents are stable several months stored in amber bottles at 3°.

Procedure. Incubation mixtures contain in a final volume of 0.35 ml: 30 μl Tris·HCl, 20 μl ribose 5-phosphate, 20 μl NH_4Cl, 20 μl glycine, 10 μl ATP, 10 μl $MgCl_2$, water, and enzyme (diluted if necessary in 0.01 M Tris·HCl). Lithium acetylphosphate (4 μmoles) can be included in assays of crude extracts but generally has little effect.

The mixtures are incubated in small round or conical centrifuge tubes, as for the clinical centrifuge, at 37° for 5 min. The reactions are stopped by adding 0.1 ml IMP-EDTA solution, and the assays are carried out with the addition of 0.1 ml of pigeon liver enzyme and incubation for 30 min at 37°. These reactions are stopped by the addition of 0.1 ml TCA-HCl solution. To these mixtures, at room temperature, are added, in order: 0.1 ml of sodium nitrite, 0.2 ml of ammonium sulfamate, and 0.2 ml of naphthylethylenediamine. Each sample is mixed and 30 sec or more are allowed between each addition. The samples are then centrifuged for 15 min at 800 g, and the absorption of the supernatants is determined at 540 nm in 1-cm semimicrocuvettes.

Controls are prepared in which the addition of IMP-EDTA precedes the addition of PRG synthetase. The assay is linear with time of incubation and amount of enzyme when purified preparations are used, but the linearity is restricted with crude enzyme extracts.[6] A possible false value may be obtained in the assay of crude extracts due to the formation of 5'-phosphoribosyl-5-amino-4-imidazolecarboxamide accompanying the initial steps of histidine biosynthesis from ATP. These enzymes are repressed by the inclusion of histidine in the bacterial growth medium. Histidine can be added as well to the assay mixtures to inhibit this reaction.

Based on a molecular extinction coefficient of 26,400 for the colored product of the above diazotization reaction with the compound 5-amino-4-imidazole-carboxamide, an absorbance of 0.1 should be given by 4.4 nmoles of PRG. In practice, a theoretical yield is not obtained although the transformylation reaction is irreversible.[2] When a chemically prepared sample of PRG (β-form was used), an absorbance of 0.1 was given by 6.9 nmoles; this value varies slightly with different avian liver preparations.[3] The preparation of PRG has been described,[4] and it can be used to standardize the assay.

Radioisotopic Assay

Reagents
[^{14}C]glycine, 0.1 μCi/μmole, 0.1 M
Reagents for PRG synthetase listed above, except glycine
Trichloroacetic acid, 30%
Ammonium formate, 0.05 M
Dowex 50 (NH$_4^+$), 200–400 mesh

Procedure. The incubations are carried out as described above for PRG synthetase with 20 μl of [^{14}C]glycine replacing the glycine used. After 15 min at 37° the reactions are stopped by adding 0.05 ml of trichloracetic acid. The samples are then centrifuged for 15 min at 800 g, and 0.1 ml portions of the supernatant are transferred to 0.5 × 5 cm Dowex 50 (NH$_4^+$) columns prepared in Pasteur pipettes. The columns are eluted with 2.0 ml of ammonium formate, and the radioactivity is determined by counting the dried samples on stainless-steel planchets[6] or directly in the liquid scintillation spectrometer with a cocktail accepting aqueous samples. The unreacted glycine is retained by the columns.

[6] D. P. Nierlich and B. Magasanik, *Biochim. Biophys. Acta* **230**, 349 (1971).

Units. Enzyme activity is expressed as μmoles of PRG formed per minute in the standard assay. For the determination of specific activity, protein determinations are performed with the Lowry procedure.[7] Bovine serum albumin (Fraction V, Armour) is used as the standard.

Purification of PRG Synthetase

The purification is carried out at 3° unless otherwise stated. Crude cell extracts are stable for 1 day of storage on ice, and enzyme at intermediate stages of purification is stable for several days. A typical purification is shown summarized in the table.

Source of Enzyme. *Klebsiella pneumoniae* (*Aerobacter aerogenes*) wild type (strain 1033)[8] and a mutant derived from it have been used. Cells suitable for enzyme purification can be obtained in several ways. The purine biosynthetic enzymes are, however, repressed by their end-products,[6] and thus any such supplements should be minimal. The wild-type strain is grown in a modified Werkman's medium containing, per liter: $Na_2HPO_4 \cdot 7\ H_2O$, 9.5 g; KH_2PO_4, 12.6 g; $(NH_4)_2SO_4$, 2 g; $MgSO_4 \cdot 7\ H_2O$, 0.2 g; $CaCl_2$, 0.01 g. The final pH is 6.3. This is supplemented after autoclaving with 20 ml glucose, 0.2 g/ml; 20 ml histidine-HCl, 1 mg/ml; and 0.5 ml thiamine-HCl, 0.1 mg/ml. Cultures are grown at 37° in 1-liter portions in 2.7-liter Fernbach flasks with vigorous rotary shaking, in 15-liter quantities in 20-liter carboys aerated with spargers, or in larger quantities in a fermenter. Cells are harvested in mid to late log phase to yield approximately 1 g cells/liter. Alternatively, a 2- to 4-fold higher initial specific activity can be obtained from cells of a purine requirer, such as strain PD-1.[6,9] This strain is blocked in the path prior to IMP, as indicated by growth on either adenine, guanine, or hypoxanthine. Such cells are grown in the above medium with the additional supplement of 9 ml adenine, 1 mg/ml (in 0.01 *N* HCl). In this medium cells are either grown overnight so as to include a period of several hours of purine limitation or grown under continuous purine limitation in a continuous-culture apparatus.[2] Dow Corning Antifoam B or Union Carbide SAG 471 can be added, to 0.2 ml/liter, to cultures grown with forced aeration.

Cell-free Extracts. Cultures are quickly chilled and harvested by centrifugation at 13,000 *g* for 15 min or in a continuous-flow centrifuge. The cells are resuspended in about 20 times their wet weight of cold 0.03

---

[7] O. H. Lowry, N. J. Rosenbough, A. L. Farr, and R. J. Randall, *J. Biol. Chem.* **193**, 265 (1951).

[8] D. Ushiba and B. Magasanik, *Proc. Soc. Exp. Biol. Med.* **80**, 626 (1952).

[9] Available from the author.

PURIFICATION OF PRG SYNTHETASE[a]

Fraction and step	Specific activity[b]	Relative specific activity	Yield (%)
Crude extract	0.028	1.0	100
I. Protamine supernatant	0.044	1.6	102
II. First ammonium sulfate fraction	0.085	2.3	70
III. DEAE-cellulose chromatography	0.19	6.8	39
IV. Second ammonium sulfate fraction	0.47	16.7	16
V. Sephadex G-75 chromatography	1.70	60.0	9

[a] From J. G. Flaks and L. N. Lukins, this series, Vol. 6 [9], with permission,
[b] Expressed as μmoles of PRG per milligram of protein per minute, in standard assay.

M phosphate buffer, pH 7.5, recentrifuged, and suspended again with 7 times their weight of the same buffer. The cells are broken by sonication in 30-ml batches in a jacketed, continuously cooled cup. Sonication is complete (in the Raytheon 10-kcycle oscillator in about 8 min) when the suspension held in the drawn tip of a Pasteur pipette appears in a very bright light translucent and straw-colored. Other methods of cell breakage have not been tried. The pooled extracts are centrifuged at 30,000 g for 30 min to remove cell debris.

Protamine Fractionation. To 150 ml of extract (from 20 g cells), being stirred at moderate rate in an ice bath, is added 0.2 volume of 2% protamine sulfate[10] (Schwartz/Mann, at room temperature) over 15 min. After 15 min additional slow stirring, the precipitate (containing PRPP amidotransferase[5]) is removed by centrifugation at 30,000 g for 15 min (fraction I).

First Ammonium Sulfate Fractionation. To the protamine supernatant is added over the period of 15 min, with stirring as above, about 60 g solid ammonium sulfate (Schwartz/Mann, enzyme grade). The final concentration is 55% saturation following the equation given by Kunitz[11]:

$$X = \frac{533(S_2 - S_1)}{1 - 0.3S_2}$$

where X is the grams of ammonium sulfate per liter of solution, and S_1

[10] Different batches of protamine sulfate should be standardized by a trial fractionation.
[11] M. Kunitz, *J. Gen. Physiol.* **35**, 423 (1952).

and S_2 are the initial and final fractions of saturation. After 15 min additional stirring the precipitate is collected as above and dissolved in 35 ml cold H_2O. This solution is dialyzed overnight against two changes of 2 liters of 0.005 M Tris-maleate buffer, pH 7.0 (fraction II).

DEAE-Cellulose Fractionation. Fraction II is applied to a 2.4 × 25 cm DEAE-cellulose column, previously equilibrated with the above Tris-maleate buffer. The column is then washed with 20 ml of the same buffer, and elution is carried out using a linear gradient of decreasing pH and increasing salt concentration. The reservoirs of the gradient-forming device contain initially 300 ml of the starting buffer in the mixing chamber and 300 ml of 0.05 M Tris-maleate buffer, pH 6.0, in the reservoir chamber. The column is eluted at a rate of 1.25 ml/min, collecting 10-ml fractions. After a total volume of 100 ml has been removed, the reservoir chamber is made 0.8 M in KCl. This is done by stirring in the appropriate amount of KCl, and removing a small volume from the chamber to equalize the hydrostatic pressures. Fractions are assayed for enzyme activity and the peak tubes pooled. The enzyme elutes in the range of 0.35 M KCl (fraction III).

Second Ammonium Sulfate Fractionation. To the pooled fractions from above (70 ml) are added 1 M Tris·HCl, pH 7.5, 0.2 M $MgSO_4$, and H_2O to bring the protein concentration to approximately 1.3 mg/ml and that of the Tris to 0.1 M and magnesium to 0.4 mM. To this solution (100 ml) is added ammonium sulfate (30 g) as above, to give 48% saturation. After centrifugation, the resulting supernatant is further adjusted to 58% ammonium sulfate saturation (7.2 g). The resulting precipitate is dissolved in the minimum volume of cold H_2O required to give a clear solution (6 ml, fraction IV).

Sephadex G-75 Fractionation. The enzyme is applied to a 4.5 × 60 cm Sephadex G-75 column which has been equilibrated with 0.03 M Tris·HCl, pH 7.5. The column is eluted with the same buffer at about 1.2 ml/min. Fractions are assayed, and those of highest specific activity are pooled.

This fraction, having less than 0.2 mg/ml, is relatively unstable at 3%, although it can be stored several months at $-18°$. To obtain more stable fractions the enzyme is concentrated 5-fold. The enzyme is transferred to a dialysis bag, and this is placed for several hours in the cold in a dish covered with flakes of polyethelylene glycol compound, 20,000 average molecular weight (Carbowax 20 M, Union Carbide). Although slower, Sephadex G75 can be used similarly. The recommended procedure has the disadvantage that a residue of polyethylene glycol enters the enzyme. Purified enzyme can be stored without loss of activity at $-18°$ in excess of a year (fraction V).

Properties

Substrate Kinetics. The purified PRG synthetase requires 5'-phosphoribosylamine, glycine, and ATP.[3] It is without activity in the absence of Mg^{2+}. ATP is consumed and equimolar amounts of PRG, ADP, and P_i are produced stoichiometrically. The K_m for glycine is $1.9 \times 10^{-4} M$; the K_m for ATP is $5.6 \times 10^{-5} M$ (at a Mg^{2+} concentration of 5.7 mM). The kinetic constants for 5'-phosphoribosylamine can only be estimated indirectly because this substrate is so labile.[3,4] Based on the estimate of the equilibrium constant for the nonenzymic formation of 5'-phosphoribosylamine from ribose 5-phosphate and NH_3, an approximate K_m of 7.8 $\times 10^{-6} M$ is obtained.

pH Optimum. Determination of the pH optimum is complicated by the effect of pH on the concentration of NH_3 (the apparent reactant in the nonenzymic formation of 5'-phosphoribosylamine) and also on the colorimetric assay. The activity has been determined using the radioisotopic assay and varying the concentration of NH_4Cl added so as to obtain constant NH_3 concentration at each pH. Under these conditions the enzyme exhibits a broad optimum between pH 7 and 8.[3]

Molecular Weight. By sucrose gradient centrifugation, using horse hemoglobin as a standard, the $s_{w,20}$ is approximately 3.2 S.[12] Based on sedimentation in a synthetic boundary cell the $s_{w,20}$ is 3.0 S (unpublished data).

Inhibitors. The enzyme is inactive in the presence of EDTA. However, even after 4 hr of treatment, enzyme activity is fully regained after restoration of the magnesium. $CuSO_4$, $BaCl_2$, $CaCl_2$, and $MnCl_2$ are not inhibitory at $10^{-4} M$ or $10^{-2} M$ concentrations.

NaF is approximately 50% inhibitory at 0.01 M; 5×10^{-4} M p-chloromercuribenzoate completely inhibits the enzymes after 10-min incubation. NaCN, 0.01 M, stimulates the reaction about 2-fold. Azaserine, 0.2 mM, sodium azide, 10 mM, mercaptoethanol, 10 mM, and gluthathione, 10 mM, are without effect.

A number of potential end-product inhibitors are without effect.[3] Thus AMP, ATP, GMP, GTP, and IMP have no effect at concentrations of 0.003 or 0.014 M. ADP at a concentration of 0.014 M inhibits the reaction by 80% when the concentration of ATP in the assay mixture is $3.2 \times 10^{-4} M$. Presumably this is due to a reversal of the biosynthetic reaction.

---

[12] Based on a value of 4.09 S for horse hemoglobin.

[24] *N*-(5-Amino-1-ribosyl-4-imidazolylcarbonyl)-L-aspartic Acid 5′-Phosphate Synthetase

By JOHN M. BUCHANAN, LEWIS N. LUKENS, and RICHARD W. MILLER

5-Amino-1-ribosyl-4-imidazole carboxylic acid 5′-phosphate (carboxy-AIR) + ATP
+ aspartate
→ *N*(5-amino-4-imidazolylcarbonyl)-L-aspartic acid 5′-phosphate (succino-AICAR)
+ ADP + P_i

Succino-AICAR[1] synthetase is one of the enzymes of the series involved in the biosynthesis of the purine nucleotides *de novo*.[2] The nitrogen of aspartic acid is the precursor of N_1 of the purine ring and the amino group of adenylic acid. The reactions for succino-AICAR formation from carboxy-AIR[3,4] and succino-AMP from IMP[5-8] are similar except that GTP is utilized for the latter and ATP for the former reaction. Adenylosuccinase is involved in the removal of carbon atoms of aspartic acid from both of the above substrates as fumarate[7-9] by a stereospecific trans elimination.[10]

Earlier studies on the preparation of succino-AICAR synthetase and succino-AICAR have been reported[11,12] in Vol. VI, articles 9 and 96, respectively, of *Methods in Enzymology*.

[1] The abbreviations used are: succino-AICAR, *N*-(5-amino-1-ribosyl-4-imidazolylcarbonyl)-L-aspartic acid 5′-phosphate; AICAR, 5-amino-1-ribosyl-4-imidazolecarboxamide 5′-phosphate; AIR, 5-amino-1-ribosylimidazole 5′-phosphate; carboxy-AIR, 5-amino-1-ribosyl-4-imidazolecarboxylic acid 5′-phosphate; succino-AMP, 6-(L-1,2-dicarboxyethylamino)-9-β-D-ribosylpurine 5′-phosphate (adenylosuccinic acid). The following trivial names have been used: AIR, 5-aminoimidazole ribonucleotide; carboxy-AIR, 5-amino-4-imidazolecarboxylic acid ribonucleotide; AICAR, 5-amino-4-imidazolecarboxamide ribonucleotide; succino-AICAR, 5-amino-4-imidazole-*N*-succinocarboxamide ribonucleotide.
[2] J. M. Buchanan, *Harvey Lect.* **54**, 104 (1960).
[3] L. N. Lukens and J. M. Buchanan, *J. Biol. Chem.* **234**, 1791 (1959).
[4] R. W. Miller and J. M. Buchanan, *J. Biol. Chem.* **237**, 485 (1962).
[5] R. Abrams and M. Bentley, *J. Am. Chem. Soc.* **77**, 4179 (1955).
[6] C. E. Carter and L. H. Cohen, *J. Am. Chem. Soc.* **77**, 499 (1955).
[7] C. E. Carter and L. H. Cohen, *J. Biol. Chem.* **222**, 17 (1956).
[8] I. Lieberman, *J. Biol. Chem.* **223**, 327 (1956).
[9] R. W. Miller, L. N. Lukens, and J. M. Buchanan, *J. Biol. Chem.* **34**, 1806 (1959).
[10] R. W. Miller and J. M. Buchanan, *J. Biol. Chem.* **237**, 491 (1962).
[11] J. G. Flaks and L. N. Lukens, this series, Vol. VI [9], p. 52.
[12] L. N. Lukens and J. G. Flaks, this series, Vol. VI [96], p. 671.

Assay Method

In view of the relative inaccessibility of the immediate substrate of the forward reaction, carboxy-AIR, an assay was developed for succino-AICAR synthetase activity that involves the arsenolysis of succino-AICAR to carboxy-AIR or AIR (5-amino-1-ribosylimidazole 5'-phosphate).[4] Since the succino-AICAR synthetase in its present state of purification[4] still contains AIR carboxylase,[13] the primary product for the reaction is AIR. In the reverse direction both ADP and orthophosphate are required for succino-AICAR metabolism. Phosphate may be replaced by arsenate. The putative product, ADP-arsenate, is believed to hydrolyze spontaneously.

AIR is determined by a modification[3] of the Bratton-Marshall method[14] for nonacetylatable arylamines. The absorbance maximum of the reaction product is 500 nm. Since crude enzyme preparations contain adenylosuccinase, AICAR may also be a product of succino-AICAR reaction. When the diazotization of succino-AICAR is carried out at room temperature, a color does not develop upon coupling with the chromogenic dye. Normal color development (absorbance maximum, 560 nm) does occur, however, at 0°. As pointed out by Lukens and Buchanan[3] heating of the products of reaction in acid solution before diazotization results in the breakdown of AIR or carboxy-AIR but not AICAR. Thus the amount of AIR can be determined by difference in solutions that contain both arylamines.

Reagents

    Tris·Cl buffer, pH 7.8
    ADP
    Sodium arsenate, pH 7.8
    Magnesium chloride
    Succino-AICAR
    Enzyme fractions

Procedure. Under certain conditions the rate of arsenolysis of succino-AICAR is proportional to enzyme concentration and thus could be used as a measure of the amount of enzyme present. The contents of each assay vessel were as follows: succino-AICAR, 0.08 μmole; Tris, 57 μmoles; ADP, 6.8 μmoles; sodium arsenate, 57 μmoles; magnesium chloride, 7.6 μmoles; succino-AICAR synthetase, 0.7–3.0 units; total

[13] L. N. Lukens and J. M. Buchanan, *J. Biol. Chem.* **234**, 1799 (1959).
[14] A. C. Bratton and E. K. Marshall, Jr., *J. Biol. Chem.* **128**, 537 (1939).

volume 0.65 ml. Vessels were incubated for 20 min at 38°, after which time the reaction was stopped by the addition of 0.1 ml of 30% trichloroacetic acid. Protein was removed by centrifugation, and a 0.40-ml aliquot was transferred to a second vessel. AIR was determined by the colorimetric method[3] after addition of reagents to make a final volume of 0.7 ml.

The arsenolysis of succino-AICAR was also followed spectrophotometrically in a constant-temperature spectrophotometer by observing the decrease in absorbancy at 270 nm, which accompanies the degradation of succino-AICAR. This method was useful as an assay in some cases and demonstrated that under the conditions employed, the reaction was proceeding at a rate that was linear with time during an incubation period of up to 40 min at 37°. Each cuvette contained the following amounts of materials in a volume of 2.9 ml: succino-AICAR, 0.14 μmole; ADP, 0.083 μmole; sodium arsenate, pH 7.8, 22 μmoles; magnesium chloride, 26 μmoles; succino-AICAR synthetase, 10–20 units. The sensitivity of this assay is much less than that of the colorimetric one. For this reason the colorimetric assay was used throughout the purification procedure as a standard measure of succino-AICAR synthetase activity. The spectrophotometric assay, however, proved useful in investigations of the properties of the enzyme and the requirements of the reverse direction.

It should be noted that the optimal conditions for use of the arsenolysis of succino-AICAR as an assay method varied from one enzyme source to another. For example, the enzyme as found in a mutant of *Neurospora crassa* was inhibited by the high levels of ADP and arsenate employed in the chicken liver assay. In the case of the neurospora enzyme, a pronounced nonlinearity in the rate of AIR formation as a function of enzyme concentration was noted.

Units. A unit of activity is defined as that amount of enzyme giving rise to the formation of sufficient AIR after 30 min incubation at 38° to yield a change in absorbancy of 0.10 at 500 nm in the modified Bratton-Marshall assay.[3] In the case of the cruder enzyme fractions, calculations of specific activity were always made on the basis of an absorbancy of around 0.1 in the colorimetric assay.

Since the molecular absorbancy of the colored product at 500 nm is 24,600 M^{-1} cm^{-1}, the excursion of 0.1 OD unit per 0.7 ml is equivalent to 0.0031 μmole of AIR or 0.0058 μmole in the contents of a vessel that had been incubated for 20 min. A standard unit of enzyme as recommended by the International Union of Biochemistry is that amount which transforms 1 μmole of substrate per minute. Thus one unit as defined by Lukens and Buchanan[3] is equivalent to 2.9 × 10^{-4} standard units.

Preparation

Preparation of Materials. Since the reactions encompassing the conversion of aminoimidazole ribonucleotide (AIR) to inosinic acid are reversible, it is possible to approach the preparation of difficulty obtainable intermediates of purine nucleotide intermediates from either direction. This choice is determined by the reactants that are commercially available.

Preparation of Succino-AICAR. Succino-AICAR, the substrate of the reaction under consideration, may be formed from formylglycinamide ribonucleotide (FGAR) in a series of reactions catalyzed by a relatively crude enzyme extract. However, for preparative purposes succino-AICAR is more conveniently produced by the reaction of fumarate with 5-amino-4-imidazolecarboxamide ribonucleotide (AICAR) catalyzed by the splitting enzyme, adenylosuccinase. AICAR may be formed[15] from the free base, 5-amino-4-imidazolecarboxamide (AICA), and 5-phosphoribosylpyrophosphate (PRPP) in a reaction catalyzed by partially purified adenine (or AICA) phosphoribosyltransferase from bovine liver or by the chemical hydrolysis of inosinic acid (IMP). The enzymic preparation of AICAR as well as reaction of the latter with fumarate to yield succino-AICAR in the presence of yeast adenylosuccinase has been described by Lukens and Flaks.[12]

Chemical Method for Preparation of AICAR. The use of a chemical method for the preparation of AICAR from IMP as described by Friedman and Gots[16] and modified by Yokota and Sevag[17] is as follows. To a 1-liter mixture of ammonium chloride (7.5 g), zinc dust (7.5 g), and HCl (0.5 *N*), 1.3 mmoles of disodium IMP were added. After stirring for exactly 5 min, the zinc was rapidly removed by filtration on a Büchner funnel. The filtrate was treated with 6 g of charcoal (Darco G-60), which had been previously washed with 1 *N* HCl. After allowing the mixture to stand for 10 min, the charcoal was collected on a Büchner funnel and washed with 500 ml of distilled water. AICAR was eluted from the charcoal by extraction with several 30-ml aliquots of a mixture of equal volumes of absolute ethanol and 1 *M* ammonium hydroxide. The extracts were combined (150 ml) and evaporated to about 10 ml. The pH of the concentrated solution was around 8.5. The concentrate was diluted to 100 ml and run directly on a Dowex-1 resin column in the bromide form. Elution and isolation of AICAR were carried out as described for the enzymic preparations.[12] The yield of AICAR after purification was

[15] J. G. Flaks, M. J. Erwin, and J. M. Buchanan, *J. Biol. Chem.* **228**, 201 (1957).
[16] S. Friedman and J. S. Gots, *Arch. Biochem. Biophys.* **39**, 254 (1952).
[17] T. Yokota and M. G. Sevag, personal communication.

110 μmoles corresponding to an 8% conversion of the IMP. However, yields as high as 50% have been claimed by Yokota and Sevag. This method has several advantages over the enzymic procedure in that it is more rapid and simpler. More importantly, however, the resulting AICAR contains no contaminating traces of ADP and consequently succino-AICAR, is completely free from ADP, ribose-5-phosphate, and other impurities. There is some evidence that AICAR formed by the reductive hydrolysis of IMP may contain other arylamines of unknown structure. These, however, do not interfere in the preparation of succino-AICAR.

Yeast Adenylosuccinase. The purest preparations of cleaving enzyme were obtained from baker's yeast by a procedure whose initial steps were similar to those employed by Carter and Cohen[7] in the purification of adenylosuccinase. The preparation of adenylosuccinase through the heat step was adequate for synthesis of succino-AICAR. The spectrophotometric assay for adenylosuccinase in which adenylosuccinate (Sigma) is split to AMP and fumarate was used. The enzyme is stabilized by sulfhydryl compounds.

Purification of Succino-AICAR Synthetase

Several sources, including *Escherichia coli, Neurospora crassa, Salmonella typhimurium,* and *Saccharomyces cerevisiae,* were examined as prospective starting materials for purification of succino-AICAR synthetase. All of these, however, were inferior in activity to extracts of chicken liver. The microbial sources had the additional disadvantage of containing high levels of succino-AICAR-cleaving activity, the presence of which made the assay for succino-AICAR synthetase by arsenolysis of succino-AICAR unfeasible.

Due to the instability of the succino-AICAR-cleaving activity (i.e., adenylosuccinase) of avian tissues, this enzyme was easily removed in the preliminary steps of the purification procedure. Fresh livers obtained from chickens (after decapitation and without sterilization) were converted to an acétone powder[18] which yielded practically no cleaving activity on extraction. The dry powder (10 g) was extracted at 2° by stirring for 30 min with 100 ml of 0.05 M potassium phosphate buffer at pH 7.3. The crude acetone powder extract was treated with pancreatic RNase, 0.2 mg of a crystalline preparation being added per milliliter of enzyme solution. The mixture was allowed to come to 38° over a period of 10–15 min in a constant-temperature bath. This procedure usually brought about the formation of a small precipitate which was removed

[18] J. M. Buchanan, S. Ohnoki, and B.-S. Hong, this volume [25].

PURIFICATION OF SUCCINO-AICAR SYNTHETASE

Fraction	Yield (units[a])	Specific activity (units/mg)
Acetone powder extract	3800	0.8–1.2
RNase supernatant solution	3600	0.8–1.2
Ammonium sulfate precipitate (A-1)	3400	1.5
Ammonium sulfate precipitate (A-2)	2900	3.5
Gel fractions		
Initial supernatant solution (G-1)	100	
Phosphate wash, 0.05 *M* (G-2)	120	
Phosphate eluate, 0.18 *M* (G-3)	1700	7
Dialyzed ammonium sulfate precipitate (A-3)	2000	8
Hydroxylapatite fractions		
Phosphate eluate, 0.12 *M*	0	
Phosphate eluates, 0.18–0.23 *M*	1100	15
Phosphate eluate, 0.35 *M*	410	7

[a] Units are expressed according to the definition of Miller and Buchanan.[4]

by centrifugation after cooling to 2°. The remaining steps of the purification were carried out at this temperature. Sufficient solid ammonium sulfate was then added to the supernatant solution to bring the salt concentration to 55% of saturation. The precipitate, containing virtually all of the desired enzymic activity, was collected by centrifugation at 13,000 rpm in the Lourdes model AA centrifuge, and the clear supernatant solution was discarded. This precipitate (fraction A-1 in the table) was taken up in 100 ml of 0.05 *M* potassium phosphate buffer, pH 7.3, and refractionated with solid ammonium sulfate. The protein precipitating between 25 and 45% of saturation, with respect to ammonium sulfate, was collected by centrifugation (fraction A-2 in the table) and was dissolved in 20 ml of 0.02 *M* potassium phosphate buffer, pH 7.3.

The clear enzyme solution was dialyzed with stirring against 6 liters of 0.01 *M* potassium phosphate buffer of the same pH. The buffer was changed after approximately 5 hr, and the dialysis was continued for an additional 9 hr. The dialyzed material was diluted 2-fold with distilled water and assayed for protein content by the standard biuret method.[19] The protein concentration was generally around 15 mg/ml at this stage of purification. A suspension of calcium phosphate gel, 21 mg/ml, which had been prepared according to the method of Keilin and Hartree,[20] was

[19] L. G. Mokrasch and R. W. McGilvery, *J. Biol. Chem.* **221**, 909 (1956).
[20] D. Keilin and E. F. Hartree, *Proc. R. Soc. London* **124**, 397 (1938).

stirred into the protein solution. The dry weight of gel added was equal to the total weight of protein present. The mixture was then allowed to equilibrate at 2° over a period of 20 min, after which the gel was collected by centrifugation. The gel, having absorbed all of the succino-AICAR synthetase and AIR-carboxylase activities, was washed with two successive 50-ml aliquots of 0.05 M potassium phosphate buffer, pH 7.2, and two similar aliquots of 0.18 M phosphate buffer at pH 7.2. Practically no activity was found in the initial supernatant solution (G-1) or in the 0.05 M phosphate wash (G-2), and these fractions were accordingly discarded. The bulk of the enzymic activity appeared in the 0.18 M phosphate eluates, which were pooled (G-3).

The G-3 fraction (80 ml) was subjected to a third ammonium sulfate fractionation. In this case, most of the activity resided in the fraction precipitating between 30 and 50% of saturation. The precipitate was collected by centrifugation and dissolved in 50 ml of 0.02 M potassium phosphate buffer, pH 6.9, and dialyzed for 10 hr against 6 liters of 0.01 M phosphate buffer at the same pH.

The dialysate was diluted 2-fold with water and poured onto a column of hydroxylapatite (2.4 cm in width, 18 cm in height), which had been prepared according to the procedure of Tiselius, Hjertén, and Levin.[21] In view of variations in the properties of hydroxylapatite, it was found expedient to elute the column with successive aliquots of potassium phosphate buffer at pH 6.9 as follows: 50 ml, 0.12 M; 100 ml, 0.18 M; 50 ml, 0.23 M; and 50 ml, 0.35 M. The pH of all such buffers was compared with the pH of standard buffers at 2°.

The enzymically active fraction obtained by eluting the hydroxylapatite column with a 0.18 M phosphate solution was either used directly in some cases or dialyzed several hours against 0.02 M phosphate buffer, pH 7.3, before it was used. The dialyzed material was diluted with water to a buffer concentration of 0.005 M and absorbed on a 1.8 × 15 cm column of DEAE-cellulose ion-exchange agent. The activity was then eluted in a small volume with 0.2 M phosphate at pH 7.3. This step was primarily one of concentration. It was found that gradient elution of the cellulose columns led to loss of activity and did not bring about separation of the enzyme from the bulk of the inactive protein. Attempts at chromatography on carboxymethyl cellulose were also unsuccessful.

The enzyme, which was somewhat unstable during purification, was not preserved or protected to any extent during or after purification by sulfhydryl-containing reagents, aspartic acid, or other dicarboxylic acids. Under conditions of storage at −20° the concentrated enzyme (>1 mg/ml) was stable.

---

[21] A. Tiselius, S. Hjertén, and Ö. Levin, *Arch. Biochem. Biophys.* **65**, 132 (1956).

Properties of the Enzyme

At concentrations of ADP of 0.05 mM, Mg^{2+} saturates the enzyme at a concentration of 2–3 mM.[4] Mn^{2+} and Co^{2+} could replace Mg^{2+} and produced maximal activation of the enzyme at concentrations of 0.06 and 0.04 mM, respectively. The activation afforded by Mg^{2+} under optimal conditions was from two to three times greater than that resulting from the other cations tested. ADP could not be replaced to a significant extent by any of the other nucleoside diphosphates. In the presence of hydroxamic acid, hydroxamates are formed during the course of the splitting of succino-AICAR to carboxy-AIR.

[25] 2-Formamido-N-ribosylacetamide 5′-Phosphate: L-Glutamine Amido-Ligase (Adenosine Diphosphate)

By John M. Buchanan, Shiro Ohnoki, and Bor Shyue Hong

2-Formamido-N-ribosylacetamide 5′-phosphate (FGAR) + glutamine + ATP + H$_2$O
→ 2-formamino-N-ribosylacetamidine 5′-phosphate (FGAM) + glutamate + ADP + P$_i$

The enzyme catalyzing the above reaction is responsible for the incorporation of the amide nitrogen of a glutamine into a precursor of the purine ring at position 3 of the latter. Two trivial names have been used in reference to the enzyme, FGAR amidotransferase and FGAM synthetase. Although the latter name has been recommended by the International Commission on Biochemical Nomenclature, the former is much preferred by the present authors because it classifies this enzyme according to enzymic function, i.e., the transfer of the amide group of glutamine. A general review of the amidotransferases has been published.[1]

FGAR amidotransferase has received special attention as one of the series of enzymes involved in purine nucleotide synthesis since it is irreversibly inactivated by the antibiotics, L-azaserine and 6-diazo-5-oxo-L-norleucine (DON), which are antimetabolites of glutamine. These two compounds were the first to be recognized as active-site directed affinity reagents by Levenberg et al.[2] in this laboratory.

Assay Method

Three assays have been developed for the determination of FGAR amidotransferase. The first, which has been used for assay of enzyme in

[1] J. M. Buchanan, Adv. Enzymol. **39**, 91 (1973).
[2] B. Levenberg, I. Melnick, and J. M. Buchanan, J. Biol. Chem. **225**, 163 (1957).

crude tissue preparation and for general enzyme purification,[3,4] depends on the enzymic conversion of the nucleotide product of the reaction, formylglycinamidine ribonucleotide (FGAM) into 5-aminoimidazole ribonucleotide (AIR), which can be determined colorimetrically. In the second and third assays two other products of the reaction, glutamate and ADP, are measured, respectively. These methods have been used primarily for studies of enzyme kinetics and mechanisms with highly purified preparations.[5]

Colorimetric Assay for FGAM Production

Principle. FGAM, a product of the FGAR amidotransferase reaction, was converted to AIR in the presence of ATP, the proper cations, and an excess of the enzyme, AIR synthetase. AIR in turn was determined by reaction with the Bratton-Marshall reagents[6] and measurement of the absorbance at 500 nm.

Reagents

AIR SYNTHETASE. 3.5 g of a fraction of an acetone powder of pigeon liver that had been obtained by precipitation with cold ethanol between 13 and 33% were dissolved in 230 ml of 0.1 M sodium phosphate buffer, pH 7.4, at 3°.[7] This protein solution was then further fractionated with ammonium sulfate between 45 and 60% saturation. This fraction contained AIR synthetase and could be used in the assay of FGAR amidotransferase. It was dissolved in 0.03 M potassium phosphate buffer, pH 7.4, to a concentration of 10 mg/ml. The enzyme remained active for 6 weeks or more at 2° in this solution. Best results are obtained, however, with recently diluted solutions of enzyme obtained from freshly prepared acetone powder.

FGAR. The preparation of this compound has been described in *Advances in Enzymology,* Vol. 6 [96][8] from a paper by Hartman *et al.*[9]

Reaction Mixture
INCUBATION MEDIUM
   Tris-Cl buffer, pH 8.0 (10 mM)
   L-glutamine, 1.33 mM

[3] T. C. French, I. B. Dawid, R. A. Day, and J. M. Buchanan, *J. Biol. Chem.* **238**, 2171 (1963).
[4] K. Mizobuchi and J. M. Buchanan, *J. Biol. Chem.* **243**, 4842 (1968).
[5] H.-C. Li and J. M. Buchanan, *J. Biol. Chem.* **246**, 4713 (1971).
[6] A. C. Bratton and E. K. Marshall, Jr., *J. Biol. Chem.* **128**, 537 (1939).
[7] B. Levenberg and J. M. Buchanan, *J. Biol. Chem.* **224**, 1019 (1957).
[8] L. N. Lukens and J. G. Flaks, *in* "Methods in Enzymology" (S. P. Colowick and N. O. Kaplan, eds.), Vol. 6, Artic. 96, p. 671. Academic Press, New York, 1963.
[9] S. C. Hartman, B. Levenberg, and J. M. Buchanan, *J. Biol. Chem.* **221**, 1057 (1956).

FGAR, 0.47 mM
ATP, 13.3 mM
MgCl$_2$, 13.3 mM
KCl, 20.0 mM
AIR synthetase, 0.04 ml
ASSAY REAGENTS
    Sodium nitrite, 0.1%
    Ammonium sulfamate, 0.05%
    N-1-(naphthyl)-ethylene diamine dihydrochloride, 0.1%
    Trichloroacetic acid, 20%, adjusted to pH 1.4 in 1.33 M potassium
    phosphate

Procedure. The reaction mixture (0.3 ml) of the above composition was incubated with the enzyme sample at 38° for 15 min and was then chilled in an ice bath. In the control, the enzyme was omitted. One-tenth milliliter of 20% trichloroacetic acid solution, adjusted to pH 1.4 in 1.33 M potassium phosphate, was added to stop the reaction, and the protein precipitate was removed by centrifugation at room temperature. The following reagents were mixed at room temperature with the supernatant solution without disturbing the protein pellet: 0.05 ml of 0.1% sodium nitrite; after 3 min, 0.05 ml of 0.05% ammonium sulfamate; and after 3 min, 0.05 ml of 0.1% N-1-(naphthyl)ethylene diamine dihydrochloride. The mixture was then allowed to stand for at least 10 min for the complete development of color at room temperature, and the absorbance at 500 nm was read with a microcell attachment of a Beckman DU spectrophotometer. Under these conditions the molecular absorbancy of the colored derivative was 24,600 M^{-1} cm^{-1}. It should be noted that the ratio of concentration of ATP to that of MgCl$_2$ is important for the performance of the assay system.[4] As the ratio of ATP to MgCl$_2$ increased beyond one, ATP inhibited the enzyme strongly. However, a concentration of MgCl$_2$ in excess of that of ATP was not particularly harmful to the best performance of the assay system.

Assay for Glutamate Production

Principle. The amount of glutamate produced in the enzymic reaction was determined by step by step coupling with the reduction of the 3-acetylpyridine analogue of NAD in the presence of excess glutamine dehydrogenase.

Reagents
INCUBATION MEDIUM
    Tris-Cl, pH 8, 20 mM
    KCl, 60 mM

MgCl$_2$, 4 mM
ATP, 14 mM
FGAR, 0.5 mM
Glutamine, 1.4 mM
ASSAY REAGENTS
3- Acetylpyridine analogue of NAD (Schwarz Bioresearch)
Crystalline glutamic dehydrogenase (Sigma)

Procedure. Added to the reaction mixture (1 ml) of the above composition was a suitable amount of FGAR amidotransferase in a negligible volume. The rate of glutamate formation was measured as follows. The reaction was terminated by immersion of the incubation vessels in boiling water for 45 sec, and the vessels were then immediately cooled to 4° in ice. Four μmoles of the 3-acetylpyridine analogue of NAD and 1.4 units of crystalline glutamic dehydrogenase were then added to the reaction mixture, which was incubated at 37° for 1.5 hr to allow complete oxidation of glutamate. At the end of the incubation the reaction mixture was centrifuged twice in a Sorvall centrifuge at 10,000 g to remove any debris in the solution. The absorption of the clear supernatant solution was measured at 363 nm in a Gilford spectrophotometer to determine the concentration of the reduced NAD analogue (ϵ_{363} = 8900) and hence the amount of glutamate formed. Under these conditions of the step by step coupling with the indicator enzyme a linear relationship between enzyme concentration and glutamate production could be obtained. In an assay in which the glutamate generated in the primary reaction is directly coupled with the reduction of the analogue of NAD in the presence of excess glutamic dehydrogenase, a lag phase seen during the first 10 min of incubation obscured the linear relationship.

Assay for ADP Production

Principle. In this assay the ADP produced during FGAM synthesis is measured by the procedure of Kornberg and Pricer.[10] The conversion of ADP to ATP by reaction with phosphoenol pyruvate in the presence of pyruvic kinase yields pyruvate, which in turn is reduced by NADH in the reaction catalyzed by lactic dehydrogenase. The reaction is followed spectrophotometrically by measurement of the change in absorbance at 340 nm (ϵ_{340} = 6220). Both pyruvic kinase and lactic dehydrogenase (ammonium sulfate-free) were added in large excess so that the reaction rate was dependent solely on the amounts of amidotransferase added.

[10] A. Kornberg and W. E. Pricer, Jr., *J. Biol. Chem.* **193**, 481 (1951).

Reagents
INCUBATION MIXTURE
  Tris-Cl, pH 8.0, 20 mM
  KCl, 60 mM
  FGAR, 0.43 mM
  ATP, 14 mM
  MgCl$_2$, 4 mM
  Glutamine, 1.4 mM
  Phosphoenolpyruvate, 1 mM
  NADH, 0.35 mM (Sigma)
  Pyruvic kinase, 50 μg (Sigma)
  Lactic dehydrogenase, 100 μg (Sigma)

Procedure. The reaction mixture (1 ml) containing the above ingredients was allowed to equilibrate at 37° for 10 min, during which time traces of ADP and pyruvate in the reactants were consumed. The reaction was initiated by the addition of FGAR amidotransferase. The initial velocity is expressed in terms of nmoles formed per minute.

Units. A standard unit of enzyme is that amount which transforms one μmole of substrate per minute under the assay conditions. This definition applies to all three methods of assay. In some publications, however, a unit has been defined as that amount which causes the excursion of 0.1 of an absorbance at 500 nm per 20-min incubation in the colorimetric assay. One such unit is equivalent to 2.9×10^{-4} standard units.

Purification of Enzyme from Chicken Liver

The enzyme has been isolated and purified (see the table) by a procedure developed by Mizobuchi and Buchanan[4] as modified by Ohnoki, Hong, and Buchanan.[11]

Composition of Buffers. All buffer solutions contained $5 \times 10^{-4} M$ L-glutamine, $1 \times 10^{-3} M$ ethylenediamine tetraacetate, and $1 \times 10^{-4} M$ dithiothreitol.

  Buffer 1: 0.01 KPO$_4$, pH 7.2
  Buffer 2: 0.002 M KPO$_4$, pH 7.2, containing 10% glycerol
  Buffer 3: 0.1 M KPO$_4$, pH 7.2, containing 10% glycerol
  Buffer 4: 0.002 M KPO$_4$, pH 6.5, containing 10% glycerol
  Buffer 5: 0.05 M KPO$_4$, pH 6.5, containing 10% glycerol

[11] S. Ohnoki, B.-S. Hong, and J. M. Buchanan, *Biochemistry* **16**, 1065 (1977).

PURIFICATION OF FGAR AMIDOTRANSFERASE

Fraction	Volume (ml)	Protein concentration (mg/ml)	Total protein (mg)	Specific activity[a] (unit/mg)
1. Acetone powder extract	2720	20.6	55,900	1.69
2. Ammonium sulfate precipitate	400	17.6	7040	12.5
3. Heat treatment	390	16.5	6420	11.1
4. Ammonium sulfate ppt.	35	108.9	3810	15.3
5. Sephadex G200	246	5.4	1323	36.9
6. DEAE-cellulose	440	0.73	321	116.5
7. $(Ca)_3(PO_4)_2$ gel	344	0.17	57	396.6

[a] Units are expressed in terms of the amount of enzyme which causes the excursion of 0.1 of an absorbance at 500 nm per 20-min incubation in the colorimetric assay. The value for the specific activity may depend on the freshness of the preparation of AIR synthetase.

Preparation of Acetone Powder. Fresh chicken livers (4.5 kg) obtained from birds that had been killed by decapitation (rather than sterilization) were ground with a large meat grinder and mixed with 19 liters of acetone previously cooled to $-20°$.[4] The mixture was homogenized in a Waring Blendor for 45 sec at top speed. The homogenate was allowed to stand at $-20°$ for about 1 hr to allow for partial sedimentation of solids. This procedure facilitated collection of the solid materials by vacuum filtration. The upper portion of the mixture was first poured onto a series of large Büchner funnels at room temperature, and, after most of the liquid had been removed by filtration at the water pump, the lower portion of the partially sedimented mixture was added to the filter cakes. The filter cakes were suspended directly on the funnels, agitated by manual stirring, and washed with cold acetone (total of 55 liters/4.5 kg of liver) until the filtrate became colorless. Anhydrous ether, previously cooled to $-20°$, was then added to the residues on the funnels, and the residues were sucked dry. During these operations the temperature of the residue was maintained below $0°$. The slightly moist residue was crumbled by hand, and ether and residual water were removed over P_2O_5 in a vacuum desiccator by application of high vacuum for several hours in the cold room. Approximately 900–1000 g of the powder were obtained from 4.5 kg of chicken liver. The powder was stored at $-20°$ under reduced pressure until used. Under these conditions, the enzyme was stable for at least 3 months.

Step 1. Extraction of Acetone Powder. Acetone powder (100 g) prepared from chicken liver was suspended in 1.5 liters of extraction buffer (buffer 1). The suspension was mixed with a magnetic stirrer for 40 min and centrifuged at about 5000 g (top speed of the Stock centrifuge) for 20 min. The supernate (fraction I) was used for the next step.

Step 2. Fractionation with Ammonium Sulfate between 33% and 45% Saturation. After the volume of the extract had been adjusted to 1.5 liters by further addition of buffer 1, 294 g of solid ammonium sulfate (33% saturation) were slowly added to the enzyme solution with stirring over a period of 90–100 min. The solution was centrifuged at about 5000 g for 20 min, and the supernate was retained. Solid ammonium sulfate (74 g/liter) was added slowly to the supernatant solution with stirring for 30 min. The precipitate was collected by centrifugation at 5000 g for 20 min and dissolved in 250 ml of buffer 1 (fraction II).

Step 3. Heat Treatment. The solution (fraction II) was placed in a water bath of 45° and was stirred mechanically at this temperature for 15 min. It was then rapidly cooled to below 15° in an ice bath and centrifuged at about 8000–13,000 g for 20 min in a Sorvall centrifuge. The supernatant fraction (fraction III) was decanted and used in the next step.

Step 4. Fractionation with Ammonium Sulfate at 40% Saturation. After the volume of fraction III has been adjusted to 250 ml by addition of buffer 1, 60.5 g of ammonium sulfate (40% saturation) were slowly added to the enzyme solution with stirring for 20 min. The precipitate was collected by centrifugation at 13,000 g for 20 min in the Sorvall centrifuge and then suspended in buffer 2, which was used to equilibrate the Sephadex G-200 column used in the next step. The volume should be kept as small as possible (fraction IV).

Step 5. Chromatrography on Sephadex G-200. Then fraction IV was applied to a Sephadex G-200 column (5.0 × 80 cm), which had been equilibrated with buffer 2. The flow rate was 24 ml/hr, and 10 ml were collected per tube. FGAR amidotransferase activity appeared between tubes 80–100 (fraction V).

Step 6. Chromatography on DEAE-Cellulose. A DEAE-cellulose column (4.5 × 50 cm) was equilibrated with buffer 2. After the application of the enzyme solution (fraction V) the column was washed with 2 bed volumes of buffer 2. Most of the contaminant was eluted before starting the gradient which was composed of 2 liters of buffer 2 and 2 liters of buffer 3. The active fractions were collected and concentrated by ultrafiltration through a PM-30 membrane (Amicon). The concentrated solution (about 30 ml) was dialyzed overnight against

100 volumes of buffer 4. The resolution achieved by a newly prepared DEAE-cellulose column is better than that obtained on a column that has been regenerated several times.

Step 7. Chromatography on Calcium Phosphate Gel. A calcium phosphate gel column (3.4 × 25 cm) prepared according to Keilin and Hartree[12] was equilibrated with buffer 4. The enzyme was applied to the column which then was washed with 2 bed volumes of buffer 4 before starting the elution of protein with a gradient composed of equal volumes of buffers 4 and 5. Fractions containing active enzyme were collected and concentrated by use of an Amicon membrane PM-30.

Purification of Enzyme from *Salmonella typhimurium*

Procedure. A method of purification of the enzyme to homogeneity has been adequately described in a previous volume of *Methods in Enzymology.*[13] Since then a paper has appeared[3] including not only the details of this method but also an alternate procedure that is more reliable but does not yield as pure a product. In the second procedure a step is introduced in which the extract of an acetone powder is treated with RNase at 45° for 20 min. After precipitation of the enzyme with ammonium sulfate and application of the dissolved precipitate on a column of DEAE-cellulose, the entire activity was obtained in the eluant of one peak in contrast to the first procedure where as many as three peaks of activity have been observed at this stage.

Properties

Molecular Weight and Molecular Activity. FGAR amidotransferase isolated from either source has a molecular weight of approximately 133,000.[3,4] Upon standing in the absence of thiol reagents the enzyme aggregates by formation of disulfide bonds.[14] Disaggregation occurs upon treatment of the enzyme with thiol reagents. The observation that enzyme may be found in more than one form may be attributed at least in part to its capacity to undergo aggregation without loss of enzyme activity. So far no subunits have been conclusively demonstrated in spite of the fact that separate active sites have been recognized for

---

[12] D. Keilin and E. F. Hartree, *Proc. R. Soc. London* **124**, 397 (1938).
[13] J. G. Flaks and L. N. Lukens, this series, Vol. VI [9], p. 52.
[14] J-M. Frère, D. D. Schroeder, and J. M. Buchanan, *J. Biol. Chem.* **246**, 4727 (1971).

glutamine and for an FGAR-ATP-Mg^{2+} complex.[15,16] The enzyme has a molecular activity of 100–120 units per μmole.[3,4]

Kinetics. As studied with the chicken liver enzyme, the K_m for FGAR is $1 \times 10^{-4}\,M$,[4] for ATP-Mg^{2+}, $4 \times 10^{-4}\,M$,[5] for glutamine $2 \times 10^{-4}\,M$,[2,4] and for ammonium chloride at pH 8.0, 8–$10 \times 10^{-2}\,M$.[3] γ Glutamyl derivatives, such as γ glutamylhydroxamate, γ glutamyl hydrazine, γ glutamyl esters, and γ glutamylthioesters, are hydrolyzed, with a strict dependence on Mg^{2+}, ATP, and FGAR.[5] The K_m of the latter varies with the individual substrates. In some cases the amount of γ glutamyl substrate hydrolyzed exceeds the amount of FGAR present.

From initial rate studies of the reaction of the three substrates of the enzyme system, it was found that the substrates add sequentially to form a quaternary complex with the enzyme before undergoing reaction with release of products.[17] Glutamine adds first to the enzyme with ATP and FGAR adding randomly thereafter (partially compulsory order mechanism).

Inhibitors. The enzyme is inhibited for reaction with glutamine by several compounds including L-azaserine, 6-diazo-5-oxo-L-norleucine, L-albizziin, iodoacetate, iodoacetamide, hydroxylamine, and cyanate.[18] The first three are competitive analogues of glutamine and bind irreversibly at the glutamine active site. By labeling the enzyme with radioactive antimetabolites of glutamine or by selective reaction with labeled iodoacetate it has been possible to identify the reactive residue at the glutamine site as a cysteinyl residue.[19,20] The amino acid composition around this site as determined in part with the bacterial and chicken liver enzymes is

where the asterisk is the residue reactive with inhibitors.

[15] K. Mizobuchi and J. M. Buchanan, *J. Biol. Chem.* **243**, 4853 (1968).
[16] K. Mizobuchi, G. L. Kenyon, and J. M. Buchanan, *J. Biol. Chem.* **243**, 4863 (1968).
[17] H.-C. Li and J. M. Buchanan, *J. Biol. Chem.* **246**, 4720 (1971).
[18] D. D. Schroeder, A. J. Allison, and J. M. Buchanan, *J. Biol. Chem.* **244**, 5856 (1969).
[19] I. B. Dawid, T. C. French, and J. M. Buchanan, *J. Biol. Chem.* **238**, 2178 (1963).
[20] S. Ohnoki, B.-S. Hong, and J. M. Buchanan, *Biochemistry* **16**, 1070 (1977).

[26] Adenylosuccinate AMP-Lyase (*Neurospora crassa*)

By Dow O. Woodward

Introduction

The enzyme adenylosuccinate AMP-lyase (previously referred to in the literature as adenylosuccinase), EC 4.3.2.2, from *Neurospora crassa* is a bifunctional enzyme and catalyzes two nonsequential reactions in purine biosynthesis. The early reaction is the conversion of 5-amino-4-imidazole-*N*-succinocarboxamide ribonucleotide (succino-AICAR) to 5-amino-4-imidazolecarboxamide ribonucleotide (AICAR) and fumaric acid (see Fig. 1).[1]

The other reaction is the terminal step in AMP biosynthesis, the cleaving of adenylosuccinate (AMP-S) to adenylic acid (AMP) and fumaric acid (see Fig. 2). The two reactions are similar in that they both involve the breakage of a carbon-nitrogen bond to yield fumaric acid. Among a large number of mutants with mutationally altered forms of adenylosuccinate AMP-lyase, an alteration in the early reaction was also observed.[2] Mutants in which one activity was completely missing always failed to show activity for the other reaction. No cofactor requirement has been demonstrated for either reaction.

Assay Method

Principle. The principal assay of adenylosuccinate-AMP-lyase is a spectrophotometric assay which takes advantage of the slightly different absorption spectra of AMP-S[3] and AMP. The absorption maximum of AMP-S is at 267 nm, and that of AMP is near 260 nm. The maximum change in absorbance occurs at 280 nm as the reaction proceeds from AMP-S to AMP; a change in optical density of 10.7/nmole/ml. Thus the assay is a measure of decrease in absorbance at 280 nm in a double-beam spectrophotometer.[4]

---

[1] R. W. Miller, L. N. Lukens, and J. M. Buchanan, *J. Biol. Chem.* **234**, 1806 (1959).

[2] D. O. Woodward, C. W. H. Partridge, and N. H. Giles, *Genetics* **45**, 535 (1959).

[3] AMP-S is prepared by enzymic synthesis. C. E. Carter and L. H. Cohen, *J. Biol. Chem.* **222**, 17 (1956).

[4] N. H. Giles, C. W. H. Partridge, and N. J. Nelson, *Proc. Natl. Acad. Sci. U.S.A.* **43**, 305 (1957).

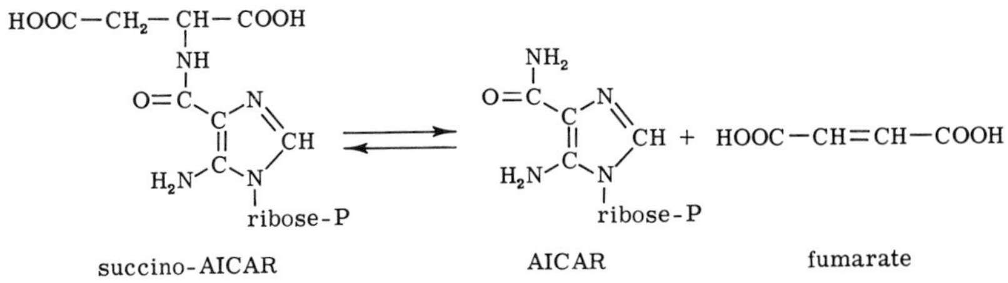

HOOC—CH₂—CH—COOH
NH
O=C
C
N
CH
H₂N
C
N
ribose-P
NH₂
O=C
C
N
CH + HOOC—CH=CH—COOH
H₂N
C
N
ribose-P
succino-AICAR
AICAR
fumarate

FIG. 1

Reagents

Sample cell: 0.05 *M* Tris·HCl buffer, pH 8.0, containing enzyme extract (can be crude or purified) and a final concentration of AMP-S at 30 μg/ml

Reference cell: 0.05 *M* Tris·HCl buffer, pH 8.0, containing the same amount of enzyme extract as the sample cell

The assay cannot be performed easily by substituting AMP-S in the reference cell in place of the enzyme extract (especially in crude extracts) because of the high 280 nm absorbance of the extract, the same wavelength as utilized in the assay.

Definition of Unit Activity. One unit of activity is defined by the disappearance of 1 μmole of adenylosuccinic acid per milligram of protein per minute.

Alternative Assay Procedure. A qualitative assay can be performed by separating AMP and AMP-S by paper chromatography[5] after completion of the reaction.

The 5-amino-4-imidazole-*N*-succinocarboxamide ribonucleotide lyase activity is measured by a change in absorbance at 267 nm.

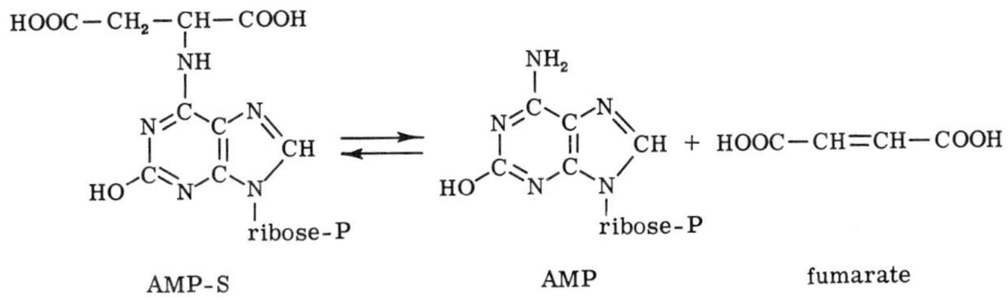

HOOC—CH₂—CH—COOH
NH
N
C
C
N
CH
HO
C
N
C
N
ribose-P
NH₂
CH + HOOC—CH=CH—COOH
HO
ribose-P
AMP-S
AMP
fumarate

FIG. 2

[5] C. E. Carter and L. H. Cohen, *J. Biol. Chem.* **222,** 17 (1956).

Preparation of Adenylosuccinate AMP-lyase[6]

A convenient way to extract adenylosuccinate AMP-lyase from *Neurospora* is by harvesting mycelium, drying it in a frozen state, grinding the dried mycelium to a fine powder, and extracting by shaking in buffer (Tris·HCl, 0.05 M, pH 8.0, or phosphate, 0.05 M, pH 7.0) for 1 hr. Centrifugation to remove visible sediment leaves a crude extract that is suitable for assay or as the starting material for the purification procedure outlined below (see the table). All operations should be carried out at 0–4°.

Extraction. One kilogram of powdered mycelium is stirred into 20 liters of 0.05 M Tris·HCl buffer, pH 8.0, for 1 hr. This mixture is filtered (e.g., with cheesecloth) to remove cell debris.

Treatment with Manganous Chloride. The supernatant from the filtration is stirred while 400 ml of 1 M MnCl$_2$ are being added. After 15 min the suspension is centrifuged at 12,500 g for 15 min. The sediment is discarded.

Ammonium Sulfate Fractionation. To each 100 ml of supernatant remaining, 28.5 g (46% saturation) of ammonium sulfate are added while stirring. After an additional 15-min stirring, the material is centrifuged at 12,500 g for 20 min. The sediment is discarded.

PURIFICATION OF ADENYLOSUCCINATE AMP-LYASE

Purification procedure	Total protein (g)	Specific activity[a] (Δ μmoles/ mg/min)	Total units $\times 10^{-5}$	Yield (%)	Purification (-fold)
1. Crude	500	9.5	46.4	100	1.0
2. MnCl$_2$	430	10.6	45.5	98	1.1
3. Ammonium sulfate	78	34.2	26.8	58	3.0
4. Hydroxylapatite	3.2	456	14.5	31	48.0
5. DEAE-cellulose (pH gradient)	0.15	4,260	6.4	14	450
6. Second DEAE-cellulose (pH gradient)	0.025	16,080	3.9	8.5	1700
7. DEAE-cellulose (phosphate gradient)	0.011	18,700	2.1	4.5	1970

[a] One unit of activity is defined by the disappearance of 1 μmole of adenylosuccinic acid per milligram of protein per minute.

[6] D. O. Woodward and H. D. Braymer, *J. Biol. Chem.* **241**, 580–587 (1966).

To each 100 ml of this supernatant fluid, 8.5 g of ammonium sulfate (58% saturation) are added while stirring. Stirring is continued for 15 min, and then the mixture is centrifuged at 12,500 g for 20 min. The supernatant fluid is discarded. The sediment can be stored at 4° at this stage as an ammonium sulfate suspension for several months with little loss of adenylosuccinate AMP-lyase activity.

Hydroxylapatite Column Chromatography. A column containing hydroxylapatite (Hypatite-C) with a bed volume of 196 cm³ is used for the next fractionation step. The ammonium sulfate precipitate (58% saturation) is dialyzed 6–8 hr against 0.0125 M Tris·HCl, pH 8.0. The hydroxylapatite column is washed first with 0.005 M phosphate buffer, pH 6.8. Three to four grams of protein can be applied to a column this size. After adsorption to the column, the enzyme is eluted with a phosphate buffer, pH 6.8, in a gradient ranging from 0.005–0.2 M.

DEAE-Cellulose Chromatography (pH Gradient). DEAE-cellulose mixed in a 1:1 ratio (w/w) is washed several times in 0.1 N NaOH containing 1 M NaCl in a Waring Blendor. It is subsequently washed several times in deionized water. The resin is then resuspended in a large volume of water and neutralized with HCl to pH 7. A column 2.5 cm in diameter is poured with a DEAE-cellulose bed volume of 90 cm.

Less than 1 g of protein is applied to a column this size after it has been dialyzed and centrifuged. Best results are obtained when it is applied to the column in the smallest volume possible. After the enzyme is adsorbed to the resin, a nonlinear gradient is used composed of 1 liter of 0.15 M KH₂PO₄, pH 4.5, running into a 500-ml mixing volume of 0.125 M Tris·HCl, pH 8.0, which feeds directly into the column. Both solutions contain 10^{-4} M β-mercaptoethanol.

If the adenylosuccinate AMP-lyase eluted is concentrated, dialyzed, and rerun on a similar but smaller (1 cm diameter, 40 cm bed volume) column, an additional 4-fold increase in specific activity can be obtained.

DEAE-Cellulose Chromatography (Phosphate Gradient). A DEAE-cellulose column prepared in the same way is used as a final step to remove small remaining amounts of foreign protein. After the dialyzed concentrated material eluted from the pH gradient is adsorbed to the column, a linear phosphate gradient is applied. One liter of 0.15 M phosphate buffer, pH 7.0, is connected to a 1-liter mixing volume containing 0.01 M phosphate, pH 7.0, which feeds directly into the column. The eluate containing adenylosuccinate AMP-lyase activity obtained from this column is estimated to contain less than 5% total foreign protein.

Properties[6,7]

Adenylosuccinate AMP-lyase has a pH optimum of 8.3 in 0.05 M Tris·HCl buffer, and its stability is higher in phosphate buffer at pH 8 than in Tris·HCl at pH 8 when stored at 0°. The K_m for this enzyme is 1.2×10^{-6} M. The molecular weight is estimated at 200,000, being composed of six to eight identical monomers.

In vitro, the enzyme exists in three forms that can be separated by ultracentrifugation or by moving boundary electrophoresis. The 10 S form and the 4.0 S form are readily interconvertible by changes in hydrogen ion concentration. Once the 3.0 S form has been produced by high hydrogen ion concentrations, it appears quite stable, and no way has been found to assemble it into either the 4.0 S or 10 S forms. A pH near 5.2 is optimal for the equilibrium between 4.0 S and 10 S to be shifted almost completely to the 10 S form. Raising the pH shifts the equilibrium to the 4.0 S form, and lowering it irreversibly converts it to the 3.0 S form. The adenylosuccinate AMP-lyase activity and the 5-amino-4-imidazole-*N*-succinocarboxamide ribonucleotide lyase activity are observed only in the 10 S form which predominates at pH 5.2, whereas the maximum activity observed is at pH 8.3.

At pH 8.3 under *in vitro* conditions, the ratio of 10 S to 4 S is about 0.7, and an adequate explanation for this apparent discrepancy has not been found.

Other Properties

One of the properties of *Neurospora* adenylsuccinate AMP-lyase that is of special interest is related to the equilibrium between the subunits described above. When this property is dealt with in the context of mutationally altered forms of the enzyme, the phenomenon termed interallelic or intragenic complementation can be observed. This simply means that two differentially altered forms of the enzyme can exchange subunits,[7,8] and the resulting "hybrid" enzyme is capable of functioning both *in vivo* and *in vitro*. The optimal conditions for producing the hybrid form of the enzyme are those conditions that generally tend to dissociate and reassociate the subunits, e.g., pH shifting or sulfhydryl reagents.[8] *In vivo* it is not known how the hybrid forms of the enzyme are produced, but the process is generally more efficient than the *in vitro* processes that have been used. The type of mutational alteration in the

[7] C. W. H. Partridge and N. H. Giles, *Nature (London)* **199**, 304 (1963); C. W. H. Partridge, *Biochem. Biophys. Res. Commun.* **3**, 613 (1960).

[8] D. O. Woodward, *Proc. Natl. Acad. Sci. U.S.A.* **45**, 846 (1959); *Q. Rev. Biol.* **35**, 313 (1960).

enzyme, and presumably the location, influence its ability to complement another enzyme. The pattern of these interactions is an ordered
pattern and has been termed a complementation map.[9]

[9] D. O. Woodward, C. W. H. Partridge, and N. H. Giles, *Proc. Natl. Acad. Sci. U.S.A.*
44, 1237 (1958).

[27] Adenylosuccinate Synthetase (Rabbit Muscle, Heart, and Liver)

By Harald E. Fischer, Katherine M. Muirhead, and Stephen H. Bishop

Adenylosuccinate synthetase (IMP: L-aspartate ligase [GDP], EC 6.3.4.4)

$$\text{IMP} + \text{GTP} + \text{L-aspartate} \xrightarrow{\text{Mg}^{2+}} \text{Adenylosuccinate} + \text{GDP} + \text{P}_i$$

Assay

Principle. The adenylosuccinate produced is assayed by recording
the initial rate of absorbance change at 280 nm, as the 6 position of the
purine ring is *N*-substituted.[1-3] The assay is carried out at room
temperature (22°) at pH 7.0 (about the pH maximum for the enzyme
activity).

Reagents
   IMP(Na)$_2$, 40 mM
   GTP(Li)$_4$, 10 mM
   L-aspartate, pH 7.0, 0.4 M
   Magnesium acetate, 0.4 M
   5 mM sodium phosphate buffer, pH 6.80
   N-2-hydroxyethyl-piperazine-*N*-2-ethane sulfonic acid (HEPES)
   0.1 M DL-dithiothreitol (DTT) stock solution
   0.5 M potassium phosphate buffer, pH 6.80

Procedure. Two milliliters of 50 mM HEPES buffer (pH 7.0) are
used in a 3-ml quartz cuvette to which 10 μl of 40 mM IMP, 10 μl of 10
mM GTP, 20 μl of 0.4 M magnesium-acetate, and an appropriate amount
of enzyme (between 5–25 μl) are added. The solution is thoroughly
mixed by inversion of the cuvette several times. The absorbance offset

[1] K. M. Muirhead and S. H. Bishop, *J. Biol. Chem.* **249**, 459–464 (1974).
[2] I. Lieberman, *J. Biol. Chem.* **223**, 327–339 (1956).
[3] F. B. Rudolph, Ph.D. Thesis, Iowa State University, Ames (1971).

is then zeroed on the recorder, and 20 μl of 0.4M L-asparate are added to start the reaction; the solution is quickly mixed and the cuvette put into the light path. The change in absorbance at 280 nm is recorded over 2–5 min (chart speed = ¼ inch/min and full-scale absorbance = 0.100). The reaction should be linear for this time period and should correspond to the amount of enzyme used. The change in absorbance from 0.01 to 0.100 can be measured accurately on most recording spectrophotometers with scale expansion. One unit of enzyme activity is defined as the amount of enzyme giving an increase in absorbance of 1.0 per minute at 280 nm (ΔE_{280} = 11.7 × 10^3 M^{-1} cm^{-1}). The specific activity is expressed as units of activity per milligram of protein. Protein concentrations were measured using 1.55 (A_{280}) − 0.76 (A_{260}) according to Layne.[4]

Purification Method

Preparation of Tissue Powder. Frozen young rabbit skeletal muscle, liver, or heart tissue (500 g) (Pel-Freez, Rogers, Arkansas) are thawed slowly. The connective tissue is removed, and the tissue is cut into small pieces with scissors and then homogenized in a Waring Blendor at medium speed for 45 sec in 3 liters of cold acetone (−20°). The mixture is filtered through a Büchner funnel (Whatman No. 54 filter paper) using a water-pump aspirator. The tissue cake is immediately resuspended in the same amount of cold acetone and again homogenized for 30 sec. After collection on the Büchner funnel the cake of tissue is spread thinly on Whatman 3-mm filter-paper sheets and air dried (room temperature) for 2 hr. The resulting powder (100–120 g) can be stored without any major enzyme activity loss for several months in an air-tight bottle at −20°.

Enzyme Extraction. All following steps of the enzyme purification procedure are performed at 0–4°.

Sixty-six grams of skeletal tissue powder (50 g of liver or heart tissue powder) are slowly added to 10 volumes (10 ml/g of powder) of a 50 mM KCl–1 mM DTT solution. All lumps should be broken so as to prepare a homogeneous mixture. After stirring for 3 hr, the tissue slurry is then centrifuged in a refrigerated centrifuge for 20 min at 12,000 g. The pellet is discarded and the supernatant fluid (536 ml) (420 ml for heart or liver preparations) saved. A small amount (1 ml) is dialyzed against 50 mM KCl–1 mM DTT for the enzyme activity assay and for the protein determination (using the 280/260 nm method). The dialysis removes purine nucleotides and other low-molecular-weight UV- absorbing compounds.

[4] See E. Layne, this series, Vol. IV [73].

Heat Step. The supernatant fluid (heart and skeletal muscle extracts) is divided into ~200-ml portions and heated in 500-ml Erlenmeyer flasks. The flasks are immersed in a 75–80° water bath with constant stirring in order to raise the temperature of the extract rapidly to 60°. After maintaining the temperature at 60° for 1 min, the solution is cooled immediately in ice water with continuous rotation of the flasks and then centrifuged for 20 min at 12,000 g. (For liver enzyme purification the extract is raised to 52° and held for 1 min at 52°.) The resulting whitish pellet is discarded and the supernatant kept. For a more precise protein determination and enzyme assay a quick dialysis (4–5 hr) of 1 ml of the supernatant solution is recommended. The preparations yield about 500 ml and 400 ml for the skeletal muscle and for the heart or liver preparations, respectively.

Ammonium Sulfate Fraction. The supernatant fluid from the heat step is then gently stirred with a magnetic stirrer, and 22.6 g of powdered $(NH_4)_2SO_4$ (Enzyme grade, Schwartz-Mann) are added for each 100 ml. After stirring for an additional 30 min, the cloudy reddish solution is centrifuged for 20 min at 12,000 g, and the pellet is discarded. Powdered $(NH_4)_2SO_4$ is slowly stirred into the supernatant solution (16.5 g/100 ml). After gentle stirring for another 30 min, the pellet is collected by centrifugation and the supernatant discarded. This pellet is then dissolved in a minimal amount of 5 mM sodium phosphate–1 mM DTT (pH 6.8) buffer and the solution dialyzed twice for 10 hr against 10 volumes of the same buffer.

Phosphocellulose Chromatography. Phosphocellulose (Sigma Chem. Co., St. Louis, Mo.) is prepared by washing with 0.5 M HCl, then 0.5 M NaOH, then titrated to pH 6.8 with 10% H_3PO_4. After removal of the fines with water washes and equilibration with 5 mM sodium phosphate–1 mM DTT (pH 6.8), the phosphocellulose suspension is degassed and poured into a glass column (1.5 × 26 cm) to form a bed volume of 45–50 ml. The 20–30 ml of dialyzed enzyme solution are then passed through the column, and the column is washed with 100 ml of the same buffer to elute some inactive protein. A linear gradient of 250 ml of the above buffer and 250 ml of 102.5 mM sodium phosphate–1 mM DTT (pH 6.8) is applied with a flow rate of 40–50 ml/hr, and fractions of 4.5 to 5 ml are collected. The enzyme peak elutes at about 60–70 mM sodium phosphate (pH 6.8) as a shoulder of a larger inactive protein peak. The most active fractions (specific activity 1.5 times higher than that of the enzyme solution added on the column) are then combined. The skeletal muscle and the heart or liver preparations are about 350 ml and 170–180 ml, respectively.

Hydroxylapatite Chromatography. Bio-gel HTP (Bio-Rad) powder (5 g) is washed repeatedly with degassed 75 mM sodium phosphate–1 mM DTT (pH 6.8) buffer; the fines are carefully decanted, and the slurry is then poured in a column (12 × 1.5 cm, bed volume, 15–20 ml). After washing with 50 ml of this 75 mM buffer the enzyme solution is applied and subsequently the column is washed with 30 ml of the degassed buffer. The enzyme is then eluted with a linear gradient of 100 ml of the same buffer and 100 ml of 0.5 M potassium phosphate–1 mM DTT buffer (pH 6.8). The flow rate should be between 10–15 ml/hr, and the fractions collected should be about 2 ml. The enzyme peak elutes at 125 mM on the high concentration side of a major inactive protein peak. The combined fractions should have a volume of 45–50 ml. The total activity recovered after this step should be about 85–90% of the applied enzyme activity.

Concentration of the Combined Fractions by (NH$_4$)$_2$SO$_4$ Precipitation. The combined fractions from the hydroxylapatite column (~45 ml) are placed in a dialysis bag and dialyzed against 10 volumes of a saturated (NH$_4$)$_2$SO$_4$ solution for 6–8 hr. The precipitated enzyme is collected by centrifugation (20 min/12,000 g), and the pellet is redissolved in 1.5–2.0 ml of 5 mM sodium phosphate–1 mM DTT–0.1 M NaCl buffer (pH 6.8).

Sephadex G-150 Gel Filtration. The G-150 Sephadex gel (Pharmacia Fine Chemicals) is prepared according to the Pharmacia manual, degassed, poured into a glass column (2 × 56 cm, bed volume 175 ml), and then equilibrated with 5 mM sodium phosphate–1 mM DTT–0.1 M NaCl degassed buffer (pH 6.8). The NaCl is included in the buffer to prevent nonspecific interaction of the enzyme with the Sephadex gel (results in a delayed elution of the enzyme). The concentrated enzyme solution (2–3 ml) is carefully applied on the column and eluted with the above buffer at a flow rate of 7–8 ml/hr. The collected fractions are usually 2 ml. The enzyme elutes as the second smaller protein peak at an elution volume of 1.8 times the V_0 as determined with Blue Dextran 2000 (Pharmacia). The combined fractions (15–18 ml) all have a specific activity of 6–7 and 3–4 units/mg protein for the skeletal muscle and the heart or liver preparations, respectively (see the table).

At this stage, the overall purification is 230- to 260-fold, and the final recovery of total activity is 20–25% of that in the original KCl extract.

Crystallization. Several milliliters of the purified G-150 fraction are dialyzed against 200 ml of 58% (NH$_4$)$_2$SO$_4$ solution, to which over 2 days powdered (NH$_4$)$_2$SO$_4$ is added gradually to result in 2% increases of the (NH$_4$)$_2$SO$_4$ solution up to a concentration of 65–67%. At this concentra-

PURIFICATION OF ADENYLOSUCCINATE SYNTHETASE FROM RABBIT SKELETAL MUSCLE
AND HEART

	Skeletal muscle[a]		Heart muscle[a]	
Step	Total activity (units)	Specific activity (units/mg protein)[b]	Total activity (units)	Specific activity (units/mg protein)[b]
Acetone powder extract	280	0.025	88	0.014
Heat step	290	0.055	80	0.017
Ammonium sulfate fraction	274	0.301	87	0.05
Phosphocellulose column	113	1.63	76	0.55
Hydroxylapatite column	79	3.50	27	1.8
Sephadex G-150	53	6.54	21	3.2

[a] In the purification for the two enzymes different amounts of acetone powder were used (skeletal muscle 67 g, heart 50 g). The isolation of the liver enzyme paralleled the heart enzyme with recoveries of 10% of the original activity.
[b] Using $\Delta E_{280} = 11.7 \times 10^3$, the final specific activities are 1.0 and 0.5 μmoles/min/mg protein.

tion diamond-shaped crystals are formed. These are centrifuged and recrystallized twice in the same manner. There is no apparent increase in specific activity with crystallization.

Purity and Stability. The purified and crystallized enzyme appears usually as a single band on a 7.5% SDS–polyacrylamide gel electrophoresis performed according to Weber et al.[5] It is stable for at least 6 months in saturated ammonium sulfate at 4° with only a minor decrease in activity; according to Spector and Miller,[6] it can be frozen and rethawed without activity loss.

Properties

Under denaturing conditions (SDS–polyacrylamide gel electrophoresis) the skeletal muscle and heart enzyme have similar molecular weights in the 53×10^3 to 54×10^3 MW range. The liver enzyme has a slightly lower molecular weight of 44×10^3 to 46×10^3. The native enzymes from these three different tissues show different molecular weights between 90×10^3 to 105×10^3 using gel filtration (Sephadex G-150). The three enzymes have a broad pH optimum between 6.6–7.0 and are

[5] See K. Weber, J. R. Pringle, and M. Osborn, this series, Vol. XXVI [1].
[6] T. Spector and R. L. Miller, *Biochim. Biophys. Acta* **445**, 509–517 (1976).

inactivated at low Zn^{2+} concentrations (K_i = 5–10 μM)[7] or by sulfhydryl group reagents.[1] Apparent K_m's for the substrates have been determined.[1,6] GDP, AMP, and fructose-1,6-diphosphate are inhibitors and considered potential physiological regulators because their K_i values approximate the concentration found in the various tissues.[1,8,9]

Anions of the Hofmeister lyotropic series inhibit strongly.[8,10] Although the inhibitory mechanism is uncertain, these anions tend to be competitive inhibitors of GTP binding. Acetate salts of buffers and reagents should be used in the reaction mixtures to avoid complications in kinetic or binding investigations.

Adenylosuccinate synthetase preparations with varying degrees of purity have been obtained from cell-free extracts of *Escherichia coli*,[2,3,11-17] *Bacillissubtilis*,[18] *Schizosaccharomyces pombe*,[19] wheat germ,[20] Ehrlich ascites tumor cells,[21] Novikoff ascites tumor cells,[22] and human placenta.[23]

Comment

The spectrophotometric assay may be sensitive to stray light effects which are dependent on the quality of the optical system of the spectrophotometer and the respective range of absorption.[24-26] This becomes especially important for inhibition studies with purine deriva-

[7] H. E. Fischer, J. M. Trujillo, and S. H. Bishop, *Fed. Proc., Fed. Am. Soc. Exp. Biol.* **36,** 715 (1977).

[8] S. H. Bishop, H. E. Fischer, K. L. Gibbs, and J. E. Stouffer, *Fed. Proc., Fed. Am. Soc. Exp. Biol.* **34,** 548 (1975).

[9] H. Ogawa, H. Shiraki, and H. Nakagawa, *Biochem. Biophys. Res. Commun.* **68,** 524–528 (1976).

[10] G. D. Markham and G. H. Reed, *FEBS Lett.* **54,** 266–268 (1975).

[11] C. E. Carter and L. H. Cohen, *J. Am. Chem. Soc.* **77,** 499–500 (1955).

[12] C. E. Carter and L. H. Cohen, *J. Biol. Chem.* **222,** 17–30 (1956).

[13] H. F. Fromm, *Biochim. Biophys. Acta* **29,** 255–262 (1958).

[14] R. W. Miller and J. M. Buchanan, *J. Biol. Chem.* **237,** 485–490 (1962).

[15] J. B. Wyngaarden and R. A. Greenland, *J. Biol. Chem.* **238,** 1054–1057 (1963).

[16] F. B. Rudolph and H. F. Fromm, *J. Biol. Chem.* **244,** 3832–3839 (1969).

[17] A. Hampton and S. Y. Chu, *Biochim. Biophys. Acta* **198,** 594–600 (1970).

[18] K. Ishii and I. Shiio, *J. Biochem. (Tokyo)* **68,** 171–178 (1970).

[19] M. Nagy, M. Djembo-Taty, and H. Heslot, *Biochim. Biophys. Acta* **309,** 1–10 (1973).

[20] M. D. Hatch, *Biochem. J.* **98,** 198–203 (1966).

[21] M. R. Atkinson, R. K. Morton, and A. W. Murray, *Biochem. J.* **92,** 398–404 (1964).

[22] S. W. Clark and F. B. Rudolph, *Biochim. Biophys. Acta* **437,** 87–93 (1976).

[23] M. B. van DerWeyden and W. N. Kelley, *J. Biol. Chem.* **249,** 7282–7289 (1974).

[24] R. B. Cook and R. Jankow, *J. Chem. Educ.* **49,** 405–408 (1972).

[25] R. L. Cavalieri and H. Z. Sable, *Anal. Biochem.* **59,** 122–128 (1974).

[26] J. Eyaguirre, *Biochem. Biophys. Res. Commun.* **60,** 35–41 (1974).

tives which absorb in the 280 nm region. Spector and Miller[6] used a coupled enzyme assay system with PEP, pyruvate kinase, and LDH and monitored the oxidation of NADH at 340 nm. This system simultaneously prevents the accumulation of GDP, a powerful inhibitor of the enzyme (K_i = 0.007 mM). Radiometric assays for the measurement of adenylosuccinate synthetase have also been developed.[2,6,23]

Acknowledgments

This work was supported by grants from the National Institutes of Health (HL 17372), the National Science Foundation (BMS 74-10433), and The Robert A. Welch Foundation. HEF was a postdoctoral fellow of the American Cancer Society (Texas Division, 1974–75).

[28] GMP Synthetase (*Escherichia coli*)[1]

By NAOTO SAKAMOTO

$$XMP + ATP + NH_3 \rightarrow GMP + AMP + PP_i \ ^2$$

$$XMP + ATP + glutamine \rightarrow GMP + AMP + PP_i + glutamate$$

Assay Method[3]

Principle.[4] The amount of GMP produced is estimated by measuring the increase in absorption at 290 nm of the reaction mixture deproteinized with perchloric acid. Under these conditions the molar extinction of GMP at 290 nm is 6.0×10^3, whereas that of XMP, ATP, and AMP, the other ultraviolet-absorbing components of the reaction mixture, is negligible.

Reagents
Tris·HCl buffer, 1.0 M, pH 8.5
ATP, 0.20 M, pH 7.0
MgCl$_2$, 0.40 M
XMP, 0.12 M

[1] Xanthosine 5'-phosphate:ammonia ligase (AMP-forming), EC 6.3.4.1.
[2] The abbreviations used are: XMP, xanthosine 5'-phosphate; ATP, adenosine 5'-triphosphate; GMP, guanosine 5'-phosphate; AMP, adenosine 5'-phosphate; PP$_i$, inorganic pyrophosphate.
[3] A slight modification of the method described by B. Magasanik, *in* "Methods in Enzymology" (S. P. Colowick and N. O. Kaplan, eds.), Vol. VI, p. 111. Academic Press, New York, 1963.
[4] H. S. Moyed and B. Magasanik, *J. Biol. Chem.* **226**, 351 (1957).

$(NH_4)_2SO_4$, 2.0 M

Perchloric acid, 3.5 %

Substrate solution: An assay uses 0.10 ml of the substrate solution consisting of 40 μl of Tris buffer, 5 μl of ATP, 10 μl of $MgCl_2$, 5 μl of XMP, and 40 μl of $(NH_4)_2SO_4$. The substrate solution is stored at $-20°$.

Procedure.[4] The reaction is started by the addition of 0.10 ml of the substrate solution to 0.15 ml of enzyme solution.[5] A similar mixture from which XMP has been omitted serves as control. The mixtures are incubated at 39° for 15 min. The reaction is stopped by the addition of 2.75 ml of 3.5% perchloric acid. The precipitated protein is removed by centrifugation, and the absorbance of the solution at 290 nm is determined in a quartz cuvette with a light path of 10 mm.

When purified enzyme preparations are used it is possible to follow XMP disappearance at pH 8.5 by measuring the decrease in absorption at 290 nm.[4,6]

Definition of Unit and Specific Activity. One unit of enzyme is defined as that amount of enzyme which catalyzes the formation of 1 μmole of GMP per minute under the above conditions. That is, 0.01 unit is equivalent to an increase in absorption of 0.30 at 290 nm.

Specific activity is expressed as units per milligram of protein. Protein concentrations are estimated spectrophotometrically by the method of Layne.[7] The protein concentration of the purified enzyme solutions can be determined by the absorbance at 280 nm using an extinction of 1.12 liter g^{-1} cm^{-1}.[8]

Purification Procedure[8]

Growth of Cells. The bacterial strain used is *Escherichia coli* B-96 (ATCC #13473), a purine-requiring mutant which lacks both inosinicase and transformylase. Derepressed synthesis of GMP synthetase is achieved in strain B-96 by aerobic growth at 37° for 6 hr in a mineral salts–glucose medium[9] supplemented with 2 mg/ml Bacto Difco casamino acids (vitamin free), 5 μg/ml thiamine, and 40 μg/ml AMP. The

[5] Initial concentrations in the reaction mixture are: Tris buffer, 0.15 M; ATP, 4 mM; $MgCl_2$, 16 mM; XMP, 2.4 mM; $(NH_4)_2SO_4$, 0.32 M; enzyme, 0.005–0.015 units.

[6] N. Zyk, N. Citri, and H. S. Moyed, *Biochemistry* **8**, 2787 (1969).

[7] E. Layne, this series, Vol. III, p. 447.

[8] N. Sakamoto, G. W. Hatfield, and H. S. Moyed, *J. Biol. Chem.* **247**, 5880 (1972).

[9] B. D. Davis and E. S. Mingioli, *J. Bacteriol.* **60**, 17 (1950).

yield of cells is approximately 1.7 g (wet weight) per liter of culture medium.

All subsequent operations in the purification procedure are carried out at 0–4°, and centrifugations are performed at 30,900 g for 30 min.

Step 1. Extraction. The cell paste from 100 liters of culture (approximately 170 g, wet weight) is suspended in 7.5 mM potassium phosphate (pH 7.4) to yield a final volume of approximately 300 ml. The cells are disrupted in a 300-ml Rosett cell[10] with a Bronwill Biosonik II sonicator at 20 kcycles for 11 min. The broken cell suspension is centrifuged, and the supernatant fluid (crude extract) is dialyzed in 8 liters of 7.5 mM potassium phosphate (pH 7.4) for 16 hr.

Step 2. Streptomycin Sulfate Precipitation. Nucleic acids are removed from the dialyzed crude extract by the slow addition, with stirring, of a neutralized 10% (w/v) solution of streptomycin sulfate to one-tenth of the volume of the extract. After 1 hr of additional stirring, the suspension is centrifuged, and the supernatant fluid is again treated in the same manner with a neutralized 5% (w/v) solution of streptomycin sulfate. After the centrifugation, the supernatant fluid is dialyzed in 8 liters of 7.5 mM potassium phosphate (pH 7.4) for 16 hr.

Step 3. Ammonium Sulfate Fractionation. The dialyzed extract from step 2 is brought to 51% of saturation with ammonium sulfate by the slow addition, with stirring, of an appropriate volume of a saturated solution of ammonium sulfate adjusted to pH 7.4 with ammonium hydroxide. After 1 hr of additional stirring, the precipitate is removed by centrifugation, and to the supernatant fluid an appropriate volume of a saturated solution of ammonium sulfate is added in the same manner to bring to 63% of saturation. The suspension is again stirred gently for 1 hr and then centrifuged. The precipitate is dissolved in 30 ml of 20 mM potassium phosphate (pH 7.4) and dialyzed in 4 liters of the same buffer for 16 hr.

Step 4. First DEAE-Sephadex Chromatography. Two batches of the dialyzed ammonium sulfate fraction are combined (to give approximately 4150 mg of protein) and dialyzed in 1 liter of 20 mM potassium phosphate (pH 7.4) containing 0.2 M potassium chloride, 0.1 mM dithiothreitol, and 0.1 mM EDTA for 16 hr before application to a DEAE-Sephadex column. The dialyzed fraction is applied to a 5× 85 cm column of DEAE-Sephadex A-50 (medium) equilibrated with the dialysis buffer. Elution of protein is accomplished by the use of a linear gradient of potassium chloride concentrations from 0.2–0.45 M in 20 mM

[10] T. Rosett, *Appl. Microbiol.* **13**, 254 (1965).

potassium phosphate (pH 7.4) containing 0.1 mM dithiothreitol and 0.1 mM EDTA. The total volume of the gradient is 5 liters. The flow rate is approximately 40 ml/hr, and the fraction size is 12.8 ml. Peak tubes containing the enzyme with specific activities greater than 3.50 units/mg of protein are pooled and concentrated in an Amicon Diaflo cell equipped with a UM-20E membrane at 60 psi.

Step 5. Second DEAE-Sephadex Chromatography. The concentrated preparation from step 4 is dialyzed in 4 liters of 20 mM Tris·HCl (pH 8.0) containing 0.2 M potassium chloride, 0.1 mM dithiothreitol, and 0.1 mM EDTA for 16 hr and applied to a 1.5 × 90 cm column of DEAE-Sephadex A-50 (medium) equilibrated with the dialysis buffer. Proteins are eluted with a linear gradient of potassium chloride concentrations from 0.2–0.45 M in 20 mM Tris·HCl (pH 8.0) containing 0.1 mM dithiothreitol and 0.1 mM EDTA. The total volume of the gradient is 800 ml. The flow rate is approximately 20 ml/hr, and the fraction size is 3.2 ml. Fractions with specific activities greater than 5.50 units/mg of protein are pooled and concentrated as in step 4. The concentrated preparation is dialyzed in 4 liters of 20 mM potassium phosphate (pH 7.4) containing 0.1 mM dithiothreitol and 0.1 mM EDTA for 16 hr. The final preparation is stored at −20°.

A summary of the purification procedure is shown in the table.

Properties

Purity.[8] The purity of GMP synthetase preparation is established by the data obtained from disc gel electrophoresis, ultracentrifugal analyses, and gel filtration. Polyacrylamide gel electrophoresis at pH values of 7.4 and 8.2 indicates the purity of enzyme preparation. Schlieren patterns of the enzyme in the sedimentation velocity experiments reveal

PURIFICATION OF GMP SYNTHETASE FROM *Escherichia coli* STRAIN B-96

Step	Volume (ml)	Protein (mg)	Activity (units)	Specific activity (units/mg protein)
1. Extraction	540	30,300	13,400	0.442
2. Streptomycin sulfate precipitation	600	24,300	11,700	0.479
3. Ammonium sulfate fractionation	74.0	4150	7150	1.72
4. First DEAE-Sephadex chromatography	17.2	547	3650	6.67
5. Second DEAE-Sephadex chromatography	5.7	343	3120	9.09

a single symmetrical boundary throughout the course of the experiment. Sedimentation equilibrium measurements suggest that the sample is pure since the Yphantis plots are linear. Gel filtration on Sephadex G-200 also reveals a single peak containing the enzyme with constant specific activity throughout.

Amino Donor. GMP synthetase from *Escherichia coli* can utilize either ammonia or glutamine as amino donor.[11] The corresponding enzyme from pigeon liver and calf thymus has a similar specificity as to amino donor.[12]

Kinetic Properties. The Michaelis constants of GMP synthetase are 2.9×10^{-5}, 5.3×10^{-4}, 1.0×10^{-3}, and 1.0×10^{-3} M for XMP, ATP, ammonia, and glutamine, respectively.[11,13]

The adenine glycoside antibiotic psicofuranine (6-amino-9-D-psicofuranosylpurine) causes a highly specific and irreversible inactivation of the enzyme. The inhibition, K_i of 1.3×10^{-7} M, is dependent on the presence of XMP and PP_i.[13-17] Hydroxylamine is a potent inhibitor of the enzyme. The inhibition is dependent on the presence of XMP and ATP and is irreversible.[4,18] GMP synthetase is inhibited irreversibly by DON (6-diazo-5-oxo-L-norleucine).[11]

GMP synthetase is completely inactivated by guanidine hydrochloride at concentrations higher than 0.6 M. After the dilution of the denaturant the enzyme treated at concentrations lower than 1.2 M recovers almost total activity. No reactivation is attained, however, when the enzyme is treated with the denaturants at concentrations between 2.0 and 3.0 M. In 3.0–6.0 M guanidine hydrochloride, the enzyme is partially (15%) reassociable to an active form by removal of the denaturant.[19]

Extinction Coefficient.[8] The extinction coefficient ($E_{280}^{1\%}$) of GMP synthetase is 11.2 ± 0.5 cm^{-1}. An absorbance of 1.00 at 280 nm is equivalent to 0.893 ± 0.035 mg/ml. The extinction coefficient is deter-

[11] N. Patel, H. S. Moyed, and J. F. Kane, *J. Biol. Chem.* **250**, 2609 (1975).
[12] S. C. Hartman and S. Prusiner, *in* "The Enzymes of Glutamine Metabolism" (S. Prusiner and E. R. Stadtman, eds.), p. 409. Academic Press, New York, 1973.
[13] S. Udaka and H. S. Moyed, *J. Biol. Chem.* **238**, 2797 (1963).
[14] T. T. Fukuyama and H. S. Moyed, *Biochemistry* **3**, 1488 (1964).
[15] H. Kuramitsu and H. S. Moyed, *J. Biol. Chem.* **241**, 1596 (1966).
[16] T. T. Fukuyama, *J. Biol. Chem.* **241**, 4745 (1966).
[17] N. Zyk, N. Citri, and H. S. Moyed, *Biochemistry* **9**, 677 (1970).
[18] T. T. Fukuyama and K. L. Donovan, *J. Biol. Chem.* **243**, 5798 (1968).
[19] N. Sakamoto, G. W. Hatfield, and H. S. Moyed, *J. Biol. Chem.* **247**, 5888 (1972).

mined by use of the method using an analytical ultracentrifuge as differential refractometer.[20]

Diffusion Coefficient.[8] The diffusion coefficient $(D^0_{20,w})$ of GMP synthetase is $(5.09 \pm 0.51) \times 10^{-7}$ cm^2 sec^{-1}. Its determination by free diffusion in an analytical ultracentrifuge is based on the method of Schachman.[21]

Partial Specific Volume.[8] The partial specific volume ($\bar{v}$) of GMP synthetase is 0.739 ± 0.026 ml g^{-1}. It is determined by the method of differential sedimentation equilibrium.[22]

Sedimentation Coefficient.[8] The sedimentation coefficient of GMP synthetase is a linear function of protein concentration in the sedimentation velocity measurements, and extrapolation to infinite dilution yields a value for $s^0_{20,w}$ of 5.91 ± 0.10 S.

Molecular Weight.[8] Sedimentation equilibrium studies following the meniscus depletion method[23] yield a molecular weight of $126,000 \pm 4,000$ for GMP synthetase, employing the determined value of 0.739 ml g^{-1} for the partial specific volume. No concentration dependence of molecular weight is observed in a number of centrifugations at various protein concentrations (0.2–0.6 mg/ml).

Gel filtration on Sephadex G-200 following the method of Andrews[24] gives a molecular weight of $128,000 \pm 12,000$ for GMP synthetase.

Subunit Structure.[8] GMP synthetase is apparently a dimer composed of identical subunits. Polyacrylamide gel electrophoresis in the presence of sodium dodecyl sulfate[25] and sedimentation equilibrium measurement in 6.0 M guanidine hydrochloride reveal that GMP synthetase dissociates into subunits of identical molecular weight. The molecular weight of the subunit is $63,000 \pm 3000$, which is one-half of that of the native enzyme. The identity of the two subunits is also supported by the finding that arginine is the sole NH$_2$-terminal residue of GMP synthetase determined by the dansylation procedure.[26]

Acknowledgment

I am indebted to Dr. H. S. Moyed for his critical reading of the manuscript.

[20] J. Barbur and E. Stellwagen, *Anal. Biochem.* **26**, 216 (1969).
[21] H. K. Schachman, this series, Vol. IV, p. 71.
[22] S. J. Edelstein and H. K. Schachman, *J. Biol. Chem.* **242**, 306 (1967).
[23] D. A. Yphantis, *Biochemistry* **3**, 297 (1964).
[24] P. Andrews, *Biochem. J.* **96**, 595 (1965).
[25] K. Weber and M. Osborn, *J. Biol. Chem.* **244**, 4406 (1969).
[26] K. R. Woods and K. T. Wang, *Biochim. Biophys. Acta* **133**, 369 (1967).

[29] GMP Synthetase[1] from Ehrlich Ascites Cells

By Thomas Spector

$$XMP + ATP + [(Gln \text{ and } H_2O) \text{ or } (NH_3)] \xrightarrow{Mg^{2+}} GMP + AMP + PP_i + [Glu]$$

Assay Methods

Principle. Four assays[2,3] were developed to meet the special requirements of these studies. Assays I–III (see Table I) couple the formation of AMP and/or GMP to the oxidation of NADH as mediated via AMP kinase and/or GMP kinase, pyruvate kinase, and lactate dehydrogenase. The oxidation is monitored spectrophotometrically at 340 or 360 nm.[4] Assay I measures the formation of AMP and therefore has the advantage of being nonspecific with respect to the substrate undergoing amination. Assay II measures GMP formation and is therefore useful for selectively monitoring the amination of XMP. Assay III combines assays I and II to produce a doubling of the sensitivity. Assay IV is radiochemical and can be performed in volumes as small as 50 μl. It is used either to conserve enzyme or to assay GMP synthetase in crude preparations.

Reagents

Standard assay mixture (reaction mixture concentration): 0.25 mM XMP; 2.0 mM ATP; 2.0 mM Gln; 10 mM MgSO$_4$; 75 mM Tris·HCl, pH 7.6

Coupling reagents (reaction mixture concentrations): 5 I.U./ml AMP kinase; 5 IU/ml GMP kinase; 5 IU/ml pyruvate kinase; 10 IU/ml lactate dehydrogenase; 0.5 mM phosphoenolpyruvate; 0.2 mM NADH; 8 mM KCl. A 5-fold concentrated stock solution of these reagents in 25 mM Tris·HCl, pH 8.5, is filtered through a 0.45 μm Millipore filter (positive pressure) and stored at 4°. The solution is stable for at least 1 month.

Procedure

Spectrophotometric Assays (I–III). Cuvettes containing the "standard reaction mixture" and the appropriate coupling reagents are incu-

---

[1] GMP synthetase, xanthosine-5'-phosphate (XMP):L-glutamine amidoligase (AMP), EC 6.3.5.2. is also known as XMP aminase, XMP:ammonia ligase (AMP) EC 6.3.4.1.

[2] T. Spector, R. L. Miller, J. A. Fyfe, and T. A. Krenitsky, *Biochim. Biophys. Acta* **370,** 585 (1974).

[3] T. Spector, *J. Biol. Chem.* **250,** 7372 (1975).

[4] Assays were monitored at 360 nm to accommodate the thiopurines which have high absorbances at 340 nm.

TABLE I
ASSAYS FOR GMP SYNTHETASE

Assay	AMP kinase	GMP kinase	Pyruvate kinase lactate dehydrogenase phosphoenolpyruvate monovalent cation NADH	$\Delta\epsilon_{340nm}$ (mM^{-1} cm^{-1})
I.	X		X	12.44
II.		X	X	12.44
III.	X	X	X	24.88
IV.	Analysis of [^{14}C]bases following acid hydrolysis and paper chromatography.			

bated for 3 min in a 37° water bath. During this equilibration period, the coupling system converts all the ADP (present in commercial preparations of ATP) to ATP. The concentration of ATP will then remain constant throughout the reaction. GMP synthetase is then added, the contents of the cuvettes are mixed, and the ΔA_{340} is monitored on the 0–0.1 A scale of a thermostated recording spectrophotometer. Linear rates are obtained after a brief lag period. Blank rates are obtained (and subtracted from reaction rates) by performing the above assay in the absence of XMP. The blank rates of the partially purified enzyme preparation are always <10% the reaction rates.

Radiochemical Assay (IV). These reactions are carried out in ½-dram vials with screw-top lids. Vials containing the "standard reaction mixture," with [^{14}C]XMP[2] (1.2–2.4 mCi/mmole) substituted for XMP, and all coupling reagents except GMP kinase, lactate dehydrogenase, and NADH, are preincubated as described above. GMP synthetase is then added, and the reactions are initiated by mixing with a vortex mixer for 3 sec. After incubating for a time period that does not extend past the initial linear portion of the reactions, HCl (0.4 M final concentration) is added to terminate the reactions. The vials are then capped and incubated at 100° for 1 hr in order to cleave the nucleotides to their corresponding bases. Carrier bases are added, and the contents are spotted on Whatman 3 MM paper and chromatographed in an ascending direction in water-saturated butanol: NH$_4$OH (99:1) for 17 hr. The spots containing the UV-absorbing bases (R_f values: xanthine, 0.03; guanine, 0.13) are cut out and counted for their radioactivity in toluene–Triton-X–Omnifluor (30:10:0.28, v/v/w) scintillation fluid. Blank values (acid added before enzyme) are obtained (approx. 0.4% of the total counts) and subtracted from the reaction values.

Definition of Unit and Specific Activity. One unit of enzyme is that amount of enzyme which will catalyze the formation of 1 nmole of product per minute under either of the assay conditions described above. Specific activity is expressed in terms of units per milligram of protein.[5]

Purification Procedure[3]

Growth and Harvest of Ehrlich Ascites Cells. One hundred white male mice (18–20 g), inoculated 6-½ days earlier with 0.1 ml Ehrlich ascites fluid (approximately 50% of the mice would have expired by 16 days), are sacrificed by cervical dislocation. The ascites fluid is collected in a flask containing 500 ml of 0.14 M heparinized NaCl at 4°. The cells are then pelleted by centrifugation, resuspended in 600 ml of 0.07 M heparinized NaCl for 5 min (to hemolyze any contaminating erythrocytes) and recollected by centrifugation. The hemolytic procedure is repeated, and the pinkish-colored packed cells are again collected by centrifugation.

Acetone Powder Preparation. The harvested cells are slowly dropped into 2 liters of vigorously stirring acetone at 20°. The precipitate is collected by filtration and resuspended three times in 1 liter of acetone. The final retentate is air-dried for 30 min and then dried *in vacuo* for 20 hr. Approximately 12 g of light-tan powder are obtained. The GMP synthetase activity (assay IV) is stable in this form when stored at −20°.

DEAE-Sephadex (A-50) Column Chromatography. A 2-g portion of the acetone powder is added to 20 ml of 20 mM potassium phosphate buffer, pH 7.5, containing 0.1 mM EDTA ("buffer A") and slowly stirred for 20 min. This step and all succeeding steps are performed at 0–4°. The extract is centrifuged at 40,000 g for 5 min. The supernatant is applied directly onto a 4.9 × 11 cm DEAE-Sephadex (A-50) anion exchange column equilibrated with buffer A, and the pellet is then resuspended in 15 ml of buffer A and stirred for 15 min. The supernatants from the second and an identical third extraction are also applied to the column which is then washed for 4-½ hr with buffer A at a flow rate of 1 ml/min. By this time, all the visibly colored protein has eluted and the A_{280} is <0.2. A linear gradient from 0–0.25 M KCL in buffer A (600 ml total volume) is used for elution. The volume of the fractions is 15 ml. GMP synthetase elutes in a colorless peak at approximately 0.18

[5] O. H. Lowry, N. J. Rosebrough, A. L. Farr, and R. J. Randall, *J. Biol. Chem.* **193**, 265 (1951).

M KCl in the tail of a broad peak of GMP kinase. The presence of the latter enzyme permits assay III to be used conveniently.

Protein Concentration. The peak fractions from the above column are pooled, and the enzyme activity is precipitated by the addition of solid ammonium sulfate to 80% saturation (51.6 g/100 ml). The precipitate is collected by centrifugation and resuspended in less than 2 ml of buffer A. The suspension is dialyzed against 1 liter of buffer A containing 0.05 *M* KCl for 1 hr and clarified by centrifugation.

Sephadex G-100 Column Chromatography. The above solution of GMP synthetase is carefully layered onto a 1.5 × 85 cm Sephadex G-100 column equilibrated with buffer A containing 0.05 *M* KCl. Elution, with the same buffer, is performed under a 25-cm pressure head collecting 3-ml fractions. GMP synthetase (assay I) appears in a sharp peak (fraction 22) separated from the GMP kinase peak (fraction 30) by a few fractions.

Comments. The results of a typical purification sequence are presented in Table II. An important aspect of the procedure is that the GMP synthetase activity of the acetone extract is rather labile. This activity is best preserved by immediately loading the extract onto the DEAE-Sephadex column. The enzyme that elutes from this column is suitable for the spectrophotometric assay III. However, since it is very dilute at this stage, 50–100 μl of enzyme are required per 0.6 ml of assay volume. GMP synthetase can be more easily located by first finding the relatively abundant activity peak for GMP kinase.[6] GMP synthetase collected from

TABLE II
PURIFICATION OF GMP SYNTHETASE

Step	Volume (ml)	Total protein (mg)	Total activity (units)	Specific activity (units/mg)	Fold purified
Acetone powder extract	43	318	798	2.5	3.3[a]
DEAE-Sephadex and concentration	1.3	8.5	132	15.5	21
G-100 gel filtration	8.2	2.0	86	42.8	57

[a] The specific activity of cells that are lysed and assayed prior to the preparation of the acetone powder is 0.75 units/mg.

[6] It is noteworthy that only 15% (approx. 1,000 units) of the total of the peak of GMP kinase is loaded (together with the GMP synthetase) onto the Sephadex G-100 column. However, the GMP kinase obtained from the gel filtration step has a specific activity of 1,500 units/mg.

the Sephadex G-100 column may be concentrated either as described above or by ultrafiltration. Occasionally, a portion of the activity is lost following concentration, but it is regained after storage at $-70°$ for 1 day.

Storage. GMP synthetase obtained from either of the above described column steps can be fast-frozen and stored at $-70°$ for at least 1 year with no loss of activity. Thawing and refreezing do not diminish the activity.

Contaminating Enzyme Activities.[3] When assayed at pH 7.6 in 75 mM Tris·HCl, the preparation obtained after the gel filtration step is free from detectable quantities of GMP kinase, glutaminase, monophosphate nucleotide phosphatase, adenosine kinase, IMP dehydrogenase, NADH oxidase, and NAD reductase. The activity of ATPase ranges from nondetectable to <10% of the GMP synthetase activity, AMP kinase and adenosine deaminase from 10–20%, and inorganic pyrophosphatase from 200–300%.

Properties

pH Optimum.[3] The catalytic activity evaluated over a pH range from 7.4–8.6 in 75 mM Tris·HCl decreases by less than 20% at the extremes and is maximal at pH 7.6.

Molecular Weight.[3] The molecular weight estimated by the method of Andrews[7] is 85,000.

Sulfhydryl Involvement in Catalysis.[3] Unlike GMP synthetase isolated from calf thymus,[8] the activity of this enzyme does not have an absolute requirement for exogenous sulfhydryl-containing compounds. However, under standard assay conditions (assay IV), the enzyme is inhibited 50% in the presence of 10 mM 2-mercaptoethanol and stimulated to 190% in the presence of 10 mM dithiothreitol. Furthermore, p-chloromercuribenzoate is a powerful inhibitor, producing 72% inhibition at 1 μM and complete inhibition above 5 μM.

Michaelis–Menten Constants.[3] The K_m values were determined for each substrate with the other substrates fixed at their "standard concentration." These values are: XMP, 3.6 μM; ATP, 0.28 mM; Gln, 0.68 mM; $(NH_4)_2SO_4$, 36 mM. These values are similar to those obtained with

[7] P. Andrews, *Biochem. J.* **91**, 222 (1968).
[8] R. Abrams and M. Bentley, *Arch. Biochem. Biophys.* **79**, 91 (1959).

GMP synthetase from other sources[2,8–12] with the exception that the K_m for XMP with the bacterial[2,10,11] and avian[9] enzymes is about one order of magnitude higher.

Substrate Specificity.[3] Six monophosphate nucleotides are aminated by this enzyme. Listed in order of their relative substrate efficiency (V_{max}/K_m), they are: XMP (1000); 2′-dXMP (43); 8-azaXMP (11); 6-thioXMP (7); β-D-arabinosylXMP (3); and 1-ribosyloxipurinol 5′-phosphate (0.02). Cyclic XMP, 3-ribosylXMP, and 7-ribosyloxipurinol 5′-phosphate are not substrates. Since the V_{max} for 6-thioXMP is only 1% that of XMP, this compound is a strong alternate-substrate inhibitor with a K_i of 5 μM. The aminated products are competitive inhibitors with respect to XMP.

Interactions with Triphosphate Nucleotides.[3] While none of the 11 triphosphate purine and pyrimidine nucleotides studied are able to substitute for ATP as the energy source of the reaction, all are capable of binding to the ATP site. The K_i values for CTP, β-D-arabinosylATP, and 1-N^6-ethenoATP are similar in magnitude to the K_m for ATP (0.28 mM).

Synergistic Inhibition by PP$_i$ and Nucleosides.[13] Many analogs of adenosine[14] are noncompetitive inhibitors with respect to ATP and are not competitive with respect to XMP. The presence of PP$_i$ (one of the reaction products) causes the pattern of the nucleoside inhibition to shift to competitive (with respect to ATP) and the K_i values to decrease 38 ± 1-fold. PP$_i$, itself, is competitive with ATP (K_i = 0.42 mM). All the nucleoside inhibitors cause an apparent lowering of this inhibition constant for PP$_i$, but do not affect the nature of the inhibition. GMP synthetase from *Escherichia coli* is also inhibited by PP$_i$ and nucleosides.[10–12,15,16] However, the bacterial enzyme does not bind the nucleosides in the absence of PP$_i$, and in the presence of PP$_i$, the double inhibitor complex undergoes reversible isomerization into a nondissociating species.

[9] L. Lagerkvist, *J. Biol. Chem.* **233**, 143 (1958).
[10] S. Udaka and H. S. Moyed, *J. Biol. Chem.* **238**, 2797 (1963).
[11] N. Zyk, N. Citri, and H. S. Moyed, *Biochemistry* **8**, 2787 (1969).
[12] T. Spector and L. M. Beacham, III, *J. Biol. Chem.* **250**, 3101 (1975).
[13] T. Spector, T. E. Jones, T. A. Krenitsky, and R. J. Harvey, *Biochim. Biophys. Acta* **452**, 597 (1976).
[14] Examples of inhibitors are adenosine, psicofuranine, decoyenine, N^6-allyladenosine and 2-fluoroadenosine.
[15] L. Slechta, *Biochem. Biophys. Res. Commun.* **3**, 596 (1960).
[16] N. Zyk, N. Citri, and H. S. Moyed, *Biochemistry* **9**, 677 (1970).

Section IV
Deoxynucleotide Synthesis

[30] Ribonucleoside Diphosphate Reductase (*Escherichia coli*)

By LARS THELANDER, BRITT-MARIE SJÖBERG, and STAFFAN ERIKSSON

$$\text{NDP} + \text{thioredoxin-(SH)}_2 \xrightarrow{\text{Mg}^{2+}} \text{dNDP} + \text{thioredoxin-S}_2$$

Deoxyribonucleotides are synthesized by a direct reduction of the corresponding ribonucleotides. The reaction is catalyzed by ribonucleotide reductase which in this way is involved in the control of DNA synthesis. The enzyme has been purified to homogeneity and studied extensively in *Escherichia coli* and *Lactobacillus leichmannii*.[1] The preparation of ribonucleotide reductase from T4-infected *E. coli* by affinity chromatography[2] and the partial purification of the enzyme from Novikoff hepatoma[3] in rats were described in this series previously.

An *E. coli* strain (KK546)[4] which is lysogenic for a defective lambda carrying the genes of ribonucleotide reductase overproduces the enzyme upon induction. The preparation of large amounts of homogeneous enzyme rests on the use of this source containing 10% of the total protein as ribonucleotide reductase.

Assay Methods

Principle. Ribonucleotide reductase from *E. coli* consists of two nonidentical subunits, proteins B1 and B2, which have no biological activity when assayed separately. In the presence of Mg^{2+} they combine to form the active enzyme. Therefore, one subunit is always assayed in the presence of an excess of the other.[5] In the reduction of ribonucleotides, stoichiometric amounts of oxidized thioredoxin are formed. This low-molecular-weight protein is reduced by NADPH in a reaction catalyzed by a flavoprotein, thioredoxin reductase.[1]

$$\text{Thioredoxin-S}_2 + \text{NADPH} \rightarrow \text{thioredoxin-(SH)}_2 + \text{NADP}^+$$

By combining the two reactions ribonucleotide reductase activity can be measured by spectrophotometric determination of NADPH oxidation.[5]

[1] H. P. C. Hogenkamp and G. N. Sando, *Struct. Bonding (Berlin)* **20**, 23f (1974).
[2] O. Berglund and F. Eckstein, this series, Vol. 34B, p. 253.
[3] E. C. Moore, Vol. 12A, p. 155.
[4] S. Eriksson, B.-M. Sjöberg, S. Hahne, and O. Karlström, *J. Biol. Chem.* **252**, 6132 (1977).
[5] N. C. Brown, Z. N. Canellakis, B. Lundin, P. Reichard, and L. Thelander, *Eur. J. Biochem.* **9**, 561 (1969).

The spectrophotometric assay cannot be used with crude extracts due to unspecific oxidation of NADPH. Instead ribonucleotide reductase activity in crude extracts is determined by measurements of the formation of [^{3}H]dCDP from [^{3}H]CDP.[5]

A. Spectrophotometric Assay

Procedure. The enzyme was incubated at 25° in a mixture containing 200 nmoles ATP, 1.6 μmoles MgCl$_2$, 80 nmoles NADPH, 5 μmoles N-2-hydroxyethylpiperazine-N'-2-ethanesulphonic acid buffer (pH 7.6), 300 pmoles thioredoxin, 40 pmoles thioredoxin reductase, 10 nmoles EDTA, and 65 nmoles dithiothreitol in a final volume of 0.13 ml. The reaction was started by the addition of 75 nmoles CDP, and the oxidation of NADPH was monitored at 340 nm with a Zeiss automatic recording spectrophotometer equipped with microcuvettes. The best determinations of activity were obtained when 1–2.5 units of one subunit were assayed in the presence of a 5-fold excess of the other subunit. Before addition of CDP the background oxidation of NADPH was recorded for some minutes, and this background was subtracted from the NADPH oxidation observed after addition of CDP.

Comments. Dithiothreitol stabilizes the B1 subunit of ribonucleotide reductase,[6] and therefore it is included in the incubation mixture in a final concentration of 0.5 mM. At higher concentrations dithiothreitol starts to chemically reduce oxidized thioredoxin and competes with NADPH as hydrogen donor.[5] EDTA is included in the stock solutions of thioredoxin and thioredoxin reductase to protect these proteins.

B. [^{3}H] CDP Assay

Procedure. The same incubation mixture was used as in the spectrophotometric assay but [^{3}H]CDP (specific activity 2 × 10^6 cpm/μmole) replaced the CDP and the reaction was started by the addition of enzyme. Thioredoxin reductase can be replaced by 10 mM dithiothreitol.[5] After 10 min at 25° the reaction was stopped by the addition of 0.5 ml of 1 M perchloric acid and 0.5 μmole of dCMP carrier. Precipitated protein was carefully removed by centrifugation, and then the nucleotides were hydrolyzed to monophosphates by heating at 100° for 10 min. Potassium hydroxide (4 M) was added to neutralize the pH using phenol red as indicator, the potassium perchlorate was precipitated on ice for 10 min and then removed by centrifugation, and the sample was chromatographed on a 1 × 3.5 cm Dowex column (Dowex 50W-X8, 200–400

[6] L. Thelander, *J. Biol. Chem.* **248**, 4591 (1973).

mesh, H^+, in Biorad Econo columns 1 × 10 cm with polypropylene funnels). On elution with 0.2 M acetic acid, CMP eluted in the first 55 ml and dCMP in the following 25 ml. The chromatographic procedure was finished in 2 hr. Aliquots of 1 ml from the dCMP eluate were analyzed for radioactivity using Instagel (Packard) and a liquid scintillation counter. The amount of dCMP carrier in the dCMP eluate was determined by reading the absorbance at 280 nm. Usually recoveries were around 60–70%. The values from the radioactivity measurements were finally corrected for the recoveries of the dCMP carrier.

Definition of Unit and Specific Activity. One unit of ribonucleotide reductase activity is defined as the amount of proteins B1 or B2 which catalyzes the formation of 1 nmole of dCDP per minute at 25° under standard conditions in the presence of an excess of the complementary subunit. This definition of a unit is one-tenth of earlier used values.[5] The specific activity of a preparation is defined as units of enzyme activity per milligram of protein.

Purification Procedure

Reagents
Strain: *E. coli* KK546 (F⁻ *thr leu thi thy tlr tonA lac supE*44 *nrdA nalA malA/λdnrd-1 λb515 519 CI857 S7 xis*6)
Culture medium (in grams/liter): Bacto-tryptone (10), yeast extract (5), NaCl (10), glucose (1); pH is adjusted to 7.0 before autoclaving. Prior to use 2 ml of sterile 0.5 M MgSO₄ are added.
Buffer A: 50 mM Tris-Cl (pH 7.6)–20% glycerol–15 mM MgCl₂
Dithiothreitol (A-grade, Calbiochem)
Streptomycin sulfate (Novo Industri A/S)
dATP-Sepharose (prepared according to Berglund and Eckstein[2])
Immuno adsorbent columns[4] consisting of anti B1 or anti B2 γ-globulin covalently bound to Sepharose 4B
Collodion bags, type SM 13200 (Sartorius-Membranfilter, 34 Göttingen, West Germany)
Ultrogel AcA 34 (LKB-Beckman)

Growth and Disintegration of Bacteria. A 250-liter culture of *E. coli* KK546 was grown with moderate aeration (125 liters/min) at 33°. At a density of 5 × 10⁸ cells/ml the temperature was raised to 43° in less than 5 min by circulation of steam through the cooling jacket of the tank. After 20 min at 43° the culture was cooled to 37° by circulation of tap water through the cooling jackets. The culture was grown at 37° for 2 hr and then rapidly cooled to 10° by circulation of ice water. The cells (1.35

kg) were harvested through centrifugation, rapidly frozen, and stored at −20°. Immediately prior to the preparation of the crude extract the cells were disintegrated at −20−−25° in a bacterial press.[5]

Crude Extract. Frozen disintegrated cells were extracted in a loosely fitting glass homogenizer with 3 volumes of buffer A containing 4 mM DTT for 10–15 min until the frozen cells were thawed. Insoluble cell debris was removed by centrifugation at 35,000 g for 40 min.

Streptomycin Precipitation. The supernatant solution was precipitated by addition of 0.2 volume of a 5% neutralized solution of streptomycin. The precipitant was slowly added (15 min) with continuous stirring, the resulting suspension was stirred for an additional 15 min, and the precipitate was removed by centrifugation.

Ammonium Sulfate Fractionation. The supernant solution was precipitated by the addition of solid ammonium sulfate to give 60% saturation (0.39 g/ml solution). The suspension was stirred for 60 min, and the precipitate was collected by centrifugation and dissolved in ¼ of the original volume in buffer A containing 4 mM DTT. Ammonium sulfate was removed by passing this solution through a column of Sephadex G-25 equilibrated with the same buffer.

Affinity Chromatography on dATP-Sepharose. The desalted protein fraction was adsorbed to a column of dATP-Sepharose[2] (2000 enzyme units/ml), which had previously been equilibrated with buffer A containing 4 mM DTT. The column was washed free of nonspecifically bound proteins with 8–9 column volumes of the same buffer. Most of the protein B2 activity in the extract was retarded on the column but 10–20% of the activity was found in the wash fraction.

The dithiothreitol concentration was then increased to 10 mM. This improved the stability of protein B1 so that the column could be left overnight. Two different elution procedures were then used. Alternative I, which yielded a complex of proteins B1 and B2, was used when separation of the two subunits from each other was unnecessary. Alternative II was used to obtain protein B1 and protein B2 in separate fractions from which either subunit could subsequently be further purified.

Elution Alternative I. The column was eluted with 3–4 volumes of buffer A containing 10 mM DTT and 10 mM ATP. The resulting eluate contained a mixture of the two subunits with 4.5 mg of protein B1 per mg of protein B2, i.e., approximately 2 molecules of protein B1 per molecule of protein B2.

There was less than 15% contaminating proteins in the mixture of proteins B1 and B2 as judged by sodium dodecyl sulfate gel electrophoresis. The enzyme prepared by this very rapid procedure can be used as helper enzyme when other components of the ribonucleotide reductase system are to be assayed, i.e., the hydrogen donor system: thioredoxin, thioredoxin reductase,[1] or glutaredoxin.[7]

Elution Alternative II. Elution of protein B2 was started with 6 volumes of 10 m*M* Tris-Cl (pH 7.6)–20% glycerol–10 m*M* DTT. The eluate contained 50% of the applied protein B2 and 2–3% of contaminating protein B1.

Some contaminating protein was then desorbed by a second wash of the column with 4 volumes of 0.1 *M* Tris-Cl (pH 7.6)–0.2 *M* NaCl–20% glycerol–10 m*M* DTT. The column was again equilibrated with buffer A containing 10 m*M* DTT, and protein B1 was eluted with 3–4 volumes of buffer A containing 10 m*M* DTT and 10 m*M* ATP. Approximately 80% of the protein B1 and 10% of the protein B2 activity eluted in this ATP pool.

Either subunit was 70–80% pure at this stage, and the elution procedure is schematically depicted in Fig. 1.

Chromatography on Anti-B1 and Anti-B2 γ-Globulin Sepharose Col-

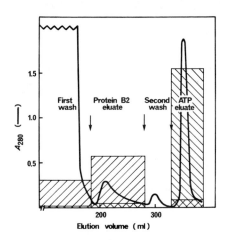

FIG. 1. Affinity chromatography on dATP-Sepharose, elution alternative II. A desalted protein fraction (607 mg in 34 ml) was applied to 15 ml of dATP-Sepharose and chromatographed as described in the text. Absorbance at 280 nm, —— (the absorbance at 290 nm was measured in the ATP eluate and then used to calculate the absorbance at 280 nm after subtraction of the ATP absorption, $A_{280} \approx 2 \times A_{290}$). Protein B2 activity, ▨; protein B1 activity, ▧.

[7] A. Holmgren, *Proc. Natl. Acad. Sci. U.S.A.* **73**, 2275 (1976).

umns. Antisera against proteins B1 and B2 were prepared from rabbit as described by Berglund[8] and precipitated with ammonium sulfate to 40% saturation. The dissolved and desalted γ-globulin fraction had a protein concentration of 8 mg/ml. This fraction was coupled to CNBr-activated Sepharose 4B by the method described by Cuatrecasas.[9] The coupled Sepharose had an antigen-binding capacity of approximately 0.05 mg/ml.

The protein B1 preparation from elution alternative II was next exposed to anti-B2 antibodies, and the protein B2 preparation was exposed to anti-B1 antibodies by immediately passing the eluates through the antibody–γ-globulin Sepharose columns. The anti-B2 column was equilibrated with buffer A containing 10 mM DTT before use and eluted with 2 volumes of buffer A containing 10 mM DTT and 0.2 M NaCl. The anti-B1 column was equilibrated with 50 mM Tris-Cl (pH 7.6)–20% glycerol–0.2 M NaCl and eluted with 2 volumes of the same buffer. By this procedure most of the contaminating subunit was removed.

The columns were regenerated by washing with 3 column volumes of 6 M guanidine hydrochloride which removed the bound antigen.

Concentration by Ultradialysis. Both eluates were concentrated by ultradialysis in collodion bags against buffer A containing 10 mM DTT (protein B1) or 50 mM Tris-Cl (pH 7.6)–20% glycerol–0.2 M NaCl (protein B2) to a concentration of approximately 40 mg protein per ml. After the concentrated protein fraction had been removed, the empty dialysis bag was left in the buffer for 1 hr. Approximately 0.1 ml of buffer diffused back into the bag and was added to the previous fraction. This gave a recovery of 80–90% of both protein and activity and up to 200-fold concentration.

Chromatography on Ultrogel AcA 34. Each concentrate from the γ-globulin Sepharose chromatography was further purified on Ultrogel AcA 34. This gel allows high enough flow rates for the chromatography to be completed within 20 hr.

PROTEIN B1. The column was equilibrated and eluted with buffer A containing 10 mM DTT. The enzyme activity emerged from the column in a peak that coincided with the major protein peak. Protein B2 activity eluted slightly ahead of the main protein B1 peak. This position in the chromatogram corresponds to the molecular weight of the complex between the two subunits. The fractions with constant specific activity were pooled and concentrated by ultradialysis. The pooled protein B1 contained no ATP as judged by the absorbance ratio of 260/280 nm.

[8] O. Berglund, *J. Biol. Chem.* **250**, 7450 (1975).
[9] P. Cuatrecasas, *J. Biol. Chem.* **245**, 3059 (1970).

TABLE I

PURIFICATION OF *E. coli* RIBONUCLEOSIDE DIPHOSPHATE REDUCTASE FROM 30 g KK546

Fraction	Protein B1			Protein B2		
	Activity (units)	Specific activity (units/mg)	Yield (%)	Activity (units)	Specific activity (units/mg)	Yield[a] (%)
Crude extract	60,000	24	100	55,000	22	—
0–60% (NH$_4$)$_2$SO$_4$	51,000	36	85	109,000	77	100
dATP-Sepharose	41,000	353	68	52,000	1560	48
(Anti-B1–γ-glob. Seph.) (Anti-B2–γ-glob. Seph.)	39,000	433	65	39,000	1850	36
Ultrogel[b]	27,000	590	45	25,000	2860	23

[a] The activity of protein B2 increased after the first steps of purification; therefore the activity of the dissolved ammonium sulfate fraction was used as 100%.

[b] The yield of pure proteins was 40 mg of protein B1 and 10 mg of protein B2.

If the antibody column step was omitted, the fraction of enzyme migrating as a complex was increased, but approximately 50% of the applied protein B1 activity could be isolated free of protein B2. This procedure gives a lower yield of protein B1 but allows isolation of a fraction with the subunits as a 1:1 complex.

PROTEIN B2. The column was equilibrated and eluted with 50 mM Tris-Cl (pH 7.6)–20% glycerol–0.2 M NaCl. The applied protein eluted in two peaks, the major one containing free protein B2 and the minor one a complex of protein B1 and B2. From the major peak, fractions with constant specific activity were pooled and concentrated by ultradialysis.

Both concentrated enzyme fractions were rapidly frozen in ethanol:Dry Ice and stored at −70°.

Yield of Homogeneous Subunit Proteins. Each purified subunit contained less than 4% contaminating protein as judged from sodium dodecyl sulfate gel electrophoresis.

Table I shows a summary of the purification of the subunits of ribonucleoside diphosphate reductase from 30 g of bacteria. Total recovery of protein B1 was 45% and of protein B2 20%.

The final output of 40 mg of protein B1 and 10 mg of protein B2 is approximately 30 times more than can be recovered from the same amount of derepressed *E. coli* B3.[5]

Physical Properties

Protein B1

Size and Subunit Structure. The B1 subunit has a molecular weight of 160,000 and consists of two polypeptide chains of similar or identical size. Both chains have the same carboxyterminal amino acid residue but they differ in the aminoterminal part. Furthermore, on polyacrylamide gel electrophoresis at pH 8.74 the protein B1 band is resolved in two very close moving components of about equal amounts.[6]

Composition. Protein B1 contains 21 half-cystine residues, and all of them are reduced in the fully active enzyme. Oxidation leads to dissociation of the B1 subunit and inactivation. Removal of Mg^{2+} by EDTA also results in a dissociation of protein B1, and addition of Mg^{2+} gives back the active 7.7 S form of the protein. The extinction coefficient ($E_{1\%}^{1cm}$) at 280 nm of the pure protein B1 is 10.8.[6]

Protein B2

Size and Subunit Structure. The protein B2 subunit has a molecular weight of 78,000 and consists of two identical or very similar polypeptide chains.[6] The $s_{20,w}$ is 5.5 S.[10]

Composition. Protein B2 contains two atoms of non-heme-bound iron per molecule and in addition one free radical.[4] The radical is recognized by a sharp peak at 410 nm in the optical spectrum and a characteristic electron paramagnetic resonance (EPR) signal centered around g = 2.0047. The spectral properties originate from an unpaired electron delocalized in the aromatic ring of a tyrosine residue in the protein.[11] The presence of the radical is directly correlated to activity.[12]

The extinction coefficient ($E_{1\%}^{1cm}$) at 280 nm of the pure protein B2 is 15.0,[6] and the A_{410}/A_{280} is 0.069 in our best preparations.[4] The ratio A_{410}/A_{280} is a sensitive measurement of the radical content of a certain protein B2 preparation, and a gradual loss of radical and activity is seen on handling of the protein.[12] The radical can be regenerated by removal of the iron and replacing it by addition of iron in the form of a ferrous-ascorbate complex.[13]

Ribonucleotide Reductase

Size and Subunit Structure. The active form of ribonucleotide reductase consists of a 1:1 complex between protein B1 and protein B2 and sediments at 10.1 S. Addition of the negative allosteric effector dATP (see below) induces inactive forms of the enzyme sedimenting at 15 and 22 S, and this aggregation is reversed by ATP.[6]

Catalytic Properties

Allosteric Regulation

Ribonucleotide reductase is allosterically controlled by deoxyribonucleoside triphosphates and ATP. The same enzyme reduces CDP, UDP, GDP, and ADP, but both the general activity and the substrate specificity is controlled by the allosteric effectors. There are both positive and negative effectors. A positive effector at the same time decreases the K_m for a given substrate and increases the V_{max}.[14]

[10] N. C. Brown and P. Reichard, *J. Mol. Biol.* **46**, 25 (1969).

[11] B.-M. Sjöberg, P. Reichard, A. Gräslund, and A. Ehrenberg, *J. Biol. Chem.* **252**, 536 (1977).

[12] A. Ehrenberg and P. Reichard, *J. Biol. Chem.* **247**, 3485 (1972).

[13] C. L. Atkin, L. Thelander, P. Reichard, and G. Lang, *J. Biol. Chem.* **248**, 7464 (1973).

[14] N. C. Brown and P. Reichard, *J. Mol. Biol.* **46**, 39 (1969).

TABLE II
ALLOSTERIC REGULATION OF RIBONUCLEOTIDE REDUCTASE
FROM *E. coli*

Substrates	CDP, UDP	GDP, ADP
Stimulation	dATP ($10^{-6} M$) ATP dTTP	dTTP dGTP (dCTP)
Inhibition	dATP ($>10^{-5} M$)	dATP ($>10^{-5} M$)

Effector Binding. All effectors bind exclusively to protein B1, and there are four effector binding sites which can be divided into two classes, each consisting of two sites. One class binds dATP, dGTP, dTTP, dCTP, and ATP (specificity sites), while the other class binds only ATP and dATP (activity sites).[14,15] The allosteric regulation of ribonucleotide reductase is summarized in Table II. Binding of effectors to the specificity sites controls the substrate specificity of the enzyme, while binding to the activity sites controls the overall activity. The dissociation constants at 2° for deoxyribonucleoside triphosphate effector binding to protein B1 vary between $0.3–5 \times 10^{-7} M$, and the corresponding value for ATP binding is about $10^{-5} M$.[14]

Substrate Binding. Also the ribonucleoside diphosphate substrates bind exclusively to protein B1. There are two substrate binding sites per protein B1 molecule, and these are distinct from the effector binding sites. The dissociation constants at 20° for substrates range from $2 \times 10^{-5} M$ to around $10^{-3} M$. A positive effector increases the affinity of protein B1 for its favored substrate; e.g., the K_{diss} of GDP binding in the absence of effector is $1.1 \times 10^{-4} M$ while addition of the positive effector dTTP decreases the K_{diss} to $2.2 \times 10^{-5} M$.[16]

Redox Properties. Protein B1 contains redox-active dithiol groups of the same type as those present in the active sites of thioredoxin and thioredoxin reductase. These redox-active dithiols are the immediate electron donors during ribonucleotide reduction. Ribonucleotide reductase acts by a ping-pong mechanism alternating between the dithiol and disulfide form of protein B1.[17]

[15] U. von Döbeln, *Biochemistry* **16**, 4368 (1977).
[16] U. von Döbeln and P. Reichard, *J. Biol. Chem.* **251**, 3616 (1976).
[17] L. Thelander, *J. Biol. Chem.* **249**, 4858 (1974).

Active Site. The active site of ribonucleotide reductase is formed both from protein B1, which contributes the redox-active dithiols, and protein B2, which contributes the free radical.[18] The conformation of the active site is regulated by the allosteric effectors.

[18] L. Thelander, B. Larsson, J. Hobbs, and F. Eckstein, *J. Biol. Chem.* **251**, 1398 (1976).

[31] Ribonucleotide Reductase of Rabbit Bone Marrow

By SARAH HOPPER

Studies on ribonucleotide reduction in partially purified extracts of Novikoff hepatoma,[1] rat embryos,[2] rat liver,[3,4] and rabbit bone marrow[5] suggest that animal cells may contain enzymes which resemble the iron-requiring ribonucleotide reductase of *Escherichia coli*[6,7] rather than the B_{12}-dependent enzyme of *Lactobacillus leichmannii*.[8] However, since it has not been possible to purify the mammalian reductases by conventional methods, studies on highly purified ribonucleotide reductases have been limited to microbial systems. The activity of ribonucleotide reductase of rabbit bone marrow which was previously shown to be dependent on two protein fractions[5] has now been highly purified by affinity chromatography on dATP-Sepharose. The purified preparation of the bone marrow enzyme catalyzes the reduction of CDP, UDP, ADP, and GDP at comparable rates and requires iron. At high dilutions of the enzyme the reaction rate is second order with respect to protein concentration.

Materials and Methods

ATP-Sepharose and dATP-Sepharose were synthesized by the procedure described by Berglund and Eckstein[9,10] utilizing cyanogen bromide-

[1] E. C. Moore and P. Reichard, *J. Biol. Chem.* **239**, 3453 (1964).
[2] S. Murphree, E. C. Moore, and P. T. Beall, *Cancer Res.* **28**, 860 (1968).
[3] C. D. King and J. L. Van Lancker, *Arch. Biochem. Biophys.* **129**, 603 (1969).
[4] A. Larsson, *Eur. J. Biochem.* **11**, 113 (1969).
[5] S. Hopper, *J. Biol. Chem.* **247**, 3336 (1972).
[6] N. C. Brown, R. Eliasson, P. Reichard, and L. Thelander, *Biochem. Biophys. Res. Commun.* **30**, 522 (1968).
[7] C. L. Atkin, L. Thelander, P. Reichard, and G. Lang, *J. Biol. Chem.* **248**, 7464 (1973).
[8] R. L. Blakely and H. A. Barker, *Biochem. Biophys. Res. Commun.* **16**, 391 (1964).
[9] O. Berglund and F. Eckstein, *Eur. J. Biochem.* **28**, 492 (1972).
[10] O. Berglund and F. Eckstein, this series, Vol. 34, p. 253.

activated Sepharose 4B obtained from Pharmacia Fine Chemicals. The amounts of ligand bound in two preparations each of ATP- and of dATP-Sepharose were estimated as 0.08 and 0.19 μmoles of ATP and 0.48 and 1.42 μmoles of dATP per cm^3 of wet packed Sepharose by the spectrophotometric measurement of adenosine or deoxyadenosine released after treatment of the ATP- or dATP-Sepharose preparations with *Crotalus terr. terr.* venom phosphodiesterase (Boehringer Mannheim) and *E. coli* alkaline phosphatase (Worthington: BAPF purified on DEAE-Sephadex). The variation in the amounts of *p*-aminophenyl-ATP or -dATP bound to Sepharose in different batches is not understood. The preparation containing 1.42 μmoles of dATP per cm^3 of Sepharose tightly binds the bone marrow reductase and was used for purification of the enzyme described in this report.

Frozen rabbit bone marrow was prepared by Pel-Freez Biologicals, Rogers, Arkansas. Six-week-old rabbits were given 4 daily intraperitoneal injections of acetophenylhydrazide in 2.0 ml of 150 mM sodium chloride and the marrow collected on the 5th day was frozen immediately on Dry Ice. Somewhat reduced amounts of enzyme of similar specific activity are obtained when the reductase is purified from the same weight of bone marrow from 6-week-old untreated rabbits.

Enzyme Assays

To assay for the reduction of CDP the enzyme was incubated at 37° for 45 min in a final volume of 0.25 ml. The solution contained 25 μmoles N-2-hydroxyethylpiperazine-N'-2-ethanesulfonate, pH 7.0, 1 μmole sodium ATP, 2 μmoles magnesium chloride, 2.5 μmoles dithiothreitol, 25 nmoles ferrous ammonium sulfate, and 34 nmoles [^{3}H]CDP (specific radioactivity, 1.7×10^7 cpm/μmole). After acid hydrolysis, dCMP was separated from CMP on Dowex 50, hydrogen form, columns, and radioactivity in dCMP was determined by liquid scintillation spectrometry as previously described.[5] Conditions for the reduction of all four ribonucleoside diphosphates were identical except that the appropriate tritium-labeled substrate replaced [^{3}H]CDP and 1.5 mM GTP or 0.5 mM dTTP replaced ATP as activators when ADP or GDP were the substrates, respectively. UDP reduction was activated by 1.6 mM ATP. After 45 min at 37°, the mixture was heated at 95° for 2 min, and 125 μg of *E. coli* alkaline phosphatase (Boehringer Mannheim) were added in 0.125 ml of 150 mM Tris·HCl, pH 8.2. The mixture was then incubated at 37° for 30 min and again heated at 95° for 2 min. The tubes were chilled, homologous ribonucleoside and deoxyribonucleoside carriers were added, and protein was removed by centrifugation. Adenosine was separated from 2'-deoxyadenosine on columns of Dowex-1, hydroxide form, as described by Dekker.[11] Radioactivity was corrected for recovery of 2'-deoxyadenosine carrier. The UDP reaction mixture was passed over a column of Dowex 50, hydrogen form, to remove adenosine, and the fractions containing radioactivity were eluted with 200 mM acetic acid. 2'-Deoxyuridine was separated from uridine and uracil on thinlayer cellulose sheets in isopropanol:concentrated HCl:water (65:16.6:18.4). For estimation of GDP reduction, guanosine and 2'-deoxyguanosine were separated on thin-layer cellulose sheets developed in 1-butanol:water:concentrated NH_4OH (86:14:5). In each case the entire lane was cut into 3-mm sections which were counted by liquid scintillation spectrometry. The fraction of the total radioactivity recovered in the 2'-deoxyribonucleoside was corrected to the radioactivity added initially as [³H]UDP or [³H]GDP.

One unit of enzyme activity is defined as the amount of protein which catalyzes the formation of 1 pmole of dCMP per minute. In experiments where ADP, UDP, and GDP were tested as substrates, a unit of enzyme activity is defined as the amount of protein which catalyzes the formation of 1 pmole of the corresponding deoxyribonucleoside per minute. Specific activity is expressed as units of activity per milligram of protein. Due to the presence of dithiothreitol and ATP in some fractions of the enzyme, protein was determined by a microprecipitation modification of the Lowry method as described by Clark and Jakoby[12] using bovine serum albumin as a reference. All reagents used in the purification and enzyme assays were prepared in deionized water maintained at a specific resistance greater than 18 Mohms-cm by Continental Water Conditioning Company.

Purification of the Enzyme

The data in Table I summarize the purification of ribonucleotide reductase from 690 g of frozen bone marrow. Steps 1–3 were performed essentially as previously described.[5] All procedures were carried out at 4° unless stated otherwise. Columns of Sephadex G-25 and G-200 were equilibrated with 20 mM Tris·HCl, pH 7.6, containing 2 mM sodium EDTA, pH 7.0, 2 mM dithiothreitol, and 10 mM $MgCl_2$, a solution referred to in this report as buffer A.

Step 1. Preparation of Crude Extract. Fifty-gram portions of frozen bone marrow were added to 50 ml of water in an ice-cold 250-ml cup. After 10 min in ice, the marrow was homogenized at full speed for two 45-sec intervals in a Sorvall Omnimixer. The homogenates were com-

[11] C. A. Dekker, *J. Am. Chem. Soc.* **87**, 4027 (1965).
[12] J. F. Clark and W. B. Jakoby, *J. Biol. Chem.* **245**, 6065 (1970).

TABLE I
PURIFICATION OF RIBONUCLEOTIDE REDUCTASE

Step	Volume (ml)	Total protein (mg)	Total activity (units)	Specific activity (units/mg)
1. Crude extract	707	16,190	21,920	1.35
2. Ammonium sulfate *	89	3,765	22,870	6.07
3. Sephadex G-200	99	2,841	40,180	14.1
4. dATP-Sepharose	1.6	4.5	37,670	8370

bined and centrifuged at 12,000 g for 30 min. The upper fatty layer was removed, and the supernatant containing the enzyme activity was centrifuged again at 20,000 g for 30 min.

Step 2. Ammonium Sulfate Precipitation. Ammonium sulfate saturated at 4° (474 ml) and 4.7 ml of 500 mM dithiothreitol were added to 707 ml of the supernatant to bring the concentrations to 0.40 saturation and 2 mM, respectively. After 30 min in ice, the suspension was centrifuged at 20,000 g for 30 min. The pellet containing the enzyme activity was washed with 40 ml of 0.50 saturated ammonium sulfate containing 2 mM dithiothreitol, and the suspension was centrifuged at 20,000 g for 20 min. This wash was repeated twice. The pellet was dissolved in enough buffer A to give approximately 80 ml. A small amount of insoluble material was removed by centrifugation at 20,000 g for 30 min. The supernatant was made 10% in sucrose and was frozen in liquid nitrogen in four equal portions which were stored at −70°.

Step 3. Sephadex G-200 Column. One portion of the enzyme solution (22 ml) from step 2 was thawed at room temperature before layering it underneath the buffer A on top of a Sephadex G-25 column (2.5 × 36 cm). Buffer A was passed down the column of Sephadex G-25 into a capillary tubing connected to the bottom of a Sephadex G-200 column (2.5 × 90 cm). After all protein eluted from the Sephadex G-25 had entered the bottom of the column of Sephadex G-200, the tubing leading from the Sephadex G-25 column was disconnected. Buffer A from a Mariotte flask was then allowed to flow upward through the Sephadex G-200 column at a rate of about 20 ml/hr, and 10-ml fractions were collected. The first UV-absorbing material (40 ml) to emerge from the Sephadex G-200 column had a milky appearance and lacked ribonucleotide reductase activity. The fractions comprising the next 200 ml were pooled, combined with an equal volume of saturated ammonium sulfate containing 2 mM dithiothreitol, and allowed to stand in ice for 2 hr before centrifuging at 20,000 g for 30 min. The pellet was dissolved in 10 mM histidine·HCl, pH 7.0, containing 2 mM dithiothreitol and was frozen and stored as before. After the remaining three portions of enzyme solution from step 2 were purified by the same procedure, the four Sephadex G-200 eluates were pooled and dialyzed against 1-liter portions of the same buffer which were changed every hour until the dialysate was free of ammonium sulfate. The enzyme preparation was frozen and stored as above.

Step 4. Chromatography on dATP-Sepharose. A solution of ribonucleotide reductase from step 3 of the purification containing approximately 40,000 units of activity and 2841 mg of protein was made 100 mM in Tris·HCl, pH 7.6, and 10 mM in MgCl$_2$, centrifuged to remove insoluble material, and added to a column (0.9 × 7.0 cm) of dATP-Sepharose equilibrated with the same buffer. The column was washed with 200 ml of buffer to remove the unbound fraction and to eliminate any material absorbing light at 280 nm. A loosely bound fraction was then eluted with 35 ml of this buffer containing 3 mM ATP adjusted to pH 7.6 with sodium hydroxide. After a small amount of a moderately bound fraction was removed with 30 ml of buffer containing 300 mM sodium chloride, the tightly bound active fraction was eluted with 37 ml of buffer containing 100 mM ATP adjusted to pH 7.6. Contents of all the tubes collected with each eluting solution were pooled, and the four separate fractions were dialyzed overnight against saturated ammonium sulfate containing 5 mM dithiothreitol. Each suspension was collected along with saturated ammonium sulfate washes of each dialysis bag. After centrifugation at 20,000 g for 30 min, the pellets were dissolved in 10 mM histidine·HCl, pH 7.0, containing 2 mM dithiothreitol and were dialyzed against 800-ml portions of the same buffer changed every hour until the dialysate was free of ammonium sulfate. Insoluble material was removed by centrifugation at 20,000 g for 20 min. The ribonucleotide reductase activity and protein in the four fractions obtained by chromatography on dATP-Sepharose are shown in Fig. 1.

Stabilization of the Enzyme Activity by Glycerol and ATP

When stored for 6 months in 10 mM histidine·HCl, pH 7.0, containing 2 mM dithiothreitol and approximately 2 mM ATP which remains in the preparation after elution of the enzyme from the dATP-Sepharose column, the purified enzyme appeared to aggregate and lost approximately 60 and 80% of its activity at $-70°$ and $-15°$, respectively. While addition of 200 mM potassium chloride to the above medium did not

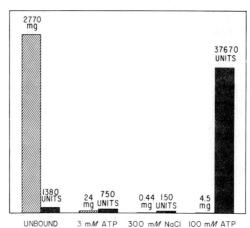

Fig. 1. Purification of bone marrow ribonucleotide reductase by affinity chromatography on dATP-Sepharose. Protein (2841 mg) from step 3 of the purification (Table I) was chromatographed on a column (0.9 × 7.0 cm) of dATP-Sepharose as described in the text. Total protein, diagonal bars; total units of enzyme activity, solid bars.

protect the activity of the enzyme during storage, the addition of glycerol to a final concentration of 40% (v/v) stabilized the enzyme activity completely when stored at either −70° or −15° for 6 months. In experiments to be described elsewhere both glycerol and ATP were required to prevent aggregation and loss of enzyme activity. It has not yet been determined whether dithiothreitol is essential.

Analytical Electrophoresis on Polyacrylamide Gels

All polyacrylamide gels were prepared in 5 × 125 mm tubes. Electrophoresis on nondenaturing gels containing 4% acrylamide was performed by the method of Davis.[13] Sodium dodecyl sulfate (SDS)–polyacrylamide electrophoresis was performed as described by Laemmli.[14] The stacking gel (8 mm) of 3% acrylamide was layered over the separation gel (85 mm) of 8% acrylamide.

Electrophoresis of the purified reductase on nondenaturing polyacrylamide gels yields one major diffuse band and one very faint band, but attempts to recover enzyme activity from the gels have not been successful. SDS–polyacrylamide electrophoresis of the tightly bound fraction separates a major band which migrates slightly faster but overlaps the band produced by the *L. leichmanii* ribonucleotide reduc-

[13] B. J. Davis, *Ann. N.Y. Acad. Sci.* **121**, 404 (1964).
[14] U. K. Laemmli, *Nature (London)* **227**, 680 (1970).

tase (M.S. 76,000)[15] and five minor bands. However, it cannot be assumed that the major band represents the bone marrow enzyme.

Properties

Effect of Enzyme Concentration

The anomalous effect of dilution on the activity of the purified bone marrow reductase illustrated in Fig. 2 is similar to that observed in extracts of several mammalian ribonucleotide reductases.[4,5,16–18] When activity is plotted against enzyme concentration, there is an upward curvature of the line at high dilutions of the enzyme. However, a linear relationship is observed when the same data are replotted as velocity against the square of the enzyme concentration, indicating a second-

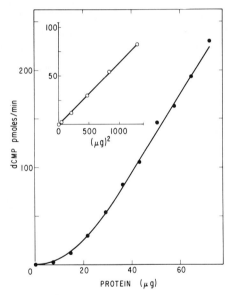

FIG. 2. The relationship of velocity to enzyme concentration. The rate of dCMP formation was assayed under standard conditions in the presence of the purified bone marrow ribonucleotide reductase. The protein concentration was varied as indicated. ●, protein expressed as μg; ○, protein expressed as (μg)².

[15] A. K. Chen, A. Bhan, S. Hopper, R. Abrams, and J. S. Franzen, *Biochemistry* **13**, 654 (1974).

[16] P. R. Walsh, Ph.D. Dissertation, University of Pittsburgh, Pittsburgh, Pennsylvania (1970).

[17] J. G. Cory, *Cancer Res.* **33**, 993 (1973).

[18] D. M. Peterson and E. C. Moore, *Biochim. Biophys. Acta* **432**, 80 (1976).

order reaction at low enzyme concentrations. The linearity between velocity and enzyme concentration at low dilutions of the enzyme denotes a first-order or pseudo first-order reaction at high enzyme concentrations. These data suggest that the purified bone marrow reductase isolated by affinity chromatography on dATP-Sepharose may consist of two components which dissociate in dilute solution. An alternative explanation of the phenomenon, inactivation of the enzyme in dilute solution, is unlikely since the effect of dilution is fully reversible.[5]

Iron Requirement

The degree of stimulation of the activity of the bone marrow reductase by iron varies considerably in different preparations of the dATP-Sepharose tightly bound fraction, i.e., from only slight stimulation to essentially complete dependence on added iron. This observation suggests that a loosely bound iron cofactor may be required for the activity. The data in Table II show that iron chelators which completely inhibit the activity of the bone marrow enzyme are without effect in the presence of excess iron.

TABLE II

THE EFFECT OF IRON AND IRON CHELATORS ON RIBONUCLEOTIDE REDUCTASE ACTIVITY[a]

Additions	dCMP formed (pmoles/45 min)
None	190
+ BPS	$<10(50)^b$
+ HQS	$<10(140)^b$
Fe (50 μM)	610
+ BPS	730
+ HQS	690
Fe (100 μM)	670
+ BPS	720
+ HQS	700

[a] Protein (12 μg) from the tightly bound fraction which was chromatographed twice on dATP-Sepharose was preincubated at 0° for 30 min in the presence of all the components except CDP which was then added to start the reaction. The concentration of ferrous ammonium sulfate (Fe) was varied as indicated in the presence and absence of either 50 μM sodium bathophenanthrolinesulfonate (BPS) or 100 μM lithium 8-hydroxy-5-quinolinesulfonate (HQS).
[b] 10 μM chelator.

TABLE III

REDUCTION OF RIBONUCLEOSIDE DIPHOSPHATES[a]

dATP-Sepharose fraction added	Total protein (mg)	CDP[b]	ADP[b]	UDP[b]	GDP[b]
Unbound	770	1.67	0.50	0.83	[c]
Loosely bound	12	11.8	5.3	9.2	[c]
Tightly bound	5.3	2667	2083	2217	1667

[a] The loosely bound and tightly bound fractions used in this experiment were eluted from the dATP-Sepharose column with 2 mM and 100 mM ATP, respectively, without the intermediate 300 mM sodium chloride elution step. All other conditions were described in the text.

[b] In units/mg protein.

[c] Small amounts of deoxyguanosine formed were difficult to measure accurately due to incomplete separation of labeled guanosine.

Ribonucleoside Diphosphate Substrates

The rate of reduction of CDP under standard assay conditions was used to measure ribonucleotide reductase activity throughout the purification procedure. However, the capacity of the bone marrow ribonucleotide reductase to utilize all four ribonucleoside diphosphate substrates was tested in the presence of three dATP-Sepharose fractions (Table III). ATP is required for the reduction of CDP,[5] and GTP, reported to produce maximal dADP formation in HeLa cells,[19] increased the reduction of ADP by the bone marrow enzyme 4-fold. As found in a Novikoff hepatoma ribonucleotide reductase preparation,[20] dTTP and ATP markedly increased the rate of reduction of GDP and UDP, respectively. The specific activities for the reduction of CDP, UDP, ADP, and GDP of the purified enzyme were comparable, while these activities in the unbound and loosely bound dATP-Sepharose fractions were lower by two orders of magnitude.

Comments

Ribonucleotide reductase of rabbit bone marrow has been purified by chromatography on dATP-Sepharose, a method used previously to

[19] D. M. Peterson and E. C. Moore, Fed. Proc., Fed. Am. Soc. Exp. Biol. 33, 1271 (1974).
[20] E. C. Moore and R. B. Hurlbert, J. Biol. Chem. 241, 4802 (1966).

purify the B_1 protein of the *E. coli* ribonucleotide reductase[21,22] and both proteins of the phage T_4-induced ribonucleotide reductase.[9,10] In this one step alone there is approximately a 600-fold purification with almost complete recovery of the enzyme activity.

The high concentration of ATP required for elution of the enzyme from dATP-Sepharose suggests either that ATP competes only weakly for a dATP binding site or even that removal of the enzyme with ATP may be due to a simple ion-exchange reaction. The latter is unlikely since the tightly bound fraction cannot be eluted with 300 mM sodium chloride. The use of dATP to elute the reductase from the dATP-Sepharose column has been avoided since it is a potent inhibitor of CDP reduction and is difficult to remove from the purified enzyme.

The availability of a purified mammalian ribonucleotide reductase should simplify separation and characterization of the components of the enzyme system.

Acknowledgment

I wish to thank David Kopp of the Life Sciences Department, University of Pittsburgh, for his assistance in the preparation of dATP-Sepharose and Joan Riley and Janice Dorman for technical assistance.

This work was supported by research grants GB-23070, BMS-70-00696, and PCM-76-05625 from the National Science Foundation and GM-20759 from the National Institutes of Health.

[21] J. A. Fuchs, H. O. Karlström, H. R. Warner, and P. Reichard, *Nature (London) New Biol.* **238**, 69 (1972).
[22] L. Thelander, *J. Biol. Chem.* **248**, 4591 (1973).

[32] Ribonucleoside Triphosphate Reductase from *Lactobacillus leichmannii*

By RAYMOND L. BLAKLEY

Ribonucleoside triphosphate

$$+ R(SH)_2 \xrightarrow{\text{adenosyl cobalamin}} \text{deoxyribonucleoside triphosphate} + RS_2 + H_2O$$

Ribonucleotide reductases that require adenosylcobalamin (coenzyme B_{12}) for activity (EC 1.17.4.2) are present in a number of lactobacilli,[1] clostridia,[2,3] rhizobia,[4] pseudomonads,[3] corynebacteria,[3]

[1] R. L. Blakley, R. K. Ghambeer, P. F. Nixon, and E. Vitols, *Biochem. Biophys. Res. Commun.* **20**, 439 (1965).
[2] R. H. Abeles and W. S. Beck, *J. Biol. Chem.* **242**, 3589 (1967).
[3] F. K. Gleason and H. P. C. Hogenkamp, *Biochim. Biophys. Acta* **277**, 466 (1972).
[4] J. Cowles, H. Evans, and S. Russell, *J. Bacteriol.* **97**, 1460 (1969).

micrococci,[3] and thermophyllic bacteria.[3,5] The only eukaryotes known to contain the cobalamin-dependent reductase are the euglenoid flagellates (*Euglena* and *Astasia*).[3,6] The reductase from *L. leichmannii* is specific for reduction of ribonucleoside triphosphates, but the enzyme from some other species reduces diphosphates[6,7] and the latter are the preferred substrates in some cases.[7] Like the non-B_{12} ribonucleotide reductase of *E. coli*[8] and mammals[9] the lactobacillus enzyme has as its physiological reducing substrate reduced thioredoxin. This is a low-molecular-weight protein which is converted from its disulfide to its dithiol form at the expense of NADPH by a specific flavoprotein, thioredoxin reductase. However, as with the reductase from other sources, 1,3- or 1,4-dithiols such as dihydrolipoate, dithiothreitol, and dithioerythitol can substitute for reduced thioredoxin as reducing substrate for the lactobacillus enzyme.[10,11]

Assay Methods

Colorimetric Method

Principle. This is the method used routinely and is applicable at all stages of purification. With ATP and dihydrolipoate as substrates, the formation of dATP is estimated colorimetrically by the use of the diphenylamine reagent.[12] Before color can be developed with the latter, however, the interfering sulfhydryl groups of dihydrolipoate must be removed. This is achieved by alkylation with chloroacetamide. Reagent blanks, which contain heat-inactivated enzyme and all other components of the assay system, are included to correct for color produced by reaction of assay components with diphenylamine. Standards, containing heat-inactivated enzyme, dAMP, and assay components, are included with each assay.

Reagents

Sodium acetate, 4 M
Potassium phosphate buffer, 1 M, pH 7.3
ATP, 0.1 M, pH 7.0

[5] G. N. Sando and H. P. C. Hogenkamp, *Biochemistry* 12, 3316 (1973).
[6] F. K. Gleason and H. P. C. Hogenkamp, *J. Biol. Chem.* 245, 4894 (1970).
[7] J. R. Cowles and H. J. Evans, *Arch. Biochem. Biophys.* 127, 770 (1968).
[8] A. Larsson and P. Reichard, *Prog. Nucleic Acid Res. Mol. Biol.* 7, 303 (1967).
[9] E. C. Moore, this series, Vol. 12A, p. 155.
[10] E. Vitols and R. L. Blakley, *Biochem. Biophys. Res. Commun.* 21, 466 (1965).
[11] R. L. Blakley, *Fed. Proc., Fed. Am. Soc. Exp. Biol.* 25, 1633 (1966).
[12] R. L. Blakley, *J. Biol. Chem.* 241, 176 (1966).

Adenosylcobalamin, 0.2 mM (protect from light)

Sodium dihydrolipoate, 0.5 M, pH 7.0, containing 16.7 mM EDTA. Preparation of dihydrolipoate is carried out as described by Gunsalus and Razzell,[13] except that distillation is unnecessary. The solution is stored frozen under nitrogen.

Chloroacetamide, 0.5 M in 0.25 M potassium phosphate buffer, pH 7.3

dAMP, 2.0 mM

Diphenylamine reagent. This is prepared by dissolving 2 g of recrystallized (ethanol-water) diphenylamine in 100 g of glacial acetic acid and adding 3.0 ml of concentrated H_2SO_4 and 1.0 ml of 0.05 M cupric acetate.

Procedure. Assays are carried out in 13 × 100 mm test tubes which are stood in an ice bath until additions have been completed. The enzyme (about 0.02 unit) is added to each tube. The enzyme in reagent-blank and standard tubes is inactivated by heating the covered tubes at 100° for 3 min. After cooling all tubes to 0°, 0.1 ml of dAMP solution is added to standards. To each tube is then added 0.125 ml of sodium acetate solution, 0.025 ml of buffer solution, 0.05 ml of ATP solution, and water so that the final volume will be 0.50 ml after dihydrolipoate and coenzyme addition. Subsequent operations (until incubation at 37° is finished) are conducted in a dim light. To each tube is added 0.02 ml of adenosylcobalamin solution and 0.03 ml of dihydrolipoate solution. The tubes are quickly flushed with nitrogen, stoppered, the contents mixed, and the rack of tubes transferred from the ice bath to a vigorously stirred bath maintained at 37°. Because of a very high temperature coefficient[14] this method of initiating the reaction is satisfactory. After 10 min at 37° the rack of tubes is returned to the ice bath and to each tube is added 0.4 ml of chloroacetamide solution. The contents are mixed well, and the tubes are covered (marbles) and heated at 100° for 40 min. After cooling the tubes, 2.0 ml of diphenylamine reagent are added to each, the contents mixed well, and the covered tubes kept at 37° for up to 20 hr. A purple color develops in the first few hours and later changes to blue. Absorbance at 595 nm can be read after 4 hr, but best results are obtained after 16 hr.[12] The mean reagent blank value is subtracted from other absorbance values, and the dATP formed is calculated from the corrected values by comparison with the corrected absorbance of the standard.

---

[13] I. C. Gunsalus and W. E. Razzell, this series, Vol. 3, p. 941.
[14] Y. Tamao and R. L. Blakley, *Biochemistry* **12**, 24 (1973).

Spectrophotometric Method

Principle. ATP and reduced thioredoxin are used as substrates, and the reduction of ribonucleotide is coupled with the thioredoxin reductase system as follows[15]:

$$\text{Thioredoxin} + \text{NADPH} + \text{H}^+ \xrightarrow{\text{thioredoxin reductase}} \text{reduced thioredoxin} + \text{NADP}^+$$

$$\text{ATP} + \text{reduced thioredoxin} \xrightarrow[\text{adenosylcobalamin}]{\text{ribonucleotide reductase}} \text{dATP} + \text{thioredoxin} + \text{H}_2\text{O}$$

With thioredoxin reductase in excess, the rate of dATP reduction is equivalent to the rate of NADPH oxidation, which is measured continuously by the decrease in absorbance at 340 nm. A reference cell, containing all components except adenosylcobalamin, is used to correct for extraneous reactions, particularly NADPH oxidase.

Reagents

Potassium phosphate buffer, 0.5 M, pH 7.5
EDTA, 40 mM, pH 7.3
Sodium acetate, 4 M
ATP, 0.1 M, pH 7.3
NADPH, 2 mM, in Tris·HCl buffer, 0.05 M, pH 7.3
Adenosylcobalamin, 0.2 mM (protect from light)
Thioredoxin, 120 μM, prepared according to Laurent *et al.*[16] The final purification on Sephadex G50 may be omitted.
Thioredoxin reductase, 6 units/ml, prepared according to Moore *et al.*[17] The final purification on Sephadex G100 may be omitted.

Procedure. The reaction is carried out in cuvettes having 5-mm path length and requiring a volume of only 0.5 ml. To reaction and reference cells the following are added: 0.025 ml of buffer, 0.125 ml of sodium acetate solution, 0.05 ml of EDTA solution, 0.05 ml of ATP solution, 0.05 ml of NADPH solution, 0.03 ml of thioredoxin, 0.01 ml of thioredoxin reductase, ribonucleotide reductase (about 0.05 unit), and water to give a volume of 0.48 ml in the reaction cell and 0.50 ml in the reference cell. The cells are quickly stoppered, the contents mixed by inversion, and the cells placed in a double-beam recording spectrophotometer thermostatted at 37°. During a 3-min period of thermal equilibration the absorbance recorded at a sensitivity of 0.1 absorbance full scale

[15] E. Vitols, C. Brownson, W. Gardiner, and R. L. Blakley, *J. Biol. Chem.* **242**, 3035 (1967).
[16] T. C. Laurent, E. C. Moore, and P. Reichard, *J. Biol. Chem.* **239**, 3436 (1964).
[17] E. C. Moore, P. Reichard, and L. Thelander, *J. Biol. Chem.* **239**, 3445 (1964).

should remain constant within 0.003 absorbance units. The instrument should be adjusted so that the initial absorbance reading is about 0.09 so that the decrease in absorbance when the reaction is initiated can be measured. Alternatively, if this is not possible on the instrument used, the reference solution can be placed in the sample compartment, the reaction cell in the reference compartment, and the initial absorbance adjusted close to 0.01. NADPH oxidation in the reaction cell then causes an absorbance increase. After thermal equilibration the reaction cell is quickly removed, 0.02 ml of adenosylcobalamin solution is added in dim light, the contents are mixed by inversion, and the rate of absorbance change is measured over the next 2 min.

Definition of Units of Activity. In the literature to date, a unit of activity has usually been defined as the enzyme catalyzing formation of 1 μmole of dATP per hour. However, there appears to be no reason why the unit of activity should not be the international unit, that is, the enzyme catalyzing formation of 1 μmole of dATP per minute. This definition is accordingly used throughout this chapter. Specific activity is defined as units of activity per milligram of protein. Protein concentration is determined by the method of Lowry *et al.*[18] as modified by Hartree.[19] For highly purified solutions the standard reference solution should be pure ribonucleotide reductase, the concentration of which has been determined refractorimetrically in the analytical ultracentrifuge with a double-sector, synthetic-boundary cell.[20] Protein concentration of highly purified solutions can also be determined spectrophotometrically[21] from the absorption coefficient $E^{1\%}_{280 \text{ nm}} = 13.3 \pm 0.1$.

Alternative Assay Procedures. In the colorimetric method removal of interfering dithiol by mercuric chloride has been described.[22] A radioactive assay using labeled CTP as substrate is employed by some workers.[23]

Purification Procedure

Growth of Organism. The organism was originally obtained as ATCC 4797, though there is some evidence that the strain currently used in our

[18] O. H. Lowry, N. J. Rosebrough, A. L. Farr, and R. J. Randall, *J. Biol. Chem.* **193**, 265 (1951).
[19] E. F. Hartree, *Anal. Biochem.* **48**, 422 (1972).
[20] J. Babul and E. Stellwagen, *Anal. Biochem.* **28**, 216 (1969).
[21] P. J. Hoffmann and R. L. Blakley, *Biochemistry* **14**, 4804 (1975).
[22] F. Stutzenberger, *Anal. Biochem.* **56**, 294 (1973).
[23] A. K. Chen, A. Bhan, S. Hopper, R. Abrams, and J. S. Franzen, *Biochemistry* **13**, 655 (1974).

laboratory has developed some differences from the parent strain.[24] The medium is prepared as follows. Glucose (20 g), sodium acetate·3 H_2O (10 g), potassium acetate (10 g), casein hydrolysate (1.5 g), and magnesium acetate (0.15 g) are dissolved in 150 ml of water. Tween 80 (1 g) is then added and dissolved with stirring. L-Tryptophan (50 mg), L-tyrosine (50 mg), and KH_2PO_4 (85 mg) are dissolved in 20 ml of water (heating), and the solution is added to the main solution. Adenine (16 mg), guanine (16 mg), uracil (16 mg), and L-cystine (100 mg) are dissolved in about 5 ml of 5 N HCl with heating and added to the main solution with mixing. Pyridoxine hydrochloride (4 mg), nicotinic acid (2 mg), folic acid (1 mg), and calcium pantothenate (1 mg) are dissolved in 5 ml of 0.01 N NaOH and added to the main solution. L-Cysteine hydrochloride (1 g) is added immediately before adjusting to pH 6.0, making up to 1 liter, and sterilizing at 15 psi for 30 min. If the medium is not to be inoculated the same day it should be stored frozen. The organism is maintained by monthly stab subculture in solid medium of the same composition. Although the organism requires vitamin B_{12} for growth, the vitamin present in the commercial preparation of casein hydrolysate is adequate for good growth. Additional cobalamin represses reductase synthesis.[25]

The pH of the fermentation is adequately controlled by the acetate buffer, and no aeration is necessary. Reductase production is a maximum as the culture enters the stationary phase which corresponds to an absorbance of 1.1 to 1.2 at 660 nm. When the culture reaches this stage it is cooled to below 20° within 15 min and harvesting is commenced. Cooling is continued down to 5° as harvesting proceeds, and harvesting should be complete within 90 min. The paste is frozen as soon as harvesting is complete—the yield is about 2 g/liter. Frozen paste may be stored at −100° for at least 3 months without significant loss of activity. Frozen paste has been supplied commercially by Grain Processing of Muscatine, Iowa, and by the New England Enzyme Center, Boston.

Step 1. Frozen cell paste (about 1 lb) is thawed and the cells suspended in 0.1 M sodium citrate buffer, pH 6.3 (5°), containing 1 mM 2-mercaptoethanol and 0.02% sodium azide. All subsequent steps are carried out at 5° unless otherwise stated. The cells are disrupted by two passages through a Manton–Gaulin mill at 7500 psi. After adjusting the pH of the extract to 5.6, the cell debris is removed by centrifugation (Sorvall GSA rotor, 12,000 rpm, 15 min). The supernatant is dialyzed against 0.1 M sodium citrate buffer at pH 5.6, containing 1 mM 2-mercaptoethanol and 0.02% sodium azide, for 16–24 hr.

[24] R. L. Blakley, *J. Biol. Chem.* **240**, 2173 (1965).
[25] R. K. Ghambeer and R. L. Blakley, *J. Biol. Chem.* **241**, 4710 (1966).

Step 2. The enzyme solution is diluted with the same buffer as used for dialysis to a protein concentration of 10 mg/ml. A protamine sulfate solution is prepared by stirring 2 g of the solid/100 ml of water and adjusting to pH 5.6 at room temperature with 5 N KOH. It is not necessary to remove undissolved material. The dialyzed enzyme solution in batches that can be centrifuged at one time (1–4 liter) is vigorously stirred at 0–5° while the freshly prepared protamine sulfate solution (at 25°) is added during a 15- to 20-min period. Stirring is continued for 15 min after all the protamine sulfate has been added, and the mixture is then centrifuged for 10 min (25,000 g). The optimum volume of protamine sulfate solution to be added must be determined in a preliminary trial with a series of 10-ml portions of enzyme solution; it varies from 0.05–0.16 volume depending on the batch of cells and the batch of protamine sulfate.

Step 3. The supernatant fraction from the previous step is then made 2 mM in EDTA before precipitation of enzyme with ammonium sulfate (360 g/liter). The precipitate is recovered by centrifugation, suspended in a minimum volume of 0.1 M sodium citrate buffer, pH 5.5, containing 1 mM 2-mercaptoethanol and 0.02% sodium azide, and dialyzed overnight against the same buffer.

Step 4. Any insoluble material is removed by centrifugation before applying the dialyzed solution (maximum volume, 500 ml) to a Sephadex G-100 column (16 × 130 cm), previously equilibrated with the same buffer used for dialysis.

Step 5. The active fractions from the column are combined and the enzyme reprecipitated by addition of solid ammonium sulfate (440 g/ liter) after preliminary addition of EDTA (to 2 mM). The precipitate is suspended in and dialyzed against 0.1 M sodium citrate buffer, pH 5.0, containing 1 mM mercaptoethanol and 0.02% sodium azide for 36–48 hr if it is to be subjected immediately to affinity chromatography. When solutions are to be stored for several days mercaptoethanol is omitted from the dialysis buffer.

Affinity Chromatography

Pure enzyme can be most conveniently prepared by subjecting the enzyme preparation from step 5 to chromatography on Sepharose to which dGTP has been attached through a hexanolamine spacer group (Fig. 1). The substituted Sepharose is prepared as follows.

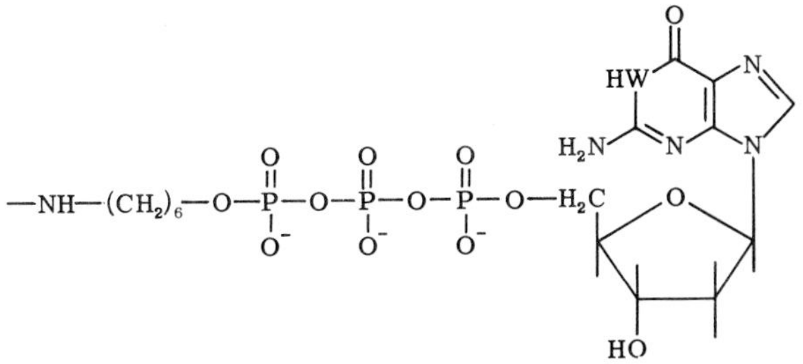

FIG. 1

FIG. 1

6-Amino-1-hexanol Phosphate (I)[26]

6-Amino-1-hexanol (84 g, 0.72 mole) is heated with crystalline phosphoric acid (73 g, 0.74 mole) to 150° under vacuum (<0.1 mm Hg) for 20 hr. The reaction mixture is cooled to room temperature and the solid residue dissolved in water (1 liter). The solution is adjusted to pH 10.5 with 5 N LiOH and chilled in an ice bath. Precipitated lithium phosphate is removed by filtration through a layer of Celite. The filtrate is adjusted to pH 3 with glacial acetic acid and passed through Dowex 50-X2 (pyridinium form, 200–400 mesh, 6.2 × 60 cm). The ninhydrin-positive eluate is evaporated to dryness under vacuum, and the residue is dissolved in water (540 ml) with warming and crystallized by the addition of ethanol (540 ml). After recrystallization in the same way the material has mp 246–250°; the yield is 69 g, 58%.

N-Trifluoroacetyl 6-Amino-1-hexanol Phosphate (II)[26]

40 g of I (0.24 mole) are dissolved in 300 ml of water, the pH adjusted to 9.5 with 5 N LiOH, and the solution made to a volume of 400 ml and chilled in an ice bath. S-Ethyl trifluorothiol acetate (20 ml, 0.156 mole) is added with vigorous stirring to produce a fine dispersion. The temperature is maintained below 4°, and 5 N LiOH is added intermittently to maintain the pH at 9.5. Proton release ceases after 45 min, and a further 20 ml of ethyl trifluorothiol acetate is added while stirring, cooling, and pH adjustment continues. After a further 45 min another 10 ml of ethyl trifluorothiol acetate are added. Progress of the reaction is followed by spotting 5-μl samples of the reaction mixture on paper, spraying with ninhydrin, and heating at 100° for 5 min. The reaction is complete (2 hr) when only a faint pink color is obtained with ninhydrin. The solution is adjusted to pH 5 with trifluoroacetic acid, and then concentrated to

[26] R. Barker, K. W. Olsen, J. H. Shaper, and R. L. Hill, J. Biol. Chem. 247, 7135 (1972).

dryness under vacuum (40°). The residue is redissolved in water (1 liter) and then concentrated to dryness several times to remove residual reagents; it is finally dissolved in 1 liter of water and adjusted to pH 1.5 with trifluoroacetic acid. The solution is passed through a column containing 1 liter of Dowex 50-X8 (H^+ form, 200–400 mesh) to remove unreacted amine. The fractions containing the product can be detected by conductivity. The eluate is taken to dryness under vacuum at 45° and dried under a high vacuum over P_2O_5 and NaOH at room temperature for at least 24 hr. The residue, which usually partially solidifies, is dissolved in 200 ml of water and converted to the tributylammonium salt by treating with tributylamine (60 ml, 200 mmoles). The mixture is evaporated to dryness under vacuum and dried by repeated solution in dry dimethylformamide and evaporation to dryness under vacuum (45°). The residue is finally dissolved in 300 ml of dry dimethylformamide.

N-Trifluoroacetyl 6-Aminohexanol 1-Pyrophosphate (III)[21]

Tributylammonium phosphate is prepared by dissolving phosphoric acid (39.2 g, 0.40 mole) in 200 ml of water, adding tributylamine (74.1 g, 0.40 mole), and mixing vigorously. The solution is allowed to stand overnight before concentrating to a syrup. Water is removed by repeated evaporation to dryness after solution in dry dimethylformamide, and the residue is finally dissolved in 300 ml of dry dimethylformamide.

The solution of II is converted to the imidazolide by addition of carbonyldiimidazole (50 g, 0.31 mole) with stirring under a drying tube. Some suspended material dissolves, and there is considerable evolution of CO_2. After 6 hr at room temperature methanol (4.5 ml, 0.11 mole) is added to decompose excess diimidazole. After stirring for 30 min the solution of tributylammonium phosphate is added with stirring. A precipitate of imidazolium phosphate begins to form at once. After 24 hr under a drying tube the precipitate is removed by filtration through a layer of Celite, and after treating the clear filtrate with an equal volume of methanol the mixture is evaporated to a thick syrup. The syrup is dissolved in 1 liter of water and chromatographed at 4°C on a 6 × 100 cm column of Dowex 1-X2 (chloride form, 100–200 mesh) with a gradient formed with 15 liters of 0.01 N HCl in the mixing chamber and 15 liters of 0.55 M LiCl in 0.01 N HCl in the reservoir. A flow rate of about 300 ml/hr is used. Each fraction (~250 ml) is neutralized with 5 N LiOH soon after collection and a sample (0.05 ml) assay for acid-labile and inorganic phosphate. Fractions in the major peak of acid hydrolyzable phosphate (at about fraction 100) are combined and concentrated to dryness on a rotary evaporator (bath 45°). LiCl is extracted from the residue with 6 × 300 ml of ice-cold methanol and discarded. The residue is collected and dried under high vacuum overnight at room temperature. Yield is 42.8 g (56%). The crude III can be crystallized by addition of methanol or ethanol to the concentrated aqueous solution. Chromatography on cellulose in isobutyric acid–concentrated NH_4OH–water (66:1:33, v/v) or on PEI-cellulose with development by 1.0 M KCl shows the presence of only one phosphate compound (R_f 0.69 and 0.34 respectively) which is ninhydrin negative but becomes ninhydrin positive after treatment with alkali (pH 11.5 for several minutes at 25°).

P^3-(6-Aminohex-1-yl)dGTP (IV)[21]

The sodium salt of dGMP and the lithium salt of III are converted to the pyridinium salts by passage of an aqueous solution of the compound through a 4.4 × 30 cm column of Dowex 50-X2 (pyridinium form, 200–400 mesh). Fractions containing the dGMP are detected by spotting 5-μl samples on a cellulose TLC sheet containing fluorescent indicator. III is detected in the eluate from acid-labile phosphate in a 0.05-ml sample. The solution of the pyridinium salt of dGMP is treated with one molar equivalent of methyl tri-N-octylammonium hydroxide,[27] with ethanol added if necessary to give a homogeneous solution, and the solution is then concentrated to dryness and dried by repeated solution in dimethylformamide and evaporation. III was converted with tributylamine to the tributylammonium salt and dried similarly.

To a solution of methyl tri-N-octylammonium dGMP (13 mmole) in 140 ml of dimethylformamide are added 12 g (75 mmole) of carbonyldiimidizole. After stirring for 8 hr under a drying tube at room temperature, 2.55 ml (72 mmole) of methanol are added and the solution stirred for a further 40 min. Then a solution of the tributylammonium salt of III (43 mmole) in 500 ml of dimethylformamide is added and the reaction kept at room temperature for 24 hr with exclusion of moisture. The dimethylformamide is removed on the rotary evaporator, and 1 liter of 67% methanol is used to dissolve the syrupy residue. The product is purified by application of this solution to a Dowex 1-X2 column and chromatography in the same way as for III. Absorbance at 252 nm and acid-labile phosphate are determined on fractions of the eluate. The desired product, P^3-(6-(N-trifluoroacetyl)aminohex-1-yl)dGTP is in the major absorbing peak. LiCl is removed from the concentrated, combined fractions by passage through Bio-Gel P2 (200–400 mesh, 6.2 × 110 cm) equilibrated and developed with 0.05 M triethylammonium bicarbonate, pH 7.0. Ultraviolet-absorbing fractions are combined and incubated at pH 11.5 at room temperature overnight to form IV. The solution of the

[27] R. Letters and A. M. Michelson, *Bull. Soc. Chim. Biol.* **45**, 1353 (1963).

latter is purified on DEAE cellulose (4.4 × 45 cm, bicarbonate form). Elution is accomplished with a gradient generated by water (3.5 liters) in the mixing chamber and 0.2 M triethylammonium bicarbonate, pH 7.5 (3.5 liters) in the reservoir. Fractions containing the major absorbing peak are combined, concentrated to a syrup, and freed from triethylammonium bicarbonate by repeated addition of ethanol and evaporation. The final residue is dissolved in water (100 ml), adjusted to pH 7.5, and stored frozen.

Coupling to Sepharose. Sepharose 4B is activated with CNBr according to the procedure of Porath for producing highly activated gel,[28,29] and 600 g of the well-washed, activated gel are stirred with 2 mmole of IV in 300 ml of water at pH 10.0 at 4°C for 18 hr. The gel is then washed extensively with water in a column at room temperature. Approximately 2.4 μmole of ligand are bound per milliliter of packed wet gel as estimated by the loss of ultraviolet-absorbing material.

Chromatography on dGTP-Sepharose. A portion of the dialyzed solution from step 5 of the enzyme preparation containing about 600 units of activity is adjusted to pH 7.3 at 20° after addition of sodium phosphate buffer to a final concentration of 0.1 M. The solution is then subjected to affinity chromatography at 25° on a 2.5 × 95 cm column equilibrated with 0.1 M sodium phosphate, pH 7.3. The column is washed with the same buffer until the absorbance at 280 nm falls to 0.2 (about 1.5 bed volumes), and then with 2 bed volumes of 0.1 M sodium citrate buffer, pH 6.3. These washes elute extraneous proteins as well as low-activity forms of the reductase in which the protein is either oxidized or has scrambled disulfide bridges.[30] The high-activity form of the enzyme is then eluted with 2.0 M urea in 0.1 M sodium citrate buffer, pH 6.3. Fractions in this peak with absorbance of 0.05 or greater are pooled and dialyzed against 500 volumes of 0.1 M sodium citrate buffer, pH 5.0, containing 0.02% sodium azide at 4° over 24 hr. The enzyme is brought to a concentration of about 10 mg/ml by pressure dialysis or ultrafiltration (UM 10 Amicon membrane) and stored under N_2 at 0°. The behavior of the enzyme during chromatography is critically dependent on the temperature, pH, and ionic strength. Deviation from the conditions specified, particularly the temperature (25°), gives much less satisfactory results. Weakly binding reductase forms can be partially converted to high-activity enzyme by concentrating, treating with 5 mM dihydrolipoate at pH 7.3 and 0° for 30 min, and rechromatography at 25°.

[28] J. Porath, K. Aspberg, H. Drevin, and R. Axén, *J. Chromatogr.* **86**, 53 (1973).
[29] J. Porath, this series, Vol. 34, p. 23.
[30] D. Singh, Y. Tamao, and R. L. Blakley, *Adv. Enzyme Regul.* **15**, 81 (1977).

PURIFICATION OF RIBONUCLEOTIDE REDUCTASE FROM *L. leichmannii*[a]

Purification step	Volume (ml)	Protein (g)	Activity (units)	Specific activity (units/mg)	Recovery (%)
Crude extract	1710	20.9	1660	0.08	
Protamine sulfate precipitation	1700	12.8	1550	0.12	94
First ammonium sulfate	170	8.1	1440	0.18	87
Sephadex G-100 fractions	1850	1.62	1290	0.80	78
Second ammonium sulfate	72	1.51	1280	0.85	77
Affinity column fractions[b]	5.8	0.24	663	2.82	40

[a] Results for 459 g of cell paste.
[b] Calculated from results with a portion of the material of the previous step. An additional 5.4% of the activity was obtained as high-activity enzyme by treating weakly bound forms from the first affinity column with dihydrolipoate and rechromatographing.

The table summarizes data for a typical preparation.

Properties

Purity. The purified enzyme is homogeneous by polyacrylamide gel electrophoresis, giving a single band at a loading of 100 µg per gel.[21] A single symmetrical schlieren peak was obtained in the ultracentrifuge, and results of Edman degradation were also consistent with homogeneity.

Stability. The reductase is fairly stable under the conditions of storage specified, losing only about 23% of its activity in 15 months.[30] On freezing there is some loss (10–20%), but this does not increase during a prolonged period in the frozen state at −100°.

Molecular Properties. The molecular weight is about 76,000, and the enzyme consists of a single polypeptide chain.[31] The amino-terminal sequence is Ser- Glu- — Ile- Ser- Leu- Ser- Ala with the uncertainty that positions 5 and 7 might be either Ser or Cys. The carboxyl terminus is Lys. The isoelectric point is 4.5.[32]

Specificity and Regulation. This reductase is specific for the reduction of ribonucleoside triphosphates and is subject to allosteric activating effects by the deoxyribonucleoside triphosphate products.[15] The magni-

[31] D. Panagou, M. D. Orr, J. R. Dunstone, and R. L. Blakley, *Biochemistry* **11**, 2378 (1972).
[32] M. D. Orr, R. L. Blakley, and D. Panagou, *Anal. Biochem.* **45**, 68 (1972).

tude of these activating effects depends on the concentration of monovalent cations,[15] the latter also having an activating effect which is nonspecific and decreases in the order $Na^+ > K^+ > Rb^+, Cs^+, NH_4^+ \gg Li^+$.[11] The activating effects of deoxyribonucleoside triphosphates are specific[15]: dGTP is a specific activator for ATP reduction; dATP for CTP; dCTP for dUTP; and dTTP for ITP. Only the deoxyribonucleoside triphosphates are able to cause activation of ribonucleotide reduction: neither the diphosphates nor monophosphates are activators. A variety of 1,3- and 1,4-dithiols can act as reducing substrates but monothiols or 1,2-dithiols are relatively inactive.[11] Dithiothreitol gives a lower V_{max} than dihydrolipoate, and other dithiols give rates similar to that for dithiothreitol or still lower.[10] Thioredoxin of $E.$ $coli$ gives a similar K_m and V_{max} to that from $L.$ $leichmannii$.[33] Of the analogues of adenosylcobalamin tested, two with altered bases were found to be active: $Co\alpha$-(aden-9-yl)- and $Co\alpha$-(benzimidazolyl)-$Co\beta$-adenosylcobamide.[15] Of many analogues altered in the nucleoside moiety the following cobalamins were active: 3-isoadenosyl, nebularyl, 2′-deoxyadenosyl, and tubercidyl.[34] Isopropylideneadenosyl- and L-adenosylcobalamin appeared to have a trace of activity. All analogues are inhibitory, however, with K_i values in the range 5–45 μM.

$Kinetic$ $Properties.$ Determination of Michaelis constants is complicated by the fact that in the absence of the activator or at low coenzyme levels or both, double reciprocal plots are nonlinear. In the case of plots of reciprocal velocity versus reciprocal GTP concentration the nonlinearity has been shown to be consistent with self-activation by GTP with a K_m value of 0.24 mM, a dissociation constant of 4 mM for the complex formed by binding GTP to the regulatory site, and an approximately 3-fold increase in V_{max} due to this complex formation.[30] In the presence of the appropriate activator, double reciprocal plots become linear whether coenzyme or ribonucleotide concentration is varied, and the apparent K_m of the ribonucleotide and of the coenzyme is lowered.[15] The apparent K_m for ATP is 0.22 mM in the presence of 1 mM dGTP, and for CTP it is 0.13 mM in the presence of 1 mM dATP. The apparent K_m for the coenzyme changes from 5.5 μM in the presence of 2 mM CTP without activator, to 0.45 μM in the presence of 0.1 mM CTP with 1 mM dATP present, and to 0.3 μM with 1 mM CTP and 1 mM dATP. K_m for $E.$ $coli$ thioredoxin is 4 μM, and for thioredoxin from $L.$ $leichmannii$ it is 3 μM.

[33] M. D. Orr and E. Vitols, $Biochem.$ $Biophys.$ $Res.$ $Commun.$ **25,** 109 (1966).

[34] G. N. Sando, R. L. Blakley, H. P. C. Hogenkamp, and P. J. Hoffmann, $J.$ $Biol.$ $Chem.$ **250,** 8774 (1975).

Ligand Binding. The allosteric activators bind relatively tightly to the regulatory site,[23,30] the dissociation constant for dGTP being about 2 μM at 4.5° and 49 μM at 37°, with the actual value obtained depending on the ionic strength. In agreement with the kinetic results the ribonucleoside triphosphate substrates bind to the regulatory site but with relatively low affinity.[23] In agreement with the kinetic results, binding of the activators to the regulatory site causes an increased affinity of the coenzyme for the catalytic site, but a similar effect on binding of ribonucleotide substrates could not be demonstrated because of the weak binding of these substrates to the enzyme or the enzyme-activation complex.[30] This presumably indicates that the reaction is ordered with ribonucleotide binding last.

Mechanism. Stopped-flow spectrophotometry[14] and ESR of freeze-quenched reactions[35] have shown that in the presence of dithiol, enzyme, and deoxyribonucleotide, the cobalt–carbon bond of enzyme-bound coenzyme is rapidly cleaved to yield a radical pair which has unusual magnetic properties.[36] This formation of a radical pair is considered to be a step in the mechanism of the reaction, since it disappears when ribonucleotide is added to the system and rates of formation and disappearance indicate kinetic competence.[14,35]

[35] W. H. Orme-Johnson, H. Beinert, and R. L. Blakley, *J. Biol. Chem.* **249**, 2338 (1974).

[36] R. E. Coffman, Y. Ishikawa, R. L. Blakley, H. Beinert, and W. H. Orme-Johnson, *Biochim. Biophys. Acta* **444**, 307 (1976).

Section V
Nucleotidases and Nucleosidases

[33] Adenosine Monophosphate Nucleosidase from *Azotobacter vinelandii* and *Escherichia coli*

By VERN L. SCHRAMM and HAZEL B. LEUNG

$$AMP + H_2O \xrightarrow{\text{MgATP}^{2-}} \text{Adenine} + \text{Ribose 5-PO}_4$$

Adenosine monophosphate nucleosidase (AMP nucleosidase) was first described by Hurwitz *et al.*[1] during studies of the polynucleotide metabolizing enzymes of *Azotobacter vinelandii*. The enzyme differs from enzymes which are usually classified as nucleotide-degrading enzymes in that it exhibits a high degree of specificity for AMP as substrate and is essentially inactive unless MgATP^{2-}, the allosteric activator, is present.[2] Allosteric inhibition of activity occurs in the presence of P_i. These controls regulate the enzyme activity which has been postulated to control intracellular AMP levels.[3,4] The resulting adenine is deaminated by adenine deaminase to form hypoxanthine, which is the end-product of AMP catabolism in *Azotobacter*. AMP nucleosidase is also present in *E. coli* and has similar kinetic properties. The end-product of AMP catabolism in *E. coli* appears to be adenine, since adenine deaminase is missing in this organism.[5]

Assay Method

Principle. Hydrolysis of AMP to ribose 5-PO$_4$ and adenine releases the anomeric carbon and therefore causes it to become a reducing sugar. Recent improvements in the reducing sugar assay[6] allow accurate and sensitive quantitation of the ribose 5-PO$_4$. Previous investigators have used *n*-butanol extraction of adenine as an assay method for AMP nucleosidase.[7] This method suffers from a lower sensitivity and a lack of specificity, since adenosine is also partially extracted into *n*-butanol. Conversely, few enzyme activities interfere with the reducing sugar assay.

---

[1] J. Hurwitz, L. A. Heppel, and B. L. Horecker, *J. Biol. Chem.* **226**, 525 (1957).
[2] V. L. Schramm, *J. Biol. Chem.* **249**, 1729 (1974).
[3] V. L. Schramm and H. Leung, *J. Biol. Chem.* **248**, 8313 (1973).
[4] V. L. Schramm and F. C. Lazorik, *J. Biol. Chem.* **250**, 1801 (1975).
[5] H. Leung and V. L. Schramm, unpublished observations (1976).
[6] S. Dygert, L. H. Li, D. Florida, and J. A. Thoma, *Anal. Biochem.* **13**, 367 (1965).
[7] M. Yoshino, *J. Biochem.* (*Tokyo*) **68**, 321 (1970).

METHODS IN ENZYMOLOGY, VOL. LI

Reagents

Assay mixture: 0.1 M triethanolamine-HCl, pH 8.0, 4 mM AMP, 7 mM ATP, and 7 mM MgCl$_2$

Alkaline Cu reagent: 16 g Na$_2$CO$_3$, 6.4 g glycine, and 0.18 g CuSO$_4$·5 H$_2$O in a final volume of 400 ml

Indicator reagent: 0.48 g 2,9-dimethyl-1,10-phenanthroline in a final volume of 400 ml; pH adjusted to 3.0 with HCl

Enzyme Dilution. Enzyme is diluted in 0.1 M Tris·HCl, pH 8.0, containing 2 mM AMP, 0.1 mM EDTA, 0.1 mM dithiothreitol, and 3 μM phenylmethyl sulfonyl fluoride. Diluted enzyme is >90% stable for 8 hr at 0°.

Procedure. Diluted enzyme (0.001–0.01 unit) is added to 0.25 ml of the assay mixture which has been equilibrated to 30°. The amount of protein added should be kept below 30 μg, since higher concentrations will cause interference in the reducing sugar assay. Blank values must always be run by adding an enzyme aliquot to assay mixture containing 0.3 ml of the alkaline Cu reagent. The reducing sugar formed in an appropriate incubation period (20–2 min for 0.001–0.01 unit, respectively) is estimated by adding 0.3 ml of alkaline Cu reagent, 0.3 ml of 2,9-dimethyl-1,10-phenanthroline reagent, and sufficient H$_2$O to give a final volume of 1.6 ml. The alkaline Cu reagent stops catalysis, and the mixture is stable for several hours at room temperature. Color development occurs when the covered tubes are placed in a boiling H$_2$O bath (95–100°) for 8 min. After cooling to 30° in a circulating H$_2$O bath, the Cu-2,9-dimethyl-1,10-phenanthroline complex is read at 450 nm. A standard curve with ribose 5-PO$_4$ should be run. The normal sensitivity under these conditions is 0.0025 μmole of ribose 5-PO$_4$ which will give an A_{450} of approximately 0.045.

Definition of Unit. One unit of enzyme activity is defined as the amount of enzyme that catalyzes the formation of 1 μmole of product per minute. Protein is determined by the Folin method.[8]

Growth of Cells

Azotobacter vinelandii OP[9]

Cultures of *A. vinelandii* OP were maintained and grown at 30° in nitrogen-free medium[10] containing the following ingredients per liter of

[8] E. Layne, this series, Vol. 3 [73].
[9] J. A. Bush and P. W. Wilson, *Nature (London)* **184**, 381 (1959).
[10] J. W. Newton, P. W. Wilson, and R. H. Barris, *J. Biol. Chem.* **204**, 445 (1953).

deionized water: K_2HPO_4, 1.4 g; KH_2PO_4, 0.35 g; $MgSO_4 \cdot 7 H_2O$, 0.2 g; NaCl, 0.1 g; $CaCl_2 \cdot 2 H_2O$, 0.1 g; $Fe_2(SO_4)_3 \cdot 3 H_2O$, 0.01 g; $FeCl_2 \cdot 4 H_2O$, 3.5 mg; $Na_2MoO_4 \cdot 2 H_2O$, 0.25 mg; and sucrose, 20 g. The phosphate buffer was autoclaved separately. For growth of large-scale cultures, 50 ml of the media in a 500-ml Erlenmeyer flask were inoculated from an agar slant and grown to mid-log phase (200–400 Klett units using a No. 42 blue filter) on a New Brunswick model VS rotory shaker. Five-milliliter portions were transferred to six 500-ml portions of media in 2-liter Erlenmeyer flasks. These were grown to mid-log phase as before and transferred to 200 liters of media contained in a growth chamber.[11] Cells were harvested at late-log or early-stationary phase to give approximately 2 kg of cell paste.

Escherichia coli K_{12}

For growth of *E. coli* K_{12}, a similar procedure was used except the cells were maintained and grown at 37° in medium containing the following ingredients per liter of deionized water: K_2HPO_4, 10 g; KH_2PO_4, 2.5 g; $MgSO_4 \cdot 7 H_2O$, 0.2 g; $CaCl_2 \cdot 2 H_2O$, 0.01 g; $Fe_2(SO_4)_3 \cdot 3 H_2O$, 0.01 g; $(NH_4)_2SO_4$, 1 g; and glucose 5 g. Medium and glucose were autoclaved separately and mixed before use. Inoculation of the 200-liter growth chamber was with 100 ml of mid-log phase cells. Cells were harvested in log phase (Klett 490) to give approximately 1 kg of cell paste.

Cell pastes or cells suspended 1:1 (w:w) in 0.4 M phosphate, pH 7.5, can be frozen and stored at −20° or −85°. At −20°, the enzyme activity is stable for 6 months followed by a slow decline of activity. Cells frozen and stored at −85° show no loss of enzyme activity for up to 3 years. This applies to both *Azotobacter* and *Escherichia* cell pastes.

Purification Procedures

AMP Nucleosidase from Azotobacter vinelandii OP

Unless otherwise mentioned, all buffers used during the purification procedure contained EDTA (10^{-4} M), dithiothreitol (10^{-4} M), and phenylmethylsulfonyl fluoride ($3 \times 10^{-6} M$). Temperatures were at 0–5° except where noted.

Step 1. Initial Extract. Frozen cells (954 g) were thawed in a 30° water bath and suspended by mixing with 1.0 liter of 0.4 M potassium phosphate buffer (pH 7.5) containing EDTA ($2 \times 10^{-4} M$), dithiothreitol ($2 \times 10^{-4} M$), and PMSF ($6 \times 10^{-6} M$). The suspension was passed

[11] V. L. Schramm, *Anal. Biochem.* **57**, 377 (1974).

through an Aminco French pressure cell at 15,000–25,000 psi and diluted to 3.3 liters with 0.2 M potassium phosphate buffer (pH 7.5). A 5-ml sample of the extract was centrifuged at 12,000 g for 15 min, and the supernatant fluid was used to determine activity and protein.

Step 2. Heat Treatment. The initial extract, placed in a 3.5-liter stainless-steel beaker, was heated to 60° while being stirred over a period of 30 min. The extract was maintained at 60° for 10 min and then cooled by placing the beaker in an ice bath. The supernatant fluid (2.5 liters) was recovered after centrifugation at 12,000 g for 8 hr. Heating rates were not critical during this step, and similar results were obtained when the temperature was increased to 60° in as little as 2 min.

Step 3. Ammonium Sulfate Fractionation. The supernatant fluid from the heat treatment was brought to 0.37 saturation at 2° by the addition of solid ammonium sulfate over a period of 13 min followed by an additional 17-min stirring. The precipitate was removed by centrifugation (12,000 g for 40 min) and suspended in 150 ml of 0.1 M Tris·HCl buffer (pH 8.0) containing 0.15 M NaCl and 2 mM AMP (the Tris-NaCl-AMP buffer) to give a total volume of 208 ml.

Step 4. Desalt on G-25 Sephadex. The ammonium sulfate fraction was applied to a large (4.6 × 56 cm) G-25 Sephadex column which had been equilibrated with the Tris-NaCl-AMP buffer, and eluted with the same buffer. Those protein fractions free of ammonium sulfate, as determined by Nessler reagent, were pooled to give 250 ml.

Step 5. First DEAE A-50 Sephadex Fractionation. The desalted protein solution was applied to a 4.7 × 12 cm DEAE A-50 Sephadex column equilibrated against the Tris-NaCl-AMP buffer. The protein was eluted with a 2-liter linear gradient of NaCl (0.15–0.30 M) in the same buffer. The enzyme peak appeared at 0.23 M NaCl and was diluted with the Tris-AMP buffer, containing no NaCl, to reduce the NaCl concentration to 0.15 M. The solution was concentrated to 230 ml on an Amicon Diaflow pressure dialysis unit.

Step 6. Second DEAE A-50 Sephadex Fractionation. The solution from the previous step was applied to a 2.5 × 50 cm DEAE A-50 Sephadex column, and the column was developed using a 2-liter gradient from 0.15 to 0.28 M NaCl. Fractions of the highest activity were pooled and concentrated to 57 ml in an Amicon Diaflow concentrator.

Step 7. Hydroxylapatite Fractionation. Without further adjustment of the NaCl concentration, the fraction from the second DEAE column was applied to a 1.5 × 7.5 cm column of hydroxylapatite:cellulose mixture (1.4:1.0 dry weight basis) which had been equilibrated by washing with 10 volumes of Tris-NaCl-AMP buffer. The column was eluted sequentially with 25-ml portions of the buffer containing 20, 40, 80, 160, and 320 mM K$_2$HPO$_4$ (pH 8.0). Fractions containing enzymic activity were pooled and concentrated on an Amicon Diaflow concentrator to 7.4 ml.

Step 8. Sephadex G-200 Gel Filtration. The solution from the previous step was passed through a (2.5 × 100 cm) column of Sephadex G-200 equilibrated with Tris·HCl buffer (0.1 M, pH 8.0) containing 2.0 mM AMP. Those fractions containing the highest specific activities were combined and concentrated by pressure dialysis (Amicon Diaflow) to 3.2 ml. Polyacrylamide gel electrophoresis of this fraction gave a single band both in the presence and absence of sodium dodecyl sulfate.

Crystallization of AMP Nucleosidase. Purified enzyme (4.6 ml; 1.7 mg/ml, specific activity of 30) in 0.1 M Tris·HCl (pH 8.0) containing 1 mM AMP was precipitated by the addition of solid ammonium sulfate to 0.75 saturation. The precipitate was collected by centrifugation at 29,000 g for 7 min and extracted sequentially with 1 ml of 0.70, 0.66, 0.62, 0.58, 0.54, 0.50, 0.46, and 0.42 saturated ammonium sulfate in 0.1 M Tris·HCl and 1 mM AMP (pH 8.0). Upon standing overnight at 4°, crystals appeared in tubes containing 0.42 and 0.46 saturated ammonium sulfate, both having specific activities of 34.3, and both giving a single band in polyacrylamide gel electrophoresis. Microscopic examination revealed the crystalline structure to be hexagonal plates with a diameter of ~0.03 mm. These crystals have been used to induce crystallization using enzyme preparations with specific activities as low as 15. The resulting crystalline material always exhibits specific activities near 34 and appears either as hexagonal plates or clusters of needles. Yield is usually 50–80% for the crystallization step. The purification procedure is summarized in Table I.

Stability of the Purified Enzyme. Following Sephadex G-200 the enzyme was stable for periods of at least 1 year after freezing in Dry Ice–ethanol mixtures and storing at −70°. The crystalline enzyme was also stable for periods of at least 1 year at 4°, stored in the ammonium sulfate solution from which it was crystallized.

AMP Nucleosidase from Escherichia coli

In general, the above purification procedure can be used for preparation of the *E. coli* enzyme. The differences in purification and yield are due primarily to *E. coli* containing less of the enzyme, and the enzyme

TABLE I
PURIFICATION OF AMP NUCLEOSIDASE FROM *Azotobacter vinelandi*[a]

Purification procedure	Volume (ml)	Total protein (mg)	Total units (μmoles/ min)	Specific activity (μmoles/ min/mg)	Yield (%)
Initial extract	3300	108,000	2700	0.025	100
Heat treatment	2500	32,000	2100	0.067	78
Ammonium sulfate fractionation	208	8100	1500	0.19	58
Sephadex G-25	250	7500	1500	0.21	58
First DEAE-Sephadex column	230	300	1400	4.6	51
Second DEAE-Sephadex column	57	130	1300	9.9	48
Hydroxylapatite column	7.4	60	1100	18	39
Sephadex G-200 fractionation[b]	3.2	21	690	34	34[d]
Crystals from G-200 side fractions[c]	1.1	6.7	230	34	—

[a] Starting material was 954 g wet weight of cells.
[b] Fractions containing enzyme of specific activity 34 were pooled, concentrated, and frozen for storage.
[c] In this purification, fractions containing substantial activity, but specific activities less than 34, were pooled, concentrated, brought to near precipitation with ammonium sulfate, and crystallization was induced by the addition of 1 μl of a solution containing seed crystals of AMP nucleosidase. These were collected by centrifugation, dissolved in Tris·HCl (0.1 M, pH 8.0) containing 2 mM AMP, dialyzed to remove ammonium sulfate, and assayed for protein and activity.
[d] Percent yield includes both fractions with specific activity of 34.

being less stable than that from *Azotobacter*. The purification scheme outlined below will emphasize the differences in the two procedures.

Step 1. Initial Extract. This is similar to the *A. vinelandii* procedure except that 4 mM MgCl$_2$ and 9 mg crystalline deoxyribonuclease (Worthington, crystalline) were added to 1.5 kg of cells during the thawing procedure. The pH was adjusted to 7.5 with 10 N KOH before the next step.

Step 2. Heat Treatment. This is similar to the *Azotobacter* preparation except that only 30 min of centrifugation at 12,000 g were required to remove the precipitate.

Step 3. Ammonium Sulfate Fractionation. The fraction precipitating between 0.28 and 0.39 saturation of ammonium sulfate was dissolved in a minimum volume of Tris-NaCl-AMP buffer and was dialyzed against 3.5 liters of the same buffer.

Step 4. First DEAE A-50 Sephadex Fractionation. The dialyzed solution was eluted from a 3.8 × 21 cm column of DEAE A-50 Sephadex with a 1.5-liter linear gradient of NaCl (0.16–0.35 M).

Step 5. Second DEAE A-50 Sephadex Fractionation. The active fraction from the first DEAE column was diluted with an equal volume of Tris-AMP buffer (no NaCl) and was placed on a 1.8 × 21 cm column of DEAE A-50 Sephadex. The enzyme was eluted with a 500-ml PO_4 gradient (0.05–0.15 M, pH 8.0) in the Tris-NaCl-AMP buffer.

Step 6. 1,6-Diaminohexane-Sepharose 4B Fractionation. The AMP nucleosidase peak (140 ml) from step 5 was dialyzed overnight against 2 liters of salt-free Tris-AMP buffer and applied to a 1.6 × 20 cm column of Sepharose 4B coupled to 1,6-diaminohexane[12] (coupling was done with 1 M 1,6-diaminohexane). The enzyme was eluted with a 800-ml NaCl gradient (0–1.0 M) followed by 200 ml of 1.1 M NaCl. The gradient was buffered with Tris-AMP buffer. Active fractions were concentrated to 4.2 ml with Amicon ultrafiltration using a PM 30 membrane.

Step 7. Sephadex G-200 Gel Filtration. Enzyme from the previous step was eluted from Sephadex G-200 with Tris-NaCl-AMP buffer in which the NaCl had been increased to 0.2 M. The active fractions were pooled and concentrated using a Sartorius vacuum ultrafiltration device. A summary of the purification is listed in Table II.

TABLE II

PURIFICATION OF AMP NUCLEOSIDASE FROM *Escherichia coli* K_{12}[a]

Purification procedure	Volume (ml)	Total protein (mg)	Total units (μmoles/ min)	Specific activity (μmoles/ min/mg)	Yield (%)
Initial extract	2950	446,000[b]	1010	0.002	100
Heat treatment	5325	35,000	685	0.019	68
Ammonium sulfate fractionation	200	2,500	580	0.19	48
First DEAE-Sephadex column	298	410	360	0.88	36
Second DEAE-Sephadex column	140	64	260	4.1	26
1,6-Diaminohexane-Sepharose fractionation	4.2	5.7	85	15	8
Sephadex G-200 fractionation	0.2	3.9	67	17	7

[a] Starting material was 1.5 kg wet weight of cells.
[b] Initial extract protein values may not be accurate.

[12] P. Cuatrecasas, M. Wilchek, and C. B. Anfinsen, *Proc. Natl. Acad. Sci. U.S.A.* **61**, 636 (1968).

TABLE III
PROPERTIES OF AMP NUCLEOSIDASE FROM *Azotobacter vinelandii* OP

Molecular weight [a,b]	320,000	Equilibrium sedimentation
Subunit molecular weight[a,b]	54,000	SDS gel electrophoresis and equilibrium sedimentation
Number of subunits[a,b]	6	Apparently identical size and charge
K_m for AMP[c]	100 μM	K_m is independent of [MgATP^{2-}], exhibits Michaelis-Menten kinetics
$M_{0.5}$ for MgATP^{2-} [c]	40 μM	Increases V_{max} 400-fold, Hill coefficient >3, $M_{0.5}$ is independent of [AMP]
$I_{0.5}$ for P$_i$ [c]	150 μM	Competitive with MgATP^{2-}, Hill coefficient of inhibition >3
Binding sites for tubercidin 5'-PO$_4$ (an AMP analog)[b]	3	Binding follows standard saturation isotherm
K_d for tubercidin 5'-PO$_4$ [b]	15 μM	MgATP^{2-} has no effect on K_d
Binding sites for MgATP^{2-} [b]	6	Binding strongly cooperative
K_d for MgATP^{2-} [b]	90 μM	K_d is [MgATP^{2-}] for 50% saturation
Binding sites for P$_i$ [b]	6	At high [P$_i$] nonspecific binding occurs, binding strongly cooperative
K_d for P$_i$ [b]	170 μM	K_d is [P$_i$] for 50% saturation, P$_i$ and MgATP^{2-} binding are mutually exclusive

[a] Copyright by the American Chemical Society. V. L. Schramm and L. I. Hochstein, *Biochemistry* **11**, 2777 (1972).
[b] V. L. Schramm, *J. Biol. Chem.* **251**, 3417 (1976).
[c] V. L. Schramm, *J. Biol. Chem.* **249**, 1729 (1974).

Comments on Purification Procedure. The *E. coli* AMP nucleosidase is present at lower levels than the *A. vinelandii* enzyme, and it is also less stable as judged by the lower yields. The reason for this instability, and the methods to stabilize the *E. coli* enzyme, have not yet been investigated. However, the enzyme retains >80% of the initial activity when stored frozen for 6 months at −80° in the buffer used for the G-200

TABLE IV
KINETIC PROPERTIES OF *E. coli* AMP NUCLEOSIDASE[a]

K_m for AMP	120 μM	K_m dependent on MgATP^{2-} V_{max} unaffected, cooperative kinetics at low [MgATP^{2-}], Hill coefficient >2
$M_{0.5}$ for MgATP^{2-}	40 μM	$M_{0.5}$ dependent on [AMP], MgATP^{2-} activation curves cooperative, Hill coefficient >1.5
$I_{0.5}$ for P$_i$	200 μM	P$_i$ competitive with MgATP^{2-}, cooperativity of MgATP^{2-} curves retained at high [P$_i$]

[a] H. Leung and V. L. Schramm, unpublished observations (1976).

Sephadex step. Analysis of purified *E. coli* AMP nucleosidase on polyacrylamide gel electrophoresis gave two bands which stained for protein with Coomassie blue. Similar unstained gels showed that AMP nucleosidase activity was present in both protein bands. This result suggests that the *E. coli* enzyme is homogeneous following the Sephadex G-200 step. No attempts have been made to crystallize the *E. coli* enzyme.

Properties of AMP Nucleosidases

The kinetic patterns of AMP nucleosidases are complex, with cooperative activation by $MgATP^{2-}$ and cooperative inhibition by P_i.[2] The kinetic constants and physical properties of the *A. vinelandii* enzyme are listed in Table III, and the kinetic properties of the *E. coli* enzyme are listed in Table IV.

[34] An Acid Nucleotidase from Rat Liver Lysosomes

By CHARALAMPOS ARSENIS and OSCAR TOUSTER

$$\text{Nucleotide} + H_2O \rightarrow \text{nucleoside} + \text{phosphate}$$

Rat liver lysosomes are capable of degrading nucleic acid to nucleosides and inorganic phosphate.[1] This process requires an acidic pH and the presence of nucleases and nucleotidases. Lysosomes have long been known to contain various enzymes exhibiting phosphatase activity toward a wide variety of substrates, including nucleotides. Resolution of these activities was achieved by DEAE-cellulose chromatography,[2,3] yielding an enzyme which acts on nucleotides but not on simple sugar phosphates.[3] It is apparently the only acid nucleotidase that has been found in mammalian cells, and it is the only nucleotidase reported to be active toward 2'-, 3'-, and 5'-nucleotides.

The following preparation of the acid nucleotidase from an extract of a crude lysosomal fraction, and the description of its properties, are based on a previous report[3] in which the preparation of the enzyme from highly purified lysosomes ("tritosomes") was also reported. The use of purified lysosomes offers the advantage of yielding a higher purity

[1] C. Arsenis, J. S. Gordon, and O. Touster, *J. Biol. Chem.* **245**, 205 (1970).
[2] C. Arsenis and O. Touster, *J. Biol. Chem.* **242**, 3400 (1967).
[3] C. Arsenis and O. Touster, *J. Biol. Chem.* **243**, 5702 (1968).

METHODS IN ENZYMOLOGY, VOL. LI

product, but lower yields are obtained because of the more complex procedure for obtaining the organelles.

Assay Method

Principle. The enzymic hydrolysis of nucleotides is followed by determining the release of inorganic phosphate by the method of Lowry and Lopez.[4] *p*-Nitrophenyl phosphate is used as substrate during the purification procedure. The *p*-nitrophenol released is determined by its absorbance at 410 nm at pH 10–11.

Reagents

Sodium acetate or sodium malonate, 0.2 M, pH 4.8
Nucleotide, 0.1 M, adjusted to pH 4.8
p-Nitrophenyl phosphate, 0.1 M, adjusted to pH 4.8
Sodium acetate buffer, 0.25 M, pH 3.1
Ammonium molybdate, 1%
Ascorbic acid, 1%
Sodium hydroxide, 0.04 M

Procedure. The standard assay mixture (1.0 ml) for determining nucleotidase activity contains 6 μmoles of nucleotide, 50 μmoles of sodium malonate or sodium acetate, pH 4.8, and an amount of enzyme which hydrolyzes less than 20% of the substrate. After 30 min of incubation at 37°, the reaction is stopped by the addition of 1.0 ml of 0.25 M sodium acetate buffer, pH 3.1, and then heated in a boiling-water bath for 1 min. Inorganic phosphate released is determined by the addition of 0.2 ml each of 1% solutions of ammonium molybdate and ascorbic acid in water. The absorbance at 700 nm is proportional to the inorganic phosphate concentration. When *p*-nitrophenyl phosphate is the substrate, the *p*-nitrophenol produced is determined by stopping the reaction with 10 ml of 0.04 M NaOH and determining the absorbance at 410 nm.

Definition of Unit and Specific Activity. One unit of enzyme corresponds to the release of 1 μmole of product per minute. Specific activity is defined as units per milligram of protein determined by the method of Lowry *et al.*[5]

---

[4] O. H. Lowry and J. A. Lopez, *J. Biol. Chem.* **162**, 421 (1946).
[5] O. H. Lowry, N. J. Rosebrough, A. L. Farr, and R. J. Randall, *J. Biol. Chem.* **193**, 265 (1951).

Purification Procedure

Step 1. Preparation of Lysosomal Extract. Male Wistar rats (150–250 g) were fasted for 18–20 hr before they were killed by decapitation; their livers were quickly removed and homogenized in cold 0.25 M sucrose. All subsequent operations are carried out at 0–4°. The lysosomal fraction is prepared essentially according to De Duve *et al.*[6] The lysosomal pellet is suspended in cold water containing 0.1% (v/v) Triton X-100 and dialyzed overnight against 0.001 M Tris·HCl buffer, pH 7.4. After removal of insoluble material by centrifugation, the supernatant solution is used for column chromatography.

Step 2. Column Chromatography. The supernatant solution obtained above (from 15 g of liver) is poured over a DEAE-cellulose column, 1.9 × 20 cm, equilibrated with 0.01 M Tris·HCl buffer, pH 7.4. The column is washed with 100 ml of the same buffer to yield the first acid phosphatase peak (I). Elution of the acid nucleotidase (II) is then accomplished by washing the column with 100 ml of 0.01 M 5′-dAMP in buffer. Finally, a sugar phosphate phosphohydrolase (III) is eluted with 100 ml of 0.01 M Tris·HCl buffer, pH 7.4, which is 0.2 M in NaCl. Fractions (10 ml) are collected in a fraction collector at 5° and assayed with *p*-nitrophenyl phosphate as substrate at pH 4.5. The most active fractions are pooled and concentrated by ultrafiltration.

Salient features of the purification are shown in the table. The buffer wash (I) and the NaCl eluate (III) contain enzymes active towards a wide variety of phosphate esters. Eluate II is the acid nucleotidase. The three eluates contain approximately 22% of the protein and 47% of the *p*-

FRACTIONATION OF PHOSPHATASE ACTIVITY FROM RAT LIVER LYSOSOMES

Fraction	Volume (ml)	Total protein (mg)	Total activity[a] (units)	Specific activity (units/mg)
Lysosomal extract	6.5	19.1	17.6	0.92
DEAE-cellulose				
Fractions 6–9 (I)	40	0.58	2.07	3.57
Fractions 16–19 (II)	40	0.17	1.37	8.06
Fractions 26–29 (III)	40	3.58	4.90	1.37

[a] Activity measured with *p*-nitrophenyl phosphate as substrate.

[6] C. De Duve, B. C. Pressman, R. Gianetto, R. Wattiaux, and F. Appelmans, *Biochem. J.* **60**, 604 (1955).

nitrophenyl phosphatase activity applied to the column. The extent of purification of the acid nucleotidase is much greater than is apparent from the table, since the isolation of the lysosomal extract involves the separation of the enzyme from most of the liver protein.

Properties

Substrate Specificity. Rat liver lysosomal acid nucleotidase hydrolyzes a wide range of nucleotides containing various purine and pyrimidine bases regardless of the position of the linkage of the phosphate group, and it is active toward p-nitrophenyl phosphate, α-naphthyl phosphate, and β-glycerophosphate. K_m values are 2′-AMP, 5.8×10^{-4} M; 3′-AMP, $3.3 \times 10^{-4}\ M$; and 5′-AMP, $1.25 \times 10^{-4}\ M$. Essentially no activity is observed with simple sugar phosphates as substrates.

Activators and Inhibitors. Unlike other nucleotidases of mammalian origin, the lysosomal acid nucleotidase is not strongly activated or inhibited by particular cations, except for mercuric ions, which are inhibitory. At 5 mM concentration, potassium fluoride inhibits 91%. Among organic compounds, the most effective inhibitors are tartrate, citrate, oxalate, malate, p-hydroxymercuriphenyl sulfonate, and p-hydroxymercuribenzoate. Tartrate is a potent competitive inhibitor; 10 μM tartrate inhibits more than 80%. Iodoacetic acid, iodoacetamide, and N-ethylmaleimide are relatively inactive. EDTA is a potent inhibitor, but a metal requirement for the enzyme has not been demonstrated.

Effect of pH. The optimum pH is 5.0, with no activity being observed below pH 2.5 or above pH 7.0.

Effect of Temperature. The optimal temperature for the hydrolysis of 5′-dAMP is between 50° and 60° for a 15-min incubation.

Effect of Substrate Concentration. Under the assay conditions described herein, the reaction is linear for at least 60 min.

Purity. Although lysosomal extracts yield more than one band of acid phosphatase activity in polyacrylamide gel electrophoresis in both basic and acidic gels, the purified acid nucleotidase shows only a single band of activity. Thermal inactivation rates employing several different substrates are very similar. The purified preparation therefore appears to be composed of a single phosphatase. The preparation shows essentially no contamination by several other lysosomal hydrolases tested, namely, aryl sulfatase, cathepsin, β-glucuronidase, and N-acetyl-β-D-glucosaminidase.

Molecular Weight. By the method of Martin and Ames,[7] employing sucrose density gradient centrifugation, the lysosomal acid nucleotidase behaves as a protein with a molecular weight of 79,500.

[7] R. G. Martin and B. N. Ames, *J. Biol. Chem.* **236**, 1372 (1961).

[35] Nucleoside Triphosphate Pyrophosphohydrolase (NTPH)

By ALLAN J. MORRIS

$$NTP + H_2O \rightarrow NMP + PP_i$$

Assay Method

The analysis of NTPH activity utilizes added yeast inorganic pyrophosphatase to hydrolyze the pyrophosphate produced during incubation with substrate (inosine triphosphate, ITP). The inorganic phosphate formed is then determined colorimetrically by the procedure of Rathbun and Betlach using KH_2PO_4 as a reference standard.[1] The reaction mixture (1.0 ml) contains 50 mM β-alanine buffer (pH 9.5), 10 mM $MgCl_2$, mM dithiothreitol, 1 unit of yeast inorganic pyrophosphatase, 0.5 mM ITP, and the NTPH-containing solution to be analyzed. The reaction may be initiated either by the addition of the ITP or the NTPH-containing solutions. Incubation is carried out for 20 min at 37°, and the reaction is terminated by the addition of 0.1 ml of 50% trichloroacetic acid. The protein precipitate which forms is removed by centrifugation (2000 g for 5 min), and the supernatant solution is decanted into an 18 × 150 mm test tube for subsequent colorimetric phosphate quantitation. Appropriate blank reaction mixtures are incubated and analyzed in order to correct for trace amounts of inorganic phosphate present in some NTPH-containing solutions and, in particular, the ITP solution used as substrate.

Protein analyses are carried out using the procedure of Lowry *et al.* using serum albumin as a reference standard.[2] For routine screening of human blood for the specific activity of NTPH, the hemoglobin of the

[1] W. B. Rathbun and M. V. Betlach, *Anal. Biochem.* **28**, 436–445 (1969).
[2] O. H. Lowry, N. J. Rosebrough, A. L. Farr, and R. J. Randall, . *Biol. Chem.* **193**, 265–275 (1951).

METHODS IN ENZYMOLOGY, VOL. LI

lysate analyzed may be determined by the method of Austin and Drabkin.[3]

The unit of NTPH activity used here is 1 μmole of inorganic pyrophosphate hydrolyzed per minute from ITP in the standard 20-min incubation. The specific activity of NTPH is the units of NTPH activity per milligram of protein.

Purification Procedures

General Considerations. All steps in the purification of NTPH from the sources described here were carried out at 4° unless otherwise indicated. The pH values of buffers are indicated for the temperatures at which the buffered solutions are to be utilized.

It has been our experience that the solubilities of rabbit hemoglobin and human hemoglobin in ammonium sulfate solutions are sufficiently different to necessitate unique isolation procedures of NTPH from red cells of the two species. Neither of these procedures is applicable to NTPH isolation from rabbit liver, and hence a third procedure for purification has been developed for that tissue.

Purification of Nucleoside Triphosphate Pyrophosphohydrolase from Rabbit Red Cells[4]

Preparation of Crude Lysate and Lysate. The blood from each of three rabbits is obtained by heart puncture following intravenous injection of 2000 units of heparin in isotonic saline and 100 mg of sodium pentabarbital (Nembutal). The cells are collected by centrifugation (3000 g for 10 min). The packed cells are washed twice by resuspension in plasma volumes of isotonic saline and sedimented as before. The packed red cells are then pooled and lysed in 4 volumes of water for 10 min with gentle stirring (*crude lysate*). The crude lysate fraction is then centrifuged (10,000 g for 20 min), and the supernatant fraction is removed (*lysate*). To the lysate fraction is added a 100 mg/ml solution of streptomycin sulfate to a final concentration of 2.8 mg/ml. After gentle mixing for 30 min the precipitate which forms is removed by centrifugation at 10,000 g for 20 min and discarded. The supernatant solution is dialyzed against 15 volumes of 50 mM Tris-Cl (pH 7.0), mM MgCl$_2$, and mM glutathione (GSH). The dialysate is replaced twice during the ensuing 24 hr of dialysis.

Ammonium Sulfate Fractionation. To the dialyzed lysate is added

[3] J. H. Austin and D. L. Drabkin, *J. Biol. Chem.* **112**, 67–88 (1935).
[4] C. J. Chern, A. B. MacDonald, and A. J. Morris, *J. Biol. Chem.* **244**, 5489–5495 (1969).

TABLE I
PURIFICATION OF NUCLEOSIDE TRIPHOSPHATE PYROPHOSPHOHYDROLASE FROM RABBIT
RED CELLS

Fraction	Protein (mg)	Total activity (units)	Specific activity (units/mg)	Recovery (%)	Relative purification
Crude lysate	14,800	167	0.01	100	1.0
Lysate	10,200	139	0.01	83	1.0
40–70% Ammonium sulfate	300	93.0	0.31	56	31
Sephadex column	86.3	71.0	0.82	43	82
DEAE-cellulose column	1.19	31.5	26.5	19	2650

Tris-Cl (pH 7.5) to a final concentration of 0.1 M. The solution is then titrated to pH 6.5 by the addition of 1 N acetic acid. Powdered (mortar-ground) ammonium sulfate is added slowly (over a period of 30 min) to 40% of saturation (22.6 g/100 ml). Stir gently for an additional 30 min and centrifuge at 10,000 g for 20 min. The volume of the supernatant solution is measured, and 18.2 g of powdered ammonium sulfate per 100 ml are added as before. After 30 min of gentle stirring the precipitate, which contains the NTPH activity, is collected by centrifugation at 10,000 g for 20 min. The pellet (*40–70% ammonium sulfate fraction*) is dissolved in a minimum volume of 50 mM Tris-Cl (pH 7.0), mM MgCl$_2$, and mM GSH.

Sephadex G-100 Gel Filtration. The *40–70% ammonium sulfate fraction* is added to a Sephadex column (5 × 100 cm) containing Sephadex G-100 previously equilibrated with 50 mM Tris-Cl (pH 7.0), mM MgCl$_2$, mM GSH and eluted with the same buffer solution. Elution is performed at a flow rate of 25 ml/hr. Aliquots of the eluate fractions are analyzed for NTPH activity (Fig. 1), pooled, and concentrated by Amicon Diaflow ultrafiltration using a PM-10 membrane (*Sephadex column fraction*).

DEAE-Cellulose Chromatography. Dialyze the Sephadex column fraction for several hours against 50 mM Tris-Cl (pH 8.0), 4 mM MgCl$_2$, and 5 mM GSH. Apply the dialyzed sample to a DEAE-cellulose column (2 × 10 cm, 0.76 meq/g) previously equilibrated with the same buffer. Wash the column with the buffer solution until the hemoglobin present in the sample begins to emerge from the column. At this point begin a linear gradient of from 0 to 0.1 M NaCl in the column buffer solution (200 ml in each of the two chambers of the gradient-delivering device) (Fig. 2). The column eluate fractions (4.0 ml) are analyzed for protein by

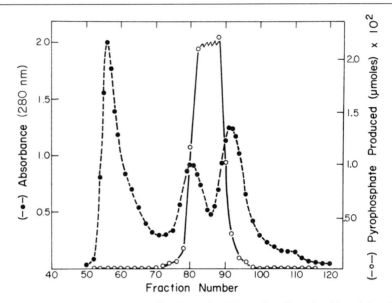

FIG. 1. Sephadex G-100 gel filtration of rabbit red cell nucleoside triphosphate pyrophosphohydrolase.

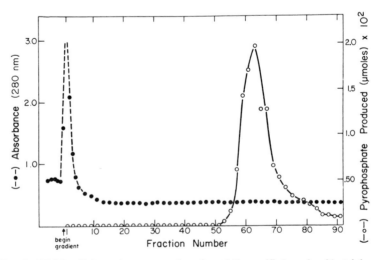

FIG. 2. DEAE-cellulose chromatography of partially purified nucleoside triphosphate pyrophosphohydrolase.

280 nm absorbance and NTPH by analysis of an aliquot as described under *Assay Method*. The amount of protein present in the fractions containing NTPH activity should be below the detectable level of this procedure. Fractions containing NTPH activity should be pooled, concentrated by ultrafiltration, and aliquots stored in liquid N_2.

Properties. Rabbit NTPH has pyrophosphohydrolytic activity with a number of nucleoside triphosphates including xanthosine triphosphate, uridine triphosphate, and guanosine triphosphate, but inosine triphosphate is cleaved most rapidly of the compounds tested. Neither adenosine triphosphate nor inosine diphosphate is hydrolyzed by NTPH.[4] The K_m of NTPH for ITP has been found to be $3.37 \times 10^{-5} M$. The pH optimum is near 9.75 in β-alanine buffer. Enzyme activity is lost rapidly in the absence of added sulfhydryl compounds, especially during the later stages of purification. $MgCl_2$ is required for NTPH activity; maximal activity is shown in the presence of 10 mM $MgCl_2$ or above. Molecular weight estimations by sucrose density centrifugation, using rabbit hemoglobin and pancreatic DNase as reference markers, indicate a molecule of 37,000 g/mole.

The NTPH prepared by the procedure described here should be free of inorganic pyrophosphatase contamination, but polyacrylamide gel electrophoresis reveals two bands of contaminating proteins.

Isolation of Nucleoside Triphosphate Pyrophosphohydrolase from
Rabbit Liver[5]

Introductory Comments. It has been observed that liver homogenates contain phosphohydrolytic activity which carries out a nonspecific cleavage of nucleoside triphosphates as well as the monophosphate and diphosphate derivatives. Consequently, approximations of NTPH activity in crude liver fractions are obtained by comparing the hydrolysis of ATP to the hydrolysis of ITP at pH 9.5 in the standard assay. The excess hydrolysis of ITP is deemed to be due to the presence of NTPH in those fractions. Partially purified NTPH fractions from liver are free of this contaminating hydrolytic activity.

Preparation of the Rabbit Liver Homogenate and Supernatant Fractions. The liver of an adult rabbit is perfused through the aorta with 1 liter of isotonic saline to remove blood cells from the tissue. A portion of the liver is excised and passed through a tissue mincer, and a wet weight is (about 100 g) obtained. The minced liver is then homogenized in 3 volumes (w/v) of 50 mM Tris-Cl (pH 7.4) and mM GSH using a

[5] J. K. Wang and A. J. Morris, *Arch. Biochem. Biophys.* **161**, 118–124 (1974).

TABLE II
PARTIAL PURIFICATION OF RABBIT LIVER NUCLEOSIDE TRIPHOSPHATE
PYROPHOSPHOHYDROLASE

Fraction	Total protein (mg)	Total activity (units)	Specific activity (units/mg)	Relative purification	Recovery (%)
Crude homogenate	14,400	141[a]	0.01	1.0	100
High-speed supernatant	3460	196[a]	0.06	6.0	139
Heat-denatured supernatant	563	95	0.17	17.0	67
Ammonium sulfate column	82.1	79	0.96	96.0	56
DEAE-cellulose column	7.10	46.5	6.55	655	33

[a] Enzyme activity estimated by comparison of ITP hydrolysis and ATP hydrolysis (see *Introductory Comments*).

Potter–Elvehjem homogenizer equipped with a Teflon pestle (*crude homogenate*). The crude homogenate fraction is centrifuged at 27,000 *g* for 20 min to remove cell debris, and the supernatant fraction is removed and subjected to centrifugation at 143,000 *g* for 2 hr (*high-speed supernatant*).

Heat Denaturation. The high-speed supernatant fraction is placed in a large flask and immersed in a 65° water bath for 5 min with continuous stirring. The materials are then rapidly cooled in an ice bath, and the precipitate which has formed is removed by centrifugation at 10,000 *g* for 20 min and discarded (*heat-denatured supernatant*).

Celite Chromatography. Further purification of liver NTPH is achieved by subjecting the heat-denatured supernatant fraction to reverse ammonium sulfate gradient elution from a Hyflo-supra celite column.[6] Celite is added to the heat-denatured supernatant fraction (1 g celite per 100 mg protein), and the materials are mixed thoroughly. The suspension is then brought to 80% of saturation by the slow addition of 51.6 g of ammonium sulfate per 100 ml of the supernatant fraction. After stirring for 30 min the slurry is used to prepare a column bed (approximately 2.8 × 3 cm). Elution is conducted with a linear gradient of decreasing ammonium sulfate concentration (80–0%) prepared by means of 450 ml of 80% saturated ammonium sulfate (51.6 g/100 ml) in 50 m*M* Tris-Cl (pH 7.4), m*M* GSH in one chamber of the gradient-delivery device, and 450 ml of the same buffered solution lacking ammonium sulfate in the second chamber. NTPH is eluted in those column fractions

[6] T. P. King, *Biochemistry* **11**, 367–371 (1972).

possessing 45–50% of ammonium sulfate saturation (Fig. 3), and the NTPH is concentrated from these pooled fractions by the addition of 15.6 g of ammonium sulfate per 100 ml (70% of saturation). The mixture is stirred gently for 30 min, and the precipitate which forms is collected by centrifugation at 10,000 g for 20 min. The precipitate is then dissolved in 20 ml of 50 mM Tris-Cl (pH 8.1) and mM GSH and dialyzed overnight against 1 liter of the same buffer solution (*ammonium sulfate column fraction*).

DEAE-Cellulose Chromatography. The dialyzed ammonium sulfate column fraction is applied onto a DEAE-cellulose column (2.5 cm in diameter and a bed height of 1.4 cm) previously equilibrated with 50 mM Tris-Cl, pH 8.1, mM GSH. The column is eluted using a linear gradient of NaCl produced by 50 ml of elution buffer in one chamber of the gradient delivery device and 50 ml of elution buffer containing 0.2 M NaCl in the second chamber. Eluate fractions containing NTPH activity are pooled and concentrated by ultrafiltration using an Amicon Diaflow concentrator equipped with a PM-10 membrane (*DEAE cellulose column fraction*). Aliquots of the concentrated enzyme solution may be stored in liquid nitrogen.

Properties. All parameters studied to date indicate that NTPH obtained from rabbit liver is identical to that obtained from rabbit red blood cells. The final enzyme preparation from liver, obtained by the

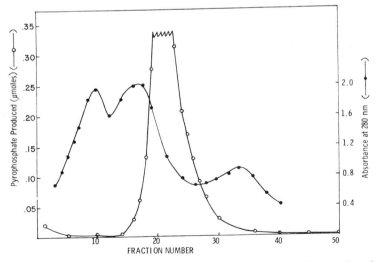

FIG. 3. Elution of rabbit liver nucleoside triphosphate pyrophosphohydrolase from a celite column using an inverse ammonium sulfate concentration gradient.

procedure described above, is considerably less pure than that obtained by the procedure described for red blood cells.

Isolation of Nucleoside Triphosphate Pyrophosphohydrolase from Human Red Blood Cells

Introductory Comments. Since the NTPH activity of human erythrocytes has been found to vary widely from one human to another (range: 0–6 nmoles of ITP hydrolyzed per milligram of hemoglobin[7]), preliminary analysis of NTPH in the blood sample to be utilized is a judicious preliminary step in this procedure. Red cell lysates with the highest specific activity available have been utilized for isolation of NTPH.

Preparation of Crude Lysate and Lysate. Outdated human blood obtained from the American Red Cross (355 ml) is centrifuged at 4000 g for 15 min and the plasma measured and discarded. The packed cells are washed twice by resuspension in a plasma volume of 0.9% NaCl and sedimented as before. The washed cells (140 ml packed cell volume) are lysed in 3 volumes of mM dithiothreitol with gentle stirring for 60 min (*crude lysate*). The crude lysate is centrifuged at 10,000 g for 20 min, and the clear supernatant solution is carefully removed (340 ml) (*lysate*). The absorbance at 280 nm of an appropriate dilution of the lysate is determined.

Calcium Phosphate Gel Elution. Calcium phosphate gel prepared by the method of Tsuboi and Hudson is suspended as an aqueous slurry, and aliquots are dried on tared planchets in order to establish gel concentration.[8] Calcium phosphate gel is added to the lysate to the extent of 0.11 mg dry weight of gel per milliliter of lysate for each unit of absorbance at 280 nm. The mixture is stirred for 1 hr and centrifuged at 10,000 g for 20 min, and the supernatant fraction is discarded. The gel pellet is washed by resuspension in 200 ml of mM GSH and sedimented as before. To the resulting gel pellet is added 200 ml of 20% ammonium sulfate solution (10.6 g ammonium sulfate per 100 ml), and the gel is suspended thoroughly in this solution. Following sedimentation of the gel the supernatant solution is removed and 38.7 g of ammonium sulfate per 100 ml are added to the supernatant solution. Following centrifugation at 10,000 g for 20 min the protein precipitate is dissolved in an imidazole-acetate buffer prepared by a 1 : 100 dilution of stock imidazole-acetate buffer to which mM GSH is added (*gel eluate fraction*). Stock

---

[7] C. Soder, J. F. Henderson, G. Zombor, E. E. McCoy, V. Verhoef, and A. J. Morris, *Can. J. Biochem.* **54**, 843–847 (1976).

[8] K. K. Tsuboi and P. B. Hudson, *J. Biol. Chem.* **224**, 879–887 (1957).

TABLE III
PARTIAL PURIFICATION OF NUCLEOSIDE TRIPHOSPHATE PYROPHOSPHOHYDROLASE FROM
HUMAN ERYTHROCYTES

Fraction	Protein (grams)	Activity (units)	Specific activity ($\times 10^3$)	Recovery (%)	Purification
Crude lysate	58.2	64.7	1.11	100	1
Lysate	33.7	55.4	1.64	86	1.5
CaPO$_4$ gel eluate	2.44	37.8	15.5	58	14.0
C-50 eluate	0.039	23.2	594	36	535
A-20 eluate	0.0046	5.9	1280	9.1	1153

imidazole-acetate buffer is conveniently prepared by titrating 1 M acetic acid with 1 M imidazole to pH 6.5 at 4° (essentially a 1:1 mixture). The gel eluate fraction is dialyzed against two 2-liter portions of 1:100 dilution imidazole-acetate buffer containing mM GSH. Protein concentration of the dialyzed gel eluate fraction is estimated by 260/280 nm absorbance ratio using the dialysate to establish the blank absorbancy.[9]

Chromatography on C-50-120 Sephadex. The CM Sephadex column is prepared by stirring a weighed amount of C-50-120 Sephadex (4.5 ± 0.5 meq/g) in a 1:10 dilution of stock imidazole-acetate buffer, pH 6.25. The CM Sephadex is collected by centrifugation at 10,000 g for 15 min and washed twice by resuspension and sedimentation in a 1:100 dilution of stock imidazole-acetate buffer, mM GSH. The final washed CM Sephadex is resuspended to an exact volume in the latter solution and a column prepared using 6.5 mg (dry weight) of the C-50-120 Sephadex per milligram of protein in the gel eluate fraction. The resulting column bed is typically 7 × 25 cm. Wash the column with the 1:100 dilution buffer, mM GSH, until the pH of the eluate is identical to input solution. Add the gel eluate fraction (29 ml) to the CM Sephadex column bed and eluate. NTPH activity is eluted at the void volume of the column while many contaminating proteins are retained on the column. The NTPH-containing eluate fractions are pooled and concentrated to approximately 20 ml by ultrafiltration and dialyzed against 0.05 M Tris-Cl, pH 8.0, mM GSH (*C-50 eluate fraction*).

DEAE-Sephadex Ion Filtration Chromatography.[10] Prepare a column

[9] O. Warburg and W. Christian, *Biochem. Z.* **310**, 384–421 (1942).
[10] L. H. Kirkegaard, T. J. A. Johnson, and R. M. Bock, *Anal. Biochem.* **50**, 122–138 (1972).

2.5 × 36 cm using A-20-150 DEAE-Sephadex previously prepared and equilibrated with 0.1 M Tris-Cl, pH 8.0, and 60 mM KCl. Wash the column with this same solution until the pH of the eluate corresponds to that of the buffer entering the column. Change the elution buffer to 0.05 M Tris-Cl, pH 8.0, mM GSH, 60 mM KCl, and 4 mM MgCl$_2$ and continue until reequilibration is assured.

To one-half of the C-50 eluate fraction add solid KCl to 60 mM and place the enzyme fraction onto the DEAE-Sephadex column and elute with the final buffer above (Fig. 4). The major NTPH-containing fractions are pooled, concentrated by ultrafiltration to about 5 ml, and a 35-ml portion of 0.05 M Tris-Cl (pH 7.0), mM GSH is added and the concentration continued. Following addition of a second 35-ml portion of the latter buffer the enzyme solution is concentrated to approximately 6 ml and aliquots are frozen in liquid nitrogen. Addition of dithiothreitol to 5 mM has been used on occasion to ensure adequate thiol during prolonged storage (*A-20 eluate fraction*).

The DEAE-Sephadex column may be reutilized to fractionate the other portion of the C-50 eluate fraction if the top of the column bed containing the colored materials is replaced with fresh DEAE-Sephadex and the column is reequilibrated with the elution buffer.

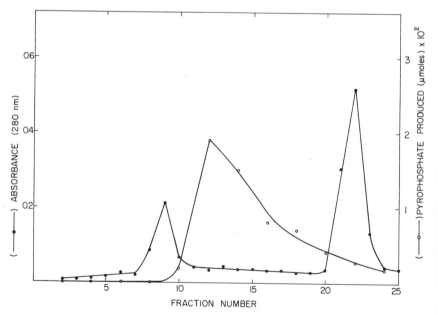

Fɪɢ. 4. DEAE-Sephadex ion filtration chromatography of human red cell nucleoside triphosphate pyrophosphohydrolase.

Properties. The K_m of the human red cell NTPH for ITP has been shown to be $3.46 \times 10^{-5}\,M$, essentially the same as that determined for the rabbit red cell enzyme.[7] By use of a procedure which permits visual localization of NTPH in polyacrylamide gels, the electrophoretic mobilities of rabbit red cell and rabbit liver NTPH at pH 8.6 were shown to be identical to that of the human red cell NTPH.[5] Similarly, pH optimum and the reagent requirements for maximal activity of human red cell NTPH are indistinguishable from those of rabbit red cell NTPH. The ratio of the rate of hydrolysis of ITP to the rate of hydrolysis of GTP by the rabbit red cell NTPH is 10, while that ratio as determined with human NTPH is 30. Hence human red cell NTPH exhibits a higher degree of specificity for ITP than the rabbit counterpart.

It has been our experience that NTPH from human red cells is considerably less stable than the NTPH isolated from rabbit red cells.

[36] Deoxythymidylate Phosphohydrolase from PBS2 Phage-Infected *Bacillus subtilis*

By ALAN R. PRICE

$$5'\text{-dTMP} + H_2O \rightarrow \text{thymidine} + P_i$$

Deoxythymidylate phosphohydrolase is an enzyme induced by bacteriophage PBS2 during infection of *Bacillus subtilis*.[1] Its greatest affinity is toward dTMP, although other 5'-deoxyribonucleotides containing 4-hydroxypyrimidine or 6-hydroxypurine moieties are also attacked. The enzyme probably functions to hydrolyze dTMP and thus to prevent the accumulation of dTTP in the cell during infection by PBS2 phage, which has uracil instead of thymine in its DNA.

Assay Method[1,2]

Principle. The colorimetric assay measures the enzymic release from the substrate of inorganic phosphate, which is quantitated as its acid–molybdate complex.[3]

Reagents

2-(*N*-morpholino)ethanesulfonate-NaOH (MES) buffer, $1\,M$, pH 6.2
MgCl$_2$, $0.1\,M$

[1] A. R. Price and S. M. Fogt, *J. Biol. Chem.* **248**, 1372 (1973).
[2] H. V. Aposhian, *Biochem. Biophys. Res. Commun.* **18**, 230 (1965).
[3] G. R. Bartlett, *J. Biol. Chem.* **234**, 466 (1959).

Disodium ethylenediaminetetraacetate (EDTA), 10 mM
dTMP, 20 mM

Procedure. Place in a tube 10 μmoles of MES buffer, 0.5 μmoles of MgCl$_2$, 0.05 μmoles of EDTA, 0.3 μmoles of dTMP, a rate-limiting amount of enzyme, and water to a volume of 0.125 ml. The reaction is initiated by the addition of enzyme to the mixed reagents at 37°. After 15 min, the tube is placed in a 100° bath for 1 min. Two control tubes are incubated: one without any enzyme; and one with enzyme, but with the substrate dTMP withheld until after boiling the reaction mixture. The contents of the tubes are diluted with water and colorimetric reagents[3] (H$_2$SO$_4$, ammonium molybdate, NaHSO$_3$, Na$_2$SO$_3$, and 1-amino-2-naphthol-4-sulfonic acid) to a volume of 1.25 ml. After heating for 7 min at 100°, the tubes were cooled and centrifuged to remove any precipitate. The absorbance at 750 nm due to the blue phosphomolybdate complex is measured (0.01 μmole of P$_i$ gives an absorbance of 0.11 in a 1-cm cuvette).

Units. One unit of enzyme catalyzes the hydrolysis at 37° of 1 μmole of dTMP per minute. Specific activity is expressed per milligram of protein as determined by the method of Lowry *et al.*[4]

Alternative Assay Procedures. Phosphohydrolase activity can also be measured by separation of substrate and product using either paper or ion-exchange column chromatography. The nucleoside product is quantitated by its ultraviolet absorbance. Radioactive substrates can also be employed, so that nucleoside release is measured by scintillation counting.[1] Enzyme preparations were diluted when necessary for assay in 10 mM Tris·HCl buffer (pH 7.5) containing 1 mM EDTA, 0.1 mM dithiothreitol, and 1 mg of bovine serum albumin per milliliter.

Purification Procedure[1]

Preparation of Infected Cells. Bacillus subtilis strain SB19 (ATCC 15575) was selected for its motility (since PBS2 phage adsorbs only to active flagella[5]). Cells were grown with vigorous shaking at 37° using 750 ml of Difco Penassy Broth in 4-liter flasks. Alternatively, a richer tryptone–yeast extract broth which gives higher phage yields can be employed.[6] At a culture absorbance at 660 nm of 1.0 (about 2 × 10^8 cells/ml), phage PBS2 (ATCC 15575-B) was added at a multiplicity of 5 per

[4] O. H. Lowry, N. J. Rosebrough, A. L. Farr, and R. J. Randall, *J. Biol. Chem.* **193**, 265 (1951).
[5] A. R. Price and S. J. Cook, *J. Virol.* **9**, 602 (1972).
[6] G. E. Katz, A. R. Price, and M. J. Pomerantz, *J. Virol.* **20**, 535 (1976).

cell. Shaking was continued for 25 min (lysis begins about 40 min after infection), and then the flasks were chilled in ice. The infected cells were then harvested at 0° by centrifugation for 15 min at 5000 g.

Preparation of Cell Extracts. The cells were resuspended in 1% of the original volume using 10 mM Tris·HCl buffer, pH 7.5, containing 1 mM EDTA. Egg lysozyme was added at 100 μg/ml, followed by incubation at 37° for about 30 min until the suspension cleared. Brief sonic oscillation was used to shear the DNA and thus reduce the viscosity. Centrifugation at 15,000 g for 20 min at 4° gave a clarified cell extract.

General. The following procedures were performed at 0–4°. A summary of the purification scheme is given in the table.

Streptomycin Treatment. While the extract was being stirred, a 5% (w/v) solution of streptomycin sulfate (pH 7) was added in drops over a 5-min period to give a final concentration of 1.15% (w/v). After 15 min, the suspension was centrifuged at 15,000 g for 15 min, and the supernatant fluid was made 1 mM in 2-mercaptoethanol.

DEAE-Cellulose Chromatography. The above fraction was applied to a column (4 × 36 cm) of DEAE-cellulose equilibrated in 10 mM Tris·HCl buffer (pH 7.5) containing 1 mM EDTA and 1 mM 2-mercaptoethanol. It was eluted with 375 ml of the same buffer at a rate of 120 ml/hr. A linear 4-liter gradient from 0 to 400 mM NaCl in this buffer was employed to elute the activity. The peak fractions (200 to 240 mM NaCl) were pooled.

Hydroxylapatite Chromatography. The above fraction was put on a

PARTIAL PURIFICATION OF DEOXYTHYMIDYLATE PHOSPHOHYDROLASE

Fraction	Volume (ml)	Total activity (units)	Specific activity (units/mg)
Cell extract	102	315	0.30
Streptomycin	123	304	0.37
DEAE-cellulose	299	184	1.4
Hydroxylapatite	4.2	192	5.7
Sephadex G-100 I	19	193	33
Electrofocusing	3.4	41	9.1[a]
DEAE-Sephadex	2.4	22	22[a]
Sephadex G-100 II	3.0	22	113

[a] These fractions are contaminated by ampholines from electrofocusing, which react in the assay for protein and thus decrease the apparent specific activity.

column (4 × 12 cm) of hydroxylapatite equilibrated in 10 mM potassium phosphate buffer (pH 7.0) containing 0.1 mM EDTA and 1 mM 2-mercaptoethanol. It was eluted with 600 ml of the same buffer at 120 ml/ hr. A linear gradient from 0 to 200 mM phosphate in this buffer eluted the activity. The peak fractions (30–90 mM phosphate) were pooled and dialyzed in 50 mM Tris·HCl buffer (pH 7.5) containing 10 mM NaCl, 0.1 mM EDTA, and 0.1 mM dithiothreitol. The volume was reduced by ultrafiltration in a collodion bag in the same buffer.

First Sephadex G-100 Chromatography. The above fraction was applied to a column (1.4 × 100 cm) of Sephadex G-100 and eluted in the above buffer at a rate of 27 ml/hr. Peak fractions (corresponding to a molecular weight of 35,000–45,000) were pooled.

Electrofocusing. The above fraction was dialyzed in 0.5% glycine containing 0.1 mM EDTA and 0.1 mM dithiothreitol. Electrofocusing was performed in 110 ml of 2% ampholines (pH 3–6) as described in the LKB 8100 manual. Fractions containing activity were pooled and then dialyzed and concentrated in 50 mM Tris·HCl buffer (pH 7.5) containing 10 mM NaCl, 0.1 mM EDTA, and 0.1 mM dithiothreitol.

DEAE-Sephadex Chromatography. The above fraction was applied to a column (2.5 × 7.5 cm) of DEAE-Sephadex and eluted with 70 ml of the above buffer at 20 ml/hr. A linear gradient from 0 to 1 M NaCl in this buffer was employed to elute the enzyme. Active fractions (0.4–0.6 M NaCl) were pooled and concentrated by ultrafiltration in the same buffer.

Second Sephadex G-100 Chromatography. The above fraction was applied to the Sephadex G-100 column described above, eluted with the same buffer, and concentrated by ultrafiltration. The enzyme was purified 380-fold with a recovery of 7% of the activity. It was not homogeneous, showing at least 5 protein bands on electrophoresis in neutral or dodecylsulfate-containing polyacrylamide gels.[1]

Properties[1]

Substrate Specificity. When assayed under the usual conditions (1–3 mM substrate), several nucleotides could replace dTMP with rates of hydrolysis equal to or greater than dTMP. These included dTMP analogues with: (1) ribose, arabinose, or 2′-fluorodeoxyribose instead of deoxyribose as the sugar residue; or (2) 2-thio, 4-thio, or 6-aza substituents on the thymine ring. Likewise, high activity was obtained with dUMP and its derivatives with the following substituents on the 5-position of the ring: fluoro, chloro, bromo, iodo, trifluoromethyl, hydroxymethyl, methylamino, and ethyl. Even a reduced ring (O^4-4,5,6-

tetrahydro-dUMP) was acceptable as a substrate, although sugar analogues (ribo, arabino, or 2'-O-methylribo) of dUMP were poor (1–3% of the rate with dTMP).

All cytosine-containing nucleotides gave little or no activity, although 5-methyl-dCMP (which is 4-amino-dTMP) and dCMP were slowly dephosphorylated (3% and 0.4%, respectively) without prior deamination. Only dAMP (3%) among adenine nucleotides had any activity. However, other purine deoxyribonucleotides (dGMP, dIMP, and dXMP) gave 40–70% of the rate with dTMP. Deoxyribose-5-phosphate was inactive.

Analogues of dTMP with 3'-phosphate substituents (pdTp, dTp, pdTpdT, dTpdT, and cyclic-dTMP) showed little or no activity. Also inactive were 5'-phosphate derivatives, such as dTDP, dTTP, dTMP-p-nitrophenylester, dUDP, dUTP, dCTP, ATP, dATP, and dGTP. In summary, the PBS2 phosphohydrolase requires a free 5'-phosphate on a nucleotide, preferably a deoxyribonucleotide derivative of 4-hydroxypyrimidine or 6-hydroxypurine (without the amino groups found in cytosine or adenine).

Substrate and Inhibitor Affinities. The apparent K_m values for dTMP, dUMP, and dGMP were 0.01, 0.8, and 0.7 mM, respectively. The K_m for each of these substrates was equal to its K_i as a competitive inhibitor of the other two substrates. Likewise, each nucleoside product had about the same K_i (1 mM for thymidine, 70 mM for deoxyuridine, and 80 mM for deoxyguanosine) versus each of the three substrates. These results (and others on enzyme inactivation by heating, trypsin treatment, or sulfhydryl reagents[1]) indicate that one enzyme hydrolyzes all three substrates.

Studies of various nucleotides as inhibitors of dTMP hydrolysis indicated that the affinity for dTMP is much greater than for any of the other nucleotides mentioned above. Only 4-thio-dTMP was a good inhibitor; most other dTMP analogues which gave high V_{max} values as substrates showed weak binding as inhibitors. However, at high concentrations, most nucleotides could inhibit the hydrolysis of sub-K_m levels of dTMP. The data support strong preference for a 2'-deoxyribonucleotide with a 4-hydroxypyrimidine ring, and a high affinity only for dTMP.

pH and Temperature Optima. The enzyme is maximally active with MES and sodium glycylglycinate buffers between pH 5.8–6.8. It is inactivated below pH 4.5 or above pH 10.5. The optimum assay temperature was 42°.

Salt Effects. The hydrolysis of dTMP was inhibited less than 20% by 2 M NaCl. In 0.4 mM EDTA, the activity was 10% of that with optimal (4 mM) Mg^{2+} concentrations; the apparent K_m for Mg^{2+} was 0.06 mM. Only Co^{2+} or Ni^{2+} could effectively substitute for Mg^{2+}. $HgCl_2$ and

CuCl$_2$ completely inhibited the enzyme, unless excess dithiothreitol was present. Potassium fluoride also inhibited (by *50%* at 7 m*M*). The reaction proceeded to completion (over 99% hydrolysis), and the P$_i$ produced was only weakly inhibitory.

Molecular Weight. Sephadex column chromatography in the presence of standard proteins gave an apparent molecular weight of 40,000. Since the enzyme appeared smaller on Sephadex chromatography and on sucrose density gradient centrifugation when 1 *M* NaCl was present, the enzyme may contain subunits.[1]

Stability. The crude cell extract or the purified fractions in 50% glycerol could be stored at −20° for at least a year without loss of activity.

Comparative Properties. The PBS2-induced deoxythymidylate phosphohydrolase in crude extracts shows very similar properties to those described here for the purified preparation, since the levels of nucleotidase activities in uninfected *B. subtilis* extracts are very low (less than 0.01 units/mg). These *B. subtilis* 3'- and 5'-nucleotidases are rather nonspecific,[7,8] like the known snake venom, bull semen, and prostate gland phosphatases. Thus, the PBS2 phosphohydrolase has a unique specificity, suited to its proposed role of degrading dTMP to prevent thymine incorporation into PBS2 phage DNA.

Acknowledgment

This work was supported by the U.S. Energy Research and Development Administration as EY-76-S-02-2101, Report No. C00-2101-30.

[7] A. L. Demain and D. Hendlin, *J. Bacteriol.* **94**, 66 (1967).
[8] R. A. Felicioli, S. Senesi, F. Marmocchi, G. Falcone, and P. L. Ipata, *Biochemistry* **12**, 547 (1973).

[37] Uridine Nucleosidase from Yeast

By GIULIO MAGNI

Uridine + H$_2$O → uracil + ribose

Uridine nucleosidase (EC 3.2.2.3) was first discovered in yeast by Carter.[1] The method given below was developed to provide for the first

[1] E. C. Carter, *J. Am. Chem. Soc.* **73**, 1508–1510 (1951).

time a homogeneous preparation of the enzyme. The details of the procedure and the properties of the enzyme have been reported previously.[2,3]

Enzyme Assay

Assay I

Principle. The method is based on differential absorption between uridine and uracil at 280 nm at pH = 7.2 (ΔE_m^M at 280 nm = 2.1). The rate of production of uracil is determined by measuring the decrease of absorption at 280 nm with the time, in the presence of an appropriate concentration of uridine.

Reagents
Sodium phosphate buffer, pH 7.2, 0.2 M
Uridine, 4 mM

Procedure

The enzyme assay is conducted at 37°, and the reaction is followed by measuring the rate of disappearance of nucleoside using a recording spectrophotometer equipped with a thermostatted cell compartment. Assay mixtures contain the following components in a final volume of 2.0 ml: 1.0 ml of the sodium phosphate buffer, 0.1 ml of uridine, an appropriate amount of enzyme, and H_2O up to 2 ml. Reactions are usually initiated by the addition of enzyme.

Assay II

Principle. The method is based on different R_f values of nucleoside and base when they are chromatographed on cellulose thin layer using as a solvent a mixture of 4 parts acetone, 2 parts acetic acid, and 1 part H_2O.

Materials

Acetone, acetic acid, cellulose thin layer sheet.

[2] G. Magni, E. Fioretti, P. L. Ipata, and P. Natalini, *J. Biol. Chem.* **250**, 9–13 (1975).
[3] G. Magni, P. Natalini, S. Ruggieri, and A. Vita, *Biochem. Biophys. Res. Commun.* **69**, 724–730 (1976).

Procedure

The 0.2 ml reaction mixture consisted of 0.2 mM uridine, 108 mM phosphate buffer, and an appropriate amount of enzyme. After incubation at 37°C, the reaction was started upon addition of 25 μl of acetic acid. After centrifugation, 18 μl of supernatant was spotted on cellulose thin layer together with an appropriate amount of uridine and uracil and were chromatographed as previously described. After the chromatograph was developed, it was dried and the spots (visualized by fluorescence) corresponding to uridine and uracil were cut out and placed in scintillation vials with 12 ml of scintillation mixture for counting in a liquid scintillation spectrometer.

Definition of Unit. One unit of enzyme activity is defined as the amount of enzyme capable of hydrolyzing 1 μmole of uridine per minute at 37°. Protein content is measured by a modification of the biuret test[4] by the procedure developed by Schacterle and Pollack.[5] For pure enzyme, the extinction coefficient (see below) can be used.

Materials

Nucleosides are obtained from Boehringer and Soehne. Commercial baker's yeast is from Vulcania, Italy. Sephadex G-100 is purchased from Pharmacia, and hydroxylapatite is from Bio-Rad.

Purification Procedure

All steps are performed at 0–4°.

Step 1. Four kilograms of commercial baker's yeast (Vulcania) are plasmolyzed according to Kunitz[6] by swelling with 2 liters of toluene at 45°C, leaving at room temperature for 2–3 hr, mixing with 4 liters of cold distilled water, and then transferring to a separatory funnel (in the cold). After 18 hr the aqueous phase containing the cellular homogenate is collected and centrifuged for 20 min at 15,000 g; the precipitate fraction is discarded and any turbidity from the supernatant fluid is eliminated by filtration through Whatman No. 3 MM paper. The filtered supernatant fluid is considered as crude extract.

Step 2. The crude extract is precipitated by adding solid ammonium

[4] A. G. Cornall, C. J. Bardawill, and M. M. David, *J. Biol. Chem.* **177**, 751 (1949).
[5] G. R. Schacterle and R. L. Pollack, *Anal. Biochem.* **51**, 654–655 (1973).
[6] M. J. Kunitz, *J. Gen. Physiol.* **29**, 393–406 (1947).

sulfate up to 50% saturation. After centrifugation the supernatant fluid is discarded, and the pellet is dissolved in 70 ml of 100 mM phosphate buffer, pH 7.2, and dialyzed overnight against the same buffer.

Step 3. The dialyzed fraction 2 is placed onto a Sephadex G-100 column (7.5 × 110 cm) equilibrated with the above buffer, and the elution is performed, with 15-ml fractions being collected at a constant flow rate of 56 ml/hr.

The enzyme activity is eluted between 1590 and 1780 ml (volume elution). The eluate is precipitated by addition of solid ammonium sulfate up to 50% saturation. After centrifugation, the pellet is redissolved in 11 ml of 100 mM phosphate buffer, pH 7.2, and dialyzed 6 hr against the same buffer.

Step 4. The dialyzed fraction 3 is placed onto a Sephadex G-100 column (3 × 120) equilibrated with the above buffer, and the elution is performed with 5-ml fractions being collected at a constant flow rate of 20 ml/hr. The uridine nucleosidase activity is eluted between 350 and 390 ml (volume elution). The eluate is saturated with solid ammonium sulfate up to 50% and centrifuged, and the pellet is redissolved in 5 ml of 10 mM phosphate buffer and dialyzed 16 hr with several changes against 10 mM phosphate buffer, pH 7.2.

Step 5. The dialyzed fraction 4 is adsorbed onto a hydroxylapatite column (Bio-Gel HT) equilibrated with 10 mM phosphate buffer, pH 7.2, using a peristaltic pump at a constant flow rate of 10 ml/hr.

The column volume is calculated to provide 1 ml of gel per 5 mg of protein. Using the same pump, the column is washed with 3 times its volume of 10 mM phosphate buffer, pH 7.2, and the enzyme is eluted step by step with increasing concentrations of the same buffer. At 30 mM buffer concentration the enzyme is eluted in a volume of approximately 36 ml and 5-ml fractions are collected.

All ammonium sulfate precipitations are performed with the pH kept at 7 by adding ammonium hydroxide. No loss of activity is observed after such precipitations. The purification procedure is summarized in Table I.

Properties of the Enzyme

Table I shows a 260-fold purification of the enzyme which is stable up to 15 days at 4° in 100 mM phosphate buffer, pH 7.2, or in 50 mM Tris·HCl, pH 7.0. At −20° the enzyme is stable for several months. A

TABLE I
PURIFICATION OF BAKER'S YEAST URIDINE NUCLEOSIDASE[a]

Purification step	Total protein (mg)	Total activity (units)	Specific activity (units/mg)	Purification (fold)	Yield (%)
1. Crude extract	36,000	200.16	0.006	—	100
2. Ammonium sulfate	9,863.4	88.770	0.009	1.5	44.3
3. First G-100 column	526.2	72.090	0.137	23	36
4. Second G-100 column	156.0	38.482	0.247	41	19.2
5. Hydroxylapatite column	9.04	14.10	1.56	260	7

[a] Taken from Magni et al.[2]

few enzyme preparations showed, under the above conditions, a remarkable instability, and a complete loss of activity was observed in less than 1 week. The preparation is homogeneous as judged by a single band observed on polyacrylamide disc gel electrophoresis in the presence and in the absence of 0.2% sodium dodecylsulfate. This last method indi-

TABLE II
SUMMARY OF SOME PROPERTIES OF BAKER'S YEAST URIDINE NUCLEOSIDASE

Substrates	Uridine (K_m = 0.86 × 10^{-3} M)
	5-Methyluridine (K_m = 1.66 × 10^{-3} M)
pH optimum	7.1
Isoelectric point	5.1
E^1 at 280 nm, pH 7.2	5.1
Molecular weight	32,500
Subunit molecular weight	16,500
Sulfhydryl groups per enzyme mole	1
Metal content (g-atom per enzyme mole)	1 Cu^{2+}
Inhibitors	Ribosylthymine (competitive K_i = 7 × 10^{-5} M)
	Glucose-6-phosphate (competitive K_i = 1.9 × 10^{-4} M)
	Ribose (competitive K_i = 7.2 × 10^{-3} M)
	Ribulose-5-phosphate (nonhyperbolic behavior: n' > 1)
	Ribose-5-phosphate (nonhyperbolic behavior: n' > 1)
	$ZnCl_2$, $CdCl_2$, $CoCl_2$, $CuCl_2$, $MnCl_2$, $NiCl_2$, o-phenanthroline, Etilenediaminetetracetate, NaCN

TABLE III
AMINO ACID COMPOSITION OF URIDINE NUCLEOSIDASE[a]

Amino acid	Amino acid content[b] (moles/mole enzyme)	Residues per mole of enzyme (nearest integer)
Aspartic acid	27.8	28
Threonine	15.1[c]	15
Serine	15.8[c]	16
Proline	11.9	12
Glutamic acid	25.1	25
Glycine	33.2	33
Alanine	20.2	20
Valine	23.3[d]	23
Half-cystine	3.8[e]	4
Isoleucine	9.9[d]	10
Leucine	12.6[d]	13
Tyrosine	2.4	2
Phenylalanine	11.9	12
Lysine	15.6	16
Histidine	8.8	9
Arginine	6.4	6
Tryptophan[f]		1
Total residues		245

[a] Taken from Magni et al.[2]
[b] All calculations based on molecular weight of 32,500.
[c] Values were extrapolated to zero time of hydrolysis.
[d] Values were extrapolated to infinite time of hydrolysis.
[e] Determined as cysteic acid after performic acid oxidation.
[f] Determined spectrophotometrically according to Edelhoch.[7]

cates that the protein is composed of two apparently identical subunits of 16,500 mol wt. The enzyme exhibits a strict specificity toward uridine and 5-methyluridine (ribosylthymine), resulting inactive on all other pyrimidine and purine nucleosides tested.[2]

The inhibition exerted on the enzyme activity by pentose phosphate pathway metabolites, and particularly by ribose-5-phosphate, appears to be of some significance. The enzyme is also inhibited by various metal ions, and even though it did not seem to require the addition of any metal for activity it is inhibited by various chelating agents. This inhibition strongly suggested that baker's yeast uridine nucleosidase could be a metalloenzyme. Metal analysis performed on different enzyme preparations by atomic absorption spectrophotometry reveals that the only metal present in significant quantity is copper, ranging in stoichiometry between 0.65 and 1.08 g-atom of copper per enzyme mole. In addition, colorimetric and electron paramagnetic resonance measurements indicate that all copper is in the cupric state.[3]

Some general properties of the enzyme are summarized in Table II, and the amino acid composition is given in Table III.[2,7]

[7] H. Edelhoch, *Biochemistry* **6**, 1948–1954 (1967).

Section VI

Pyrimidine Metabolizing Enzymes

A. Kinases
Articles 38 through 50

B. Deaminases
Articles 51 through 55

C. Phosphorylases and *Trans*-Deoxyribosylase
Articles 56 through 60

[38] Uridine-Cytidine Kinase from Novikoff Ascites Rat Tumor and *Bacillus stearothermophilus*

By ANTONIO ORENGO and SHU-HEI KOBAYASHI

$$UR(CR) + ATP \xrightarrow{Mg^{2+}} UMP(CMP) + ADP$$

The terminal product of pyrimidine ribonucleotide biosynthesis, CTP, appears to regulate its own synthesis, acting as a specific inhibitor of the first step of the enzymic sequence. By such a mechanism, both the synthesis *de novo* of pyrimidine nucleotides from aspartate, carbon dioxide, ammonia, and 5-phosphoribosyl pyrophosphate[1] and the salvage pathway[2-4] are controlled by the inhibitor. Pyrimidine ribonucleoside kinase (uridine kinase, EC 2.7.1.48) is the first enzyme of the salvage pathway.

The kinase was first described by Canellakis in rat liver extracts.[5] It was shown by others to phosphorylate cytidine also, and to be present in a variety of tissues and tumors.[6-9] Rapidly dividing cells use the kinase to a significant extent.[8]

The partial purification and some of the general properties of the kinase from Novikoff ascites rat tumor and the obligate thermophile *Bacillus stearothermophilus* (EC 2.7.1.48) are reported. The thermophile kinase appears to be particularly suitable for studies of mechanisms of regulation of catalytic activity.

Assay Methods

Two types of assays are used. The radiochemical assay measures the conversion of the nucleoside to nucleotide by chromatographic separation of the reactants and products on DEAE-cellulose paper.

The optical assay is a compound assay in which pyruvate kinase is coupled with lactate dehydrogenase in order to measure ADP. In the presence of an excess of lactate dehydrogenase, pyruvate kinase,

[1] J. C. Gerhart and A. B. Pardee, *J. Biol. Chem.* **237**, 891 (1962).
[2] A. Orengo, *Exp. Cell Res.* **41**, 338 (1966).
[3] A. Orengo, *J. Biol. Chem.* **244**, 2204 (1969).
[4] A. Orengo and G. F. Saunders, *Biochemistry* **11**, 1761 (1972).
[5] E. S. Canellakis, *J. Biol. Chem.* **227**, 329 (1957).
[6] P. Reichard and O. Sköld, *Acta Chem. Scand.* **11**, 17 (1957).
[7] O. Sköld, *J. Biol. Chem.* **235**, 3273 (1960).
[8] O. Sköld, *Biochim. Biophys. Acta* **44**, 1 (1960).
[9] E. P. Anderson and R. W. Brockman, *Biochim. Biophys. Acta* **91**, 380 (1964).

phosphoenolpyruvate, and DPNH, ADP is stoichiometrically converted to ATP with the simultaneous oxidation of DPNH. Thus, disappearance of DPNH, as measured by a decrease in absorbance at 340 nm, is a measure of uridine kinase activity.

Reagents
    Tris·Cl buffer, pH 7.54, 83 mM
    $MgCl_2$, 7 mM
    ATPNa$_4$, 2.85 mM
    [2-^{14}C]uridine or cytidine, 0.83 mM

An appropriate amount of enzyme and/or H_2O is added to the above reaction mixture at 37° to give a final volume of 0.060 ml. The reaction is carried out at 37° for a length of time which converts less than 10% of substrate to product. Then 0.025 ml of the incubation mixture is taken in duplicate and subjected to descending chromatography on a DEAE-cellulose paper (Whatman No. DE81). Paper strips 2.9 × 14 cm are used, and the samples are applied as spots at 5 cm from one end together with 0.1 μmole of uridine or cytidine as carriers. The solvent used is 85% ethanol. Whereas the nucleosides move with the solvent and are found near the front, the nucleotides remain at the origin. The nucleotide spots are viewed with an ultraviolet light, cut out, and placed in vials containing 15 ml of Packard Permafluor liquid scintillation fluid diluted 25-fold with toluene. Samples are counted in a liquid scintillation system.

Optical Assay
    Tris·Cl buffer, pH 7.54, 0.1 M
    ATPNa$_4$, 3 mM
    $MgCl_2$, 7 mM
    PEPNa, 3 mM
    KCl, 15 mM
    NADH, 0.128 mM
    Pyruvate kinase, 7 μg (1930 units/mg)
    Lactate dehydrogenase, 10 μg (300 units/mg)
    Uridine or cytidine, 1 mM

The enzyme and/or water is added to a final volume of 1.0 ml. The reaction is initiated by adding the nucleoside, and the decrease in absorbance at 340 nm is followed with the Cary 15 spectrophotometer. The reaction is measured at 37°, and particular care is taken in using batches of ATP and phosphoenolpyruvate with minimal amounts of ADP or pyruvate. Blank runs without uridine are stable at 37°, showing no decrease in absorbance at 340 nm over a period of at least 30 min.

Definition of Unit and Specific Activity. One unit of enzyme is defined as the amount catalyzing the conversion of 1 μmole of uridine or cytidine to UMP or CMP per minute under the condition of the standard assay. The specific activity is determined by dividing units by milligrams of protein as measured by the method of Lowry et al.[10] using bovine serum albumin as a standard.

Purification Procedures

Enzyme from Rat Novikoff Ascites Hepatoma

The Novikoff ascites cells are transplanted and grown for 6 days in the peritoneal cavity of young female Holtzmann Sprague-Dawley rats (120–150 g in weight). The ascitic fluid is removed from the abdomen, diluted 1:2 with 0.25 M sucrose–1 mM MgCl$_2$, and the tumor cells are collected by low-speed centrifugation. To remove most of the red cells, alternate suspension in the sucrose solution and low-speed centrifugation are used. Finally, to estimate the volume, the cells are packed by centrifugation and suspended in 0.01 M Tris–0.25 M sucrose, pH 7.7, at a ratio of 1:4 (cells to buffer).

The suspension is homogenized in an Emanuel–Chaikoff orifice-type homogenizer.[11] In order to break most of the cells, it is necessary to pass the suspension through the homogenizer twice. The homogenate is centrifuged at 20,000 g for 20 min. The supernatant fluid is collected and centrifuged again for 2 hr at 147,000 g. The sediment is discarded, and the supernatant fluid yielded fraction 1 (Table I).

TABLE I
PURIFICATION OF URIDINE-CYTIDINE KINASE FROM NOVIKOFF ASCITES RAT TUMOR[a]

Fractions and steps	Activity (units/ml)	Total activity (units)	Protein (mg/ml)	Specific activity (units/mg protein)
1. Extract	0.10	99.5	10.80	0.009
2. Streptomycin	0.07	131.3	3.88	0.018
3. Ammonium sulfate	0.30	59.7	1.30	0.23
4. Calcium phosphate gel	0.72	21.5	1.56	0.46
5. Sephadex G-200 chromatography	0.53	11.2	0.60	0.88

[a] The radiochemical assay as described in the text was used, with uridine as substrate.

[10] O. H. Lowry, N. J. Rosebrough, A. L. Farr, and R. J. Randall, *J. Biol. Chem.* **193**, 265 (1951).

[11] C. F. Emanuel and I. L. Chaikoff, *Biochim. Biophys. Acta* **24**, 254 (1957).

The protein concentration is then adjusted to 7.0 mg/ml with 0.02 M Tris·Cl (pH 7.5)–1 × 10^{-4} M dithiothreitol.

Streptomycin Treatment. To 1 liter of fraction 1, 200 ml of 5% streptomycin sulfate are added with continuous stirring. The suspensions are stored overnight and centrifuged at 20,000 g for 20 min (fraction 2).

Ammonium Sulfate Treatment. Fraction 2 was brought to 33% saturation with ammonium sulfate. Ammonium sulfate powder is added at a rate of 1 g/min with continuous stirrings.

The suspension is stored for 1 hr and then centrifuged at 20,000 g for 20 min. The precipitate is discarded. The supernatant fluid is then brought to 38% saturation with ammonium sulfate and stored for 1 hr. After centrifuging at 20,000 g for 20 min the supernatant fluid is discarded and the sediment is dissolved in 200 ml of 0.01 M Tris·Cl, pH 7.54, containing 1 × 10^{-4} M dithiothreitol (fraction 3). The protein concentration of fraction 3 is then adjusted to 1.0 mg/ml.

Calcium Phosphate Gel Treatment. For each 100 ml of fraction 3, 7.5 ml of calcium phosphate gel (32 mg/ml) are added slowly with constant stirring. After 45 min of stirring, the suspension is centrifuged. The sediment is then rinsed with 10% ammonium sulfate–0.01 M Tris·Cl–1 × 10^{-4} M dithiothreitol, pH 7.54, at a ratio of 2 times the volume of fraction 3. The sediment is collected by centrifugation and dissolved in 0.1 M Versene–1 × 10^{-3} M uridine–1 × 10^{-4} M dithiothreitol, pH 7.2, in a volume equal to the volume of fraction 3. After a short centrifugation to remove some insoluble material, the clear supernatant fluid is brought to 70% saturation with a solution of saturated ammonium sulfate. After storage for 3 hr the sediment is removed by centrifugation and dissolved in 15 ml of 0.01 M Tris·Cl–1 × 10^{-4} M dithiothreitol, pH 7.54 (fraction 4). In some instances the calcium phosphate gel step is repeated yielding fraction 4a.

Sephadex G-200 Chromatography. A 30-ml portion of fraction 4 (1–2 mg/ml) is carefully pipetted on the top of a Sephadex G-200 column, 3.1 × 48.5 cm, equilibrated with 0.01 M Tris·Cl–1 × 10^{-4} M dithiothreitol, pH 7.7. The flow rate is usually 15 ml/hr, and fractions of 3 ml were collected. The enzyme is eluted with the equilibration buffer. The fractions are read at 280 nm to follow the protein elution. The enzymic activity is determined with the spectrophotometric assay. The fractions with the highest specific activities are pooled and yielded fraction 5.

Temperature was maintained at 2–4° throughout the entire procedure.

Stability. Fraction 4 can be kept at 4° without appreciable loss of catalytic activity for at least a month. It can also be stored in 50% glycerol at −20° for at least 6 months.

Enzyme from Bacillus Stearothermophilus

Growth of Bacterial Cells. Bacillus stearothermophilus strain 10 is grown in a trypticase medium[12] at 60° in a 24-liter laboratory fermentor to an optical density of 0.9–1.2 at 540 nm. The cells are chilled with ice, harvested by continuous-flow centrifugation, washed once with 0.2 M NH₄Cl, and stored at −20°.

Preparation of Bacterial Crude Extracts. Frozen cells are suspended in 0.05 M Tris·Cl (pH 7.5), 2 ml/g, and disrupted by four 15-sec bursts with a Branson sonifier, Model W-185C. Cell debris is removed by centrifugation at 12,000 g for 30 min. The preparation is adjusted to 0.2% in deoxycholate and then centrifuged at 123,000 g for 2 hr to prepare the crude enzyme. Protein concentration, usually around 20 mg/ml, is adjusted to 7.0 mg/ml with distilled water (fraction 1).

Streptomycin Treatment. The procedure is identical to the one described for the Novikoff kinase and yields fraction 2.

Ammonium Sulfate Treatment. Fraction 2 is brought to 35% saturation by the slow addition with continuous stirring of ammonium sulfate. The suspension is stored for 1 hr and then centrifuged at 20,000 g for 20 min. The precipitate is dissolved in 0.01 M Tris·Cl (pH 7.6) as soon as possible. Protein concentration is adjusted to 1.5 mg/ml.

For each 100 ml, 10.85 g of ammonium sulfate are added with continuous stirring. The suspension is stored for 30 min and then centrifuged at 20,000 g for 10 min. To the supernatant fluid, 9.6 g of ammonium sulfate are added, and the suspension is allowed to stand for 10 min and centrifuged again at 20,000 g for 10 min. The pellet is dissolved in 0.01 M Tris·Cl (pH 7.6) (fraction 3) and the protein concentration adjusted to 1.0 mg/ml.

For each 100 ml of fraction 3, 7.5 ml of calcium phosphate gel (32 mg/ml) are added slowly with constant stirring. After 45 min of stirring, the suspension is centrifuged for 10 min, and the sediment is discarded. To the supernatant fluid an additional 15 ml of calcium phosphate gel are added. After 45-min stirring, the sediment is collected by centrifugation and dissolved in 0.5 M Versene–2 mM uridine–20% glycerol (pH 7.6) by stirring it gently overnight in the cold room (Ca ppt. 2).

[12] G. F. Saunders and L. L. Campbell, *J. Bacteriol.* **91**, 332 (1966).

To improve the recovery an additional 12 mg of calcium phosphate gel (32 mg/ml) are added for each 100 ml of the original fraction 3. The adsorption, collection, and solubilization of the precipitate are carried out as above described (Ca ppt. 3). The two fractions (Ca ppt. 2 and 3) are combined, and after a brief centrifugation to remove some insoluble material, the clear supernatant fraction is brought to 60% saturation with saturated ammonium sulfate solutions in 0.01 M Tris·Cl (pH 7.6)–1 mM uridine. After overnight storage, the sediment is collected by centrifugation and dissolved in a minimal volume of 0.01 M Tris·Cl–10% glycerol–1 mM uridine, pH 8.0 (fraction 4).

Fraction 4 is charged on a Sephadex G-25 column (2.5 × 75 cm) which had been equilibrated with 0.01 M Tris·Cl (pH 7.25)–1 mM uridine–10% glycerol. The same buffer is used to elute the kinase. The active fractions are pooled.

DEAE-Cellulose Chromatography I. DEAE-cellulose DE52 is washed with 0.1 M EDTA (pH 7.0) and equilibrated with 0.01 M Tris·Cl (pH 7.25)–1 mM uridine–10% glycerol. A column (2.5 × 6.0 cm) is prepared, and a volume of fraction 4 containing about 40 mg of protein is applied to the column at a rate of 0.65 ml/min. The column is then washed with 70 ml of the equilibration buffer followed by 100 ml of 0.1 M Tris·Cl (pH 7.25)–1 mM uridine–10% glycerol. The kinase is then eluted by 0.1 M Tris·Cl (pH 7.25)–1 mM uridine–10% glycerol–0.1 M ammonium sulfate (fraction 5). The active fractions are pooled and brought to 60% saturation with a saturated ammonium sulfate solution in 0.01 M Tris·Cl (pH 7.6)–1 mM uridine. After overnight storage in cold the sediment is collected and dissolved in a minimum volume of 0.005 M Tris–glycine (pH 8.3)–1 mM uridine.

Sephadex G-200 Chromatography. A 4-ml portion of fraction 5 (20–25 mg/ml) is loaded on a Sephadex G-200 column (2.5 × 95 cm) previously equilibrated with 0.005 M Tris–glycine–1 mM uridine (pH 8.3). The enzyme is eluted with the same buffer at a flow rate of 0.15 ml/min. The fractions with the highest specific activities are pooled and yield fraction 6.

DEAE-Cellulose Chromatography II. A column (1 × 52 cm) is equilibrated with 0.005 M Tris–glycine (pH 8.3)–1 mM uridine–10% glycerol. A portion of fraction 6 (7–8 mg of protein) is applied to the column at a rate of 0.1 ml/min. A 700-ml linear gradient of 0–0.2 M ammonium sulfate in 0.005 M Tris–glycine (pH 8.3)–1 mM uridine–10% glycerol is used to elute the enzyme (0.08–0.09 M). The active fractions are pooled, yielding fraction 7.

TABLE II
PURIFICATION OF URIDINE-CYTIDINE KINASE FROM B. *stearothermophilus*[a]

Fractions and steps	Activity (units/ml)	Total activity (units)	Protein (mg/ml)	Specific activity (units/mg of protein)
1. Extract	0.02	182.3	7.0	0.003
2. Streptomycin	0.04	172.9	2.5	0.016
3. Ammonium sulfate	0.29	110.1	1.0	0.29
4. Calcium phosphate gel	2.02	64.0	3.8	0.53
5. I, DEAE-cellulose chromatography	0.80	53.8	0.7	1.14
6. Sephadex G-200 chromatography	0.27	33.8	0.1	2.70
7. II, DEAE-cellulose chromatography	0.13	19.7	0.03	4.33

[a] The radiochemical assay was used with uridine as substrate.

In all the chromatographies described here protein elution was monitored by measuring absorbance at 290 nm since uridine has practically no absorbance at this wavelength (290/260 = 0.04).

Stability. Fraction 7 can be stored in sterile conditions at 4° for at least 2 years without loss of activity.

Properties

Specificity

Phosphate Acceptor—Tumor Kinase. The enzyme can utilize uridine or cytidine equally well. The ratio of activity with uridine to activity with cytidine was constant (1.14) during the purification of the enzyme. Phosphorylations of [2-^{14}C]cytidine in the presence of increasing amounts of unlabeled uridine, and vice versa, are clearly in accordance with the assumption that only one enzyme is involved.

Specificity for Phosphate Donor and Magnesium Requirement. The reaction depends strictly on ATP and Mg^{2+}; in choosing the concentration of Mg^{2+}, we were guided by the stability constants of $MgATP^{2-}$ and $MgADP^-$ measured by O'Sullivan and Perrin.[13] For practical purposes all the enzymic assays were done in the presence of a concentration of at least 1 mM free Mg^{2+}. A concentration of free Mg^{2+} 20 times the optimal did not affect the reaction rate.

All of the physiological nucleoside triphosphates were tested for their

[13] W. J. O'Sullivan and D. D. Perrin, *Biochemistry* **3**, 18 (1954).

capacity for phosphorylating uridine or cytidine; ATP (13.8 nmoles of UMP formed per 10 min) is the most effective phosphate donor. As substitutes for ATP, dATP (8.7), dGTP (9.2), and dUTP (9.4) can also be considered good phosphate donors. Less effective were GTP (6.9), dCTP (4.7), dTTP (5.7), and ITP (4.6); UTP (1.4) and CTP (0.6) appeared to be very poor phosphate donors in the reaction. All the nucleoside triphosphates were used in a final concentration of 2.8 mM.

Initial Velocity Patterns. The apparent affinity constant of the kinase for ATP appears to increase with increasing concentrations of uridine. The ratio of the apparent Michaelis constant (K_m) to the apparent maximum velocity remains practically unchanged in the range of uridine concentrations studied. The values of the apparent K_m and V_{max} are listed in Table III. The kinetic data indicate a "ping-pong" mechanism, i.e., the first product leaves the enzyme before the second substrate combines.[14] The kinase is thereby postulated to bind first ATP with release of ADP and then uridine or cytidine with the subsequent release of UMP or CMP.

Thermophile Kinase

Effect of Temperature on the K_m of the Phosphate Acceptor, Uridine. A higher concentration of uridine is necessary to saturate the thermophile enzyme at 60° than at either 37° or 10°. The maximum

TABLE III
KINETIC CONSTANTS OF URIDINE-CYTIDINE KINASE FOR URIDINE AND ATP[a]

Compound	Concentration second substrate (mM)	Apparent K_m (mM)	Apparent V_{max}	K_m / V_{max}
ATP	Uridine, 0.185	0.15 ± 0.04	6.37 ± 0.26[b]	0.024 ± 0.005
	Uridine, 0.371	0.27 ± 0.02	9.91 ± 0.15[b]	0.027 ± 0.002
	Uridine, 0.742	0.31 ± 0.09	11.69 ± 0.74[b]	0.027 ± 0.006
Uridine	ATP, 1.40	0.26 ± 0.02	17.89 ± 0.53[c]	0.015 ± 0.001
	ATP, 2.80	0.27 ± 0.02	18.87 ± 0.44[c]	0.014 ± 0.001

[a] Standard conditions were as described in the text for radiochemical assay.
[b] Measured as nmoles of UMP per 10 min per 5.2 μg of protein.
[c] Measured as nmoles of UMP per 10 min per 4.3 μg of protein.

[14] W. W. Cleland, *Biochim. Biophys. Acta* **67**, 104 (1963).

velocities in μmoles of UMP hr^{-1} mg^{-1} were: 2 at 10°, 25 at 37°, and 66 at 60°. The Michaelis constants for uridine, calculated from Hofstee plots, were 2.1 × 10^{-4}, 6.4 × 10^{-4}, and 16.0 × 10^{-4} M, respectively. Since the concentration of uridine was not high enough to attain a plateau at 60°, the values of V_{max} and K_m at this temperature may be less accurate.

Effects of Concentration of ATP at Several Temperatures. At temperatures below the temperature of optimal growth of the thermophilic bacterium, it appears that there is more than one level of saturation for the enzyme. At growth temperature, i.e., 60°, the saturation curve tends to be sigmoidal.[4]

Feedback Regulation

The kinase is inhibited by CTP and UTP. The tumor kinase is inhibited 50% by CTP at 0.56 mM when the concentration of ATP is 2.8 mM and that of uridine 0.80 mM.

Studies on kinetics show that the inhibition by CTP is competitive with respect to concentration of ATP and not competitive with respect to concentration of phosphate acceptor. Upon aging of the enzyme, the inhibitory effect of CTP is partially lost and CTP became effective as a phosphate donor; the catalytic activity of the enzyme is retained.

Thermophile Kinase. In contrast to the tumor, the thermophile kinase shows a temperature dependency of the CTP inhibition. As the temperature is lowered, a progressively larger fraction of the thermophile enzyme becomes insensitive to inhibition by CTP.[4]

Since it is possible to dissociate regulation from catalysis, the existence of a distinct site for CTP binding has been postulated. Thermodenaturation studies were carried out with chelate oxyderivatives of boron involving the two cis-(OH) groups of the ribose moiety of nucleoside triphosphates. They revealed that the regulatory site is primarily coded for the recognition of the two vicinal cis-(OH) groups of ribonucleotides. A pyrimidine is preferred to a purine base, and the triphosphate has a stronger influence on protein conformation.[15]

[15] A. Orengo, S. H. Kobayashi, and H. Thames, unpublished results.

[39] Uridine-Cytidine Kinase from *Escherichia coli*

By POUL VALENTIN-HANSEN

$$\text{Uridine} + \text{GTP} \overset{\text{Mg}^{2+}}{\rightleftharpoons} \text{UMP} + \text{GDP}$$

$$\text{Cytidine} + \text{GTP} \overset{\text{Mg}^{2+}}{\rightleftharpoons} \text{CMP} + \text{GDP}$$

The method given below describes the preparation and use of 5'amino-5'- deoxyuridine coupled to bromoacetaminoethyl-Sepharose-4B for the purification of uridine-cytidine kinase by affinity chromatography.

Assay Method

Method 1

Principle. Enzyme activity is measured as the GTP-dependent formation of UMP from ^{14}C-labeled uridine.

Reagents. Reaction mixture contains:
Tris-hydrochloride, 0.1 M, pH 7.8
$MgCl_2$, 12 mM
GTP, 6 mM
[2-^{14}C]Uridine, 2 mM (1 mCi/mmole)
Enzyme dilution buffer: Tris-hydrochloride, 40 mM, pH 7.8

Assay Procedure. The reaction mixture (30 μl) is incubated at 37° in small test tubes. The reaction is started by the addition of 30 μl of enzyme (approximately 0.001 units). At 2-4-8-16-min 10-μl samples are spotted on thin-layer plates coated with poly(ethylene-imine) impregnated cellulose (PEI)[1] and dried with hot air. Plates are developed step by step in methanol (3 cm) and water (10 cm). The start spots, containing UMP, are cut out and counted in a liquid scintillation spectrometer using a toluene-based scintillation mixture. The specific radioactivity of the substrate is determined by applying 5-μl samples of each reaction mixture to small pieces of PEI and counting them.

Method 2

Principle. With partially purified and pure enzyme it is possible to use a coupled spectrophotometric assay. The GDP production is meas-

[1] K. Randerath and E. Randerath, *J. Chromatogr.* **16**, 111 (1964).

ured by the change in absorbance at 340 nm produced by the oxidation of NADH in a coupled enzyme system using uridine and GTP as substrates and PEP, pyruvate kinase, and lactate dehydrogenase as coupling factors.

Reagents
Tris-hydrochloride, 0.25 M, pH 7.8
$MgCl_2$, 50 mM
KCl, 1 M
NADH, 7.5 mM
GTP, 15 mM
PEP, 5 mM
Uridine, 10 mM
Pyruvate kinase, 10 mg/ml
Lactate dehydrogenase, 10 mg/ml (The crystalline enzyme preparations are obtained from Boehringer, Mannheim Corporation.)

Assay Procedure. The incubation mixture contains in a total volume of 500 μl: 100 μl Tris·HCl, 50 μl $MgCl_2$, 50 μl KCl, 50 μl GTP, 50 μl PEP, 50 μl uridine, 10 μl NADH, 5 μl lactate dehydrogenase, 5 μl pyruvate kinase, enzyme 0.005–0.030 units, and water to make the correct final volume. Reactions are initiated by the addition of uridine to the reaction mixture at 37° in a 10-mm quartz cuvette. The decrease in optical density at 340 nm is followed in a Zeiss spectrophotometer connected to a WW Recorder 3012. Blanks give no change in optical density. Initial rates are converted to micromoles of UMP formed per minute by dividing the change in absorbance per minute by 12.44.

Definition of Units and Specific Activity

One unit is the amount of enzyme that catalyzes the formation of 1 μmole of UMP per minute at 37°. Specific activity is expressed as units per milligram of protein. Protein is determined by the method of Lowry *et al.*[2]

Preparation of Affinity Adsorbent

The sequence of reactions used to prepare the affinity column is shown in Fig. 1. 5′amino-5′-deoxyuridine, a competitive inhibitor of

---

[2] O. H. Lowry, N. J. Rosebrough, A. L. Farr, and R. J. Randall, *J. Biol. Chem.* **193**, 265 (1951).

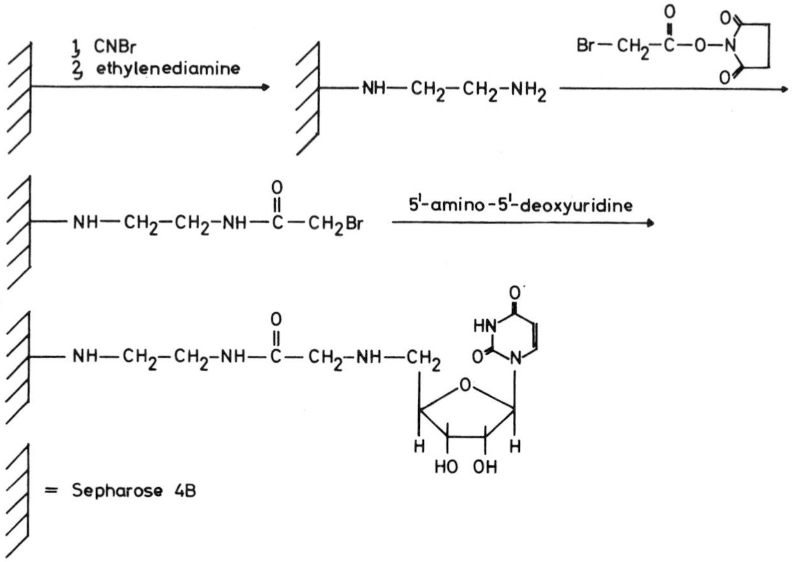

FIG. 1. Steps in the preparation of the affinity column.

uridine-cytidine kinase, is prepared from 2′,3′-isopropylidene-uridine (pharma Waldhof, Mannheim) as described by Horwitz et al.[3]

Activation. Commercially available Sepharose 4B is activated as described by Porath et al.[4] Washed agarose (60 ml) is added to an equal volume of 5 M phosphate buffer, pH 11.9, and mixed by stirring slowly. The rate of stirring is increased, and 150 ml of a cyanogen-bromide water solution (100 mg CNBr per ml H_2O) are added all at once. The slurry is stirred for 10 min, and the temperature is maintained at 8–10° by the addition of ice. The slurry is washed, with suction on a coarse sintered-glass funnel (500 ml each of cold H_2O and cold 0.1 M NaHCO$_3$, pH 9.5). After the last wash, the slurry is filtered under vacuum to a compact cake and transferred to a bottle containing 60 ml of a 10% (vol) ethylene-diamine solution in H_2O, pH 11, at 0°. The coupling is done at 4° for 60 hr. After coupling the gel is washed with 1 liter of H_2O, 1 liter 1 M NaCl, and 1 liter 0.1 M sodium phosphate buffer, pH 7.5.

Bromoacetamidoethyl-Sepharose is prepared as described by Cuatrecasas.[5] Bromoacetic acid (3 mmoles) and N-hydroxysuccinimide (3.6

[3] J. P. Horwitz, A. J. Tomson, J. A. Urbanski, and J. Chua, *J. Org. Chem.* **27**, 3045 (1962).

[4] J. Porath, K. Aspberg, H. Drevin, and R. Axén, *J. Chromatogr.* **86**, 53 (1973).

[5] P. Cuatrecasas, *J. Biol. Chem.* **245**, 3059 (1970).

mmoles) are dissolved in 24 ml of dioxane. To this solution 1.1 mmoles of dicyclohexylcarbodiimide are added. After 70 min the dicyclohexylurea formed is removed by filtration. The filtrate is added to an aminoethyl–Sepharose slurry (60 ml gel and 60 ml 0.1 M sodium phosphate buffer, pH 7.5) at 5°. After 30 min the gel is washed with 3 liters of cold 0.1 M NaCl and added to 60 ml of 0.1 M NaHCO$_3$, pH 8.5, containing 1.5 mmoles of 5'-amino-5' deoxyuridine. This mixture is left for 5 days at room temperature. The gel is removed by filtration and washed with 2 liters of water and 1 liter of 1 M NaCl. This amount of resin is sufficient for making two columns of 1.6 × 15 cm.

Comments. For good efficiency of the method the ligand concentration on the polymer should be high. This is achieved by activation of the gel with high amounts of CNBr.

Purification of Enzyme

Growth of Bacteria. Pyrimidine starvation of *E. coli* increases the level of uridine-cytidine kinase 4–5-fold.[6] Thus a pyrimidine-requiring mutant of *E. coli* K 12 is used as enzyme source.

Escherichia coli K 12 (*pyrC, metB*) is grown with aeration at 37° in the following medium (10 liters): 20 g (NH$_4$)$_2$SO$_4$, 60 g Na$_2$HPO$_4$·2 H$_2$O, 20 g KH$_2$PO$_4$, 20 g NaCl, 70 g NH$_4$Cl, 4 g MgCl$_2$·6 H$_2$O, 100 g D(+)glucose·H$_2$O, 4.5 g DL-methionine, 100 g bactotryptone, 50 g yeast extract, and 4 g uracil. At a cell density of about 5 × 10^9 cells/ml growth stops due to pyrimidine limitation. After cessation of growth, incubation of the culture, with aeration at 37°, is continued for 4 hr. Then the culture is rapidly chilled and harvested by centrifugation. The cell paste is washed with 0.1 M Tris·HCl, pH 7.8, and stored frozen at −20°.

Step 1. Preparation of Cell Extract. Fifty grams of frozen cells are resuspended in 240 ml Tris·HCl, pH 7.8, 4 mM in EDTA, and sonicated in a Branson Sonifier. The temperature of the mixture is kept below 10° during sonication. Following addition of 4 mg deoxyribonuclease and 4 mg ribonuclease the sonicate is centrifuged for 45 min in a Sorvall SS-34 rotor at 16,000 rpm.

Step 2. Ultracentrifugation. The supernatant fluid from step 1 is centrifuged 4.5 hr at 45,000 rpm in a Spinco number 50.1 preparative rotor (use polycarbonate bottles) to remove the ribosomes. The supernatant fluid is desalted on a Sephadex G-25 column (70 × 4 cm) equilibrated with 0.1 M Tris·HCl, pH 7.8 (flow rate 1 liter/hr).

[6] J. Neuhard, personal communication.

Step 3. DEAE-Cellulose Chromatography. The desalted solution is applied to a DEAE-52 cellulose column (2.5 × 40 cm) equilibrated with 0.1 M Tris·HCl, pH 7.8. After washing with two column volumes of the same buffer, the column is eluted with a linear gradient of KCl (0–250 mM) in 0.1 M Tris·HCl, pH 7.8 (total volume of gradient is 1.4 liter; flow rate is 120 ml/hr). Uridine-cytidine kinase elutes from the column at about 0.18–0.22 M KCl. The most active fractions are pooled (70 ml).

Step 4. Affinity Chromatography. Ten milliliters of the pooled fractions from step 3 are desalted on a Sephadex G-25 column (1.6 × 15 cm) equilibrated with 20 mM Tris·HCl, pH 7.8. The desalted fraction (10 ml) is applied to the affinity column (1.6 × 15 cm) equilibrated with 20 mM Tris·HCl, pH 7.8, and the column is eluted with 50 mM Tris·HCl, pH 7.8, with a flow rate of 12 ml/hr (4-ml fractions are collected). Figure 2 presents the elution profile. The column was reutilized, with identical results, after regeneration by treatment with 100 mM Tris·HCl, pH 7.8, followed by equilibration with the starting buffer. The enzyme from seven runs is concentrated on a small DEAE-cellulose column (2 ml) equilibrated with 0.1 M Tris·HCl, pH 7.8. It is eluted from this column with 0.1 M Tris·HCl (pH 7.8)–300 mM KCl.

The purification procedure is summarized in the table.

Comments on Purification Procedure

The enzyme is unstable in crude extracts, especially in the presence of Mg^{2+}. Therefore, the purification up to the DEAE-cellulose step should be done as fast as possible. Due to overlapping of bulk protein

PURIFICATION PROCEDURE FOR URIDINE-CYTIDINE KINASE FROM *Escherichia coli* K12[a]

Fraction	Total volume (ml)	Total units	Total protein (mg)	Specific activity (units/mg)	Recovery (%)
1. Crude extract	280	392	9800	0.040[b]	—
2. Ultracentrifugation	277	378	6094	0.062	96
3. DEAE-cellulose chromatography					
Top fractions	70	204	300	0.68	83
Side fractions	70	120	307	0.39	
4. Affinity[c] chromatography + concentration	6	120	0.6	200	59

[a] Data given for pyrimidine-starved cells.
[b] It is possible to get twice the enzyme level reported here.
[c] Only the DEAE–top fractions are used in the affinity chromatography step.

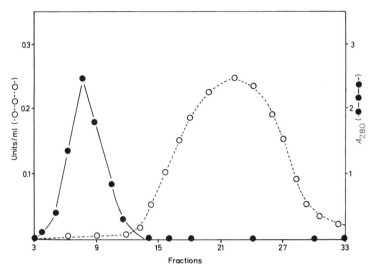

Fig. 2. Affinity chromatography of uridine-cytidine kinase on a 5'-amino-*N*-(acetamidoethyl) 5'-deoxyuridine-Sepharose column. Volume of the column is 30 ml (1.6 cm internal diameter). Equilibrating buffer: 20 m*M* Tris·HCl, pH 7.8. Load: 10 ml of active DEAE-cellulose fraction. Elute with 50 m*M* Tris·HCl, pH 7.8. Fraction volume: 4 ml. Flow rate: 12 ml/hr. All other details are given in the text. O- - -O, uridine-cytidine kinase activity; ●——●, absorbance at 280 nm.

into the enzyme fractions, it is not recommended that affinity chromatography be used on crude extracts. The purified enzyme has been found to be homogeneous when examined by polyacrylamide gel electrophoresis.

Properties

Specificity. Uridine-cytidine kinase is specific for the ribonucleosides uridine and cytidine. It does not phosphorylate 2'-deoxyribonucleosides. The relative rates of phosphorylation of uridine, cytidine, 5-fluorouridine, *N*-acetylcytidine, 5-fluorocytidine, 6-azauridine, 5-azacytidine, 5-hydroxyuridine, *N*-benzoylcytidine, 5-aminouridine, and 5-methyluridine are 100, 86, 66, 57, 40, 31, 4, 3, 1, 0.4, and 0.3, respectively. The nucleoside analogs 5'-amino-5'-deoxyuridine, 5'-deoxyuridine, 5'-iodo-5'-deoxyuridine, and 5'-azido-5'-deoxyuridine are competitive inhibitors. GTP and dGTP are the most efficient phosphate donors.[7] Activity with ATP and dATP is 10%, and with dUTP, dCTP, and dTTP 5% of that

[7] S. F. Hawking, J. R. Baird, and L. R. Finch, *Proc. Aust. Biochem. Soc.* **5,** 40 (1972).

with GTP or dGTP. CTP and UTP are inhibitors of the enzyme (CTP much stronger than UTP) and do not serve as phosphate donors.

Stability. The purified enzyme shows no loss of activity when kept in 0.1 M Tris·HCl, pH 7.8, at 4° for 6 months.

Effect of pH. The enzyme shows a broad pH optimum around 7.8.

Other Properties. The apparent K_m values for cytidine and uridine are 0.13 mM and 0.35 mM, respectively. The molecular weight of uridine-cytidine kinase is about 90,000, determined by gel filtration on a Sephadex G-200 column.

[40] Uridine-Cytidine Kinase from a Murine Neoplasm

By ELIZABETH P. ANDERSON

$$\text{Uridine} + \text{NTP} \xrightarrow{\text{Mg}^{2+}} \text{UMP} + \text{NDP}$$

$$\text{Cytidine} + \text{NTP} \xrightarrow{\text{Mg}^{2+}} \text{CMP} + \text{NDP}$$

Uridine-cytidine kinase (ATP:uridine 5′-phosphotransferase, EC 2.7.1.48) is part of the anabolic pathway by which the preformed pyrimidine nucleosides are salvaged for nucleic acid biosynthesis. In mammalian cells, at least, the enzymic activity is especially high in cells exhibiting a rapid rate of growth.[1-5] The enzyme has been reported in a variety of cells[4] and, wherever tested, phosphorylated both uridine and cytidine. This kinase is apparently the rate-limiting step in the conversion of the nucleosides to their respective triphosphates[4,5] and is susceptible to feedback regulation by both CTP and UTP.[5] In *Salmonella typhimurium,* at least, uridine-cytidine kinase does not appear to be an inducible enzyme.[6]

Assay Methods

The enzyme is assayed in one of two ways.[7]

[1] P. Reichard and O. Sköld, *Biochim. Biophys. Acta* **28,** 376 (1958).
[2] O. Sköld, *Biochim. Biophys. Acta* **44,** 1 (1960).
[3] P. G. W. Plagemann, G. A. Ward, B. W. J. Mahy, and M. Korbecki, *J. Cell. Physiol.* **73,** 233 (1969).
[4] E. P. Anderson, *in* "The Enzymes" (P. D. Boyer, ed.), 3rd ed., Vol. 9, p. 49. Academic Press, New York, 1973.
[5] E. P. Anderson and R. W. Brockman, *Biochim. Biophys. Acta* **91,** 380 (1964).
[6] J. L. Ingraham, private communication (1976).
[7] A. S. Liacouras and E. P. Anderson, *Arch. Biochem. Biophys.* **168,** 66 (1975).

Method A

Principle. In less pure preparations a specific assay employs the conversion of radioactive substrate to product, with chromatographic isolation of the products formed.

Reagents

Tris·HCl buffer, 1 M, pH 7.5, or imidazole-HCl buffer, 0.5 M, pH 6.5
Uridine, 0.1 M, or cytidine, 0.1 M
[2-^{14}C]Uridine (50 mCi/mmole), or [2-^{14}C]cytidine (25 mCi/mmole)
ATP, disodium salt, 0.1 M, adjusted to pH of assay
MgCl$_2$, 0.2 M
P-enolpyruvate, sodium salt, 0.225 M, adjusted to pH of assay
Pyruvate kinase (150 U/mg)

Procedure. The standard reaction mixture contains 3 μl Tris·HCl or imidazole-HCl buffer, 3 μl uridine or cytidine, 20 nCi [2-^{14}C]uridine or 30 nCi [2-^{14}C]cytidine, 3 μl ATP, 2 μl MgCl$_2$, 3 μl P-enolpyruvate, 1 μg pyruvate kinase, enzyme protein (in 5–15 μl 10 mM potassium phosphate buffer, pH 7.5), and water to give a final volume of 32 μl. The ATP-regenerating system, P-enolpyruvate and pyruvate kinase, is unnecessary with purified fractions of the enzyme and can be eliminated from the reaction mixture. The assay is normally run in duplicate or triplicate. The reaction is initiated by the addition of enzyme and carried out at 37° for periods of time up to 10 min, 15 μl being removed at each of two time points. The period of linearity with time should be confirmed for the preparation being assayed. The reaction is stopped by heating in boiling water for 1 min. Labeled substrate and products are then separated by ascending chromatography on thin-layer PEI-cellulose on plastic sheets, in 1-inch or 0.5-inch wide channels. The PEI-cellulose sheets[8] are stored at 0–4°. Before use they are washed at least twice by ascending chromatography in water to remove any degraded PEI. To assist in localization, 0.1 μmole each of pyrimidine base, nucleoside, and nucleoside mono-, di-, and triphosphate are applied to the origin of every channel and cochromatographed with the sample. Development with distilled water for 1 hr separates uridine and uracil and leaves nucleotides at the origin. If UMP is further phosphorylated to UDP and UTP, for example in crude fractions, and quantitation of these is desired, the three nucleotides can be separated by a second development for 2.5 hr in 1 M LiCl. The pyrimidine compounds are localized by ultraviolet fluorescence and counted in a scintillation counter. (R_fs:

[8] Available from J. T. Baker Chemical Co., Phillipsburg, New Jersey.

uridine, 0.87; uracil, 0.79; cytidine, 0.91; cytosine, 0.87, in water; UMP, 0.80; UDP, 0.65; UTP, 0.37; CMP, 0.63; CDP, 0.36; and CTP, 0.12, in LiCl.)

Method B

Principle. In more highly purified preparations, the enzymic activity can be assayed by coupling ADP formation to the pyruvate kinase–lactate dehydrogenase system.

$$\frac{\text{Uridine}}{\text{Cytidine}} + \text{ATP} \rightarrow \frac{\text{UMP}}{\text{CMP}} + \text{ADP}$$
$$\text{ADP} + \text{P-enolpyruvate} \rightarrow \text{ATP} + \text{pyruvate}$$
$$\text{Pyruvate} + \text{NADH} + \text{H}^+ \rightarrow \text{lactate} + \text{NAD}^+$$

The reaction is followed continuously by spectrophotometrically monitoring the absorbance decrease at 340 nm as NADH is oxidized to NAD^+. With nonlimiting amounts of pyruvate kinase and lactate dehydrogenase, the rate of nucleoside phosphorylation is equivalent to that of NADH oxidation.

Reagents

Imidazole-HCl buffer, 0.5 M, pH 6.5
KCl, 1 M
Uridine, 0.1 M, or cytidine, 0.1 M
ATP, disodium salt, 0.1 M, neutralized to pH 6.5
$MgCl_2$, 0.1 M
P-enolpyruvate, sodium salt, 0.06 M, neutralized to pH 6.5
NADH, 0.04 M
Pyruvate kinase (150 U/mg)
Lactate dehydrogenase (360 U/mg)

Procedure. The standard reaction mixture contains 0.1 ml imidazole-HCl buffer, 0.1 ml KCl, 0.1 ml uridine or cytidine, 0.1 ml ATP, 0.11 ml $MgCl_2$, 0.1 ml P-enolpyruvate, 0.01 ml NADH, 50 μg pyruvate kinase, 100 μg lactate dehydrogenase, and water to give a final volume of 1 ml in a cuvette (1-cm light path). The reaction is initiated by the addition of 5–20 μl of enzyme (in 0.01 M potassium phosphate buffer, pH 7.5) and followed at 25° for periods of time up to 15 min with a recording spectrophotometer (either a Cary or Gilford system, with a range span of 0.2 OD units). The rate-limiting quantity of uridine-cytidine kinase to be added is adjusted so that the initial velocity can be determined accurately over a reasonable period during which the reaction rate is linear with time. Proportionality with kinase concentration should also be confirmed. A control cuvette without nucleoside substrate measures adenosine triphosphatase and NADH oxidase activities, which must be subtracted from the total rate.

Definition of Unit and Specific Activity. With either method, one unit of enzymic activity is defined as the amount phosphorylating 1 μmole of nucleoside per minute. Specific activity is expressed as units per milligram of protein. Protein is determined either by the ratio of absorbancies at 280 and 260 nm[9] or by the method of Lowry *et al.*,[10] using bovine serum albumin as standard.

Purification Procedure[7]

Enzyme Source. The following procedure is described for the purification from acetone powders from the murine mast cell tumor P815. Acetone powders are essentially free of enzymic activities that degrade uridine to uracil, thus facilitating the assay of uridine phosphorylation,[11] and the kinase activity is stable in these powders at $-18°$ for several months. Throughout the purification, kinase activity is assayed by method A, in Tris buffer (pH 7.5), using [2-^{14}C]uridine as substrate. All purification operations are performed at 0–4°.

Step 1. Crude Extract. Tumor cells are harvested in late log-phase growth (6 days). Approximately 10 g of acetone powder are extracted twice in a mortar and pestle with 25 ml of 0.01 M potassium phosphate buffer, pH 7.5, and the slurry is centrifuged at 14,000 g for 15 min at 4°. The two supernatant volumes, totaling about 40 ml, constitute the crude extract, which is stable at $-18°$ for at least 24 hr.

Step 2. Fractionation on DEAE-Cellulose. The ionic strength of the crude enzyme is lowered with cold, distilled water to that of 0.01 M potassium phosphate buffer, and the preparation is applied to a DEAE-cellulose column (2 × 60 cm) equilibrated with 0.01 M potassium phosphate buffer, pH 7.5. The column is washed with this buffer until no further protein is removed, judged by scanning at 280 nm; it is then eluted with approximately 350 ml of 0.1 M potassium phosphate buffer, pH 7.5. Each fraction (7–8 ml) is assayed for pH, conductivity, protein content, and kinase activity. The method of Warburg and Christian[9] can also be used to determine the protein in each fraction containing enzymic activity. The kinase activity is eluted with the front of the 0.1 M buffer. The fractions with high specific activity are pooled and

[9] O. Warburg and W. Christian, *Biochem. Z.* **310**, 384 (1941).
[10] O. Lowry, N. J. Rosebrough, A. L. Farr, and R. J. Randall, *J. Biol. Chem.* **193**, 265 (1951).
[11] J. E. Ciardi, Ph.D. Thesis, George Washington University, Washington, D.C. (1968).

again assayed for enzymic activity and for protein by the method of Lowry *et al.*[10]

Step 3. Ammonium Sulfate Fractionation. To the pooled DEAE-cellulose fraction, solid, enzyme-grade ammonium sulfate is slowly added until a 33% saturated solution is obtained. After equilibration, the precipitate is removed by centrifugation at 11,000 g for 15 min and discarded. The supernatant solution is brought to 50% saturation with ammonium sulfate, equilibrated, and again centrifuged. This precipitate is dissolved in a minimum volume of 0.01 M potassium phosphate buffer, pH 7.5 (1 ml). This ammonium sulfate fraction is stable at $-18°$ for at least 1 month without significant loss of activity.

Step 4. Hydroxyapatite Fractionation. After removal of ammonium sulfate by passage through Sephadex G-50, the enzyme is applied to a column of hydroxyapatite (2.0 × 2.0 cm), prepared under 23 cm or less of hydrostatic pressure (flow rate near 3.0 ml/hr), and equilibrated with 0.01 M potassium phosphate buffer, pH 7.5. The column is washed with the buffer and then successively eluted with 0.02, 0.04, 0.06, and 0.08 M potassium phosphate buffer, all at pH 7.5. The major portion of the kinase activity is eluted with the 0.04 M buffer. These fractions are pooled, reassayed, and analyzed for protein (Lowry method).[10,12] Fractionation on hydroxyapatite gives a 10-fold increase in specific activity (see the table). At this stage, the kinase activity remains stable at $-18°$ for several months.

PURIFICATION OF URIDINE KINASE FROM MURINE MAST CELL NEOPLASM P815[a]

Fraction	Total protein (mg)	Total units (μmole/min)	Yield (%)	Specific activity (units/mg)	Ratio of uridine kinase to cytidine kinase activity
Crude extract	1420	17.0	100	0.012	2.18
DEAE-cellulose eluate	145	13.5	80	0.094	2.00
Ammonium sulfate precipitate (33–50% saturation)	36.2	11.4	67	0.315	2.00
Hydroxyapatite eluate	1.73	5.3	31	3.03	2.07

[a] Enzymic activity was assayed by method A in Tris·HCl buffer, pH 7.5, with [¹⁴C]uridine or [¹⁴C]cytidine as nucleoside substrate. The purification was based on assays with uridine as substrate, but the ratio of the two kinase activities remained essentially constant throughout the purification (last column).

[12] For these protein determinations, the sensitivity of the Lowry method was increased by the use of a Technicon AutoAnalyzer with a range expander.

The table gives the data obtained from a typical purification procedure, which achieved a 253-fold increase in specific activity over the crude extract of the acetone powder. This represents approximately 1800-fold purification from the whole tumor tissue, based on the dry weight of the acetone powder (specific activity 0.0017 U/mg dry weight of powder).

In the typical final fraction reaction velocity is linear with enzyme concentration up to at least 6 μg of protein. The ammonium sulfate fraction is essentially free of UMP kinase and of enzymes that degrade UMP to uridine or uracil. The final eluate from hydroxyapatite is also free of cytidine aminohydrolase, nucleoside diphosphate kinase, and adenosine triphosphatase, all of which are present in the crude extract.

Stability. Each step of the purification procedure is reproducible, the enzymic activity is reasonably stable at each stage, and the final fraction exhibits good stability. However, the enzyme is much more labile in 0.1 M Tris·HCl, pH 7.5, and, to achieve stability, phosphate buffers are used throughout the purification and for storage of the enzyme fractions. Even when relatively crude (DEAE-cellulose fraction), the enzyme is labile to dialysis against buffers of low ionic strength. The final hydroxyapatite eluate exhibits some loss of activity after repeated freezing and thawing.

Properties[7]

Cation Requirement of the Enzymic Reaction. With either nucleoside as substrate, Mg^{2+} can be partially (80%) replaced by Mn^{2+} and to a lesser extent by Co^{2+} (66%) and Fe^{2+} (65%). There is even less activity with Ni^{2+}, Cd^{2+}, or Zn^{2+} and negligible activity with Ca^{2+} or Cu^{2+} (hydroxyapatite eluate, assay method A; metal concentration 0.013 M in each case; phosphate donor 0.01 M ATP).

Optimum pH. The enzyme exhibits a broad pH optimum around pH 6.5–7.0 (0.1 M buffers—Tris-maleate, pH 5.5–8.0; triethanolamine-HCl, pH 6.2–7.8; Tris·HCl, pH 6.75–9.0; divalent cation 0.013 M Mg^{2+}).

Specificity of the Phosphate Acceptor. Only uridine and cytidine, among the normal, physiological nucleosides, appear to be phosphorylated by this enzyme. However, the pyrimidine ribonucleoside analogs 5-fluorouridine, 6-azauridine, 6-azacytidine, and 5-azacytidine demonstrate significant capacity to substitute as phosphate acceptor. In contrast, thymine riboside, the deoxyribosides, and cytosine arabinoside are not substrates. Adenosine, purine deoxyribosides, UMP, and CMP are also not phosphorylated. (These data are from experiments using assay method B with ATP as phosphate donor.)

The activity of the kinase with uridine is greater than that with cytidine; the ratio of the two activities remains essentially constant throughout the purification (see the table). The concurrent purification of separate but closely related isozymes might, however, be considered a likely possibility for a feedback-regulated enzyme. Further kinetic studies on uridine and cytidine as alternate substrates for the enzyme have therefore recently been carried out with the purified preparation; they indicate that a single enzyme, and probably a single catalytic site, is indeed responsible for the phosphorylation of both nucleosides.[13] Genetic evidence for the substrate specificity of uridine-cytidine kinase, from experiments with *S. typhimurium,* has shown that, at least in that organism, both uridine and cytidine are phosphorylated by a single enzyme protein encoded by a single gene.[14,15]

Specificity of the Phosphate Donor and Feedback Inhibition of the Enzyme. Like other nucleoside kinases, the enzyme exhibits a broad specificity with regard to phosphate donor. ATP can be quite effectively replaced by dATP (88%), and phosphate donor activity is also observed with dUTP (67%), dCTP (50%), GTP (23%), and dGTP (21%). In our system, GTP and dGTP are appreciably less effective than the other donors. In the absence of ATP, there is essentially no activity in the presence of the end-product inhibitors CTP and UTP. In striking contrast, the deoxyribonucleotides dUTP and dCTP are both good phosphate donors. Thus, although the deoxyribonucleoside triphosphates do exert feedback regulation over the phosphorylation of deoxyuridine and deoxycytidine, reactions which are catalyzed by other kinases, they do not inhibit the phosphorylation of uridine or cytidine by this kinase. Sensitivity of the kinase to the feedback inhibition by CTP and UTP[5] is retained throughout the purification procedure, and inhibition of the hydroxyapatite eluate by both CTP and UTP is at least as great as that of the crude enzyme (percent inhibition at equivalent ratios of inhibitor to substrate, with ATP as phosphate donor).

Molecular Weight. On Sephadex G-150 calibrated with ovalbumin (mol wt 45,000), aldolase (mol wt 158,000) and Blue dextran-2000 (mol wt 2,000,000), the peak of kinase activity is nearly coincident with that of aldolase (ammonium sulfate fraction; see the table).

Kinetic Constants. With either uridine or cytidine as phosphate acceptor, and ATP as phosphate donor, the initial velocity pattern

[13] A. S. Liacouras and E. P. Anderson, *Mol. Cell. Biochem.* **17,** 141 (1977).
[14] C. F. Beck, J. L. Ingraham, J. Neuhard, and E. Thomassen, *J. Bacteriol.* **110,** 219 (1972).
[15] C. F. Beck, J. L. Ingraham, and J. Neuhard, *Mol. Gen. Genet.* **115,** 208 (1972).

(determined with assay B) is one of intersecting rather than parallel lines, ruling out a ping-pong reaction mechanism, and suggesting that the reaction probably proceeds by the sequential addition of both substrates to the enzyme to form a ternary complex, followed by the sequential release of the two products.[16] The pattern of alternate-substrate kinetics further suggests that the sequential addition is random.[13] When analyzed by Cleland's Sequen computer program,[17] the data indicate an apparent K_m of the enzyme for uridine of 0.15 mM, one for cytidine of 0.045 mM, and for ATP, with uridine or cytidine as phosphate acceptor, a K_m of 3.6 mM or 2.1 mM, respectively. The V_{max} for the uridine-kinase reaction is twice that for cytidine phosphorylation (in these experiments, 1.83 and 0.91 μmole phosphorylated/min/mg protein).[16]

Using this purification procedure, uridine-cytidine kinase has also been partially purified (through the ammonium sulfate step) from a fluorouridine-resistant subline of the P815 tumor. This enzyme differs from the kinase of the parent cell line in being more labile, but, more important, in having a K_m for uridine of 1.5 mM, an order of magnitude higher than that of the sensitive line enzyme.[18] This alteration in the mutant cell line results in an enzyme of very low activity, as previously observed with extracts of the resistant line.[19]

[16] A. S. Liacouras, T. Q. Garvey, III, F. K. Millar, and E. P. Anderson, *Arch. Biochem. Biophys.* **168**, 74 (1975).
[17] W. W. Cleland, *Nature (London)* **198**, 463 (1963).
[18] J. M. Glick and E. P. Anderson, unpublished results.
[19] E. P. Anderson, *Cancer Res.* **23**, 1270 (1963).

[41] Pyrimidine Nucleoside Monophosphate Kinase from Rat Liver and Rat Novikoff Ascites Hepatoma (EC 2.7.4.14)

By Antonio Orengo and Patricia Maness

$$\begin{matrix} UMP & & UDP \\ CMP + ATP & \xrightarrow{(Mg^{2+})} & CDP + ADP \\ dCMP & & dCDP \end{matrix}$$

The enzyme occupies a strategic position in biosynthesis of pyrimidine nucleotides, since its phosphate-acceptor substrates are products of both the *de novo* and the salvage pathways. The phosphate-acceptor specificity is peculiar since the enzyme catalyzes the phosphorylation of

CMP, dCMP, and UMP, but not dUMP which differs from UMP by only one atom of oxygen just as dCMP differs from CMP. All three activities are strictly dependent upon the presence of a sulfhydryl reducing agent for activity. A close association of CMP and UMP kinase was found by Strominger et al.[1] in preparations from calf liver. dCMP, however, was not tested. Later Sugino et al.[2] reported a preparation purified from calf thymus which catalyzed phosphorylation of UMP, CMP, and dCMP. The purification of pyrimidine nucleoside monophosphate kinase from rat liver[3] and rat Novikoff ascites hepatoma[4] will be presented here.

Assay Methods

Two methods of assay are used. The radiochemical assay measures the conversion of nucleoside monophosphate to nucleoside diphosphate, or the transfer of γ-phosphate from ATP to the nucleoside monophosphate by separation of reactants and products on paper electrophoresis. The optical assay measures the oxidation of NADH. In the latter assay, pyruvate kinase is coupled with lactate dehydrogenase in order to measure ADP formation. The disappearance of NADH, measured by the decrease in absorbance at 340 nm, is a measure of the kinase activity. With dCMP kinase as substrate, NADH oxidation is stoichiometric with dCDP formation. However, when CMP or UMP are used as substrates, higher values are obtained, since ADP, and to a lesser degree UDP and CDP, are substrates of pyruvate kinase.[5]

Reagents
   Tris-Cl buffer, pH 7.5, 40 mM
   ATPNa$_4$, 6 mM
   MgCl$_2$, 22 mM
   [^{14}C]CMP, 2.8 mM (1.3 μCi/μmole)

The enzyme is preincubated for 60 min at 37° in 5×10^{-2} M dithiothreitol, and an appropriate amount is added to the above reaction mixture at 37° to give a final volume of 0.05 ml. The reaction is carried out at 37° for a length of time which converts less than 10% of substrate to product. The reaction is stopped by freezing in an acetone-Dry Ice bath following the addition of 10 μl of a solution which contains CMP, CDP, and CTP, each at a concentration of 1×10^{-2} M.

[1] J. L. Strominger, L. A. Hippel, and E. S. Maxwell, *Biochim. Biophys. Acta* **32**, 412 (1959).
[2] Y. Sugino, H. Teraoka, and H. Shimono, *J. Biol. Chem.* **241**, 961 (1966).
[3] P. Maness and A. Orengo, *Biochemistry* **14**, 1484 (1975).
[4] P. Maness and A. Orengo, *Cancer Res.* **36**, 2312 (1976).
[5] K. M. Plowman and A. R. Krall, *Biochemistry* **4**, 2809 (1965).

The separation of the mono-, di-, and triphosphates is accomplished by spotting 50 μl of the sample on a sheet of Whatman No. 3 MM paper (57 × 20.5 cm) which has been dampened with 5 × 10^{-2} M citric acid–citrate buffer (pH 3.5). The electrophoresis is run at 4000 V, 50 mA, for 2 hr at 10°. The nucleotide spots are viewed with an ultraviolet (UV) light, cut out, and placed in vials containing 20 ml of Packard Permafluor liquid scintillation fluid diluted 25-fold with toluene. Samples are counted in a liquid scintillation system.

Spectrophotometric Assay
  Tris-Cl, pH 7.6, 70 mM
  ATPNa$_4$, 9 mM
  MgCl$_2$, 20 mM
  PEPNa, 3 mM
  KCl, 20 mM
  NADH, 0.13 mM
  Pyruvate kinase, 7 μg (1930 units/mg)
  Lactate dehydrogenase, 10 μg (300 units/mg)
  dCMP, 8 mM
  Dithiothreitol, 50 mM

The enzyme and/or water is added to a final volume of 1.0 ml. Particular care is taken in using ATP and sodium phosphoenolpyruvate with minimal amounts of ADP and pyruvate. Enzyme activity was measured at 37° on a Cary 15 recording spectrophotometer. A small rate obtained with the enzyme in the absence of nucleoside monophosphate is subtracted.

Definition of Unit and Specific Activity. A unit is defined as the number of μmoles of CDP formed in 1 minute at 37° under the above-described conditions. The specific activity is determined by dividing units by milligrams of protein, as measured by the method of Lowry *et al.*[6] using bovine serum albumin as a standard.

Purification Procedures

Enzyme from Rat Liver

Young female Sprague-Dawley rats (albino) are used. The rats are etherized, and the livers dissected out, frozen on a block of Dry Ice, and stored until use.

[6] O. H. Lowry, N. J. Rosebrough, A. L. Farr, and R. J. Randall, *J. Biol. Chem.* **193**, 265 (1951).

Preparation of Liver Extract. Purification of the enzyme is carried out at 4°. Frozen rat liver (150 g) is minced and homogenized under nitrogen in a Tec-Mar homogenizer for approximately 2 min in 25 mM potassium phosphate (pH 7.5)–250 mM sucrose–30 mM KCl–10 mM dithiothreitol (250 ml). Phenylmethylsulfonyl fluoride (1 M, 1.5 ml) dissolved in isopropyl alcohol is added during homogenization. The homogenate (fraction 1) is centrifuged at 8000 g for 30 min. The supernatant solution is then centrifuged at 123,000 g for 2 hr. This supernatant solution yields fraction 2.

Streptomycin Treatment. Streptomycin sulfate (5%, 0.25 volume of fraction 2) is added to fraction 2 with continuous stirring under nitrogen. After 30 min of standing, the suspension is centrifuged at 8000 g for 30 min. The supernatant solution yields fraction 3.

pH Treatment. Acetic acid (7.5%) is added drop by drop to fraction 3 with continuous stirring until the pH of the suspension reaches 5.0. After standing for 10 min, the suspension is centrifuged at 8000 g for 15 min. The supernatant solution is brought to pH 7.5 with ammonium hydroxide yielding fraction 4.

Ammonium Sulfate Treatment. Fraction 4 is brought to 50% saturation by the slow addition of solid ammonium sulfate with continuous stirring under nitrogen. After 1 hr of standing, the suspension is centrifuged at 8000 g for 30 min. The supernatant solution is brought to 85% saturation, stored for 1 hr, and centrifuged at 8000 g for 30 min. The pellet is dissolved in 25 mM potassium phosphate (pH 7.5)–10 mM dithiothreitol. Phenylmethylsulfonyl fluoride (1 M, 0.12 ml) in isopropyl alcohol is added for every 25 ml of redissolved pellet.

Sephadex G-25 Dialysis I. The redissolved 70–85% ammonium sulfate fraction is applied to a Sephadex G-25 column (2.5 × 73 cm) previously equilibrated with 1 mM potassium phosphate (pH 7.5)–5 mM 2-mercaptoethanol. Fractions of 5 ml are collected and an aliquot assayed by the optical method. The enzymically active fractions which appeared in the void volume are pooled and diluted in 1 mM potassium phosphate (pH 7.5)–5 mM 2-mercaptoethanol to a protein concentration of 3.0 mg/ml, yielding fraction 5.

Calcium Phosphate Gel Treatment. For every 100 ml of fraction 5, 30 ml of calcium phosphate gel (32 mg/ml) are added with continuous stirring. The suspension is stirred for 15 min, allowed to stand for 15 min, and then centrifuged at 8000 g for 10 min. The pellet is suspended in a volume of 1 mM potassium phosphate (pH 7.5)–150 mM NaCl–5 mM 2-mercaptoethanol equal to the volume of fraction 5. This suspension is allowed to stir for 45 min and then is centrifuged at 8000 g for 10 min. The pellet is suspended in 5 mM potassium phosphate (pH 7.5)–8% ammonium sulfate–10 mM dithiothreitol (21 ml for every 100 ml of fraction 5). This suspension is stirred for 30 min and then centrifuged at 8000 g for 10 min. The supernatant solution is concentrated to 30 ml in an Amicon Diaflo ultrafiltration apparatus using an Amicon PM 10 membrane at 40 psi of nitrogen.

Sephadex G-25 Dialysis II. Concentrated enzyme is applied to a Sephadex G-25 column (2.5 × 73 cm) equilibrated with 5 mM potassium phosphate (pH 7.5)–5 mM 2-mercaptoethanol. Fractions of 5 ml are collected and assayed by the optical method. The enzymically active fractions, which appeared in the void volume, are pooled, yielding fraction 6.

DEAE-Cellulose Chromatography. DEAE-cellulose DE-52 is washed with distilled water and then with 50 mM Tris-acetate (pH 7.5). A column (2.5 × 29 cm) is packed with the resin and washed with 5 mM Tris-acetate (pH 7.5)–5 mM 2-mercaptoethanol until the effluent is of constant pH (7.5) and conductance (2.4 × 10⁻³ mho). Fraction 6, containing 300 mg of protein, is applied to the column at a rate of 0.18 ml/min. Fractions of 3.3 ml are collected, monitored at 280 nm to follow the protein elution, and an aliquot assayed by the optical method. The column is washed with 5 mM Tris-acetate (pH 7.5)–5 mM 2-mercapto-ethanol (300 ml). A 700-ml linear gradient of 0–0.25 M KCl in 5 mM Tris-acetate (pH 7.5)–5 mM 2-mercaptoethanol is used to elute the enzyme. The enzyme elutes between 0.14 and 0.18 M KCl. The enzymically active fractions are pooled, yielding fraction 7.

Cellulose Phosphate Chromatography. Cellulose phosphate is washed with 0.5 M KOH, water, 0.5 M HCl, and again with water. These washes are followed with disodium (ethylenedinitrilo)tetraacetate (5 mM) and 5 mM Tris-acetate (pH 7.5). A cellulose phosphate column (1 × 12 cm) is packed with the resin and equilibrated with 5 mM Tris-acetate (pH 7.5)–5 mM 2-mercaptoethanol until the effluent is of constant pH (7.5) and conductance (2.4 × 10⁻³ mho). Fraction 7 (50 mg of protein) is diluted into 3 volumes of 5 mM Tris-acetate (pH 7.5)–5 mM 2-mercaptoethanol and applied to the column at a rate of 0.34 ml/min. The column is washed with 5 mM Tris-acetate (pH 7.5)–5 mM 2-

TABLE I
PURIFICATION OF CMP KINASE FROM RAT LIVER[a]

Fractions and steps	Activity (units/ml)	Total activity (units)	Protein (mg/ml)	Specific activity (units/mg)
1. Homogenate	2.52	1070	176	0.014
2. Extract	2.43	852	23.9	0.102
3. Streptomycin	2.17	921	18.6	0.12
4. pH 5	2.23	949	17.4	0.13
5. 70–85% Ammonium sulfate	0.72	587	2.51	0.29
6. Calcium phosphate gel	5.25	483	3.31	1.59
7. DEAE-cellulose chromatography	7.15	236	1.44	4.97
8. Cellulose phosphate chromatography	0.78	155	0.026	30.00

[a] The radiochemical assay was used to measure the enzymic activity. Proteins were determined by the procedure of Lowry et al.[6] after precipitation with 10% $Cl_3CCOOOH$.

mercaptoethanol (160–200 ml) until protein no longer elutes. Activity is measured by the optical assay. A small amount of activity appears in the washing. A 300-ml linear gradient of 0–0.5 M KCl in 5 mM Tris-acetate (pH 7.5)–5 mM 2-mercaptoethanol is used to elute the enzyme (0.2 M KCl). The active fractions are pooled, yielding fraction 8, and are concentrated to 30 ml in an Amicon Diaflo ultrafiltration apparatus using a Pellicon type PS membrane at 40 psi of nitrogen.

When 2-mercaptoethanol removal is necessary, concentrated fraction 8 may be dialyzed for 16 hr against 6 liters of 5 mM Tris-acetate (pH 7.5), yielding fraction 8d. The results of a typical purification are presented in Table I.

Enzyme from Rat Novikoff Ascites Hepatoma

Novikoff ascites cells are grown for 5–6 days in the peritoneal cavity of young female Holtzman Sprague-Dawley rats (120–150 g). The size of the inoculum is 1×10^9 cells. Tumor-bearing rats are decapitated, and the ascitic fluid is collected and diluted 1:2 with 250 mM sucrose:1 mM $MgCl_2$ and then cooled. The material is maintained at 2–4° throughout the entire procedure. The fluid is centrifuged at 200 g, and the tumor cells are collected as a sediment. This sediment is repeatedly suspended in fresh sucrose solution and centrifuged at 200 g to remove most of the erythrocytes. The cells are finally packed by centrifugation at 1000 g to estimate their volume (130 ml), suspended in 10 mM Tris-Cl (pH 7.7)–

TABLE II

PURIFICATION OF PYRIMIDINE NUCLEOSIDE MONOPHOSPHATE KINASE FROM RAT NOVIKOFF ASCITES HEPATOMA

Fraction and steps	Volume (ml)	Total activity[a] (units)			Protein (mg/ml)	Specific activity (units/mg)			Activity ratio	
		CMP	UMP	dCMP		CMP	UMP	dCMP	dCMP: CMP	UMP: CMP
1. Homogenate	300	38.25	70.35	44.90	21.54	0.006	0.011	0.007	1.17	1.84
2. Extract	250	39.75	79.12	43.30	12.86	0.012	0.025	0.013	1.09	1.99
3. Streptomycin	300	43.80	92.85	51.75	7.72	0.019	0.040	0.022	1.18	2.12
4. pH 5	250	47.08	96.17	47.83	6.98	0.027	0.055	0.027	1.02	2.04
5. 70–85% (NH$_4$)$_2$SO$_4$	126	22.73	41.37	26.63	3.90	0.046	0.084	0.054	1.17	1.82
6. DEAE-cellulose chromatography	75	20.60	33.75	20.95	0.328	0.837	1.371	0.852	1.02	1.64
7. Cellulose phosphate chromatography, KCl elution	84	15.52	32.37	18.40	0.044	4.2	8.76	4.98	1.19	2.09

[a] The radiochemical assay was used to measure enzymic activities.

250 mM sucrose (20 ml), and stored frozen. When ready for use, the thawed cell suspension (150 ml) is diluted with an equal volume of 25 mM potassium phosphate (pH 7.5)–250 mM sucrose–30 mM KCl–20 mM dithiothreitol. The suspension is homogenized in a Tec-Mar homogenizer for 3 min. Phenylmethylsulfonyl fluoride (1 M, 1.0 ml) dissolved in isopropyl alcohol is added during homogenization. The homogenate (fraction 1) is then subjected to the same procedure of purification of pyrimidine nucleoside monophosphate kinase as that reported for rat liver, with the omission of calcium phosphate gel treatment, a step that achieved no increase in specific activity of the hepatoma kinase. Each fraction is assayed for CMP, dCMP, and UMP kinase activities by the radiochemical assay. The results of a typical purification are presented in Table II. The hepatoma kinase precipitates at 70–85% saturation of ammonium sulfate, elutes from DEAE-cellulose at 110–140 mM KCl, and elutes from cellulose phosphate at 370 mM KCl.

When required, fraction 7 is concentrated in an Amicon Diaflo ultrafiltration apparatus with a Pellicon type PS membrane under 40 psi of nitrogen. When 2-mercaptoethanol removal is necessary, concentrated enzyme is dialyzed for 24–48 hr against 5 mM Tris-acetate (pH 7.5)–150 mM KCl (10 liters) and monitored for complete but reversible loss of enzymic activity by the spectrophotometric method of assay.

Cellulose Phosphate Chromatography with Nucleoside Triphosphate Elution

Solutions of nucleoside triphosphates (1.0 mM) can substitute for KCl in eluting the kinase from cellulose phosphate. The most efficient eluents are dCTP, CTP, and ATP. Fraction 7 (Tables I and II) is applied to a column of cellulose phosphate (1 × 6 cm), previously equilibrated with 5 mM Tris-acetate (pH 7.5)–5 mM 2-mercaptoethanol. Four milligrams of protein are applied. The column is washed with equilibration buffer (20 ml) and then with 5 mM Tris-acetate–5 mM 2-mercaptoethanol–1 mM ATP (pH 7.5). The kinase is eluted sharply with the ATP front.

As an alternative, the step can be applied directly on fraction 3 prepared as previously described with the exception that 25 mM Tris-acetate–250 mM sucrose–50 mM 2-mercaptoethanol is used as homogenizing buffer. Up to 60 mg of protein are applied to a column (1 × 12 cm) of cellulose phosphate equilibrated with 25 mM Tris-acetate–50 mM 2-mercaptoethanol, pH 7.5. After a wash with 50 ml of equilibration buffer the kinase is eluted with 1.0 mM dCTP, CTP, or ATP. Using dCTP as eluent, the kinase can be purified with this step alone 90-fold.[7] This cytosol preparation is diluted 1:1 with glycerol and used as such or stored at $-20°$ until use. There is no loss of enzymic activities for at least 2 months of storage at $-20°$.

Properties

Rat Liver Kinase Specificity

Phosphate Acceptor and Donor. CMP, dCMP, and UMP are the only nucleotides capable of functioning as substrate. The radiochemical assay (ATP, 1.38 mM; nucleoside monophosphate, 0.20 mM) was used to measure production of nucleoside diphosphate from [^{14}C]CMP (32 μmole hr^{-1} mg^{-1}), [^{14}C]dCMP (6 μmole hr^{-1} mg^{-1}), and [^{14}C]UMP (24 μmole hr^{-1} mg^{-1}). [γ-^{32}P]ATP was used in testing dUMP (6.6 mM), dTMP (6.0 mM), GMP (2.6 mM), dGMP (3.6 mM), and ara-CMP (0.9 mM). Of these, only ara-CMP serves as phosphate acceptor (6 μmole hr^{-1} mg^{-1}).

Phosphate donor specificity was examined for CMP, UMP, and dCMP kinase activities. ATP and dATP are phosphate donors for each of the three kinase activities. When CMP serves as phosphate acceptor, many nucleoside triphosphates act as donor to a significant extent, particularly dCTP. However, CTP cannot donate phosphate as well, and ara-CTP cannot donate at all. When UMP acts as phosphate acceptor only ATP and dATP serve as efficient phosphate donors. dCTP acts as donor to a slight, but significant, degree. ITP can also phosphorylate UMP. When dCMP serves as phosphate acceptor, only ATP and dATP are effective substrates. None of the other nucleotides act as donor to a significant extent.

Ions. Mg^{2+} was found to be necessary for the phosphorylation of CMP, UMP, and dCMP by either ATP or dCTP. With CMP (2.66 mM) as phosphate acceptor and ATP (3.00 mM) as phosphate donor, Mn^{2+}, Ni^{2+}, and Ca^{2+} are able to substitute for Mg^{2+} but are less effective. The relative rates obtained are Mg^{2+} (100%), Mn^{2+} (42%), Ni^{2+} (16%), and Ca^{2+} (13%). No activity could be observed with Zn^{2+}, Co^{2+}, Ba^{2+}, or Cu^{2+}, or with the following monovalent cations: NH_4^+, Li^+, Na^+, or K^+. With CMP (9.34 mM) as phosphate acceptor and dCTP (8.67 mM) as phosphate donor, Mn^{2+} (100%) and Mg^{2+} (94%) are almost equally effective, followed by Co^{2+} (20%) and Ni^{2+} (17%). No activity was observed with Zn^{2+}, Ca^{2+}, Ba^{2+}, or Cu^{2+}. All cations were used at a concentration of 7.14 mM and were chloride salts.

Kinetic Constants. Kinetic parameters for CMP, UMP, and dCMP with $MgATP^{2-}$, $MgdATP^{2-}$, or $MgdCTP^{2-}$ as phosphate donors were determined from initial velocity measurements using the radiochemical assay. With ATP (1.47 mM) as phosphate donor, K_m values are for CMP

0.030 mM ± 0.007, for dCMP 2.77 mM ± 0.39, and for UMP 0.040 mM ± 0.009. With dATP (1.47 mM) K_m values are for CMP 0.027 mM ± 0.004, for dCMP 1.10 mM ± 0.18, and for UMP 0.053 mM ± 0.005. With dCTP (5.60 mM) the K_m value for CMP is 0.98 mM ± 0.07. The range of concentrations of the phosphate acceptors used were as follows: CMP (ATP or dATP), 0.004–0.048 mM; CMP (dCTP), 0.012–0.598 mM; UMP, 0.017–0.180 mM; and dCMP, 0.165–1.725 mM. With CMP (0.13 mM) as phosphate acceptor, K_m values are for ATP 0.32 mM ± 0.13, for dATP 0.074 mM ± 0.004, and for dCTP 0.82 mM ± 0.07. With dCMP (1.17 mM), K_m values are for ATP 0.68 mM ± 0.09 and for dATP 0.61 mM ± 0.06. With UMP (0.58 mM), K_m values are for ATP 0.067 mM ± 0.022 and for dATP 0.42 mM ± 0.04. The ranges of concentrations of the phosphate donors were as follows: ATP, 0.022–0.256 mM; dATP, 0.014–0.586 mM; and dCTP, 0.330–3.080 mM.

When ATP or dATP serve as phosphate donor, substrate inhibition occurs at concentrations of CMP greater than 0.13 mM. No substrate inhibition is observed with dCMP (3 mM) or UMP (1 mM).

The tumor kinase exhibits the same specificity; however, the following kinetic constants were found. With ATP (1.59 mM) as phosphate donor, the K_m values are for CMP, 0.0053 ± 0.0008 mM; for UMP, 0.043 ± 0.0009 mM; and for dCMP, 0.715 ± 0.068 mM. With dCMP (1.66 mM) as phosphate acceptor, the K_m value for ATP is 0.134 ± 0.008 mM. Fraction 7 was used as the source of enzyme with ranges of concentrations of substrates as follows: CMP, 0.0095–0.0603 mM; UMP, 0.0182–0.184 mM; dCMP, 0.208–2.250 mM; and ATP, 0.0575–0.2876 mM. No substrate inhibition was observed with CMP, dCMP, UMP, or ATP at the concentrations used in the assay.

Activation by Sulfhydryl Reducing Agents

All four enzymic activities depend strictly upon sulfhydryl reducing agents. Reduced thioredoxin is by far the most effective. The naturally occurring thiols reduced glutathione, reduced $d,1$-α-lipoic acid, and L-cysteine can substitute for thioredoxin. Dithiothreitol and 2-mercaptoethanol are also effective. Equating the specific activity of CMP(ATP) kinase after activation by dithiothreitol (85 mM) to 100, one obtains the following relative values for activation by other sulfhydryl reducing agents: 2-mercaptoethanol (85 mM) 67, reduced glutathione (85 mM) 77, L-cysteine (85 mM) 59, reduced $d,1$-α-lipoic acid (85 mM) 55, and reduced thioredoxin (17 μM) 160. The purity of the thioredoxin preparation was judged to be 50% on the basis of polyacrylamide gel electrophoresis.

Addition of 2-mercaptoethanol (5 mM) to the inactive enzyme effects a reduction in molecular weight from approximately 53,000 to 17,000, as measured by molecular sieve chromatography. This low-molecular-weight form is partially active in the presence of 5 mM 2-mercaptoethanol but becomes inactive upon removal of the reducing agent. Furthermore, higher concentrations of 2-mercaptoethanol (50 mM) fully reactivate the CMP(ATP) kinase activity, followed by dCMP(ATP) and CMP(dCTP) kinase activities in a sequential manner, without further change in molecular weight. Alkylation by iodoacetamide of the enzyme at different stages of reactivation in dithiothreitol suggests an ordered appearance of the various enzyme activities. Furthermore, iodoacetamide inactivates the fully active enzyme. Thioredoxin was found to activate the enzyme in a manner similar to 2-mercaptoethanol and dithiothreitol.[8]

Inhibitors and Activators

Fluoride was found to inhibit enzymic activity completely at a concentration of approximately 25 mM. Inhibition was also observed with NaSCN and NaClO$_4$, although at significantly higher concentrations (250 mM) NaC$_2$H$_3$O$_2$, NaCl, and Na$_2$SO$_4$ stimulated CMP-kinase activity in the latter concentration range.

Feedback Regulation

The enzyme does not appear to be subject to inhibition by CTP, dCTP, UTP, or dTTP (1.0 mM), regardless of the phosphate acceptor. It is possible that the activity of the enzyme *in vivo* is regulated by the concentration of thiols in the cell.

[7] M. Lombardi and A. Orengo, unpublished results.
[8] P. Maness and A. Orengo, *Biochim. Biophys. Acta* **429**, 182 (1976).

[42] UMP-CMP Kinase from *Tetrahymena pyriformis*

By ELIZABETH P. ANDERSON

$$UMP + ATP \xrightarrow{\text{Mg}^{2+}} UDP + ADP$$
$$CMP + ATP \xrightarrow{\text{Mg}^{2+}} CDP + ADP$$

As purified from *Tetrahymena pyriformis*, this kinase phosphorylates both UMP and CMP, and the ratio of these two activities remains constant during the purification.[1] The purified enzyme also retains appreciable kinase activity for dCMP.[1] It is therefore presently classified as ATP:(d)CMP phosphotransferase, EC 2.7.4.14,[2] which has also been purified from other sources, sometimes for activity with different substrates. For example, it may be the same as the dCMP kinase purified from calf thymus.[3,4] The present enzyme preparation was purified for phosphorylation of UMP and, as a corollary, of CMP. In contrast to this substrate specificity, CMP-dCMP and UMP kinase activities can be separated in *Escherichia coli* extracts,[5] and in *Salmonella typhimurium*, UMP and CMP kinases do not appear to be encoded by the same structural gene.[4,6]

Assay Method

Principle. The assay measures the conversion of radioactive UMP to labeled UDP (or UDP plus UTP), with chromatographic isolation of the product(s) formed. It is specific enough for use in either crude or purified fractions of the enzyme.

Reagents

Potassium phosphate buffer, 0.5 M, pH 7.5
UMP, 0.1 M, or CMP, 0.1 M
[2-^{14}C]UMP (25 mCi/mmole), or [2-^{14}C]CMP (25 mCi/mmole)
ATP, disodium salt, 0.1 M, adjusted to pH 7.5
MgCl$_2$, 0.2 M

Procedure.[1,7] The standard reaction mixture contains 3 μl potassium phosphate buffer, 3 μl UMP or CMP, 100 nCi [2-^{14}C]UMP or [2-^{14}C]CMP, 3 μl ATP, 2 μl MgCl$_2$, enzyme protein (in 5–20 μl 0.1 M potassium phosphate buffer, pH 7.5), and water to give a final volume of 32 μl. For best linearity, the enzyme protein concentration is kept below 0.15 μg. The assay is normally run in duplicate or triplicate. The

[1] B. W. Ruffner, Jr. and E. P. Anderson, *J. Biol. Chem.* **244**, 5994 (1969).
[2] "Enzyme Nomenclature," Recommendations (1972) of the Commission of Biochemical Nomenclature of the International Union of Pure and Applied Chemistry and the International Union of Biochemistry on the Nomenclature and Classification of Enzymes. Elsevier, Amsterdam, 1973.
[3] Y. Sugino, H. Teraoka, and H. Shimono, *J. Biol. Chem.* **241**, 961 (1966).
[4] E. P. Anderson, *in* "The Enzymes" (P. D. Boyer, ed.), 3rd ed., Vol. 9, p. 49. Academic Press, New York, 1973.
[5] S. Hiraga and Y. Sugino, *Biochim. Biophys. Acta* **114**, 416 (1966).
[6] J. L. Ingraham and J. Neuhard, *J. Biol. Chem.* **247**, 6259 (1972).
[7] T. Q. Garvey, III, F. K. Millar, and E. P. Anderson, *Biochim. Biophys. Acta* **302**, 38 (1973).

reaction is initiated by the addition of enzyme and carried out at 25° for periods of time up to 10 min, 15 μl being removed at each of two time points. The period of linearity with time should be confirmed for the preparation being assayed. The aliquot removed is pipetted into a tube preheated in boiling water, and the reaction is terminated by heating in boiling water for 1 min. Labeled substrate and products are then separated by ascending chromatography for 2.5 hr in 1 M LiCl on thin-layer PEI-cellulose on plastic sheets (1-inch or 0.5-inch channels). The PEI-cellulose sheets[8] are stored at 0–4°. Before use they are washed at least twice by ascending chromatography in water to remove any degraded PEI. To assist in localization, 0.1 μmole each of the nucleoside mono-, di-, and triphosphate are applied to the origin of every channel and cochromatographed with the sample. The pyrimidine compounds are localized by ultraviolet fluorescence and counted in a scintillation counter. (R_fs: UMP, 0.80; UDP, 0.65; UTP, 0.37; CMP, 0.63; CDP, 0.36; and CTP, 0.12.) Separate chromatography of the starting substrate can indicate any necessary corrections for radioactive contamination of the substrate as supplied. The ultraviolet localization has been checked against more time-consuming localization by radioautography of the chromatogram on X-ray film, with excellent agreement; the radioautogram can be useful if additional or unknown by-products may be expected.

Definition of Unit and Specific Activity. A unit of enzymic activity is defined as the amount phosphorylating 1 μmole of UMP in 1 min. Specific activity is expressed as units per milligram of protein. Protein concentration is estimated by the method of Lowry *et al.*[9] with bovine serum albumin as standard.

Alternative Assay Procedures. The enzymic activity can alternatively be assayed as conversion of radioactive ATP to ADP using ^{14}C-labeled ATP and unlabeled nucleoside monophosphate. In double-label experiments, [2-14C]UMP and [3H]ATP have both been included in the incubation. The incubation mixture is then chromatographed in duplicate, once with the set of uridine nucleotide standards and once with adenosine nucleotide standards. In addition, better resolution is attained by using serial ascents (without drying between changes of solvent) in 0.5 M, 1.0 M, and 1.5 M LiCl, for 5, 50, and 40 min, respectively.

An additional assay that couples ADP formation to the pyruvate kinase–lactate dehydrogenase system is especially useful for initial

[8] Available from J. T. Baker Chemical Co., Phillipsburg, New Jersey.
[9] O. Lowry, N. J. Rosebrough, A. L. Farr, and R. J. Randall, *J. Biol. Chem.* **193**, 265 (1951).

velocity studies since it follows the reaction continuously, spectrophoto-metrically monitoring the oxidation of NADH at 340 nm. This has proven feasible for studies of CMP phosphorylation, since CDP is a very poor substrate of pyruvate kinase[10]; it is less satisfactory for the assay of UMP phosphorylation, at least for initial velocity studies, because UDP is an appreciable substrate and the V_{max} with UDP is very different from that with ADP. Since this assay removes ADP product as it is formed, it has the added advantage of eliminating inhibition seen in the radioactiv-ity assay at moderate to high levels of ATP.[7]

Purification Procedure[1]

Step 1. Crude Lysate. The enzyme is purified from late log-phase cultures of *T. pyriformis*. All operations are performed at 0–4° unless otherwise stated. Cells are harvested on the fifth day of culture growth by centrifugation at 400 g at 4°; the usual yield is 10 ml of packed cells per 4 liters of culture. After most of the medium is decanted, the cells are resuspended and then packed by centrifugation at 10,000 g for 10 min. The packed cells are resuspended in an equal volume of 0.1 M potassium phosphate buffer, pH 7.5, and alternately frozen in Dry Ice–acetone and thawed in ice water three times. The frozen-thawed lysate is centrifuged at 100,000 g for 90 min, and the precipitate is discarded.

Step 2. Fractionation by Acid Precipitation. To the supernatant solution from the crude lysate, sufficient 1 M acetic acid is added drop by drop, with stirring, to lower the pH to 5.0. This preparation is allowed to stand at 0° for 20 min, the precipitate is removed by centrifugation at 10,000 g for 10 min, and the supernatant solution is neutralized to pH 7.5 with 1 M ammonia.

Step 3. Ammonium Sulfate Fractionation. Saturated ammonium sulfate solution (4°) is added to the preparation over a period of 1–2 min with stirring, in sufficient quantity to bring the final concentration to 60% saturation. After 20 min of equilibration, the precipitate is removed by centrifugation at 10,000 g for 10 min. This is discarded, and a volume of saturated ammonium sulfate solution equal to that of the ammonium sulfate supernatant solution is added over a period of 1–2 min to bring the preparation to 80% saturation. After standing for 20 min, the mixture is again centrifuged at 10,000 g for 10 min to collect the precipitated protein; the supernatant fraction is discarded. A minimal amount (us-

[10] T. Q. Garvey, III and E. P. Anderson, unpublished results.

ually 4 ml) of potassium phosphate buffer, pH 7.5, is added to dissolve the orange pellet.

Step 4. Fractionation on Sephadex G-75. The ammonium sulfate fraction (60–80%) is applied to a column (2.2 × 43.0 cm) of Sephadex G-75 (superfine, previously allowed to swell overnight at room temperature in 0.1 M potassium phosphate buffer, pH 7.5, and then cooled to 4°). The column is then eluted with the same phosphate buffer at a flow rate of 4 ml/hr. All fractions are assayed for protein and enzyme activity. The enzyme is eluted from the column immediately ahead of a prominent, yellow-orange band which serves as a convenient marker. The active enzyme as eluted from the column is generally contained in approximately 16 ml of buffer (the peak activity tubes typically contain a total of 4 ml). The protein concentration of these tubes is approximately 0.2 mg/ml. Experimental studies on the enzyme are usually carried out with preparations of maximal activity from the Sephadex columns. The pooled fractions are stored at 4° for further use. When stored in this way, the enzyme loses approximately 20% of its activity in 2 weeks. Experiments are therefore carried out, if possible, within 1 week after preparation of the enzyme.

A typical purification procedure (see the table) achieved approxi-

PURIFICATION OF UMP KINASE FROM *Tetrahymena pyriformis*[a]

Fraction	Total protein (mg)	Total units (μmole/min)	Specific activity (units/mg)	Yield (%)	Ratio of UMP kinase to CMP kinase activity
Frozen-thawed lysate	1315	118	0.090	—	0.72
Supernatant from 100,000 g	620	121	0.195	100	0.75
pH 5 supernatant	416	111	0.266	94	0.78
Ammonium sulfate fractionation (60–80% saturation)	51	54	1.06	46	0.77
Sephadex G-75					
Pooled fractions	3.04	38	12.5	32	
Peak fractions	0.60	16.5	27.5	14	0.68

[a] Enzymic activity was assayed with either [14C]UMP or [14C]CMP as substrate. The purification was based on assays with UMP as substrate, but the ratio of UMP kinase activity to CMP kinase remained approximately constant throughout the purification (last column). Each fraction was assayed at several protein concentrations, and specific activity was calculated from tubes containing approximately the same amount of enzyme activity (0.01 unit).

mately 300-fold purification over the crude lysate with 14% yield in the tubes of peak activity from Sephadex G-75, or about 140-fold purification with 32% yield in the total preparation after gel filtration.[11]

The fractions from Sephadex are essentially free of UDP kinase, adenosine triphosphatase, and uridine diphosphatase; all of these activities are present in the original lysate. In the presence of equimolar UDP and ADP, however, the purified preparation catalyzes the reverse of the UMP kinase reaction, at a velocity about one-third that of the forward reaction.

Stability. The purified enzyme is very labile to dilution; 3-fold dilution with either water or phosphate buffer results in 50% loss of activity in 1 hr. Loss of CMP kinase activity parallels that of UMP kinase. Bovine serum albumin (100 μg/ml) prevents this loss of activity in diluted preparations. The enzyme is also labile to elution from Sephadex in 0.1 M Tris·HCl (pH 7.5) instead of potassium phosphate; under these conditions the enzyme is eluted with a partition coefficient twice that observed in the phosphate buffer. If albumin (100 μg/ml) or NaCl (2 M) is added to the Tris buffer, the elution pattern is like that observed in phosphate.

Properties[1]

Optimum pH. The enzyme exhibits a broad optimum between pH 7.0 and 8.0 (fraction E; 0.1 M buffers: Tris-maleate, pH 4.2–8.0; potassium phosphate, pH 6.0–7.8; Tris·HCl, pH 7.2–9.0).

Specificity of the Phosphate Donor. Of the nucleoside triphosphates tested, only dATP can substitute for ATP as phosphate donor in the UMP kinase reaction, and it is only about one-tenth as active as ATP under standard assay conditions. GTP, UTP, CTP, dCTP, and dTTP are essentially inactive in this respect (fraction E).

Specificity of the Phosphate Acceptor. With [8-^{14}C]ATP and unlabeled nucleoside monophosphates as substrates, the enzyme phosphorylates CMP, UMP, and dCMP, in order of decreasing activity. AMP, GMP, dGMP, and dTMP are not phosphorylated. CMP kinase activity parallels UMP kinase activity throughout purification (see the table), and also through further fractionation on ion-exchange cellulose, as well as during inactivation. The activity with dCMP has not been followed through the purification procedure.

[11] Preliminary experiments have indicated that it is possible to purify UMP kinase activity further from the Sephadex G-75 fraction using hydroxyapatite columns. This procedure gave about 2-fold purification over that already achieved.[7]

Kinetic Constants and Reaction Mechanism. The initial velocity pattern observed with CMP as substrate (using the coupled spectrophotometric assay) is one of intersecting rather than parallel lines, ruling out a ping-pong reaction mechanism, and suggesting that the reaction proceeds by the sequential addition of both substrates to the enzyme to form a ternary complex.[10] With UMP as substrate, a similar conclusion is indicated by data on isotope exchange in the partial reactions. The enzyme catalyzes neither half-reaction, i.e., exchange of phosphate between one substrate–product pair in the absence of the second substrate, again ruling out a ping-pong mechanism of addition of one substrate and release of its product prior to addition of the second substrate.[7] The enzyme has a K_m for ATP of 0.5 mM, for CMP of 0.8 mM, and for UMP near 1.25 mM. CDP inhibition is competitive with CMP. The initial velocity pattern for the reverse reaction likewise indicates a sequential reaction mechanism.

[43] Deoxycytidine Kinase from Calf Thymus[1]

By DAVID H. IVES and SUE-MAY WANG

$$2'\text{-Deoxycytidine} + \text{ATP} \rightarrow 5'\text{-dCMP} + \text{ADP}$$

This enzyme from the cytosol of calf thymus is so named because deoxycytidine (dCyd) is the preferred substrate at low concentrations,[2] although its specificity is rather broad.[3] It has been partially purified in several laboratories[4–6] through the use of very similar procedures. The method described here is a modification of the technique used in this laboratory.[5]

Assay Method

Principle. The assay method used is based on the ability of anion-exchange paper to retain the labeled deoxynucleotide product, while

[1] This work was supported in part by Public Health Service Grant CA-06913 from the National Cancer Institute.
[2] J. P. Durham and D. H. Ives, *Mol. Pharmacol.* **5**, 358 (1969).
[3] T. Krenitsky, J. Tuttle, G. Koszalka, I. Chen, L. Beacham, III, J. Rideout, and G. Elion, *J. Biol. Chem.* **251**, 4055 (1976).
[4] R. Momparler and G. A. Fischer, *J. Biol. Chem.* **243**, 4298 (1968).
[5] J. P. Durham and D. H. Ives, *J. Biol. Chem.* **245**, 2276 (1970).
[6] Y. Kozai, S. Sonoda, S. Kobayashi, and Y. Sugino, *J. Biochem. (Tokyo)* **71**, 485 (1972).

excess nucleoside is removed by washing. The deoxynucleotide is then eluted in liquid scintillation vials and counted in the continuous phase.

Reagents

A. ATP-MgCl$_2$ solution, 40 and 48 mM, respectively. The disodium salt of ATP is stored frozen as a 200 mM stock solution after adjusting to pH 7.5 with Tris base. The working mixture contains a 20% excess of Mg^{2+} to ensure that all of the ATP will be complexed with magnesium.

B. Buffered dithiothreitol and bovine serum albumin solution; 50 mM and 1 mg/ml, respectively, in 0.1 M Tris-Cl, pH 8.0$_{37}$.[7] Crystalline albumin (Pentex) is used rather than fraction 5 powder.

C. Tritiated deoxynucleoside solution, 0.08 mM, 25 μCi/ml. This solution is prepared fresh daily from carrier-free labeled deoxynucleoside stored unfrozen in 50% ethanol at $-20°$, and from frozen stock solutions of unlabeled deoxynucleoside standardized by ultraviolet absorbance.

Procedure. For routine assays equal portions (usually 20 μl) of solutions A, B, and C are placed in a capped, disposable, 1.5-ml polypropylene centrifuge tube (Walter Sarstedt, Inc.), at 37°, and the reaction is begun by adding 20 μl of enzyme diluted with 0.1 M Tris-Cl, pH 8.0$_{37}$, and mixing with a vortex stirrer. After 30 min at 37° the reaction is stopped by adding 0.2 ml of 0.1 N formic acid, or by heating for 2 min in a boiling water bath and then diluting with 0.2 ml of water. If the concentration of enzyme is chosen such that no more than 15% of the deoxynucleoside is phosphorylated, the reaction can be expected to give a linear time-course for 30 min or more. Enzyme which has been stored should be reactivated by dilution with reagent B and incubation at 22° for 1 hr. This is particularly desirable if kinetic studies are being carried out. Since the enzyme aliquot will then contain the components of reagent B, 20 μl of 0.1 M Tris-Cl, pH 8.0$_{37}$, should replace reagent B in the above reaction mixture.

The amount of nucleotide formed is determined by spotting 50-μl aliquots of the diluted reaction mixture onto 15 × 15 mm squares of SB-2 anion-exchange paper numbered with a graphite pencil. Unreacted nucleoside is removed by washing the pieces for 20 min each in 0.001 M ammonium formate and water, with gentle stirring to keep them from settling into a heap. The washed papers are then placed in individual liquid scintillation vials, eluted with 1 ml of 0.1 M HCl–0.4 M KCl, and the radioactive solutions are counted in a Triton-toluene scintillation

---

[7] The subscript denotes the temperature at which the stated pH will be attained.

solvent, as detailed previously.[8] It is necessary to correct for absorption of impurities contained in the radioactive nucleoside preparations by determining "blank" values with an unreacted sample. Normally, this will be about 1% of the total radioactivity, but it increases with aging of the tritiated nucleoside, and may vary with different batches of SB-2 paper. If high blanks are found to be a problem, improved results may be obtained by substituting Whatman DE-81 paper, although this paper must be handled much more carefully when wet to avoid macerating it.

Definition of Unit. The activity is expressed as nmoles of deoxynucleoside phosphorylated per minute at 37°. The specific activity is defined as nmoles of deoxynucleoside phosphorylated per minute per milligram of protein, as determined by the method of Lowry *et al.,*[9] using crystalline bovine serum albumin as a standard, or by UV absorbance[10] when β-mercaptoethanol is present.

Purification Procedures

Step 1. Crude Extract. Pooled thymus glands from freshly killed calves are chilled in cracked ice and trimmed free of muscle, excess fat, and connective tissue. The glands taken from the neck contain less fat than those from near the heart, but both appear to contain similar concentrations of deoxycytidine kinase. The preparation is maintained at no more than 4° during this and all subsequent procedures. Portions weighing 625 g are minced with scissors and homogenized in a 3800-ml Waring Blendor with 2500 ml of 0.125 M KCl–0.0125 M potassium phosphate (pH 7)–0.005 M β-mercaptoethanol for 5 sec at low speed and 15 sec at high speed. The homogenate is then centrifuged for 30 min at 27,500 g in a Sorvall GSA rotor. The supernatant fluid is normally frozen in 250-ml portions in polyethylene bottles, using Dry Ice and acetone. The frozen extracts slowly decrease in activity, but have a useful shelf-life of about 12 months. Limited attempts to extract enzyme from commercial frozen thymus have, in our hands, yielded extracts of relatively low specific activities.

Step 2. Streptomycin Precipitation. The protein concentration of crude extract is estimated by measuring absorbancies at 260 and 280 nm.[10] A 10% solution of streptomycin sulfate, adjusted to pH 7 with KOH, is added drop by drop while being stirred, to a final ratio of 1 g of streptomycin sulfate per gram of protein. After stirring 20 min more, the

[8] D. H. Ives, J. P. Durham, and V. Tucker, *Anal. Biochem.* **28**, 192 (1969).
[9] O. Lowry, N. Rosebrough, N. Farr, and R. Randall, *J. Biol. Chem.* **193**, 265 (1951).
[10] O. Warburg and W. Christian, *Biochem. Z.* **310**, 384 (1941).

suspension is centrifuged for 30 min at 20,000 g, and the precipitate is discarded.

Step 3. Ammonium Sulfate Fractionation. Solid ammonium sulfate (Schwarz-Mann, enzyme grade) is added slowly, with stirring, to the supernatant fluid from step 2, to a concentration of 0.222 g/ml of supernatant solution (about 40% saturated). After stirring 20 min more, the sediment is collected by centrifugation at 20,000 g and discarded. Additional ammonium sulfate is added to the supernatant fluid (0.137 g/ml), raising the level of saturation to about 63%. The precipitate is collected as before, but the supernatant liquid is discarded. After draining the centrifuge bottles, the excess ammonium sulfate solution is blotted from the shoulders, and the precipitate is dissolved in a minimal volume of 0.01 M Tris-Cl, pH 7.0$_{20}$, containing 5 mM β-mercaptoethanol. This preparation is quite stable and may be stored frozen at $-20°$, if desired.

Step 4. Fraction by Calcium Phosphate Gel Absorption. The ammonium sulfate fraction from step 3 is diluted to about 5 mg of protein per milliliter, using the same buffer that the precipitate was dissolved in. Aged calcium phosphate gel paste (Sigma Chemical Co., used directly as supplied) is stirred into the diluted enzyme solution to a ratio of 2.24 g of gel solid per gram of protein. After 15 min of additional stirring, the gel is collected by centrifugation for 10 min at 4200 g in a Sorvall GS-3 rotor. The gel is washed first with 2 M Tris-Cl (pH 9.0$_4$)–5 mM β-mercaptoethanol, then with 0.2 M Tris-Cl (pH 9.0$_4$)–5 mM β-mercaptoethanol, using roughly 150 ml of each buffer per gram of protein from step 3, by suspending, stirring for 15 min, and centrifuging. The task of repeated resuspension of the calcium phosphate gel paste is aided greatly by a rubber resuspending cone (Dupont-Sorvall, Cat. No. 17103) attached to the shaft of a stirrer such as an "Omni-Mixer," controlled by a variable transformer. The enzyme is then eluted from the gel by two extractions of 15 min each, with 0.2 M Tris-Cl–0.025 M potassium phosphate–5 mM β-mercaptoethanol (pH 9.0$_4$), using about 85 ml of buffer per gram of protein, and centrifuging after each extraction. Enzyme is recovered from the extract by stirring in 0.39 g of solid ammonium sulfate per milliliter, stirring for total of 30 min, and centrifuging at 20,000 g. The precipitate containing the enzyme is dissolved in 0.01 M Tris-Cl (pH 9.0$_4$)–5 mM β-mercaptoethanol, using a volume about equal to one-tenth of the total volume of the gel extract. Although an earlier published procedure[5] included washing the gel with 0.2 M Tris-Cl containing 1 mM phosphate, this operation has been deleted because of a tendency for premature elution of some activity even by such low concentrations of phosphate.

Step 5. Ion-exchange Chromatography on DEAE-Cellulose. The ion-exchanger (Whatman DE-23) is thoroughly equilibrated with the same buffer used to dissolve the protein in step 4. To hasten this equilibration, the cellulose is suspended in 0.5 M Tris-Cl (pH 9.0$_4$)–5 mM β-mercaptoethanol and washed on a Büchner funnel. A column of about 3.5 × 10 cm is prepared under atmospheric pressure (this size is appropriate for 500–700 mg of protein from step 4), and then is washed with the 0.01 M Tris-Cl (pH 9.0$_4$)–5 mM β-mercaptoethanol until the pH and conductivity of the column effluent are identical to the buffer itself. The extract from step 4 is then allowed to pass slowly onto the column, followed by 1 column volume of the equilibrating buffer. The column is then washed with 3 column volumes of 0.5 M Tris-Cl (pH 9.0$_4$)–5 mM β-mercaptoethanol, and the enzyme is eluted with reduced pH, using 0.5 M Tris-Cl (pH 7.75$_4$)–5 mM β-mercaptoethanol. Depending on the characteristics of the particular lot of DEAE-cellulose, the elution volume varies from 1–2 column volumes of the eluting buffer. The enzyme is then concentrated by ultrafiltration on a PM-30 membrane filter (Amicon Corporation) in a 43-mm stirred cell. After the enzyme is concentrated to about 20 ml, a reservoir containing 0.1 M Tris-Cl (pH 8.0$_4$)–2 mM β-mercaptoethanol is connected to the cell to permit reduction of buffer concentration without further dilution. The exchange process can be monitered in the cell run-off with a conductivity meter. Finally, the concentrated enzyme (about 10 ml) is adjusted to 5 mM dithiothreitol and diluted with an equal volume of reagent glycerol. In this form the activity is relatively stable for months when stored unfrozen at −20°. Dithiothreitol is considerably more effective than β-mercaptoethanol in maintaining enzyme activity during storage, although the latter is used during purification procedures for economy.

Optional Preparative Polyacrylamide Gel Electrophoresis

Analytical polyacrylamide gel electrophoresis (PAGE) of the product of step 5 reveals that the enzyme activity lies between two major protein bands which are not active. Moreover, that preparation may contain traces of mitochondrial deoxynucleoside kinase also.[11] All of these impurities are readily removed by means of preparative PAGE carried

---

[11] Thymus cytosol and mitochondrial deoxycytidine kinases exhibit electrophoretic R_f's of 0.75 and 0.43 when developed in continuous Tris-glycine buffer at pH 9.3, using 6% polyacrylamide gel (W. R. Gower, unpublished).

out in 7% gel and continuous 40 mM Tris-maleate buffer, pH 8.0$_4$. We use a Shandon-Southern unit (No. 2782) which consists of an upper (cathode) reservoir, an annular column of separating gel surrounding a cold finger, a cross-flow elution chamber, a lower column cone serving only to provide electrical continuity, and a lower (anode) reservoir. Except for the modifications noted here, the manufacturer's operating instructions should be followed. A general discussion of preparative electrophoresis by Shuster can be found in an earlier volume.[12]

Buffers
  A. Tris-maleate stock buffer, 0.1 M: Dissolve 24.2 g Tris and 9.54 g maleic acid in a final volume of 2 liters. The pH, when diluted, is 8.0$_4$.
  B. Tris-Cl stock buffer, 0.2 M, pH 8.0$_4$
  C. Upper tank buffer, 40 mM Tris-maleate–2 mM sodium thioglycolate: Dissolve 0.456 g of sodium thioglycolate in 40 mM buffer diluted from stock A.
  D. Elution buffer, 0.1 M Tris-Cl (pH 8.0$_4$)–5 mM β-mercaptoethanol–30% glycerol. Prepare 1 lister, using stock buffer B.
  E. Lower tank buffer, 0.1 M Tris-Cl, pH 8.0$_4$: Dilute from stock B.

Preparation of Gel Column. The separating gel column is prepared for filling while noting the following additions to the operating instructions. The porous plastic ring which supports the gel should be firmly wedged in place, flush with the bottom flange of the column, by means of small sections of silicone tubing inserted between the dry plastic ring and the cold finger. The lower flange is greased with silicone vacuum grease and pressed against a piece of Parafilm on the counterbored plastic plate supplied with the instrument. This plate is held tightly in place by means of the clamping ring ordinarily used to hold the column assembly together. The gel mixture is prepared in a suction flask, and consists of 24 ml of Tris-maleate stock A, 18 ml of glycerol (ACS grade), 4.2 g of acrylamide, 0.110 g of N,N'-methylene-bis-acrylamide (BIS), and H$_2$O sufficient to make 60 ml. The solution is degassed under a vacuum, and the polymerization is initiated by adding 0.2 ml each of 0.35 M N,N,N',N'-tetramethylethylenediamine (TEMED) and 0.35 M ammonium persulfate. (All polymer components are electrophoresis grade chemicals from BioRad Laboratories). Using a funnel, about 10 ml are decanted into the separating gel column. Cold tapwater is circulated through the jacket during polymerization. Using a large one-hole stopper fitted to the top of the column, a partial vacuum is applied and released several times to permeate the fitted plastic ring with gel mixture. The

---

[12] L. Schuster, this series, Vol. 22, p. 412.

column can now be filled to a depth of 10 cm with the remaining solution and carefully topped with a layer of water to ensure a flat surface when the gel sets.

The lower column cone is filled with a stiffer 10% gel to support the elution cell membrane. In a suction flask mix 100 ml of Tris-Cl stock B, 60 ml glycerol (to maintain osmolarity with elution chamber), 20 g of acrylamide, 0.5 g of BIS, and H_2O sufficient to make 200 ml. After degassing the solution, 0.5 ml each of 0.35 M TEMED and ammonium persulfate are added, and the mixture is poured into the inverted glass cone. In the meantime, the cone is prepared by greasing the flange and clamping it upside down against a piece of Parafilm on a small glass plate, using "Bulldog" spring paper clips. The bubble-collecting pocket on the end of the cone can be eliminated by temporarily extending the end with a cylinder of plastic cut from a polyethylene bottle. The cone is then filled with gel mixture to beyond the end of the glass; the plastic sleeve is removed after the gel solidifies. The electrophoresis unit may now be assembled; take care to remove all fragments of gel below the fritted ring of the separation column with a scalpel blade, and to eliminate any air bubbles in the elution chamber. The column is cooled with a refrigerated circulator (Forma Scientific) to 4°.

Electrophoretic Removal of Ammonium Persulfate. Excess catalyst must be removed from the column by a pre-run with thioglycolate.[13] Using a glass syringe fitted with a long plastic capillary tube, 4 ml of 35 mM sodium thioglycolate in 30% glycerol (and tinted with Bromphenol blue tracking dye) are laid on the surface of the gel beneath the upper tank buffer. A constant current of 50 mA is applied until the dye emerges in the elution buffer. The glycerol layer is aspirated with a syringe before enzyme is applied.

Electrophoretic Separation of Enzyme. Since no stacking gel or buffer discontinuities are used in conjunction with the separation gel, it is necessary to achieve sample-zone sharpening by reducing the buffer concentration in the sample layer[14] to 8–10 mM. This can be done during the ultrafiltration procedure of step 5 or by buffer exchange with a BioGel P-6 column equilibrated with 8 mM Tris-maleate (pH 8.0$_4$)–5 mM β-mercaptoethanol in 30% glycerol. Unfortunately, the latter results inexplicably in the loss of 25–50% of the activity, so the ultrafiltration method is preferable. From 10–30 mg of protein from step 5, in a maximum of 4 ml of the above buffer (tinted with Bromphenol blue), are laid on the gel surface, as before. The elution buffer should flow across

[13] J. M. Brewer, *Science* **156**, 256 (1967).
[14] S. Hjerten, S. Jerstedt, and A. Tiselius, *Anal. Biochem.* **11**, 219 (1965).

PURIFICATION OF DEOXYCYTIDINE KINASE FROM CALF THYMUS

Step and fraction	Volume (ml)	Total protein (mg)	Total activity (units)	Specific activity (units/mg)	Yield (%)
1. Crude extract	1800	25,700	1440	0.056	100
2. Streptomycin	1960	21,000	1700	0.081	118
3. Ammonium sulfate	270	5500	1730	0.315	120
4. Calcium phosphate	116	487	1130	2.32	78
5. DEAE-cellulose	43	70	760	10.9	53
Preparative PAGE[a]	4	7.9	163	20.6	26[a]

[a] Based on 30 mg of protein from step 5. Yield is corrected accordingly.

the bottom of the separation gel at a rate of about 20 ml/hr. This can be controlled by means of a needle valve or peristaltic pump attached to the outlet. A slight positive pressure should be maintained on the inlet by elevation of the elution buffer reservoir. A constant current of 25 mA is applied while the protein is entering the gel. After the tracking dye is compressed into a layer 1–2 mm deep and has penetrated about 10 mm into the gel, the current is increased to 50 mA (about 700 V with the unit described) for the duration of the procedure. The main enzyme peak should emerge in 6–8 hr, usually in a total of about five 5-ml fractions, but a minor peak of activity usually follows several hours later. The upper tank buffer should be replaced whenever the total period of electrophoresis exceeds 12 hr. The enzyme fractions should be assayed as soon as possible, and the main peak is concentrated by membrane ultrafiltration and stored in 5 mM dithiothreitol–50% glycerol at $-20°$. A summary of a typical purification is shown in the table.

Properties

Substrate Specificity and Kinetic Constants. At concentrations of deoxynucleosides likely to be found in thymus cells, deoxycytidine is phosphorylated the most rapidly, but it exhibits bimodal double-reciprocal plots, such as are seen in cases of negative cooperativity.[6,15] The apparent K_m for dCyd at concentrations below 20 μM is estimated to be about 5 μM, but increases to about 50 μM at high dCyd concentrations. Surprisingly, all of the other nucleosides phosphorylated by this enzyme give normal, linear kinetics, though with considerably higher apparent

[15] D. H. Ives and J. P. Durham, *J. Biol. Chem.* **245**, 2285 (1970).

K_m values (μM): α-D-arabinosyl cytosine (40),[2,4] deoxyguanosine (310),[2] deoxyadenosine (330),[3] and cytidine (1400).[6] All of these alternative substrates produce greater maximum velocities than dCyd,[3] but all evidence available indicates that they compete with dCyd for the same enzymic site.[2,3,15] A large variety of nucleoside analogs related to dCyd, dAdo, or dGuo have also been identified as alternative substrates.[3] The enzyme is not particularly fastidious in its choice of nucleoside triphosphates, either; with the single exception of dCTP, all of the ribose and deoxyribose nucleoside triphosphates may serve as phosphate donors in the reaction catalyzed by deoxycytidine kinase.[5]

Regulatory Properties. The phosphorylation of dCyd, and of the other substrates as well, is strongly inhibited by dCTP. As might be anticipated, the inhibition pattern with dCyd is very complex, while with the other substrates such as arabinosyl cytosine or dGuo, dCTP gives simple competitive inhibition patterns, with apparent K_i values of 5.4 and 1.1 μM, respectively.[2,15] The end-product also appears to be competitive with the phosphate donor, such as ATP-Mg, with reciprocal plots converging on the ordinate (with curvature, since bimodal plots are seen with variable phosphate donor also). Competition by dCTP for the donor nucleotide site is also indicated by the fact that a more efficiently bound phosphate donor, dTTP, largely reverses the inhibition, while the β-γ-methylene diphosphonate analog dTDPCP, which is not a phosphate donor, does not.[5] The enzyme is also strongly inhibited by pyrimidine nucleoside diphosphates such as dTDP and UDP, also apparently in competition for the triphosphate donor site on the enzyme.[15]

Other Physical and Chemical Characteristics.[5] The enzyme exhibits a nearly flat response to pH variations between 6 and 10 and has an isoelectric pH of 4.85. The molecular weight is 56,000, and the frictional ratio is 1.36. An absolute requirement for divalent metal ion is best satisfied by Mg^{2+}, although a number of other divalent cations give partial activity. In order of preference, they are: Mn^{2+}, Ca^{2+}, Fe^{2+}, Co^{2+}, Zn^{2+}, Ni^{2+}, and Sr^{2+}. Activity is lost quickly in the absence of sulfhydryl reducing agents, but partial reactivation can be accomplished by incubating with 50 mM dithiothreitol, along with some bovine serum albumin. The enzyme is only moderately stable to heat. The thermostability is greatest at pH 8, less at pH 6, and much less at pH 10. The half-life in pH 8 buffer at 60° is about 8 min, but substantial protection is afforded by the addition of ATP, dCTP, or glycerol.

[44] Deoxynucleoside Kinases from *Lactobacillus acidophilus* R-26[1]

By Martin R. Deibel, Jr. and David H. Ives

2'-Deoxyadenosine + ATP → 5'-dAMP + ADP

2'-Deoxycytidine + ATP → 5'-dCMP + ADP

2'-Deoxyguanosine + ATP → 5'-dGMP + ADP

The preparation of an enzyme fraction responsible for the phosphorylation of deoxyadenosine (dAdo), deoxycytidine (dCyd), and deoxyguanosine (dGuo) has been previously reported from cell-free extracts of *Lactobacillus acidophilus* R-26.[2] These enzyme activities are further purified and resolved into two closely related deoxynucleoside kinases, one specific for the phosphorylation of dCyd and dAdo, and the other specific for dGuo and dAdo, by an affinity chromatography step using Blue Sepharose CL-6B.[3]

Assay Method

Principle. The rate of deoxynucleoside 5'-monophosphate production is determined by an ion-exchange paper method[4] in which isotopically labeled nucleoside monophosphates are retained on anion-exchange resin-impregnated paper, while labeled precursors are washed away.

Procedure. The standard assay system (0.080 ml) contains 0.1 M Tris·HCl, pH 8.0_{20},[5] 10 mM ATP, 12 mM MgCl$_2$, and 0.02 mM [^{3}H]deoxynucleoside (0.1–1.0 μCi per assay). Reactions are initiated by the addition of 20 μl of enzyme preparation (preincubated at 20° for 15 min, followed by storage on ice) to the reaction mixture at 20°, in 1.5-ml disposable plastic centrifuge tubes, and terminated after 30 min, either by the addition of 0.2 ml of 0.1 N formic acid, or by immersion in boiling water for 2 min, followed by the addition of 0.2 ml of water. Portions of the diluted radioactive assay solutions (50 μl) are spotted on 15 × 15 mm pieces of SB-2 ion-exchange paper (H. Reeve Angel), which are washed with 0.001 N ammonium formate for 20 min, followed by water alone for

---

[1] This research was supported by Public Health Service Grant CA-06913 from the National Cancer Institute.

[2] J. P. Durham and D. H. Ives, *Biochim. Biophys. Acta* **228**, 9 (1971).

[3] M. R. Deibel, Jr. and D. H. Ives, *J. Biol. Chem.* **252**, 8235 (1977).

[4] D. H. Ives, J. P. Durham, and V. S. Tucker, *Anal. Biochem.* **28**, 192 (1969).

[5] The subscript denotes the temperature at which the buffer pH was equilibrated.

20 min. The washed ion-exchange papers are prepared for scintillation counting as described by Ives et al.[4]

Definition of Units and Specific Activity. The unit of enzyme activity is defined as the amount of enzyme required to produce 1 nmole of deoxynucleoside 5'-monophosphate per minute at 20°. Specific activity is expressed as units per milligram of protein, as determined by the method of Lowry et al.[6]

Culturing and Preparation of Cells. L. acidophilus R-26 (ATCC 11506) cells are maintained on 2% agar slants containing growth medium, or are grown at 37° in fermentations of 30 liters, using a medium containing a lower sulfur content than used by Durham and Ives,[2] since the omission of thioglycolate and cysteine from this medium results in over a 300% increase in the yield of cells. When the cells reach late log growth (A_{550} = 2.2), the cell suspension is chilled to under 20° and harvested by a Sharples centrifuge. The yield of packed wet cells is usually about 3.3 g/liter of medium.

Enzyme Purification Procedure

Step 1. Preparation of Cell Extract. The wet cells are either frozen immediately for storage, or are suspended directly in 2 volumes of 0.1 M Tris·HCl, pH 8.0₄ and disrupted by two passages through a chilled French press at 10,000 lb/in². The suspension of cracked cells is centrifuged for 30 min in a Sorvalll GS-3 rotor at 13,700 g at 4°, and the supernatant fluid is stored at −20° until required. Frozen cell extracts are centrifuged at 13,700 g, as before, for 1 hr, to remove cell debris or denatured protein from the solution, and a total of 1510 ml are collected, representing over 400 g of cells, wet weight.[7] While the ambient pH of this supernatant fraction is 4.5–5.5, readjustment of the pH to between 7 and 8 results in the production of a viscous suspension which is extremely difficult to process further. However, no significant loss of enzymic activity is observed if the preparation is carried rapidly through step 3 without pH adjustment.

Step 2. Streptomycin Fractionation. To 1510 ml of extract from step 1, 650 ml of a 10% solution of neutralized streptomycin sulfate are added drop by drop (with stirring) over a period of 20 min. The solution is

---

[6] O. H. Lowry, N. J. Rosebrough, A. L. Farr, and R. J. Randall, *J. Biol. Chem.* **193**, 265 (1951).

[7] In this, and all subsequent steps, the temperature is maintained at 4°C, and partially purified preparations to be stored are kept unfrozen in 50% glycerol at −20°C to minimize loss of activity.

stirred for an additional 20 min, and then is centrifuged at 20,200 g for 45 min, yielding 2160 ml of supernatant solution.

Step 3. Ammonium Sulfate Fractionation. Solid ammonium sulfate is slowly added with stirring to the streptomycin supernatant solution to give a solution nominally 65% saturated (398 g/liter). Very little protein precipitates at lower ammonium sulfate concentrations. After stirring for 20 min more, the mixture is centrifuged at 16,300 g in a Sorvall GSA rotor for 30 min. The pellet is resuspended in 0.1 M Tris·HCl, pH 8.0$_4$, to a volume of 400 ml, and solid ammonium sulfate is added again to give a solution saturated to 65% as before. This second precipitation results in a substantial removal of deoxynucleoside 5'-monophosphate and diphosphate kinases, while having a negligible effect on the recovery of the deoxynucleoside kinase activities. The precipitate is collected by centrifugation as before and is resuspended in 56 ml of 0.1 M Tris·HCl, pH 8.0$_4$, containing 50% glycerol.

Step 4. Affinity Chromatography—I. Blue Sepharose CL-6B (4.65 μmoles of dye per milliliter of swollen gel, Pharmacia Fine Chemicals) is an affinity chromatographic support consisting of Cibacron Blue 3GA covalently attached to cross-linked, desulfated Sepharose CL-6B, which forms complexes with most enzymes which possess ATP or NAD binding pockets.[8,9] The unbound dye strongly competes with ATP for sites on the deoxynucleoside kinases, and therefore Blue Sepharose columns form strong complexes with the enzymes as well. A pH range of 6.5–7.0 in potassium phosphate is optimal for enzyme binding to Blue Sepharose, and it is likewise suitable for the separation of deoxynucleoside kinase activities, when eluted by a linear 0–10 mM gradient of ATP-Mg.[3]

For preparative fractionation, however, the solution prepared by step 3 is divided into four parts, and each aliquot is diluted at the time of chromatography to a conductivity of less than twice that of the equilibration buffer (15 mM potassium phosphate, pH 6.6, 15% glycerol). In the absence of a conductivity meter, the dilution of 15 ml of step 3 solution to 500 ml with buffer will suffice. One of the four enzyme portions is then added to a 5 × 6 cm column of Blue Sepharose, followed by extensive washing with buffer until no further absorption at 280 nm can

[8] R. L. Easterday and I. M. Easterday, *in* "Immobilized Biochemicals and Affinity Chromatography" (R. B. Dunlap, ed.), p. 123. Plenum, New York, 1974.
[9] S. T. Thompson, K. H. Cass, and E. Stellwagen, *Proc. Natl. Acad. Sci. U.S.A.* **72**, 669 (1975).

be detected.[10] All three deoxynucleoside kinase activities are eluted together with 2 mM ATP-Mg. The other three aliquots are treated identically, and the combined solutions are concentrated in the presence of 50% glycerol by ultrafiltration with a PM-10 membrane filter (Amicon) to a final volume of 5.0 ml.[11] The major purpose of the initial preparative column step is to eliminate a large quantity of protein which could presumably interfere with subsequent binding and separation of the deoxynucleoside kinase activities from the affinity column.

Step 5. Affinity Chromatography—II. Aliquots of protein from the previous step are diluted from 0.5 ml to 50 ml with equilibration buffer and applied to a 2.5 × 5.5 cm Blue Sepharose CL-6B column. The smaller column is used to increase the resolution of dCyd/dAdo kinase and dGuo/dAdo kinase, since trailing is minimized under these conditions. The column is developed by stepped increases in the concentration of ATP-Mg: 1.0 mM, for the elution of the dCyd/dAdo kinase (5A), and 3.0 mM ATP-Mg, for the elution of a dGuo/dAdo kinase fraction (5B), as shown in Fig. 1. The fractions enriched with dCyd kinase are cycled again through a 2.5 × 5.5 cm column to remove the leading edge of the dGuo kinase peak, while the dGuo kinase preparation is recycled instead through a longer column of Blue Sepharose (2.5 × 9.0 cm).

Step 6. Affinity Chromatography—III. Fraction A of step 5 is further purified, 0.5 mg at a time, by an additional passage through Blue Sepharose CL-6B (2.5 × 4.0 cm), in which a wash with 1 mM NADH is utilized to remove any residual contamination of lactate dehydrogenase, a persistent impurity not completely separated by previous steps. The dCyd kinase activity is then eluted sharply with 5.0 mM ATP-Mg, to minimize trailing of the activity, and concentrated in the presence of 50% glycerol to 4.0 ml. This additional fractionation by affinity chromatography, while increasing the specific activity of the enzyme, results in substantial losses of total activity as a result of the enzyme's enhanced lability in the highly purified state.

[10] The Blue Sepharose nonbinding fractions (approximately 16 liters altogether) are rich sources of uridine (cytidine) kinase, lactate dehydrogenase, and deoxyribosyl transferase, and these enzymes can be isolated in good yield by ammonium sulfate precipitation from the column runoff.

[11] Additional small quantities of enzyme (low specific activity) are obtained by elution of the column with 1000 ml of 0.3 M potassium phosphate, pH 6.6, 5% glycerol, for each aliquot applied originally. This procedure is also utilized for regeneration of the column for further chromatography. In addition, the relatively high cost of this affinity medium precludes carrying out this procedure at one time on a larger column.

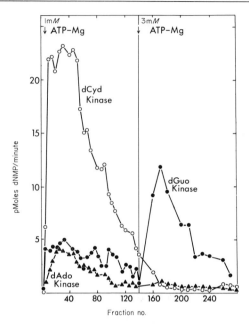

FIG. 1. Blue Sepharose CL-6B affinity chromatography of deoxynucleoside kinase activities (fraction 4); isolation of dCyd/dAdo and dGuo/dAdo kinase activities. Assays of each fraction (7.5 ml) were performed as described in *Methods*.

Notes on Further Purification. The purification procedure described above is summarized in the table. While the methods used do not result in the isolation of homogeneous deoxynucleoside kinases, it is possible to prepare pure, though highly labile, dCyd/dAdo kinase by means of preparative polyacrylamide gel electrophoresis, followed by isoelectric focusing (pH 4–6).[3]

Alternative Purification Scheme

Conventional techniques have been completely incapable of separating dCyd/dAdo kinase from dGuo/dAdo kinase activity. The obvious structural similarity of these enzymes may explain the need for repeated affinity chromatography steps to achieve their substantial separation. Often, however, such resolution may not be required, and the following scheme permits the copurification of the activities which phosphorylate dAdo, dCyd, and dGuo in fairly good yield, and without the need to generate large volumes of eluate inherent in the preceding method.

Coupled Hydrophobic and Affinity Chromatography. Protein from

PURIFICATION OF DEOXYNUCLEOSIDE KINASES FROM *Lactobacillus acidophilus* R-26

Step	Fraction (mg/ml)	Protein (mg/ml)	Volume (ml)	Specific activity (units/mg) dCyd	dAdo	dGuo	Yield[a] dCyd (%)
1	Extract supernatant	27.5	1510	0.32	0.06	0.33	100
2	Streptomycin supernatant	17.0	2160	0.32	0.09	0.49	90
3	0–65% $(NH_4)_2SO_4$ ppt.	73.5	56	2.12	0.76	3.12	66
4	Affinity chromatography—I	32.0	5	41.7	15.2	34.0	50
5	Affinity chromatography—II[b]						
	Fraction A	0.99	10.3	177.0	16.3	17.0	19.5
	Fraction B	0.35	11.0	17.8	29.7	237.0	0.7
6	Affinity chromatography—III[c]	0.027	4.0	387.0	38.0	28.5	8.5
4A	Coupled hydrophobic and affinity chromatography[d]	0.32	4.1	91.8	23.0	108.0	25.7

[a] Yields corrected to step 1.
[b] Represents 112 mg of step 4 protein.
[c] Represents 0.5 mg of step 5A protein.
[d] Represents 147 mg of step 3 protein.

step 3 (2.0 ml of dissolved ammonium sulfate precipitate) is added undiluted to a 1.9 × 7.5 cm *O*-octyl Sepharose column (Pharmacia Fine Chemicals) equilibrated with 15 mM potassium phosphate, pH 6.6, and 15% glycerol. The column is washed extensively with buffer until the absorbance at 280 nm reaches a baseline value. All three deoxynucleoside kinase activities are then eluted with 150 ml of 0.5% (v/v) Triton X-100 in the same buffer onto a 1.9 × 9.0 cm column of Blue Sepharose CL-6B. The latter column is washed extensively with 250 ml of buffer lacking detergent to remove nonbinding protein and Triton X-100. All three deoxynucleoside kinase activities are then eluted with 180 ml of 5 mM ATP-Mg, and the solution is concentrated to a volume of 4.1 ml by ultrafiltration. Up to 30% of the activity passes through PM-10 membrane filters, a phenomenon observed only with deoxynucleoside kinase preparations pretreated with Triton X-100. Thus, the activity yield reported in the table for this purification step reflects only the concentrated portion of the preparation. This method has been shown to be entirely reproducible, although the characteristics of the enzyme following exposure to Triton X-100 have not been fully determined. While the protein obtained by this procedure consists of both the dCyd/dAdo kinase and the dGuo/dAdo kinase, these activities may be resolved simply by initiating step 5.

Properties

Substrate Specificity. The phosphorylation of dCyd/dAdo and dGuo/ dAdo is accomplished primarily by ATP-Mg (1:1), although all other nucleotide triphosphates, other than the homologous end-products of each kinase activity, can function as substrates in the reactions. Under physiological conditions deoxycytidine and deoxyadenosine are the principal phosphate acceptors for the dCyd/dAdo kinase protein, while deoxyguanosine and deoxyadenosine are the preferred substrates of the dGuo/dAdo protein. The phosphorylation of dGuo in fraction 5A, and dCyd in fraction 5B, was shown by kinetic analysis to be produced by low-affinity catalytic sites on the dCyd/dAdo and dGuo/dAdo enzymes, respectively, and was not a result of the incomplete resolution of the two activities by affinity chromatography, in which a 2–5% cross-contamination of dCyd-specific kinase in fraction 5B, or dGuo-specific kinase in fraction 5A, was observed. Thus, under conditions in which ATP-Mg is present at near-saturating concentrations, and dCyd is varied at subsaturation, $K_{m(app)}$ values of 5 μM and 490 μM are obtained for the dCyd and dGuo kinase enriched fractions, while the corresponding parameters obtained with variable deoxyguanosine are reversed, with values of 80 μM and 5 μM for the respective dCyd and dGuo kinase enriched fractions. Furthermore, the apparent $K_{m(app)}$'s for dAdo phosphorylation by both fractions are 5 μM at low deoxynucleoside concentrations and greater than 1 mM at high substrate concentrations.

Stability. The enzyme, dCyd/dAdo kinase, is very unstable, especially when highly purified, and it is advisable to work with fairly

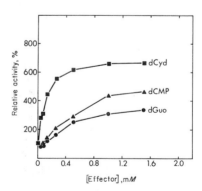

FIG. 2. The effect of varying the concentrations of dCyd, dGuo, and dCMP on the phosphorylation of dAdo. Assays were conducted as described in *Methods,* except that dCyd, dGuo, and dCMP were varied as indicated. The enzyme was 0.17 μg of fraction 5A protein.

concentrated solutions, since the enzyme is sensitive to dilution. This instability is also the reason why the assays must be carried out at 20°, rather than at higher temperatures, where the time course of the reactions tends to be nonlinear. The addition of 50% glycerol to all protein preparations, and storage at −20°, protect the enzyme against both dilution and thermal denaturation. Incubation with metal ions, sulfhydryl reagents, bovine serum albumin, or nucleosides are ineffective in preventing or reversing enzymic inactivation.

Additional Characteristics. The molecular weights of both dCyd/dAdo and dGuo/dAdo kinases are 50,000 ± 3000, as determined by Biogel P-150 gel filtration.[3] The isoelectric points both correspond to a pH of 5.3, and while the pH optimum of dCyd kinase is 7.8, the dAdo kinase function exhibits a maximal activity at pH 9.5.

Kinetic Characterization[12]

The evidence from kinetic studies favors a rapid equilibrium, random type of reaction mechanism for dCyd kinase activity, and a random sequential mechanism for dAdo kinase activity. The true dissociation constants and Michaelis constants for the dCyd kinase function are as follows: K_{dCyd}, 85 μM; K_{ATP}, 1000 μM; αK_{dCyd}, 2.2 μM; αK_{ATP}, 25.8 μM; V_{max}, 46.3 nmoles/min/mg. The constants for the dAdo kinase function are (α = 1.0): K_{dAdo}, 2.8 μM; K_{ATP}, 90 μM; V_{max}, 6.6 nmoles/min/mg.

Deoxyadenosine kinase activity exhibits apparent activation at increased substrate concentrations, which is attributed to the interaction of dAdo with both a catalytic and allosteric site on the enzyme, although the homotropic cooperativity is abolished in the presence of dCyd. Independent active sites for dCyd and dAdo kinase activities are demonstrated by kinetic analysis of the interaction of both substrates, such that dCyd noncompetitively stimulates dAdo kinase up to 500%, while dAdo is a weak competitive inhibitor of dCyd phosphorylation. As shown in Fig. 2, dAdo kinase activity is likewise stimulated by dGuo and dCMP, the latter demonstrating that mere binding to the effector site alone is sufficient for activation.

Deoxycytidine and deoxyadenosine kinase activities are inseparable by methods based on size, charge, or affinity for Blue Sepharose, although both activities can be differentially inactivated by thermal denaturation and photochemical modification by rose bengal, and both display differential sensitivity to anions in the Hofmeister series.[3] It is thus postulated that one protein containing two independently folded

[12] M. R. Deibel, Jr., R. B. Reznik, and D. H. Ives, *J. Biol. Chem.* **252**, 8240 (1977).

polypeptide regions, one for the active site of dCyd kinase (and the effector site for dAdo kinase) and the other for dAdo kinase, could account for these observations.

Characteristics of dGuo/dAdo Kinase

The properties of this enzyme preparation are not yet determined, although preliminary studies suggest that they parallel the characteristics of the dCyd/dAdo kinase enzyme.

[45] Thymidine Kinase from *Escherichia coli*

By MING S. CHEN and WILLIAM H. PRUSOFF

$$\text{Deoxythymidine} + \text{NTP} \rightarrow \text{dTMP} + \text{NDP}$$

Thymidine kinase (EC 2.7.1.75) from *Escherichia coli* is an unusual allosteric enzyme whose most striking characteristic is its regulation not only by the end-product dTTP, but also by a number of nucleoside di- and triphosphates.[1] Dimerization of the enzyme molecule accounts for the regulatory properties, and this "dimer" molecule has either an active or an inactive conformation, depending on the specific nature of the effector nucleotide with which it interacts.[2]

Assay Method

Principle. Enzymic activity has been assayed by conversion of labeled deoxythymidine to deoxythymidine monophosphate, with separation of the substrate and product by either high-voltage paper electrophoresis,[3] column chromatography,[4] thin-layer chromatography,[5] paper chromatography,[6] or disc DEAE-cellulose.[7]

[1] R. Okazaki and A. Kornberg, *J. Biol. Chem.* **239**, 275 (1964).
[2] N. Iwatsuki and R. Okazaki, *J. Mol. Biol.* **29**, 139 (1967).
[3] R. Okazaki and A. Kornberg, *J. Biol. Chem.* **239**, 269 (1964).
[4] P. S. Fitt, P. I. Peterkin, and V. L. Grey, *J. Chromatogr.* **124**, 137 (1976).
[5] D. L. Greenman, R. C. Huang, M. Smith, and L. M. Furrow, *Anal. Biochem.* **31**, 348 (1969).
[6] W. J. Reeves, Jr., A. S. Seid, and D. M. Greenberg, *Anal. Biochem.* **30**, 474 (1969).
[7] M. B. Furlong, *Anal. Biochem.* **5**, 515 (1963).

Reagents

Tris·HCl, buffer 0.2 M, pH 7.8
ATP, sodium salt, 50 mM
MgCl$_2$, 50 mM
Bovine serum albumin (BSA)
[2-^{14}C]Thymidine, 4.8 mM, sp act = 0.5–1.5 Ci/mole
The stock assay solution consists of 0.7 ml Tris buffer, 0.4 ml of ATP, 0.4 ml MgCl$_2$, 0.6 mg BSA, and 0.5 ml [2-^{14}C]thymidine. This mixture can be stored in the cold for up to 1 week. For assaying crude cell homogenates, 20 mM NaF is included to inhibit phosphatase activity.

Procedure. The reaction mixture contained in 0.1 ml of final volume is: 75 μl of the above stock assay solution plus 25 μl of enzyme solution. The conversion of deoxythymidine to deoxythymidine monophosphate was measured by one of the following two procedures:

(1) By adsorption onto DEAE discs as described by Furlong.[7] Portions of the reaction mixtures (approximately 20–30 μl) are spotted onto DEAE discs (2 cm in diameter) which are then dropped into a beaker containing about 10 ml of 95% alcohol per disc. The alcohol is decanted, and the discs are resuspended in the same volume of alcohol for 10 min. This washing procedure in 95% alcohol is repeated two times. The discs are then placed on a paper towel, dried in air, and counted in a liquid scintillation counter with toluene-POPOP.

(2) By spotting portions of the reaction mixture (approximately 2–5 μl) onto a cellulose TLC sheet with development in 0.5 M LiCl to a distance of 15 cm. The spot that corresponds to a deoxythymidine monophosphate marker is cut out and counted as in (1).

The assay of deoxythymidine monophosphate by TLC has the advantage of detecting potential contamination of the enzyme during purification with deoxythymidine monophosphate kinase, and if also present nucleoside diphosphokinase, since deoxythymidine-diphosphate and -triphosphate have different R_f values in the above TLC system. Another advantage of TLC is when [γ-^{32}P]ATP is used to study the potential of an agent to be an alternate substrate. The formed monophosphate of a substrate nucleoside analog in this system is readily separated from the phosphate donor (e.g., [γ-^{32}P]ATP, R_f = 0.06; nucleoside monophosphate, R_f = 0.47).

Definition of Units of Activity and Specific Activity. One unit of thymidine kinase activity is defined as the amount of enzyme catalyzing the formation of one μmole of deoxythymidine monophosphate per minute at 37°. Specific activity is expressed as the number of units of thymidine kinase activity per milligram of protein. Proteins were determined according to Lowry et al.[8]

Purification Procedure

The purification procedure has been described by Voytek et al.[9] The enzyme solution is maintained at 0–4° unless otherwise indicated.

Step 1. Preparation of Crude Extract. Frozen *E. coli* B cells which had been grown in Kornberg's medium and harvested in late log phase were purchased from General Biochemicals and stored at −70°. The extraction procedure is a modification of the method reported by Okazaki and Kornberg.[3] Cells (500 g) were thawed for 12 hr at 0° and then placed in a commercial Waring Blendor that contained 300–400 ml of chilled Tris·HCl buffer (20 mM, pH 7.8) plus EDTA (2 mM). The blender was started at a low speed, and 500 g of glass beads (Superbrite No. 100, Minnesota Mining and Manufacturing Company, Inc.) which had been previously chilled to −70° were added slowly. The speed of the blender was increased (blender speed low, rheostat setting 75 V) and maintained for 25 min. During the extraction the temperature remained below 0°. The suspension was then centrifuged for 30 min at 13,000 g. The supernatant fluid was decanted and subjected to a second centrifugation at 100,000 g for 3 hr. The supernatant liquid obtained was clear and yellow and is designated as fraction 1 in the table.

PURIFICATION DATA FOR *Escherichia coli* THYMIDINE KINASE[a]

Fraction and step	Volume (ml)	Units	Protein (mg/ml)	Specific activity (units/mg) × 10³	Purification factor
1. Extract	440	9.27	15.8	1.3	1.0
2. Streptomycin	415	8.27	16.6	1.0	0.75
3. Ammonium sulfate	16	2.63	34.0	5.0	4
4. Sephadex	10	0.77	4.4	17.0	12
5. First gel electrophoresis	16	0.23	0.2	78.0	59
6. Second gel electrophoresis	6	0.11	0.025	720.0	531

[a] Reproduced from Voytek et al.[9] (Table I). The units are expressed as the conversion of 1 μmole of deoxythymidine to deoxythymidine monophosphate in 1 min at 37° rather than in 1 hr as presented in the original table.[11]

[8] O. H. Lowry, N. J. Rosebrough, A. L. Farr, and R. J. Randall, *J. Biol. Chem.* **193**, 265 (1951).

[9] P. Voytek, P. K. Chang, and W. H. Prusoff, *J. Biol. Chem.* **246**, 1432 (1971).

Step 2. Streptomycin Sulfate Fractionation. To every 100 ml of fraction 1, 2.63 ml of a 38% streptomycin sulfate solution (Grade B, Calbiochem) were added drop by drop over a 30-min period with stirring. The solution was stirred for an additional 30 min, frozen, stored at −20° overnight, thawed, and then centrifuged at 13,000 *g* for 30 min. The supernatant fluid (fraction 2) was clear yellow.

Step 3. Ammonium Sulfate Fractionation. Solid ammonium sulfate was slowly added to fraction 2 to a final saturation of 40%. After additional stirring for 30 min, the mixture was centrifuged at 13,000 *g*, and the precipitate was resuspended in 16 ml of 50 m*M* Tris·HCl buffer (pH 7.8) containing EDTA (0.5 m*M*) and 2-mercaptoethanol (9 m*M*). The mixture was dialyzed overnight against the same buffer (fraction 3).

Step 4. Sephadex G-150 Chromatography. Fraction 3 was added to a Sephadex G-150 column (2.5 × 100 cm) which had been previously washed with 2 liters of the dialysis buffer. The proteins were eluted from the column by the upward flow technique. The major fractions containing 75% of deoxythymidine kinase activity were pooled and concentrated 8-fold in an Amicon ultrafiltration cell model 202 with a UM-20E filter (fraction 4).

Step 5. Preparative Polyacrylamide Disc Gel Electrophoresis I. A Buchler preparative polyacrylamide gel electrophoresis apparatus was used. Sucrose and Bromphenol blue were added to fraction 4 making the final concentrations 3 and 0.001%, respectively. The protein solution was then layered on top of a 4-cm 7.5% gel which was prepared with the use of the Tris-glycine buffer system at pH 8.9 according to Davis.[10] MgCl₂ (1 m*M*) was also included in the gels. The upper buffer (pH 8.9) consisted of Tris (53 m*M*), glycine (53 m*M*), cysteine (1.2 m*M*), MgCl₂ (1 m*M*), and 2-mercaptoethanol (20 m*M*). The composition of the lower buffer was Tris·HCl (100 m*M*) at pH 8.0 and MgCl₂ (1 m*M*). The elution buffer contained Tris·HCl (100 m*M*) at pH 8.0, MgCl₂ (1 m*M*), cysteine (1.2 m*M*), and 2-mercaptoethanol (20 m*M*). The membrane-holder buffer was Tris·HCl (400 m*M*) at pH 8.1 and MgCl₂ (1 m*M*). A constant current of 60 mA was applied across the gel for approximately 18 hr by which time the deoxythymidine kinase had passed through the gel. The flow rate of the elution buffer was 1 ml per minute. The coolant temperature was maintained at 1°. Eluted fractions containing deoxythymidine kinase activity were pooled and concentrated approximately 8-fold by ultrafiltration (fraction 5).

Step 6. Preparative Polyacrylamide Disc Gel Electrophoresis II. The

---

[10] B. J. Davis, *Ann. N. Y. Acad. Sci.* **121,** 404 (1964).

second preparative electrophoresis was performed in a manner identical to the first but modified by use of a 5% polyacrylamide gel at a height of 5 cm. Fractions that contained the enzymic activity were pooled and concentrated as described above (fraction 6). Fraction 6 (25–40 μg of protein per milliliter) was divided into 0.3-ml aliquots and stored at −70°. No loss in activity occurred within a 1-month storage period, and there was no change in the response to the nucleoside effectors.

The results of a typical preparation are summarized in the table,[9,11] and the enzyme obtained was found to be electrophoretically homogeneous.

Properties

Substrate Specificity. The enzyme can use the following as a phosphate acceptor: deoxythymidine, deoxyuridine, 5-halogenated analogs (5-fluoro-, 5-chloro-, 5-bromo-, and 5-iodo-deoxyuridine[3]), and 5-mercaptodeoxyuridine.[12] Like other nucleoside kinases, the enzyme has a broad specificity with regard to the phosphate donor; most nucleotide triphosphates with the exception of dTTP have been found able to donate phosphate in the reaction. The ability to serve as a phosphate donor follows the following order; ATP, dGTP, dATP, dCTP, GTP, and ITP. Other nucleoside triphosphates and all nucleoside diphosphates are inert as phosphate donors. The metal ion requirement can be fulfilled by either Mg^{2+} or Mn^{2+}.

pH Optimum. Activity as a function of pH is maximal at about pH 7.5.

Stability. The enzyme appears to be stable for several months if kept at −70° in the presence of 10 mM β-mercaptoethanol and added BSA (0.5 mg/ml).

Molecular and Allosteric Regulatory Properties. E. coli thymidine kinase is an allosteric enzyme.[1] Its activity is not only regulated by the end-product dTTP, but also by a number of nucleoside di- and triphosphate. The end-product dTTP is an inhibitor, while the following nucleotides are activators: dCTP > dATP > dGTP > dCDP > hydroxymethyl-dCTP > hydroxymethyl-dCDP > GTP > dADP > dGDP > GDP.[1] The halogenated analogs 5-iodo-dCTP and 5-bromo-dCTP are more potent allosteric activators than the naturally occurring dCTP.[9] These nucleoside di- and triphosphates, whether inhibitory or activating,

[11] P. Voytek, P. K. Chang, and W. H. Prusoff, *J. Biol. Chem.* **247**, 367 (1972).
[12] T. I. Kalman and T. J. Bardos, *Mol. Pharmacol.* **6**, 621 (1970).

exert their regulatory control of the enzyme activity by an allosteric phenomenon in which the initial event appears to be the dimerization of the enzyme.[2] Those allosteric regulators that are activators increase the sedimentation coefficient from 3.4–3.5 S to 5.3–5.6 S, whereas those that inhibit increase the sedimentation coefficient to 5.9–6.0.[9] Whereas 5-iodo-dUTP is a very potent activator at pH 7.8, replacement of the oxygen in the 5'-position of IdUTP with an imino (-NH-) moiety produced not only an inhibitory effector but also one that is 60-fold more potent than dTTP as allosteric regulator.[13]

In general, the effect of activating effectors is to decrease the K_m for substrate and increase the V_m of the reaction, while the inhibiting effectors increase the K_m and decrease the V_m. The molecular weight of the "monomer" and "dimer" is 42,000 and 90,000, respectively.[2] In addition to its role as a phosphate donor, ATP at high concentrations behaves as an activator, and hence its presence normalizes the plot of Michaelis-Menten kinetics of thymidine phosphorylation from sigmoidal to hyperbolic in shape.[1]

The monomer form of the enzyme is more sensitive to temperature[14] and ultraviolet (UV) irradiation[15] than when in the dimer form (in the presence of a naturally occurring regulatory nucleotide). When the dimer is induced by 5-iodo-dUTP or 5-iodo-dCTP the enzyme is more sensitive to UV irradiation than when in the monomer form (Chen and Prusoff, unpublished result). The enzyme is protected against UV inactivation by the normal substrate deoxythymidine, whereas the halogenated analog of thymidine, 5-iododeoxyuridine (an alternate substrate), enhances the rate of UV inactivation.[11,15] The formation of a free radical in the 5-position of the pyrimidine moiety during radiation-induced dehalogenation of 5-iodo analogs of either the alternate substrate or regulatory nucleotides accounts for such enhanced UV sensitization of the enzyme. [14C]-Labeled IdUrd or IdUTP when irradiated in the presence of the enzyme forms a covalent linkage with the protein which is thus inactivated.

Kinetic Constants.[9,15] The K_m for deoxythymidine is $1.7 \times 10^{-5} M$ and for 5-iodo-2'-deoxyuridine $1.2 \times 10^{-5} M$. The K_i for 3-*N*-methyl-5-iodo-2'-deoxyuridine is $1.7 \times 10^{-3} M$. Since ATP also regulates the enzyme activity as an allosteric activator at high concentration, the above kinetic constants will depend on the ATP concentration.[1] K_m values of the activating regulators are $5 \times 10^{-6} M$ (5-iodo-dCTP, pH 5.5), $22 \times 10^{-6} M$ (5-bromo-dCTP, pH 5.5), $29 \times 10^{-6} M$ (dCTP, pH

[13] M. S. Chen, D. C. Ward, and W. H. Prusoff, *J. Biol. Chem.* **251**, 4839 (1976).
[14] N. Iwatsuki and R. Okazaki, *J. Mol. Biol.* **29**, 155 (1967).
[15] R. Cysyk and W. H. Prusoff, *J. Biol. Chem.* **247**, 2522 (1972).

5.5), $4 \times 10^{-6} M$ (5-iodo-dCTP, pH 7.4), $14 \times 10^{-6} M$ (5-bromo-dCTP, pH 7.4), $27 \times 10^{-6} M$ (dCTP, pH 7.4), and $10 \times 10^{-6} M$ (5-iodo-dUTP, pH 7.4). The K_m value for the inhibitory effector dTTP is $16 \times 10^{-5} M$, and for 5′-triphosphate of 5-iodo-5′-amino-2′,5′-dideoxyuridine (AIdUTP) it is 0.67 μM.

[46] Deoxythymidine Kinase in Regenerating Rat Liver

By EDWARD BRESNICK

$$\text{Deoxythymidine} + \text{ATP} \xrightarrow{\text{Mg}^{2+}} \text{d-TMP} + \text{ATP}$$

Introduction

Deoxythymidine kinase (TdR kinase; ATP-thymidine 5′-phosphotransferase, EC 2.7.1.21) catalyzes the phosphorylation of deoxythymidine and a number of its analogs to form the corresponding deoxyribonucleotides, e.g., d-TMP, at the expense of ATP; this reaction requires a divalent cation, Mg^{2+}.

The enzyme is important in that it is responsible for recycling endogenous deoxythymidine via the pyrimidine salvage pathway and for utilization of exogenous deoxythymidine within tissues. In addition, TdR kinase fulfills an important role in the activation of a number of pyrimidine analogs for chemotherapeutic efficacy, particularly the halogenated pyrimidine deoxyribonucleosides.[1]

TdR kinase activity within mammalian cells is closely correlated with their proliferation capacity. In this respect, enzyme activity is markedly elevated in regenerating liver,[2] neoplastic tissues,[3-8] in virally infected

[1] E. Bresnick, *in* "The Molecular Biology of Cancer" (H. Busch, ed.), p. 277. Academic Press, New York, 1973.
[2] E. Bresnick, *Methods Cancer Res.* **6**, 347 (1971).
[3] T. W. Sneider, V. R. Potter, and H. P. Morris, *Cancer Res.* **29**, 40 (1969).
[4] E. Bresnick and U. B. Thompson, *J. Biol. Chem.* **240**, 3967 (1965).
[5] J. Bukovsky and J. S. Roth, *Cancer Res.* **25**, 358 (1965).
[6] T. Hashimoto, T. Arima, H. Okuda, and S. Fujii, *Cancer Res.* **32**, 67 (1972).
[7] E. Bresnick, U. B. Thompson, H. P. Morris, and A. G. Liebelt, *Biochem. Biophys. Res. Commun.* **16**, 278 (1964).
[8] E. Bresnick and R. J. Karjala, *Cancer Res.* **24**, 841 (1964).

cells,[9-12] in cells entering the S-phase of the cell cycle,[13] in tissues undergoing dietary and diurnal variations,[14,15] and in embryonic liver.[7,16,17]

The enzyme is not only affected by induction at the transcriptional level, but is markedly influenced by the concentration of deoxyribonucleotides. Thus, TdR kinase is end-product inhibited by d-TTP. The latter inhibition is manifested in a wide variety of systems.[18,19]

In addition to the cytosolic TdR kinase described above, a second set of enzymes have been reported to be associated with mitochondria. This TdR kinase apparently functions in the incorporation of the deoxyribonucleotide precursor into mitochondrial DNA.[20,21] However, in the present report, only the cytosolic enzyme will be considered.

Assay Method

Principle. The assay method, which is largely a modification of the procedures used by Bollum and Potter[22] and Breitman,[23] is based upon the binding potential of the cationic support, DEAE-cellulose, for anions, e.g., d-TMP. The labeled precursor, [14C]TdR, is rapidly washed from the DEAE-cellulose, and the resultant radioactivity which represents the product of the reaction, d-TMP, may easily be quantitated by liquid scintillation techniques.

Reagents

[2-14C]TdR (or [methyl-14C]TdR), 3 μCi/μmole
ATP, 25 mM, pH 8.0
MgCl$_2$, 25 mM

[9] B. R. McAuslan, *Virology* **20**, 162 (1963).
[10] S. Kit, D. R. Dubbs, and P. M. Frearson, *Cancer Res.* **26**, 638 (1966).
[11] J. Kara and R. Weil, *Proc. Natl. Acad. Sci. U.S.A.* **57**, 63 (1967).
[12] R. Sheinin, *Virology* **28**, 47 (1966).
[13] T. P. Brent, *Cell Tissue Kinet.* **4**, 297 (1971).
[14] R. E. Beltz, *Arch. Biochem. Biophys.* **99**, 304 (1962).
[15] H. A. Hopkins, H. A. Campbell, B. Barbiroli, and V. R. Potter, *Biochem. J.* **136**, 955 (1973).
[16] H. G. Klemperer and G. R. Haynes, *Biochem. J.* **108**, 541 (1968).
[17] A. T. Taylor, M. A. Stafford, and O. W. Jones, *J. Biol. Chem.* **247**, 1930 (1972).
[18] R. Okazaki and A. Kornberg, *J. Biol. Chem.* **239**, 275 (1964).
[19] D. H. Ives, P. A. Morse, Jr., and V. R. Potter, *J. Biol. Chem.* **238**, 1467 (1963).
[20] S. Kit and Y. Minekawa, *Cancer Res.* **32**, 2277 (1972).
[21] A. J. Berk and D. A. Clayton, *J. Biol. Chem.* **248**, 2722 (1973).
[22] F. J. Bollum and V. R. Potter, *J. Biol. Chem.* **233**, 478 (1958).
[23] T. R. Breitman, *Biochim. Biophys. Acta* **67**, 153 (1963).

Sucrose, 0.25 M
Saline, 0.9%
Calcium phosphate gel
Methanol
Tris buffer, 0.05 M, pH 8.0
DEAE-cellulose, 1.5-cm discs, cut with a cork borer
Ammonium formate, 1 mM
Scintillation fluor, 0.5% PPO–0.03% dimethyl POPOP in toluene
Heating buffer, 0.05 M Tris (pH 8.0)–0.25 M sucrose–6 mM β-mercaptoethanol–1 mM TdR
Storage buffer, 0.05 M Tris (pH 8.0)–3 mM β-mercaptoethanol–50% glycerol

Procedure.[8] The reaction mixture included in a total volume of 0.25 ml is: enzyme; [^{14}C]TdR, 0.1 μCi; ATP, 5 mM; MgCl$_2$, 2.5 mM; and 0.05 M Tris buffer, pH 8.0. The incubation is conducted at 37° for 15 min. The reaction is stopped by plunging the test tubes in a 100° water bath for 3 min. Aliquots, 25 μl, of the cooled reaction mixture are applied to numbered 1.5-cm discs of DEAE-cellulose paper which are held upright on the shaft of a hypodermic needle. After the aqueous solution has permeated the disc, the latter is immersed in approximately 30 ml of 1 mM ammonium formate, washed 5 times in the latter, and 3 times in methanol. The discs are dried at 80°, then placed in scintillation vials containing 5 ml of PPO-dimethyl-POPO in toluene and residual ^{14}C is counted. Since the [^{14}C]d-TMP is insoluble in toluene, the vials with their contents (minus the discs) may be reused a number of times.

Definition of Units. One unit of thymidine kinase is defined as the amount of enzyme that catalyzes the formation of 1 nmole of thymidine monophosphate per minute.

Purification Procedure

All steps in the partial purification are conducted at 4° and are easily accomplished in 5 hr.

Preparation of Animals. The source of the cytosolic TdR kinase is 24-hr regenerating rat liver. Male Sprague-Dawley rats, 180–200 g, are partially hepatectomized at 8:00 A.M. according to the procedure of Higgins and Anderson.[24] After 24 hr, the rats are killed by exsanquination and the liver remnants are removed, washed in cold 0.9% NaCl, and homogenized in cold 0.25 M sucrose (1 g liver per 4.5 ml) in a glass–

[24] G. M. Higgins and R. M. Anderson, *AMA Arch. Pathol.* **12**, 186 (1931).

Teflon homogenizer. The TdR kinase is partially purified from this homogenate.

Partial Purification of Enzyme. The liver homogenate is centrifuged at 100,000 g for 90 min to yield a supernatant fraction which contains approximately 90% of the homogenate activity.

Solid $(NH_4)_2SO_4$ is added slowly to this high-speed supernatant fraction to 35% saturation, and the mixture should be mechanically stirred at 4° for 10 min. After centrifugation at 20,000 g for 15 min, the supernatant fraction is discarded. The precipitate is redissolved in $\frac{1}{2}$ of the original volume of cold 0.05 M Tris (pH 8.0)–0.25 M sucrose–6 mM β-mercaptoethanol–1 mM TdR. The resultant solution is shaken at 50° for 5 min (55° water bath) and then immediately cooled. Centrifugation at 15,000 g for 10 min is required to remove any denatured protein.

Calcium phosphate gel is added to the supernate, 2 mg gel/mg protein, and the suspension is stirred at 4° for 15 min. After centrifugation at 10,000 g for 10 min, the gel is discarded. The enzyme can be precipitated from the supernatant fraction upon the addition of solid $(NH_4)_2SO_4$ to 40% saturation. The precipitate, obtained after centrifugation at 20,000 g for 15 min, is dissolved in 0.05 M Tris(pH 8.0)–3 mM β-mercaptoethanol–50% glycerol and may be stored at $-10°$ for at least several weeks. The specific activity of the final preparation has varied between 60 and 100 units/mg protein over the course of many purifications when conducted under the assay conditions described above. A typical purification from regenerating rat liver is shown in the table.

Further Purification. Several methods have subsequently been published[25,26] in which considerable further purification has been achieved

PARTIAL PURIFICATION OF TdR KINASE FROM REGENERATING RAT LIVER

Fraction	Volume (ml)	Protein (mg)	Total activity (units)[a]	Specific activity (units/mg)	Yield (%)
Liver homogenate	320	6000	1400	0.23	100
$10^5 \times$ g supernate	290	3500	1400	0.40	100
$(NH_4)_2SO_4$, 0–30%	150	210	840	4.0	60
50° treatment	150	105	800	8.0	57
Negative gel	140	31	733	23.7	52
Second $(NH_4)_2SO_4$	10	7	420	60.0	30

[a] 1 unit = 1 nmole product per minute.

[25] E. P. Kowal and G. Markus, *Prep. Biochem.* **6**, 369 (1976).
[26] L. S. Lee and Y. C. Cheng, *J. Biol. Chem.* **251**, 2600 (1976).

by affinity chromatography. In these procedures, thymidine-3'-(4-nitrophenyl phosphate) is reduced catalytically to the 4-amino derivative, and the latter is coupled to carboxyhexyl-sepharose in the presence of 1-ethyl-3-(3'-dimethyl-aminopropyl) carbodiimide. Unreacted carboxylic acid groups on the support are blocked either with galactosamine[26] or glycinamide.[25] Partially purified TdR kinase is applied to the affinity gel column which has been equilibrated with 10 mM Tris(pH 7.5)–10% glycerol–2 mM dithiothreitol–0.5 mM EDTA. The column is washed with this buffer, and the enzyme is eluted with 0.2 M Tris(pH 7.5)–2 mM dithiothreitol–10% glycerol–0.1 mM TdR. This yields a preparation which is essentially homogenous with a specific activity of approximately 200 units/mg protein.

Properties

Specificity of the Reaction. The deoxyribonucleoside specificity of the enzyme was investigated using [^{14}C]ATP as the phosphorylating agent.[4] Equally effective substrates were TdR and the halogenated pyrimidine deoxyribonucleosides, 5-bromo-, 5-chloro-, and 5-iododeoxyuridine as well as 5-trifluoromethyl deoxyuridine. Deoxyuridine and 5-fluorodeoxyuridine were only 60% as effective; 6-azadeoxythymidine exhibited only 25% of the substrate efficacy as TdR. The following compounds were without substrate activity: deoxycytidine, 5-Br-deoxycytidine, 5-I-deoxycytidine, Br-uridine, I-uridine, purine ribonucleosides, uridine, and 5-F-deoxycytidine. The K_m for TdR has been determined to be 5.6 μM at pH 8.0 and was markedly dependent upon pH.[27]

TdR kinase required Mg^{2+} almost exclusively,[4] with Ca^{2+}, Co^{2+}, or Mn^{2+} exhibiting only 30% activity at comparable concentrations. The K_m for MgCl$_2$ is 0.8 mM.[4] The activity of TdR kinase is dependent upon the ATP concentration in such a fashion that Lineweaver-Burk plots exhibit bimodality.[27] At an ATP concentration <10^{-3} M, the K_m is 20 mM; at higher concentrations, the K_m for ATP is 2.6 mM.

pH Optimum. TdR kinase activity exhibited a maximum at pH 8.0[4] with a sharp decline at more acid pHs. The pK_m for TdR has been determined as a function of pH and a pK_E of 7.5 has been observed.[27]

Sedimentation Characteristics and Molecular Weight. The molecular weight of TdR kinase from regenerating rat liver has been determined

---

[27] E. Bresnick, K. D. Mainigi, R. Buccino, and S. S. Burleson, *Cancer Res.* **30**, 2502 (1970).

from its sedimentation in sucrose density gradients.[27] The enzyme has a sedimentation of 5.0 S and a molecular weight of approximately 81,000.

Inhibition of TdR Kinase. Enzyme activity[4] is markedly inhibited by Mn^{2+}, Co^{2+}, Ca^{2+}, Ni^{2+}, Sr^{2+}, Fe^{2+}, Ba^{2+}, or Zn^{2+}. TdR kinase is also inhibited by Be^{2+}, and this inhibition has been studied in some detail.[28] The K_i for Be^{2+} was 1.0 μM. Enzyme activity[4] was reduced by sulfhydryl reagents, p-hydroxymercuribenzoate, N-ethyl maleimide, iodoacetamide, and Hg^{2+}. In addition, the enzyme was sensitive to dioxane, ethylene glycol, urea, SDS, and methylguanidine sulfate.[4]

Enzyme activity with TdR as substrate was reduced in competitive fashion by a number of the other substrates. The K_i for the halogenated pyrimidine deoxyribonucleosides, 5-bromo-, 5-chloro-, and 5-iododeoxyuridine, was approximately 3–5 μM[4]; that of 5-trifluoromethyl deoxyuridine was also in this range.[29] The K_i for deoxyuridine and 5-fluorodeoxyuridine was 110 and 22 μM, respectively.[4]

TdR kinase is end-product inhibited by d-TTP in a noncompetitive fashion which is markedly dependent upon the ATP concentration.[4] At low concentrations of ATP, d-TTP inhibition was most profound.[4]

[28] K. D. Mainigi and E. Bresnick, *Biochem. Pharmacol.* **18**, 2003 (1969).
[29] E. Bresnick and S. S. Williams, *Biochem. Pharmacol.* **16**, 503 (1967).

[47] Thymidine Kinase from Blast Cells of Myelocytic Leukemia

By Y.-C. CHENG

$$dThd + ATP \xrightarrow{\text{Mg}^{2+}} dTMP + ADP$$

Assay Method

Principle. This is a radioactive assay based on the conversion of [^{14}C]dThd to [^{14}C]dTMP. Substrate and product have different adsorption properties on an anion-exchange paper disc and thus can be readily separated.

Reagents

REACTION MIXTURE. It contains 250 μM [^{14}C]dThd (specific activity of 12 μCi/nm), 2.5 mM ATP, 2.5 mM $MgCl_2$, 12.5 mM NaF, 2 mM

dithiothreitol, 4.5 mM phosphocreatine, 6 units/ml creatine kinase, and 1% bovine serum albumin in 0.19 M Tris·HCl (pH 7.5).

ENZYME. It is prepared in a 0.05 M Tris·HCl (pH 7.5) buffer containing 2 mM dithiothreitol, 50 μM dThd, and 10% glycerol (buffer S).

Procedure. Seventy-five microliters of reaction mixture are used in each assay. Five to 25 μl of enzyme solution, depending on the activity of the enzyme, are added and supplemented with buffer S to give a final volume of 0.1 ml for each assay. This is incubated at 37° for a half hour. If longer incubation is needed, then the tube is sealed with parafilm. The reaction is linear for at least 4 hr, providing at least 10% of dThd is still present in the reaction mixture. The reaction is stopped by spotting 50 μl of the mixture on a Whatman DE-81 paper disc; the disc is then dropped immediately into 95% alcohol (10 ml/disc) and washed 3 times with alcohol. Each washing should be carried out at least 5 min apart. The disc is then dried and placed in a vial containing 7.5 ml of a scintillation solution (PCS, Amersham/Searle). dThd is removed from the disc during the washing procedure. The blank value is obtained by carrying out the assay in the absence of the enzyme.

Definition of Unit. A unit is defined as the amount of enzyme which converts 1 nmole of dThd to dTMP per minute under our standard conditions.

Preparation of Affinity Gel for dThd Kinase. Cheng and Prusoff[1] in studies of mouse S-180 dThd kinase first suggested that gel with dThd linked through the 3′ position may be useful for the purification of mammalian dThd kinase. Kowal and Markus[2] coupled deoxythymidine-3′(4-amino-phenylphosphate) carboxyhexyl-sepharose and found this affinity column useful in purifying dThd kinase from rat colon adenocarcinoma; we used this system with minor modifications for the purification of dThd kinase from acute myelocytic leukemic cells. Seventy-five milligrams of deoxythymidine-3′(4-nitrophenylphosphate)[3] was dissolved in 30 ml of anhydrous methanol and reduced to deoxythymidine-3′(4-aminophenylphosphate) under 1 atm of PH$_2$ in the presence of 30 mg of 10% Palladium on carbon. The reduction was allowed to proceed for 4 hr, and then the catalyst was removed by filtration. About 90% of the starting material was converted to the reduced product. The filtrate was concentrated by lyophilization to 5 ml and loaded on a Dowex 50 column

[1] Yung-chi Cheng and W. H. Prusoff, *Biochemistry* **13**, 1179 (1974).
[2] E. P. Kowal and G. Markus, *Prep. Biochem.* **6**, 369 (1976).
[3] Obtained from Raylo Chemicals, Ltd., Alberta, Canada.

(H$^+$ form) with H$_2$O as the eluting solvent; this separated the unreacted material (not retained on the column) from the product. This purification step is not essential. The reduced product was linked to CH-sepharose gel (7 gm dry weight) in the presence of 0.1 M 1-ethyl-3-(3'-dimethyl aminopropyl) carbodimide by incubation at room temperature for 10 hr. During this period, the medium was readjusted frequently to pH 5.2. The gel was then washed with 0.05 M Tris·HCl (pH 7.5) containing 1 M NaCl and resuspended in 50% dimethylformamide aqueous solution containing 0.1 M galactosamine and 0.1 M 1-ethyl-3-(3'-dimethylamino-propyl) carbodimide for another 3 hr in order to block any unreacted carboxyl groups on the matrix. Finally, the gel was washed with 0.05 M Tris·HCl (pH 7.5) and 1 M NaCl and stored in the same buffer containing 0.1% NaN$_3$. There were 2.7 μmole of the dThd derivative bound per milliliter of wet gel. The gel could be reused.

Purification Procedure

Step 1. Acute myelocytic leukemia blast cells were obtained from patients after leukophoresis. The cells were exposed to hypotonic shock[1] to remove the contaminating red blood cells and then stored at $-70°$ until ready for use. To prepare the crude homogenate, the cells were resuspended in 4 volumes of an extraction buffer containing 0.01 M Tris·HCl (pH 7.5)–10% glycerol–0.15 M NaCl–20 μM dThd and 2 mM dithiothreitol, sonicated, and centrifuged at 39,000 rpm for 1 hr with a Beckman type-50 rotor. This is designated as the cell homogenate. There are two types of dThd kinase present in this preparation; one is cytoplasmic dThd kinase, and the other is mitochondrial dThd kinase. All the operations were performed at 4° unless a different temperature is indicated.

Step 2. Streptomycin Sulfate Fractionation. A 20% streptomycin sulfate solution was added drop by drop to the cell homogenate with stirring over a period of 30 min at 4°. After further stirring for 15 min, the solution was centrifuged at 12,000 rpm for 10 min, and the supernatant fluid was collected.

Step 3. Ammonium Sulfate Fractionation. Solid ammonium sulfate was slowly added to the above supernatant fluid with constant stirring to 20% saturation. After additional stirring for 15 min, the mixture was centrifuged at 12,000 rpm for 10 min. Additional ammonium sulfate was added to the supernatant fluid until a concentration of 50% saturation was reached. The solution was stirred for an additional 30 min, centrifuged at 12,000 rpm for 10 min, and the precipitate was resuspended in

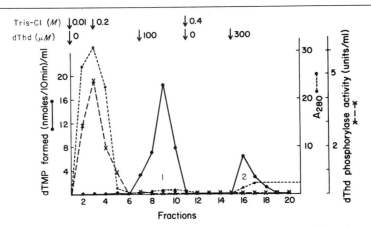

FIG. 1. Elution patterns of affinity column chromatography of dThd kinase from acute myelocytic leukemic blast cells. Fractions of 1.5 ml were collected. The column was eluted with Tris·HCl buffer (pH 7.5) containing dThd in concentrations as shown in the figure. Glycerol (10%) and dithiothreitol (5 mM) were added as stabilizers. Elution procedure is shown in the figure.

buffer A (0.01·M Tris·HCl (pH 7.5)–10% glycerol–2 mM dithiothreitol–0.5 mM EDTA).

Step 4. Affinity Column Chromatography. The above resuspended solution was applied to an affinity gel column (0.5 × 5 cm) which was equilibrated with buffer A. The column was eluted with buffers of different ionic strength containing different concentrations of thymidine which were supplemented with 10% glycerol and 5 mM dithiothreitol as indicated in Fig. 1; 1.5-ml fractions were collected. After loading the enzyme solution, the column was eluted with 0.1 M Tris·HCl with no thymidine. All unabsorbed proteins were eluted in the first four fractions. The first peak of dThd kinase was eluted at 0.2 M Tris·HCl and 100 μM dThd, which is the cytoplasmic enzyme, and the second at 0.4 M Tris·HCl and 300 μM dThd, which is the mitochondrial dThd kinase. These elutions were carried out at room temperature.[4] A summary of the purification is given in Table I.

Comments on Purification. The use of affinity column chromatography for purifying dThd kinase has three advantages: (1) there is a high recovery of the enzyme; (2) the procedure is rapid; and (3) it separates the isozymes. It is important that the ammonium sulfate precipitation procedure and affinity column chromatography should be done on the

[4] L. S. Lee and Yung-chi Cheng, *J. Biol. Chem.* **251**, 2600 (1976).

TABLE I
PURIFICATION OF dThd KINASE[a]

Procedure	Volume (ml)	Units/ml	Protein (units/mg)	Yield (%)	Purification (-fold)
1. Homogenate	20	0.874	0.076	100	1
2. Streptomycin sulfate fractionation	20	0.847	0.051	97	0.67
3. Ammonium sulfate fractionation (20–50%)	3	2.648	0.125	46	1.64
4. Affinity column chromatography					
Peak 1	1.5	1.837	183.7	36	2417
Peak 2	1.5	0.621	124.2	9	1634

[a] Enzyme activities of peaks 1 and 2 in the affinity column were measured at the peak position, and the yield was calculated by considering the total activities of the peaks.

same day; this is because there is no dThd to protect enzyme activity in the buffer in which the ammonium sulfate precipitate was dissolved. The enzyme can be stored on the column overnight without losing activity. This column can also be used to purify human cytoplasmic dCyd kinase which is retained on the column and eluted from the column by 0.2 M Tris·HCl (pH 7.5) buffer.[5] The same column can be used for purification of dThd kinase from mouse or rat cells as well as herpes simplex virus type I and II specific dThd kinase.[6]

Properties

Physical Properties. Molecular weight determinations were done by the glycerol gradient centrifugation technique. Electrophoresis was carried out by the method described previously.[4] The activation energy was determined by studying the reaction velocity between 25° and 40°. Table II summarizes the properties of cytoplasmic and mitochondrial dThd kinases.

Stability. Purified mitochondrial and cytoplasmic dThd kinases were quite stable in the absence of their substrates, and either substrate was only partially effective in stabilizing the enzyme. Dithiothreitol was essential for the stability of both dThd kinases in the presence of the substrate. Addition of 0.5 mg/ml bovine serum albumin to the purified enzyme also helped to stabilize the enzyme.

[5] Yung-chi Cheng, B. Domin, and L. S. Lee, *Biochim. Biophys. Acta* **481,** 481 (1977).
[6] Yung-chi Cheng and M. Ostrander, *J. Biol. Chem.* **251,** 2605 (1976).

TABLE II
PHYSICAL PROPERTIES OF THYMIDINE KINASES ISOLATED FROM DIFFERENT
SUBCELLULAR FRACTIONS

	Cytoplasmic	Mitochondrial
Molecular weight	90,000	70,000
Activation energy (kcal/mole)	15.17	10.93
Electrophoretic mobility (R_f)	0.0	0.6
pH optima	7.6–7.8	7.4–7.8

Divalent Salt Requirements. There is an absolute requirement of a divalent cation for the action of both enzymes. Mg^{2+} was the best divalent cation tested. In order of activity $Mg^{2+} > Ca^{2+} > Mn^{2+} > Cu^{2+} > Fe^{2+} > Zn^{2+}$ worked for the cytoplasmic dThd kinase, and $Mg^{2+} > Cu^{2+} > Mn^{2+} > Ca^{2+} > Fe^{2+} = Zn^{2+}$ worked for the mitochondrial dThd kinase. The divalent cation chelates with the phosphate donor, and this complex serves as the true substrate of the reaction instead of free ATP.

Substrate Specificity. (a) Phosphate donor: ATP and dATP are the best substrates among all of the triphosphate nucleotides tested for the cytoplasmic deoxythymidine kinase. Both substrates have the same K_m (0.2 mM) and V_{max}. They act as cooperative-type substrates with a Hill constant of 2. For mitochondrial deoxythymidine kinase, ATP is the best substrate with a K_m of 0.1 mM. It acted as a Michaelis-Menten type substrate. dATP has the same K_m but can give only 56% of the V_{max}. All the triphosphate ribonucleotides gave a higher V_{max} for the reaction catalyzed by the mitochondrial enzyme than the corresponding triphosphate deoxyribonucleotides. (b) Phosphate receptor: Mitochondrial deoxythymidine kinase could use either deoxythymidine (K_m = 5.2 μM) or deoxycytidine (K_m = 3 μM) as a substrate; in contrast, cytoplasmic deoxythymidine kinase could use deoxythymidine (K_m = 2.6 μM) but not deoxycytidine as substrate. (c) Analog studies: A number of deoxythymidine and deoxycytidine analogs modified at the 5 position were studied for their binding affinity to both isozymes. It was observed that both isozymes lose their binding affinity to deoxyuridine with an alkyl group larger than methyl substituted at this position. I^5-dCyd, Br^5-dCyd, vinyl^5-dUrd and ethynyl^5-dUrd could act as substrates for mitochondrial deoxythymidine kinase, but not for cytoplasmic deoxythymidine kinase. Ara-cyd could not be phosphorylated by the mitochondrial enzyme. I^5-dUrd and Br^5-dUrd were substrates for both enzymes.

Reaction Mechanism. Initial velocity and product inhibition studies suggested that the major mechanism for the reaction catalyzed by cytoplasmic deoxythymidine kinase was "sequential," whereas for the mitochondrial enzyme the "ping-pong" mechanism seems to be operative when deoxythymidine was used as the substrate, and the "sequential" mechanism when deoxycytidine was used as the substrate.[5,7]

Inhibitor. Cytoplasmic deoxythymidine kinase could be inhibited by dTTP, which served as competitive inhibitor with respect to ATP. The mitochondrial enzyme could be inhibited by either dTTP (K_I = 10 μM) or dCTP (K_I = 2 μM).

Acknowledgments

This work was supported, in part, by USPHS Grants CA-13038 and CA-05298 and American Cancer Society Grant CH-29. Y. C. Cheng has been an American Leukemia Society Scholar.

[7] L. S. Lee and Yung-chi Cheng, *Biochemistry* **15**, 3686 (1976).

[48] Nucleoside Diphosphokinase from *Salmonella typhimurium*

By JOHN L. INGRAHAM and CHARLES L. GINTHER

$$\text{Nucleoside}_1\text{ triphosphate} + \text{nucleoside}_2\text{ diphosphate} \xrightarrow{\text{Mg}^{2+}}$$
$$\text{nucleoside}_1\text{ diphosphate} + \text{nucleoside}_2\text{ triphosphate}$$

Nucleoside diphosphokinase plays a central role in the synthesis of nucleoside triphosphates from corresponding nucleoside diphosphates. With the exception of ATP and a small portion of GTP, all other triphosphates are synthesized by this enzyme.[1]

Assay Methods

Principle. Two sensitive radiochemical assays for enzyme activity have recently been developed. One (method A) measures the transfer of radioactivity from [γ-^{32}P]ATP to nucleoside diphosphates and is convenient when relative activities of the enzyme towards a number of diphosphate substrates are compared. The other (method B) measures one activity of the enzyme: the conversion of [2-^{14}C]UDP to [2-^{14}C]UTP

[1] C. L. Ginther and J. L. Ingraham, *J. Bacteriol.* **118**, 1020 (1974).

at the expense of the phosphoryl group of ATP. Method B offers the advantages associated with employing a radiochemical substrate with longer half-life.

In spite of the presence of interfering reactions, both methods of assay can be used to estimate activities of nucleoside diphosphokinase in crude extracts of *Salmonella typhimurium* because the reaction product is purified prior to measurement.

METHOD A

> *Buffer A*: 100 mM triethanolamine-HCl, pH 8.0, containing 10 mM MgCl$_2$ and 2 mM mercaptoethanol
> *Reagents* (prepared in triethanolamine buffer):
> [γ-^{32}P]ATP, 25 mM ($\sim$2 μCi/μmole)
> Nucleoside diphosphate, 4 mM
> *Reagents* (for chromatography):
> absolute methanol
> KH$_2$PO$_4$, 0.85 M (pH 3.4)

Procedure. Small test tubes containing 10 μl of [γ-^{32}P]ATP, 10 μl nucleoside diphosphate, and 10 μl triethanolamine buffer are placed in a 37° constant-temperature bath for 2 min; then the reaction is initiated by adding 10 μl of an appropriate dilution of enzyme in triethanolamine buffer. After 3 min of incubation at 37°, the reaction is stopped by applying 25 μl of the mixture to poly(ethyleneimine) thin-layer plates[2] and immediately drying the spots. The plates are washed with absolute methanol for 20 min, dried, and developed with: (1) absolute methanol to the origin; (2) 0.85 M KH$_2$PO$_4$ to 15 cm above the origin.

Following drying in a stream of warm air, the plates are placed against X-ray film (Kodak Royal Blue) for 4–8 hr.

The developed X-ray film is used to locate [γ-^{32}P]nucleoside triphosphate. These areas are cut from the chromatograms (plastic-backed thin-layer plates facilitate removal of appropriate areas), placed in a vial containing 7 ml of Bray's solution,[3] and counted in a scintillation spectrometer.

METHOD B

> *Buffer*: Same as method A.
> *Reagents* (prepared in triethanolamine buffer):
> ATP, 25 mM
> [2-^{14}C]UDP, 4 mM (0.5–1.25 μCi/μmole)

[2] K. Randerath and E. Randerath, *J. Chromatogr.* **22**, 110 (1966).
[3] G. A. Bray, *Anal. Biochem.* **1**, 279 (1960).

Reagents (for chromatography:
Absolute methanol
Sodium formate, 2 M, pH 3.4
Sodium formate, 4 M, pH 3.4

Procedure. The reaction is run as in method A substituting 25 mM ATP for 25 mM [γ-^{32}P]ATP and 4 mM [2-^{14}C]UDP for 4 mM nucleoside diphosphate. After spotting and drying, poly(ethyleneimine) plates are developed: (1) to the origin with absolute methanol; (2) to 2 cm above the origin with 2 M sodium formate; (3) to 10 cm above the origin with 4 M sodium formate. Radioautography and counting were done as in method A except exposure was for 8–18 hr.

Definition of Unit and Specific Activity. One unit of nucleoside diphosphokinase is defined as the amount of enzyme that catalyzes the formation of 1 μmole of UTP from UDP and ATP in 1 min.

Purification Procedure[4]

Reagents

Culture medium: basal salts medium (007)[5] containing 0.2% glucose and 0.15% vitamin-free casein hydrolysate (Nutritional Biochemical Corp.)
Standard buffer (buffer B): 100 mM Tris·HCl (pH 7.8)–10 mM MgCl$_2$–2 mM mercaptoethanol
Aged calcium phosphate (Calbiochem)
DEAE-cellulose (Whatman DE52)
Sephadex G-200 (Pharmacia)
DEAE-Sephadex (Pharmacia)

Growth of Bacteria and Preparation of Extracts. Starter cultures of *Salmonella typhimurium* grown overnight at 37° in 125 Erlenmeyer flasks on a rotary shaker are used to inoculate 4-liter Erlenmeyer flasks containing 2 liters of culture medium. Care must be taken to maintain cultures in a highly aerobic state because specific activity of nucleoside diphosphokinase is decreased over 10-fold in cells grown under conditions of fermentation rather than respiration. After 8–10 hr of growth (when the culture has reached stationary stage) the culture is harvested in a refrigerated Sharpless supercentrifuge. About 77 g of cells (wet weight) are obtained from 40 liters of culture. Cells are then washed with 200 ml of chilled 0.9% NaCl, resuspended in 200 ml of buffer B,

[4] C. L. Ginther and J. L. Ingraham, *J. Biol. Chem.* **249**, 3406 (1974).
[5] D. J. Clark and O. Maaløe, *J. Mol. Biol.* **23**, 99 (1967).

sonicated in a Biosonic II sonic oscillator in 50-ml portions for 10 × 30 sec, and clarified by centrifugation at 15,000 g for 10 min.

Calcium Phosphate Precipitation. Aged calcium phosphate buffer B (2%) is added in four 80–100 ml portions to the crude extract. After stirring 30 min at 0°, the mixture is centrifuged at 5000 g for 10 min.

Ammonium Sulfate Fractionation. The supernatant solution after calcium phosphate precipitation is fractionated by the addition of solid ammonium sulfate. The material which precipitates between 40–60% saturation is collected by centrifugation at 30,000 g for 20 min, dissolved in about one-tenth the original volume of buffer B, and dialyzed for 5 hr against 20 volumes of the same buffer.

Chromatography on DEAE-Cellulose. The dialyzed ammonium sulfate fraction is applied to a 2.5 × 40 cm column of DEAE-cellulose which was previously equilibrated against buffer B. The column is developed with a concave gradient of KCl in buffer B at a flow rate of 18 ml/hr. Fractions eluting between 0.25–0.27 M KCl are pooled.

Second Ammonium Sulfate Fractionation. Material from the DEAE-cellulose column is fractionated again by adding solid ammonium sulfate. The protein that precipitates between 50–85% saturation is collected by centrifugation at 30,000 g for 20 min and dissolved in 2 ml of buffer B.

Chromatography on Sephadex G-200. The ammonium sulfate fraction is applied to a 1.6 × 100 cm column of Sephadex G-200 and eluted with buffer B. The elution volume is 112 ml. Active fractions were pooled.

DEAE-Sephadex. Pooled fractions from Sephadex G-200 are applied

TABLE I

PURIFICATION OF NUCLEOSIDE DIPHOSPHOKINASE

Fraction	Volume (ml)	Total protein (mg)	Total activity (units)	Specific activity (units/mg)
Crude extract	210	8400	24,100	2.87
Calcium phosphate supernatant	530	1330	16,300	12.3
$(NH_4)_2SO_4$ fraction	40	500	9700	19.4
DEAE-cellulose eluate	51	28	3240	115
$(NH_4)_2SO_4$ fraction	1.8	5	2020	330
Sephadex eluate	11	1.3	1320	1010
DEAE-Sephadex eluate	28	0.28	551	1970

TABLE II
ACTIVITY OF NUCLEOSIDE DIPHOSPHOKINASE (ASSAY METHOD A) WITH VARIOUS
NUCLEOSIDE DIPHOSPHATES[4]

Nucleoside diphosphate	Specific activity (moles/min/mg protein)
ADP	1690
UDP	1790
GDP	1110
CDP	1650
dADP	1050
dGDP	990
dCDP	1100
XDP[a]	670
IDP[b]	760
dUDP	640
dTDP	480

[a] Xanthosine diphosphate.
[b] Inosine diphosphate.

to a 2.6 × 26 cm column of DEAE-Sephadex and eluted with a concave gradient of 0–0.275 M KCl in buffer B. Nucleoside diphosphokinase elutes with 0.23 M KCl.

The results of this purification are summarized in Table I.

Properties

Physical Properties. The purified enzyme does not appear to lose activity at −20° over a period of months but at 25° its half-life in the absence of substrates is only 30 min. As judged by migration on Sephadex G-200 the molecular weight of the native enzyme is estimated to be 85,000.

Catalytic Properties. Mg^{2+} is required for activity; 1 mM free Mg^{2+} appears to be optimal. Apparently the true substrate is nucleotide Mg^{2+}.[6]

The pH optimum is quite broad, the activity being approximately equal over the range of pH 6.5–9.0. A thiol group is required for activity.

The mechanism of reaction is ping-pong: phosphorylated enzyme is an intermediate of the reaction.

The striking property of the enzyme is its broad substrate spectrum. When assay method A is used at least 11 nucleoside diphosphates are active as phosphoryl acceptors with similar activities (Table II).

[6] N. Mourad and R. E. Parks, *J. Biol. Chem.* **241**, 271 (1966).

[49] Nucleoside Diphosphokinase from Human Erythrocytes

By R. P. AGARWAL, BONNIE ROBISON, and R. E. PARKS, JR.

The term "nucleoside diphosphokinase" (NDP kinase)[1] is applied to a family of enzymes that catalyze the transfer of the terminal phosphate groups of 5′-triphosphate nucleotides to 5′-diphosphate nucleotides as follows:

$$N_1TP + E \rightleftharpoons N_1DP + E \sim P$$
$$N_2DP + E \sim P \rightleftharpoons N_2TP + E$$
$$\overline{N_1TP + N_2DP \rightleftharpoons N_1DP + N_2TP}$$

where N_1 and N_2 are purine or pyrimidine ribo- or deoxyribonucleosides. All NDP kinases function through the formation of enzyme-bound high-energy phosphate intermediates.[2-4]

Assay Method

Principle. The enzyme may be assayed spectrophotometrically by following the formation of ADP from ATP in a coupled pyruvate–lactate dehydrogenase system[5,6] containing dTDP,[7] NADH, phosphoenol pyruvate (PEP), lactate dehydrogenase (LDH), and pyruvate kinase (PK) according to the following reactions:

$$dTDP + ATP \xrightarrow{\text{NDP kinase}} dTTP + ADP$$
$$ADP + PEP \xrightarrow{\text{PK}} ATP + pyruvate$$
$$Pyruvate + NADH \xrightarrow{\text{LDH}} lactate + NAD^+$$

The rate of oxidation of NADH is followed by measuring the decrease in absorbancy at 340 nm.[8]

[1] NDP kinase, nucleoside diphosphokinase, ATP:nucleoside diphosphate phosphotransferase, EC 2.7.4.6.

[2] N. Mourad and R. E. Parks, Jr., *Biochem. Biophys. Res. Commun.* **19**, 312 (1965).

[3] N. Mourad and R. E. Parks, Jr., *J. Biol. Chem.* **241**, 3838 (1966).

[4] R. E. Parks, Jr. and R. P. Agarwal, *in* "The Enzymes" (P. D. Boyer, ed.), 3rd ed., Vol. 8, p. 307. Academic Press, New York, 1973.

[5] N. Mourad and R. E. Parks, Jr., *J. Biol. Chem.* **241**, 271 (1966).

[6] R. P. Agarwal and R. E. Parks, Jr., *J. Biol. Chem.* **246**, 2258 (1971).

[7] The relative low reactivity of dTDP with pyruvate kinase makes this assay possible.

[8] To minimize the errors due to stray light, a blue filter may be used when a tungsten lamp is employed at 340 nm [see R. L. Cavalieri and H. Z. Sable, *Anal. Biochem.* **59**, 122 (1974); R. L. Miller, D. L. Adamczyk, T. Spector, K. C. Agarwal, R. P. Miech, and R. E. Parks, Jr., *Biochem. Pharmacol.* **26**, 1573 (1977).

Reagents[6]

1. Tris-acetate buffer, 0.8 M, pH 7.5
2. PEP, 0.03 M, adjusted to pH 7.5
3. ATP, 0.02 M, adjusted to pH 7.5
4. Pyruvate kinase, 12.5 units/ml, dissolved in 1% bovine plasma albumin solution in 0.1 M Tris-acetate, pH 7.5
5. Lactate dehydrogenase, 22.5 units/ml, dissolved in 1% bovine plasma albumin solution in 0.1 M Tris-acetate, pH 7.5
6. MgCl$_2$, 0.1 M
7. KCl, 0.25 M
8. NADH, 0.003 M, prepared fresh
9. dTDP, 0.004 M
10. NDP kinase, appropriately diluted with 0.1 M Tris-acetate, pH 7.5

The Assay Mixture. An assay mixture is prepared from equal volumes of solutions 1–7. This mixture may be stored frozen for several months. One volume of freshly prepared NADH solution (number 8) is added to 7 volumes of the above mixture on the day of the experiment. This reaction mixture is stable for at least 2 days if kept at 4°.

Procedure. A 1.5-ml quartz cuvette (diam = 1.0 cm) containing 0.8 ml of assay mixture and 0.1 ml NDP kinase (NDP kinase and H$_2$O) is placed in the cell compartment of a recording UV spectrophotometer for 4–5 min for temperature equilibration and to reach a constant background reaction rate.[9] After this incubation time the reaction is started by addition of 0.1 ml dTDP solution, and the rate of decrease in A_{340} is followed for 3–4 min.

A linear rate is recorded and corrected for background rates due to the enzyme[9] and the substrate.[10]

Unit of Enzyme. One unit is the amount of the enzyme that catalyzes the formation of 1 μmole of ADP ($-A_{340}$ = 6.2 cm^2) per minute under the standard conditions of assay. Specific activity is defined as units per

---

[9] Contaminating enzymes which break down ATP, PEP, or NADH may result in increased background rates. With the purified enzyme the background rate is almost zero. However, it must be determined at least once.

[10] This background is caused by the slight substrate activity of a diphosphate nucleotide substrate (e.g., dTDP) with pyruvate kinase. Therefore, careful attention must be given to addition of correct concentration of a diphosphate nucleotide and pyruvate kinase. Under the conditions of standard assay, a decrease in absorbance of 0.01 to 0.02 per min is observed.

milligram of protein. The protein is determined by UV absorption at 280 nm.[11]

Alternative methods for the measurement of NDP kinase activity include isotopic, staining, and coupled enzymic procedures.[4]

Purification Procedure

Relatively recently, the marked isoelectric variation of human erythrocytic NDP kinases was discovered.[12] Prior to this finding, however, methods of purification of the enzyme were developed which, as is now apparent, yielded different results due to isoelectric variability.[6] Therefore, to appreciate the problems involved in the purification of human erythrocytic NDP kinases, a brief review seems warranted.

The earliest attempts to purify human erythrocytic NDP kinase involved use of DEAE-cellulose (phosphate) as the initial step to separate enzyme from hemoglobin.[5] This resulted in approximately 50% recovery of the original activity. Further purification of this preparation yielded a homogeneous NDP kinase with a specific activity of about 70.[5] Subsequent to the above studies, adsorption of NDP kinase on calcium phosphate gel was introduced, which enabled separation of the enzyme from hemoglobin with recovery of greater than 95% of the original activity. When NDP kinase isolated by the calcium phosphate gel procedure was purified further to apparent homogeneity, an enzyme fraction with a specific activity of about 1000 was obtained.[6] This marked discrepancy in the specific activities of two apparently homogeneous preparations of erythrocytic NDP kinase was explained when the technique of isoelectric focusing was introduced. Here it was shown that erythrocytic NDP kinase is comprised of at least six distinct peaks of enzymic activity ranging from pI values of 5.4–8.3 (Fig. 1).[12] Upon reexamination it was learned that the NDP kinase first isolated (specific activity 70) was principally the pI 5.8 variant, and the later preparation (specific activity about 1000) was principally the pI 7.3 variant.[6] It should be noted that marked isoelectric variation has been observed with all NDP kinases examined to date in erythrocytes from a wide range of animal species, in various organs, and in the subcellular fractions of rat liver.[12-14] Therefore, it appears likely that any attempt to purify NDP kinase from any animal source will be complicated by the

---

[11] O. Warburg and W. Christian, *Biochem. Z.* **310**, 384 (1941); also see Vol. 3 [73].

[12] Y.-C. Cheng, R. P. Agarwal, and R. E. Parks, Jr., *Biochemistry* **10**, 2139 (1971).

[13] R. E. Parks, Jr., P. R. Brown, Y.-C. Cheng, K. C. Agarwal, C. M. Kong, R. P. Agarwal, and C. C. Parks, *Comp. Biochem. Physiol. B* **45**, 355 (1973).

[14] Y.-C. Cheng, B. Robison, and R. E. Parks, Jr., *Biochemistry* **12**, 5 (1973).

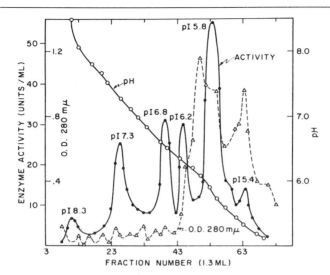

Fig. 1. Electrofocusing profile of human erythrocytic nucleoside diphosphokinase isolated from pooled blood. From Cheng *et al.*[12] Reprinted with permission from *Biochemistry* **10**, 2139 (1971). Copyright by the American Chemical Society.

occurrence of marked isoelectric variability. Another point worth noting is that with all tissues examined by this laboratory, the use of calcium phosphate gel adsorption as an initial step in purification has yielded virtually quantitative recovery of enzymic activity with excellent purification.

For the purification of human erythrocytic NDP kinases, the general method of Chapter 79 has proved highly satisfactory when steps 1 through 3 are followed.[15] This procedure has been employed at least four times on quantities of human blood of approximately 25 liters.[16] Overall recoveries of NDP kinase activity in the order of 60% with approximately 300-fold purification were achieved in most instances. In two preparations, it was possible to isolate from step 3 (calcium phosphate gel:cellulose column chromatography) three distinct fractions of enzymic activity that consisted of different mixtures of isozymic variants as follows:

Fraction 1, pIs 5.8 (major), 5.4, and 6.3
Fraction 2, pIs 6.3 (major), 5.8, and 6.8
Fraction 3, pIs 7.3 (major), 6.8, and 8.3

Further purification of fractions 2 and 3 is illustrated in the flow

[15] R. P. Agarwal, K. C. Agarwal, and R. E. Parks. Jr., this volume [79].
[16] The procedure has been performed successfully by the Enzyme Center at Tufts University School of Medicine, Boston, Massachusetts.

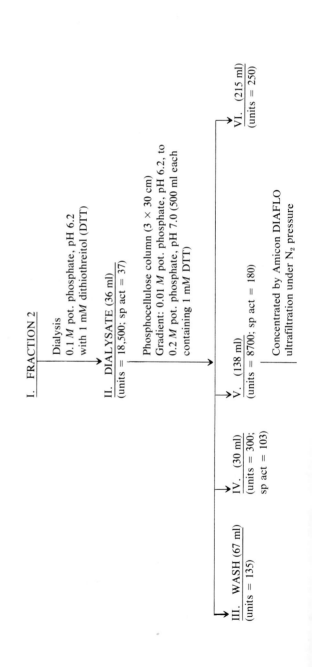

I. FRACTION 2

Dialysis
0.1 M pot. phosphate, pH 6.2
with 1 mM dithiothreitol (DTT)

II. DIALYSATE (36 ml)
(units = 18,500; sp act = 37)

Phosphocellulose column (3 × 30 cm)
Gradient: 0.01 M pot. phosphate, pH 6.2, to
0.2 M pot. phosphate, pH 7.0 (500 ml each
containing 1 mM DTT)

III. WASH (67 ml)
(units = 135)

IV. (30 ml)
(units = 300;
sp act = 103)

V. (138 ml)
(units = 8700; sp act = 180)

Concentrated by Amicon DIAFLO
ultrafiltration under N₂ pressure

VI. (215 ml)
(units = 250)

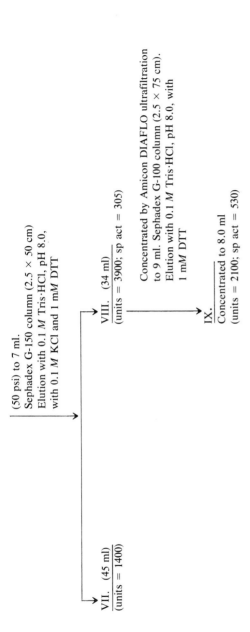

FIG. 2. Further purification of the "fraction 2" of human erythrocytic nucleoside diphosphokinase. The fraction at step IX consists of pI 6.3 variant with a small contamination of pI 5.8 variant.

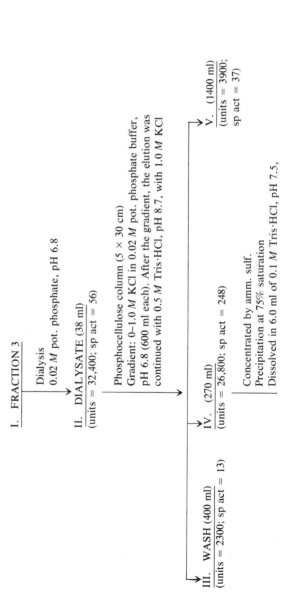

I. FRACTION 3

Dialysis
0.02 M pot. phosphate, pH 6.8

II. DIALYSATE (38 ml)
(units = 32,400; sp act = 56)

Phosphocellulose column (5 × 30 cm)
Gradient: 0–1.0 M KCl in 0.02 M pot. phosphate buffer, pH 6.8 (600 ml each). After the gradient, the elution was continued with 0.5 M Tris·HCl, pH 8.7, with 1.0 M KCl

III. WASH (400 ml)
(units = 2300; sp act = 13)

IV. (270 ml)
(units = 26,800; sp act = 248)

Concentrated by amm. sulf.
Precipitation at 75% saturation
Dissolved in 6.0 ml of 0.1 M Tris·HCl, pH 7.5,

V. (1400 ml)
(units = 3900; sp act = 37)

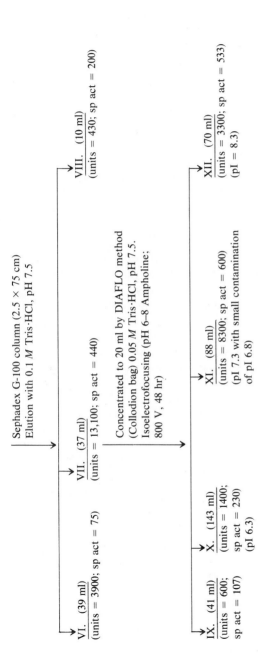

VI. (39 ml)
(units = 3900; sp act = 75)

Sephadex G-100 column (2.5 × 75 cm)
Elution with 0.1 M Tris·HCl, pH 7.5

VII. (37 ml)
(units = 13,100; sp act = 440)

VIII. (10 ml)
(units = 430; sp act = 200)

Concentrated to 20 ml by DIAFLO method
(Collodion bag) 0.05 M Tris·HCl, pH 7.5.
Isoelectrofocusing (pH 6–8 Ampholine;
800 V, 48 hr)

IX. (41 ml)
(units = 600;
sp act = 107)

X. (143 ml)
(units = 1400;
sp act = 230)
(pI 6.3)

XI. (88 ml)
(units = 8300; sp act = 600)
(pI 7.3 with small contamination
of pI 6.8)

XII. (70 ml)
(units = 3300; sp act = 533)
(pI = 8.3)

FIG. 3. Purification of the "fraction 3" of human erythrocytic nucleoside diphosphokinase.

diagrams of Figs. 2 and 3. As illustrated in these diagrams, the use of isoelectric focusing as a final step in purification made possible isolation of substantial amounts of NDP kinase variants of pI values of 8.3, 7.3, and 6.3 at specific activities of 500 or greater. Most of the procedures employed are described elsewhere in this volume (Chapters 64, 67, 72) with the exception of phosphocellulose chromatography.

Phosphocellulose Column Chromatography. Phosphocellulose is washed free of fines and impurities and converted to the K^+ form.[17] The slurry is packed in a column and equilibrated with 0.02 M potassium phosphate, pH 6.8. The dialyzed enzyme solution is added to the column, washed with the same buffer (approx. 300 ml), and the gradient elution is performed as described in Figs. 2 and 3. Ten-milliliter fractions are collected, and the enzymic activity is determined. The fractions with highest enzymic activity are pooled.

Properties

Molecular Weight, Substrate Specificity, and Response to Temperature. All six isoelectric variants of human erythrocytic NDP kinase follow ping-pong reaction sequences.[4,18] The formation of a high-energy, phosphorylated-enzyme intermediate was first demonstrated by incubating erythrocytic NDP kinase with ^{32}P-labeled ATP.[2] Subsequently, all NDP kinase preparations examined followed a similar ping-pong reaction sequence with the formation of phosphorylated enzyme intermediates. Histidine residues, phosphorylated at the 1 or 3 positions, have been isolated by column chromatography of amino acid digests of these

---

[17] Some commercial preparations of phosphocellulose that are colored completely inactivate NDP kinase. Therefore, prior to use the phosphocellulose must be cleaned. Twenty-five grams phosphocellulose (0.82 meq/g; Sigma Chemical Co., St. Louis, Missouri) is washed free of fines by resuspending in distilled water. An unidentified dark yellow colored material is removed during this procedure. The slurry is poured onto a Buchner funnel and filtered with suction until a semisolid filter cake is obtained which is then suspended in a solution of 0.5 N NaOH containing 0.5 M NaCl (approximately 2 liters), filtered through a Buchner funnel, and washed repeatedly with the same solution to remove any remaining yellow color (about 6 liters of solution are required). The filter cake is washed with distilled water until filtrate approaches neutrality. The filter cake is then suspended in 0.1 N HCl (about 2 liters), filtered, and washed repeatedly with 0.1 N HCl (6 liters) until the pH of the filtrate drops to 1.0. The filter cake is then suspended in distilled water and the pH is adjusted to about 7.0 by addition of KOH solution. The slurry is filtered through a Buchner funnel and washed 3–4 times with 0.5 M KCl solution, followed by washing three times with 0.01 M KCl, and finally 3 times with 0.02 M potassium phosphate buffer, pH 6.8. The semisolid cake is suspended in 0.02 M potassium phosphate buffer, pH 6.8, for further use.

[18] W. W. Cleland, *Biochim. Biophys. Acta* **67**, 173 and 188 (1963).

phosphorylated enzymes.[19,20] When the molecular weights of the isoelectric variants of the human erythrocytic NDP kinase were estimated by molecular sieving, values ranging from 80,000–100,000 were obtained.[12] Furthermore, although each isoelectric variant was shown capable of reacting with either purine or pyrimidine ribonucleotides or deoxyribonucleotides, substantial differences in K_m values were observed.[12] In addition to the many natural nucleotides that serve as substrates, the pI 5.8 variant has been shown capable of reacting with the analog nucleotides 8-azaUDP and 5-F-UTP.[5] Examination of temperature dependence revealed linear Arrhenius plots with isoelectric variants pI 5.4, 5.8, 6.3, and 6.8. By contrast, the pI 7.3 and 8.3 isoelectric variants displayed biphasic Arrhenius plots with downward curvatures and break points at about 31° (see the table).[6,12] These observations and others have suggested strongly that the NDP kinases are in fact a family of related enzymes that function via a similar reaction mechanism but are not classic isozymes,[4] e.g., of the lactate dehydrogenase type.

Metal Requirements. The erythrocytic NDP kinases require divalent cations. With the pI 7.3 variant, Mg^{2+}, Mn^{2+}, and Co^{2+} are approximately equally effective, whereas Ca^{2+} is about 50% as active. On the other hand, Zn^{2+} and Cu^{2+} were without activity.[6] No evidence for a monovalent cation requirement could be detected.

Thiol Group Activity. The erythrocytic NDP kinases have essential sulfhydryl groups as demonstrated by titration with PCMB, which is

SOME BIOCHEMICAL AND PHYSICAL PROPERTIES OF ERYTHROCYTIC NUCLEOSIDE
DIPHOSPHOKINASE VARIANTS[a]

Variant (pI)	K_m value $\times 10^3 M$				Mol wt	Arrhenius plot
	ATP	GTP	dTDP	CDP		
5.4	0.20	0.20	0.11	0.21	80,000	Linear
5.8	1.00	0.14	0.55	1.00	93,000	Linear
6.3	3.00	0.10	0.22	0.18	84,000	Linear
6.8	0.25	0.05	0.20	1.10	80,000	Linear
7.3	0.08	0.16	0.30	0.25	84,000	Diphasic
8.3	0.17	0.08	0.12	0.50	100,000	Diphasic

[a] From Cheng et al.[12] Reprinted with permission from *Biochemistry* **10**, 2139 (1971). Copyright by the American Chemical Society.

[19] O. Wålinder, *J. Biol. Chem.* **243**, 3947 (1968).
[20] Our unpublished results.

capable of completely inactivating the enzyme. However, complete reactivation occurs with the addition of dithiothreitol.[5,6] With both the pI 5.8 and 7.3 variants, marked protection against inactivation by PCMB was effected by the addition of ATP or dTDP in the presence or absence of Mg^{2+}.[5,6] When inactivation of the pI 7.3 variant by PCMB was carried out in the presence of 2 M urea, the enzyme could not be reactivated by the addition of dithiothreitol.[6]

Effect of pH. The effect of pH on apparent K_m and V_{max} values was determined with the pI 7.3 variant. Negligible effects on K_m values were observed between pH 5.5–9.0. The V_{max} values remained constant between pH 6.5–9.0 with decreased activities at lower and higher pH values.[6]

Comment. The physiological role of NDP kinases in human erythrocytes remains a mystery. The activity of this enzyme is among the highest found in this cell. However, the nucleotide profile of the adult human erythrocyte is perhaps the simplest of any animal cell examined to date, consisting predominantly of the mono-, di-, and triphosphate ribonucleotides of adenine, with very small quantities of guanine nucleotides.[21] In most extracts of human erythrocytes examined by high-pressure liquid chromatography, UTP and CTP are not detectable.[21] Therefore, one must question the need for multiple NDP kinase isoelectric variants that have different responses to nucleotide substrates in cells that contain few nucleotides other than those of adenine. It has been speculated that NDP kinases may represent vestigial enzymes that served vital metabolic functions at an early stage of erythrocytic formation and maturation. On the other hand, it is also possible that these NDP kinase isoelectric variants play an as yet undetected metabolic role in such vital functions as membrane transport, glutathione synthesis, etc. It must be appreciated that any enzyme that functions via a high-energy phosphate-enzyme intermediate that has low specificity for nucleotide substrates is capable of catalyzing the NDP kinase reaction.[4]

[21] E. M. Scholar, P. R. Brown, R. E. Parks, Jr., and P. Calabresi, *Blood* **41**, 927 (1973).

[50] Nucleoside Phosphotransferase from Carrot

By ELINOR F. BRUNNGRABER

3'AMP + Adenosine $\rightarrow$ Adenosine + 5'AMP
3'AMP + H_2O $\rightarrow$ Adenosine + P_i
Phenylphosphate + uridine $\rightarrow$ 5'UMP + phenol
Phenylphosphate + H_2O $\rightarrow$ Phenol + P_i

Assay Method

The nucleoside phosphotransferase (EC 2.7.1.77) isolated from carrot is a protein carrying both transfer and hydrolytic functions.[1,2] It catalyzes the phosphate transfer from a suitable donor such as a 3' or 5' mononucleotide to a nucleoside acceptor forming a new 5' nucleotide. In the absence of a suitable acceptor, the phosphate is released as inorganic phosphate. A second independent hydrolytic function is also present. The transfer function is assayed by determining the amount of 5' nucleotide formed when the enzyme is incubated with a nucleoside and a suitable phosphate donor. Uridine was originally chosen as the acceptor and phenylphosphate as the donor in the standard assay procedure. In later studies p-nitrophenylphosphate was used since it was as efficient a donor as phenylphosphate and had the advantage of providing a visible indicator with the formation of the yellow p-nitrophenol. However, for accurate measurement of the transferase function it is important to determine the nucleotide rather than the formation of the phenol because of the independent hydrolase function of the nucleoside phosphotransferase and the contaminating phosphatases in the cruder preparations. The enzymically formed nucleotide is separated on paper and measured spectrophotometrically. The phosphatase activity is determined by the estimation of phosphorus.[3] With the more highly concentrated enzyme solutions, the enzyme should be diluted with buffer to give a concentration hydrolyzing less than 25% of the donor.

Reagents

Sodium phenylphosphate
Uridine

A stock solution is prepared containing 40 mM uridine and 200 mM sodium phenylphosphate in 0.2 M sodium acetate at pH 5.0.

[1] E. F. Brunngraber and E. Chargaff, *J. Biol. Chem.* **242**, 4834 (1967).
[2] E. F. Brunngraber and E. Chargaff, *J. Biol. Chem.* **245**, 4825 (1970).
[3] E. J. King, *Biochem. J.* **26**, 292 (1932).

Procedure. Two-tenths milliliter each of enzyme and stock solution are incubated at 37° for 30 min. Aliquots of 0.02 ml or 0.04 ml are spotted on Whatman No. 1 paper and developed in a solvent system of 1-propanol, concentrated ammonium hydroxide, and water (11:7:2, v/v/v). The enzymically synthesized 5'UMP is eluted from the paper and estimated spectrophotometrically in the ultraviolet. Inorganic phosphorus is estimated in 0.05- or 0.1-ml aliquots.

Expression of Activity. The transferase activity is expressed as μmoles per minute of nucleotide formed, and the phosphatase activity is expressed as μmoles per minute of inorganic phosphate released. The specific activity is the activity per milligram protein. The phosphorylation ratio is the molar ratio of phosphate transferred to inorganic phosphate released.

Purification Procedure

All operations are performed in the cold.

Step 1. Ten pounds of washed carrot root are converted to juice either in a vegetable juicer or Waring Blendor. Approximately 2.5 liters are obtained. This extract is centrifuged (16,300 g, 30 min) and the supernatant fluid decanted through glass wool.

Step 2. The filtrate is applied to a DEAE-cellulose (15 g standard type, Schleicher & Schuell) column previously equilibrated with 0.05 M sodium acetate solution of pH 5.0. The dimensions of the column are not critical although a short, wide column is favored over a longer, narrow one in order to hasten the application of filtrate. After the column has been washed with 500 ml of the same buffered solution, protein is eluted with 0.4 M sodium acetate (pH 5.0). Fractions of 15 ml, each during a period of 45 min, are collected and assayed for protein and phosphotransferase. Transferase activity usually begins to appear in the eluates after approximately 150 ml are collected. The fractions that contain the enzyme are combined. Alternatively, a batch-wise procedure could be adopted for this step.

Step 3. The solution (325 ml) is brought to 90% saturation with ammonium sulfate by the gradual addition, with stirring, of 216 g of ammonium sulfate. The mixture is left overnight and then centrifuged (16,300 g, 15 min). The suspension of the sediment in 50 ml of water is subjected to dialysis overnight against several changes of distilled water (3 liters) and clarified by centrifugation (500 g, 5 min).

Step 4. The solution is adjusted to a 0.05 M sodium acetate concentration by the addition of molar sodium acetate (pH 5.0) and applied to a column (1.2 × 14 cm) of 2 g of DEAE-cellulose (type 20, Schleicher & Schuell) previously equilibrated with 0.05 M sodium acetate (pH 5.0). The column is washed with the same solution until the protein concentration of the effluent falls below 0.02 mg/ml. A combined salt and pH gradient is used for subsequent elution of phosphotransferase activity. The mixing chamber contains 100 ml of 0.05 M sodium acetate, pH 5.0, and the reservoir contains 100 ml of 0.1 M sodium acetate, pH 4.0. A flow rate of 5 ml every 25 min is maintained. Phosphotransferase is eluted at a pH of about 4.3. The fractions which contain the peak activity of this enzyme are combined and dialyzed against distilled water overnight. At this stage of purification all contaminating phosphatases have been removed as shown by the constant phosphorylation ratios recorded in steps 4, 5, and 6. (Table I). The step 4 enzyme can be satisfactorily used as a convenient preparatory tool for the synthesis of nucleotides that are otherwise accessible only with difficulty. Yields of product up to 45% of the nucleoside acceptor can be achieved under the conditions studied here. Further purification of the enzyme is concerned with the removal of inactive protein but entails the loss of enzyme.

Step 5. If further purification of the enzyme is required, the solution from the second DEAE-cellulose column is saturated with ammonium

TABLE I
ISOLATION OF NUCLEOSIDE PHOSPHOTRANSFERASE[a]

| | | | Phosphotransferase | | |
| | Volume (ml) | Total protein (mg) | Recovery (%) | Specific activity[b] | Phospho-rylation ratio[c] |
Step and procedure					
1. Crude extract	2500	12,500	100	0.0057	0.03
2. DEAE-cellulose column 1	330	468	100	0.15	0.7
3. (NH₄)₂SO₄ precipitate 1	60	128	99	0.55	1.9
4. DEAE-cellulose column 2	160	7.5	55	5.2	2.6
5. (NH₄)₂SO₄ precipitate 2	2	7.0	53	5.3	2.6
6. Sephadex G-100 column	16	1.0	45	32	2.6

[a] In this preparation, 10 pounds of carrots were processed.
[b] The specific activity is expressed as μmoles, per minute per milligram protein, of uridylic acid produced under the assay conditions defined in the text.
[c] The phosphorylation ratio is defined as the molar ratio of phosphate transferred to inorganic phosphate released.

sulfate. The precipitate, collected by centrifugation (16,300 g, 15 min), is dissolved in 2 ml of distilled water and subjected to dialysis against three changes of 500 ml of distilled water and once again against 1 liter of 0.05 M sodium acetate, pH 5.0.

Step 6. Two milliliters of enzyme solution are applied to a Sephadex G-100 column (1.5 × 40 cm) previously equilibrated with 0.1 M KCl in 0.05 M sodium acetate, pH 5.0. The enzyme is eluted with this buffer at a flow rate of 25 ml/hr; 4-ml fractions are collected. The active fractions after assay were dialyzed against 0.05 M sodium acetate, pH 5, and stored in the frozen state at −20°.

Properties

Chemical Characterization. Rodgers and Chargaff[4] have reported that a variety of criteria indicate that the preparation of nucleoside phosphotransferase described here may be regarded as representing a homogeneous protein. Evidence for homogeneity is provided by data from gel filtration, sedimentation equilibrium studies, sucrose density gradient studies, polyacrylamide gel electrophoresis, and isoelectric focusing.

The molecular weight of the protein is found to be near 44,000 by three different methods (gel filtration, sedimentation equilibrium, and sucrose density gradient studies). A value of 38,000 is estimated from the amino acid composition. It is assumed that the protein is highly hydrated since there is no evidence in the enzyme of anything but protein. The partial specific volume of the protein is reported as 0.722.

The amino acid analysis of the protein demonstrates a single residue of histidine and no tyrosine. The amino terminus is shown to be aspartic acid; the carboxyl termini, glycine and serine.

When the enzyme is treated with 6 M urea, two subunits may be separated by electrofocusing. Neither fraction has enzymic activity, nor is activity recovered by mixing the two. The amino acid composition of these two subunits is substantially different.

The isoelectric point of the intact enzyme is pI = 4.1. Subunit A has pI = 4.6; subunit B, pI = 5.1.

Catalytic Functions. The enzyme is considered a nucleoside phosphotransferase which also exhibits hydrolase activity rather than a nucleotidase with transfer properties because: (1) phosphate transfer is favored over hydrolysis; (2) transfer occurs specifically to nucleosides; and (3) high yields of nucleotide are recorded. Kinetic studies[2] have

[4] R. Rodgers and E. Chargaff, *J. Biol. Chem.* **247**, 5448 (1972).

demonstrated two hydrolytic sites, binding the phosphate donor, and one transfer site, engaging the nucleoside acceptor. The kinetic constants and pK values are summarized in Table II. Two ionizing groups (pK 7.0, 7.6) in the free enzyme bind the phosphate donor in the course of its hydrolysis, whereas only one (pK 7.6) serves in this capacity in the transfer function. A second group (pK 7.0) seen in the transferase reaction apparently does not function in the binding of the phosphate donor or of the nucleoside acceptor although its pK is changed in the presence of uridine.

Characterization of the Transfer Reaction. Four factors must be considered: (1) the nature of the phosphate donor; (2) the aglycone of the nucleoside acceptor; (3) the sugar moiety of the nucleoside; (4) the donor-acceptor pair. Of the several organic phosphates tested, 3'AMP, 5'AMP, phenylphosphate, and p-nitrophenylphosphate are nearly equally effective as phosphate donors under standard conditions with uridine as the acceptor. With ribose 5-phosphate as the donor a small amount of 5'UMP is formed while the 1- and 6-phosphates of glucose and β-glycerophosphate are nearly or entirely inactive in phosphate transfer. In regard to the phosphate acceptor when phenylphosphate is used as the phosphate donor, the nucleosides can be listed in descending order of efficiency: cytidine, 3.6; uridine, 2.7; deoxycytidine, 2.1; ribothymidine, 2.0; thymidine, 1.5; deoxyuridine, 1.4; deoxyguanosine, 1.4; guanosine, 1.0; deoxyadenosine, 0.6; and adenosine, 0.5. Thus, in general, when phenylphosphate serves as the donor, the pyrimidine nucleosides are better acceptors than the purine nucleosides. Within the pyrimidine series, the ribonucleosides are more efficient acceptors than

TABLE II

KINETIC CONSTANTS AND pK VALUES FOR TRANSFERASE AND HYDROLASE FUNCTIONS OF
NUCLEOSIDE PHOSPHOTRANSFERASE

	Phosphotransferase	Phosphatase
K_m		
Phenylphosphate (mM)	3.5 ± 0.7	0.7; 3.0
Uridine (mM)	3.5 ± 0.8	
K_i		
Uridine (mM)		4.3 ± 0.4
pK		
Free enzyme	7.0; 7.6	7.0; 7.6
Enzyme-phenylphosphate		4.0; 6.6
Enzyme-phenylphosphate-uridine	4.2; 5.1; 6.4	

the deoxyribonucleoside derivatives, whereas the opposite appears to hold for the purines. The importance of the sugar moiety is demonstrated by comparing adenosine, 2'-deoxyadenosine, and 3'-deoxyadenosine as phosphate acceptors. The 3'-deoxyadenosine is more efficiently phosphorylated than 2'-deoxyadenosine, which in turn surpasses adenosine. The importance of the donor-acceptor pair is demonstrated with adenosine as the acceptor. Six times more 5'AMP is produced by 3'AMP as the donor than by phenylphosphate.

Specificity. The enzymic product seen under normal conditions is exclusively a 5'-nucleotide. With highly purified enzyme preparations and long periods of incubation, however, smaller proportions of the 2' and 3' isomers are also formed. The phosphate transfer is limited to one type of acceptor, namely, nucleosides. These include such unusual ones as the ribosides of purine, 2,6-diaminopurine, and 5-bromouracil, and the glucoside or arabinoside of adenine.[5] The transferase is also able to use cyclic and 2' nucleotides as acceptors. 3'AMP is hydrolyzed producing the acceptor adenosine which in turn accepts phosphate from a second molecule of 3'AMP forming 5'AMP.

Characteristics of the Hydrolase Activity. Those substrates that are efficient phosphate donors in the transfer reaction are hydrolyzed rapidly. For example, phenylphosphate, p-nitrophenylphosphate, and 3'AMP are hydrolyzed whereas glucose 1-phosphate is degraded only slightly. However, not all hydrolyzable substrates are efficient phosphate donors in the transferase reaction. Ribose 5-phosphate is hydrolyzed initially at the same rate as phenylphosphate but is only one-fourth as effective a donor for the transferase. Acceptors for the transferase function serve as inhibitors of the hydrolase. Uridine acts as a noncompetitive inhibitor, with $K_i = 4.3$ mM.

Effect of pH. The pH optimum of the purified transferase is about pH 5.0 with either uridine or deoxycytidine as the acceptor and phenylphosphate as the donor. There is a variation of the ratio of 5'UMP to 5'dCMP formed at different pH values. Thus, this ratio at pH 4, 5, 6, and 7 is 0.7, 1.2, 1.1, 0.5, respectively. When the release of inorganic phosphate is measured in the two systems, maxima are observed not only at pH 5, but also at 6–6.5.

When the enzyme is maintained at pH 9.5 without protecting agent only about one-third of the initial transfer activity is retained whereas the hydrolase function is almost entirely lost after 22 hr. When the solutions are returned to pH 5, a slight recovery of enzymic activity is

[5] M. Tunis and E. Chargaff, *Biochim. Biophys. Acta* **40**, 206 (1960).

observed. The presence of a reducing agent during the treatment at high pH provides some protection from the irreversible loss of enzyme activity. Dithiothreitol, the most effective, permits the recovery of more than 80% of the activity after the solutions are brought back to pH 5. The formation of several species of inactive aggregates at high pH which was prevented by the presence of a strong reducing agent has been demonstrated.[4]

Site-specific Reagents. Iodoacetamide, 2-hydroxy-5-nitrobenzyl bromide, N-ethylmaleimide, and N-acetylimidazole, all at final concentrations of 0.001 M, had no effect on the activity of the enzyme. The transferase function was slowly reduced to 75% within 2 hr and to nearly 50% in 72 hr by 2,4-dinitrofluorobenzene, while the hydrolase activity of the enzyme remained unchanged. Iodine causes a drastic loss of enzyme activity. This inactivation could not be reversed by the addition of an excess of dithiothreitol. Loss in activity is accompanied by the rapid incorporation of 1 g-atom of iodine per mole of protein. The enzyme activity completely disappears after 1 hr. In a much slower reaction requiring several hours, a second atom of iodine appears to be taken up by the protein. It is probable that the inactivation by iodine involves a modification of the histidine residue via iodohistidine since histidine could no longer be demonstrated in the hydrolysate of the iodinated protein.

Effect of Cations and Anions. Metals were tested in 0.01–10 mM concentrations. The following cations had no or almost no effect on either the transferase or the phosphatase functions of the enzyme: Na^+, K^+, $(NH_4)^+$, and Zn^{2+}. Ammonium molybdate also was ineffective. Ca^{2+}, Mn^{2+}, and Pb^{2+} were equally inhibitory to both functions, and Fe^{2+} and Fe^{3+} were more inhibitory to the transfer reaction. Mg^{2+}, Cu^{2+}, and Co^{2+} enhance the phosphatase function considerably, the transferase to a small extent. Sodium fluoride has no effect on either activity at concentrations below 10 mM. At 0.25 M NaF, the phosphatase is inhibited by less than 10%, the transferase enhanced by about 45%.

Inactivation and Reactivation. Glutathione (2.5 mM) or cysteine (2.5 mM) depress both transferase and phosphatase activities by about 25%. EDTA or 8-hydroxyquinoline (oxine), both at 0.625 mM concentration, exert a slightly stronger inhibition which is reversed by 5 mM $MgCl_2$. When the enzyme solution is subjected to dialysis overnight against distilled water, it loses no activity. Dialysis, first against EDTA or oxine and then against water, produces, however, an inhibition which results in the recovery of only 36% of the initial transferase activity and 65% of the initial phosphatase activity. Copper and cobalt ions (1 mM)

reactivate the enzyme system. Magnesium ions also have a limited effect in the case of the treatment with EDTA.

Stability. The enzyme at step 4 is stable for at least 1 month when stored at −20°. No measureable loss of either transferase or phosphatase activity was observed during this period. In solutions stored at 4° for 1 week, 80% of the transferase activity was retained with no loss of hydrolase activity.

Phosphorylated Protein Intermediate. Experiments with [5′-³²P]AMP as well as kinetic studies gave no evidence for the implication of a phosphoprotein intermediate during the phosphotransferase reaction.

[51] Cytosine and Cytidine Deaminase from Yeast

By PIER LUIGI IPATA and GIOVANNI CERCIGNANI

The hydrolytic deamination of cytidine to uridine and of cytosine to uracil is catalyzed in yeast by two distinct enzyme proteins.[1] The presence of cytidine deaminase (cytidine aminohydrolase, EC 3.5.4.5) activity, distinct from cytosine deaminase (cytosine aminohydrolase, EC 3.5.4.1), was first described in crude extracts from yeast by Wang *et al.*,[2] while the latter enzyme was originally found in yeast, as well as in *Escherichia coli,* by Hahn and Schäfer[3] in 1925.

Assay Methods

Principle. Both cytosine and cytidine deaminase activities can be determined by direct spectrophotometric assay from the fall in absorbance at 286 nm following the conversion of the 4-amino to the 4-keto compounds. Table I shows the differential absorptivities for four couples of pyrimidine derivatives at 286 nm.

Reagents

Tris-Cl buffer, 0.1 *M*, pH 7
Cytidine, 1 m*M* in water
Cytosine, 1 m*M* in water

[1] P. L. Ipata, G. Cercignani, and E. Balestreri, *Biochemistry* 9, 3390 (1970).
[2] T. P. Wang, H. Z. Sable, and J. O. Lampén, *J. Biol. Chem.* 184, 17 (1950).
[3] A. Hahn and L. Schäfer, *Z. Biol. (Munich)* 83, 511 (1925).

TABLE I
DIFFERENTIAL ABSORPTIVITIES FOR SOME PYRIMIDINE DERIVATIVES AT pH 7[a]

Substrate	Product	$-\Delta\epsilon_{286}$ at pH 7
Cytosine	Uracil	0.937
5-Methylcytosine	Thymine	1.520
Cytidine	Uridine	3.140
2'-Deoxycytidine	2'-Deoxyuridine	2.650

[a] Calculated from data of W. E. Cohn, this series, Vol. 3 [107], p. 740.

Procedure

(a) ASSAY OF CYTOSINE DEAMINASE. Into a quartz microcuvette with 1-cm light path pipette 0.3 ml of Tris-Cl buffer, 0.5 ml of cytosine solution, and 0.2 ml of enzyme preparation containing a maximum of 80 munits (about 0.08 absorbance unit/min) of cytosine deaminase (dilute with water, if required). Mix rapidly and incubate at 27° in a recording spectrophotometer set at 286 nm, against a reference cuvette in which cytosine is replaced by water. Record the absorbance change for 5–10 min.

(b) ASSAY OF CYTIDINE DEAMINASE. Into a quartz microcuvette with 1-cm light path pipette 0.6 ml of Tris-Cl buffer, 0.3 ml of cytidine solution, and 0.1 ml of enzyme preparation containing a maximum of 50 munits (about 0.15 absorbance unit/min) of cytidine deaminase. Mix rapidly and incubate at 27° in a recording spectrophotometer set at 286 nm, against a reference cuvette in which cytidine is replaced by water. Record the absorbance change for 5–10 min.

Enzyme Units. One unit of activity is that amount of enzyme catalyzing the deamination of 1 μmole of substrate per minute at 27°. A change in absorbance at 286 nm of 1 absorbance unit/min corresponds to 0.318 unit of cytidine deaminase and to 1.067 unit of cytosine deaminase. Specific activity is given in milliunits per milligram protein.

Purification Procedures

All steps are carried out at 0–6° unless stated otherwise. Protein is determined according to Warburg and Christian[4] throughout the purification procedure, while the protein content of crude extracts is evaluated by a biuret method.[5]

[4] O. Warburg and W. Christian, *Biochem. Z.* **310**, 35 (1942).
[5] A. G. Gornall and A. Hunter, *J. Biol. Chem.* **147**, 593 (1943).

Purification of Cytosine Deaminase

Step 1. Preparation of Yeast Plasmolysate. First 3.5 kg of Vulcania baker's yeast are plasmolyzed according to Kunitz:[6] after swelling with toluene (0.5 liter/kg) at 45°, the slurry is kept at room temperature for 2–3 hr, then mixed with cold distilled water (1 liter/kg of yeast) and eventually transferred to separatory funnels in the cold. After 18 hr, the aqueous phase is collected and centrifuged at 10,000 rpm for 20 min; the supernatant fluid is filtered through filter paper to eliminate any floating material.

Step 2. Ammonium Sulfate Fractionation. The supernatant fluid from step 1 is fractionated by adding a saturated ammonium sulfate solution at pH 7.0. Cytosine deaminase activity is precipitated between 60–70% saturation.

Step 3. First G-100 Sephadex Chromatography. The pellet from the above step is solubilized in 0.05 M Tris-Cl buffer, pH 7.2, to a final volume of about 150 ml and applied to the top of a G-100 Sephadex column (7 × 120 cm), equilibrated with the same buffer. The proteins are eluted with the equilibration buffer; 18-ml fractions are collected with a flow rate of about 50 ml/hr. The most active fractions (between 2000 and 2200 ml of effluent) are pooled and precipitated with neutral ammonium sulfate between 65–80% saturation.

Step 4. Second G-100 Sephadex Chromatography. The pellet from the preceding step is dissolved in the minimal amount of 0.05 M Tris-Cl buffer, pH 7.2, and chromatographed through a second G-100 Sephadex column (3 × 60 cm), equilibrated as above. Proteins are eluted with the equilibration buffer; 5.5-ml fractions are collected with a flow rate of about 10 ml/hr. Cytosine deaminase is eluted between 350–430 ml of effluent.

Step 5. First DEAE-Cellulose Chromatography. The active fractions from the above step are pooled and applied to a DEAE-cellulose (acid form) column (2.5 × 18 cm). After the sample has run into the column bed, the proteins are eluted with 50 ml of 0.05 M Tris-Cl, pH 7.2, followed by 300 ml of a linear gradient between 0.05–0.25 M Tris-Cl, pH 7.2; the flow rate is kept from the beginning at about 30 ml/hr. More than 90% of the cytosine deaminase activity is eluted between 0.10–0.16 M Tris buffer.

Step 6. Second DEAE-Cellulose Chromatography. The pooled active fractions (60 ml) from the DEAE-cellulose column are dialyzed against

[6] M. J. Kunitz, *J. Gen. Physiol.* **29**, 393 (1947).

TABLE II
PURIFICATION OF CYTOSINE DEAMINASE FROM BAKER'S YEAST[a]

Step	Volume (ml)	Protein (mg)	Specific activity (munits/mg)	Yield (%)
1	4230	36,400	9.2	100
2	154	12,370	14.3	53
3	183	311	233	21
4	102	128	494	19
5	60	23.7	1973	14
6	111	11.4	3128	11

[a] Reproduced from Ipata et al.[7] Reprinted with permission from Biochemistry 10, 4270 (1971). Copyright by the American Chemical Society.

water (adjusted at pH 7.2) for 5 hr, absorbed to an identical DEAE-cellulose (acid form) column, and eluted in 8-ml fractions with 500 ml of a linear gradient between 0–0.3 M Tris-Cl, pH 7.2. The enzyme acticity is recovered, around a concentration of 0.15 M Tris-Cl, in a single peak exactly coinciding with the sole protein peak which can be eluted up to 0.3 M Tris-Cl concentration.

A typical purification is summarized in Table II.

The pooled active fractions from step 6 can be stored at $-20°$ divided into small aliquots. The preparation thus stored shows no appreciable loss of activity after 5 months, and it was shown to be free from the following enzyme activities: adenosine deaminase, AMP deaminase, cytidine deaminase, guanase, 5′-nucleotidase, ribonuclease, and RNA phosphodiesterase.[7]

Purification of Cytidine Deaminase

Step 1. Preparation of Yeast Plasmolysate. This step is carried out as described above for cytosine deaminase, starting from 1.5 kg of baker's yeast.

Step 2. Ammonium Sulfate Precipitation. The supernatant fluid from step 1 is brought to 67% saturation with respect to ammonium sulfate by slowly adding the solid salt under continuous stirring. After centrifugation, the supernatant fluid is discarded; the pellets can be stored at $-20°$ for months without any loss of activity.

Step 3. First G-100 Sephadex Chromatography. About one-fourth of

[7] P. L. Ipata, F. Marmocchi, G. Magni, R. Felicioli, and G. Polidoro, Biochemistry 10, 4270 (1971).

the pellets from the above step (corresponding to 400 g of baker's yeast cake) is suspended in the minimal volume of 0.05 M Tris-Cl buffer, pH 7.2, and applied to the top of a G-100 Sephadex column (6 × 120 cm), equilibrated with the same buffer. Proteins are eluted with the equilibration buffer, 8-ml fractions are collected with a flow rate of about 15 ml/hr. The most active fractions are eluted between 610–670 ml of effluent. In contrast with the mouse kidney enzyme,[8] inactivation of yeast cytidine deaminase never occurs during gel filtration steps.

Step 4. Second G-100 Sephadex Chromatography. The protein in the pooled active fractions is precipitated by adding solid ammonium sulphate up to 67% saturation. The pellet is solubilized in a minimal volume of 0.05 M Tris-Cl buffer, pH 7.2, and applied to a G-100 Sephadex column (3 × 95 cm), equilibrated as above. Proteins are eluted with the same buffer; 6-ml fractions are collected with a flow rate of about 9 ml/hr. Elution of cytidine deaminase is achieved between 260–300 ml of effluent.

Step 5. Third G-100 Sephadex Chromatography. The above step is repeated, with the same column. Fractions between 270–290 ml of effluent are pooled. If lower purity, but higher enzyme yield, is desired, fractions between 260–300 ml are pooled.

Step 6. DEAE-Cellulose Chromatography. The enzyme preparation from the above step is applied to a DEAE-cellulose (acid form) column (2 × 27 cm). After the sample has run into the column bed, the enzyme is eluted with 50 ml of 0.05 M Tris-Cl, pH 7.2, followed by 50 ml of 0.05 M potassium phosphate, pH 7.2, and eventually with 500 ml of a linear gradient between 0.05–0.2 M potassium phosphate, pH 7.2. Six-milliliter fractions are collected at a flow rate of 23 ml/hr. Cytidine deaminase is eluted between 0.09–0.15 M phosphate. Although most contaminant proteins are removed by this step, the purification factor is not as high as would be expected, and about 50% of the enzyme activity is lost. Since the protein concentration of active fractions is low (about 0.1 mg/ml) and cytidine deaminase is highly sensitive to mercurial agents (see below under *Properties*), inactivation is thought to occur. Indeed, a higher specific activity is obtained in this step if 1 mM glutathione is added to the eluting buffers.

A typical purification scheme is shown in Table III.

The pooled active fractions from step 6 show no appreciable loss of activity within 1 month when stored at −20°. The enzyme preparation

[8] R. Tomchick, L. D. Saslaw, and V. S. Waravdekar, *J. Biol. Chem.* **243**, 2534 (1968).

TABLE III
PURIFICATION OF CYTIDINE DEAMINASE FROM BAKER'S YEAST[a]

Step	Volume (ml)	Protein (mg)	Total activity (munits)	Specific activity (munits/mg)
1	1582	17,700	87,770	5.0
2	74	8,000	93,170	11.7
3[b]	55	765	25,120	32.8
4	37	186	13,450	72.2
5	27	47.5	2,770	125.6
6	63	6.3	1,300	206.7

[a] Adapted from Ipata et al.[1]
[b] From this step on, about one-fourth of the step 2 fraction is used.

does not deaminate the following: cytosine, guanine, adenosine, AMP, CMP, and dCMP.

Properties

Properties of Cytosine Deaminase

Molecular Weight. Estimation of the molecular weight by the method of Andrews[9] gives a value of 34,000.

Kinetic Properties. Cytosine deaminase shows a single activity optimum at pH 6.5; the enzyme preparation is stable for at least 48 hr when kept at 4° in the pH range from 5–9. The reaction rate is proportional to enzyme concentration up to 70 μg of protein/ml and obeys the Arrhenius equation in the temperature range from 20–52°; the time course of the reaction is linear up to 15 min at 27°. Both cytosine and 5-methylcytosine are good substrates, with identical K_m values (2.5 mM).[7] With either substrate, cytosine deaminase is inhibited by a number of nucleosides and nucleotides: Table IV shows the concentrations of 10 purine and pyrimidine compounds required to produce a 50% inhibition of the enzyme activity at pH 7.

Sensitivity to p-Mercuribenzoate. When 0.1 mM p-mercuribenzoate is added simultaneously with substrate, it has no effect on enzyme activity; however, 50% inactivation is produced in the presence of 0.3 mM p-mercuribenzoate, whether cytosine or 5-methylcytosine is used as substrate.[7]

[9] P. Andrews, *Biochem. J.* **91**, 222 (1964).

TABLE IV
THE INHIBITORY EFFECT OF SOME NUCLEOSIDES AND NUCLEOTIDES ON CYTOSINE
DEAMINASE FROM BAKER'S YEAST[a]

Inhibitor	Concentration (mM) required for 50% inhibition
Cytidine	0.517
Thymidine	0.690
Guanosine	0.400
CMP	0.450
CDP	0.520
CTP	0.442
GMP	0.385
GDP	0.412
GTP	0.405
dTTP	0.580

[a] Modified from Ipata et al.[7] (p. 4273).

Properties of Cytidine Deaminase

Molecular Weight. Estimation of the molecular weight by the method of Andrews[9] gives a value of 57,000.

Kinetic Properties. The pH optimum for cytidine deaminase is sharply defined around 7.2. The enzyme preparation is stable at 4° for some hours at pH values between 5.5–9.5; rapid inactivation and precipitation of the enzyme protein occurs at pH values lower than 5. Both cytidine and 2'-deoxycytidine are good substrates for the yeast enzyme,[10] but cytidine produces strong substrate inhibition at concentrations higher than 0.5 mM;[1] the apparent K_m for cytidine is $2.5 \times 10^{-4} M$, while the K_m value for 2'-deoxycytidine (which displays Michaelis kinetics) is $9.1 \times 10^{-5} M$.[10] It must be noted that the ratio of specific activities toward the two substrates is constant throughout the purification procedure. Neither 5-methylcytidine nor its 2'-deoxy derivative is deaminated when tested at 0.25 mM concentration.[10] A number of nucleotides exert an inhibitory effect on cytidine deaminase from yeast: CMP, at 0.300 and 0.750 mM concentration, produces 50% and 90% inhibition, respectively; 85% inhibition is given by 0.750 mM GTP; either dTMP or dTTP, at 0.400 mM concentration, produces about 50% inhibition; CDP, CTP, and UTP are weaker inhibitors.[10]

[10] G. Magni, P. L. Ipata, P. Natalini, R. Felicioli, and G. Cercignani, *Experientia* **30**, 861 (1974).

Sensitivity to p-mercuribenzoate. Cytidine deaminase from yeast is highly sensitive to *p*-mercuribenzoate. When added at 0.2 μM concentration simultaneously with substrate, this agent causes a 50% inactivation, while the reaction rate is immediately reduced to zero at 2 μM *p*-mercuribenzoate concentration.[1]

Heat Sensitivity. When the enzyme preparation is held for 5 min at temperatures between 40–62°, no reduction is observed in the reaction rate tested at 27° in the presence of 0.3 mM cytidine; heat treatment for 5 min at 50–60°, however, completely abolishes CMP and substrate inhibition.[1]

[52] Cytidine Deaminases (from *Escherichia coli* and Human Liver)

By DAVID F. WENTWORTH and RICHARD WOLFENDEN

Cytidine + H₂O ⇌ uridine + NH₃

The equilibrium for the hydrolytic deamination of cytidine, catalyzed by cytidine deaminase, lies far in the direction of hydrolysis, with $K_{eq} =$ [uridine][NH₃]/[cytidine][H₂O] = 78, expressed in terms of the molar concentration of uncharged reactants and products, with water activity taken as unity.[1]

A. Cytidine Deaminase from *E. coli.*[2]

Cytidine deaminase, originally detected in extracts of *E. coli* by Wang *et al.,*[3,4] has been purified extensively but not to homogeneity.

Assay Method

Principle. Enzyme activity can be determined by a direct spectrophotometric assay based on the loss of absorbance when cytidine is converted to uridine, at 282 nm where $\Delta\epsilon = -3600$ for a 1-cm light path at pH 7.5. When high levels of extraneous protein or the presence of certain nucleoside inhibitors cause the background absorbance to be too

[1] R. M. Cohen and R. Wolfenden, *J. Biol. Chem.* **246,** 7566 (1971).
[2] R. M. Cohen and R. Wolfenden, *J. Biol. Chem.* **246,** 7561 (1971).
[3] T. P. Wang, H. Z. Sable, and J. O. Lampén, *J. Biol. Chem.* **184,** 17 (1950).
[4] T. P. Wang, this series, Vol. 2, p. 478.

high at 282 nm, the reaction may be followed at 290 nm where $\Delta\epsilon$ for cytidine transformation is -2100 at pH 8.0. When using cytidine at concentrations much above K_m, the reaction may be followed at 290 nm or at 282 nm by using cuvettes of shorter path length (0.1–0.5 cm).

Reagents

0.1 M Tris·HCl, pH 7.5
0.00167 M cytidine

Procedure. To 0.5 ml of buffer and 0.1 ml of cytidine, add that amount of water that will result in a final volume of 1 ml after the addition of enzyme. Start the reaction by adding enzyme and record the early linear portion of the reaction.

Definition of Unit and Specific Activity. One unit of activity is the amount of enzyme required to deaminate 1 μmole of cytidine per minute in the above standard reaction mixture at 25°. Specific activity is units per milligram of protein as determined by the method of Lowry *et al.*[5]

Purification Procedure

Step 1. Preparation of Crude Extract. Frozen *E. coli* B cells (mid-log phase, obtained from General Biochemicals, Inc.) are thawed overnight. The resulting paste (150 g) is mixed for 10 min in a Waring Blendor with Superbrite glass beads (450 g) (obtained from 3M Co.) in Tris·HCl buffer (210 ml, 0.01 M, pH 7.4, containing 0.01 M magnesium acetate). The cell extract, after low-speed centrifugation, is centrifuged for 3 hr at 105,000 g in a Spinco model J, ultracentrifuge. To the resulting supernatant solution is added, with stirring, 0.2 volume of streptomycin B sulfate (5% solution in water), over a period of 30 min. After 20 min of additional stirring, the supernatant fluid is removed by centrifugation, divided into small portions (10 ml), placed in a water bath at 60° for 1 min, and then rapidly cooled. The precipitate is removed by centrifugation and discarded.

Step 2. Ammonium Sulfate Fractionation. To the supernatant solution is added, in a ratio of 31.5 g/100 ml, a solid mixture containing ammonium sulfate and potassium bicarbonate in a weight ratio of 99:1, respectively. The mixture is added slowly over a period of 50 min, and the solution is stirred for an additional 30 min. The resulting precipitate is removed by centrifugation and discarded, and the supernatant fluid is

[5] O. H. Lowry, N. J. Rosebrough, A. L. Farr, and R. J. Randall, *J. Biol. Chem.* **193**, 265 (1951).

similarly treated with an additional quantity of solid ammonium sulfate mixture (17.5 g/100 ml). The resulting precipitate, containing most of the deaminase activity, is recovered by centrifugation, redissolved in a minimal volume of distilled water, and dialyzed overnight against a large volume (1000 ml/10 ml of enzyme solution) of potassium phosphate buffer (0.01 M, pH 8.8).

Step 3. DEAE Column Chromatography. After dialysis, the enzyme solution is applied to a column of DEAE-cellulose (bed volume 230 ml) equilibrated with potassium phosphate buffer (0.01 M, pH 8.8) containing 0.01 M KCl. Elution is performed with a linear gradient, the first reservoir containing 1200 ml of the buffer used for equilibration, and the second reservoir containing 1200 ml of potassium phosphate buffer (0.01 M, pH 7.5) containing 0.5 M KCl.

Step 4. To the pooled fractions containing maximal activity (eluted in the neighborhood of 0.25 M KCl) is added the solid ammonium sulfate potassium bicarbonate mixture described above (49 g of mixture per 100 ml of enzyme solution). The resulting precipitate is dissolved in a minimal volume of distilled water, dialyzed exhaustively against Tris·HCl buffer (0.01 M, pH 7.5), and applied to a column of carboxymethylcellulose (75-ml bed volume) equilibrated with the same buffer. Deaminase activity is not retained appreciably on this column, which removes some inactive protein. The eluted enzyme solution is applied to a column of hydroxylapatite (100-ml bed volume) and is eluted with a linear gradient, the first reservoir containing Tris·HCl buffer (0.05 M, pH 7.5), and the second reservoir containing Tris·HCl buffer (0.5 M, pH 7.5). Active fractions are pooled, dialyzed against potassium phosphate buffer (0.01 M, pH 8.8), and concentrated on a column of DEAE-cellulose (5-ml bed volume) equilibrated with the same buffer. The enzyme, in its final state of purification, is eluted with potassium phosphate buffer (1 M, pH 6.5). Overall purification, summarized in Table I, is approximately 175-fold.

SUMMARY OF PURIFICATION OF CYTIDINE DEAMINASE FROM *E. coli*

	Total units	Specific activity (units/mg)	Recovery (%)
Extract	209	0.051	(100)
75% (NH$_4$)$_2$SO$_4$ preparation	142	0.12	68
DEAE-cellulose, peak concentration	57	0.69	27
Hydroxylapatite, peak concentration	6.7	8.9	3.2

Properties

Stability. Cytidine deaminase, purified as described, is reasonably stable, retaining 55% of its activity after 3 weeks at 4°.

Specificity. The enzyme is specific for cytosine nucleoside derivatives. At pH 7.5 cytosine deoxyriboside[3] (K_m = 8.9 × 10^{-5} M) is converted 3–4 times more rapidly than cytidine (K_m = 2.1 × 10^{-4} M), whereas 5,6-dihydrocytidine[6] (K_m = 1.1 × 10^{-4} M) is transformed 0.1 as rapidly as cytidine. 4-N-Methylcytidine is a poor substrate with K_m: V_{max} at least three orders of magnitude higher than that of cytidine. No significant change in activity is encountered when uracil, adenine, hypoxanthine, guanine, or their respective nucleosides and 5'-nucleotides are included in the standard assay at concentrations equivalent to that of cytidine (2 × 10^{-4} M). The enzyme can add methylamine to uridine to form 4-N-methylcytidine, whereas dimethylamine, trimethylamine, tris(hydroxymethyl)aminomethane, hydroxylamine, or sodium sulfide do not serve as cosubstrates in the backward reaction.[1]

Metal Involvement. No loss of activity is encountered when the enzyme is exhaustively dialyzed against EDTA (10^{-2} M) nor does EDTA at this concentration affect activity when included in the assay.

Borohydride Reduction. The enzyme is not inactivated by incubation with sodium borohydride (10^{-3} M) for short periods in the presence or absence of substrate.

Approximate Molecular Weight. The elution behavior of the enzyme on a calibrated column of Sephadex G-100, determined according to the procedure of Siegel and Monty,[7] indicates an apparent Stokes' radius of the enzyme of approximately 40 Å. Sucrose gradient centrifugation of the enzyme, according to the procedure of Martin and Ames,[8] yields a sedimentation coefficient of 4.4 S. Together, these results suggest an approximate molecular weight of 73,000, assuming a partial specific volume of 0.725 cm^3/g.

Effects of pH and D_2O. There is little variation in V_{max} and K_m in the range from pH 6.5–10.7. Below pH 6.5 there is a gradual reduction in V_{max}. An irreversible loss of activity occurs at pH values below 4. The substitution of deuterium oxide (90 atom-% excess) for water as solvent at pH 7.1, 7.5, and 8.4 results in a slight increase (~10%) in V_{max} and no significant change in K_m. The complete hydrolysis of cytidine in D_2O in

[6] B. E. Evans, G. N. Mitchell, and R. Wolfenden, *Biochemistry* **14,** 621 (1975).

[7] L. M. Siegel and K. J. Monty, *Biochim. Biophys. Acta* **112,** 346 (1966).

[8] R. G. Martin and B. N. Ames, *J. Biol. Chem.* **236,** 1372 (1961).

the presence of cytidine deaminase leads to no isotope incorporation at position 5, as indicated by the nuclear magnetic resonance spectrum of the product uridine.

Inhibitors. The product uridine and its reduced derivatives, 5,6-dihydrouridine and 3,4,5,6-tetrahydrouridine, serve as competitive inhibitors of cytidine deamination under the usual assay conditions. Tetrahydrouridine (K_i = 2.4 × 10^{-7} M) is more than four orders of magnitude more effective as an inhibitor than uridine (K_i = 2.5 × 10^{-3} M) or dihydrouridine (K_i = 3.4 × 10^{-3} M). Inhibition by tetrahydrouridine is instantaneous, purely competitive, and fully reversed by dilution within the time (approximately 5 sec) required to initiate the standard assay. The unusual affinity of cytidine deaminase for tetrahydrouridine may be due to its resemblance to a transition-state intermediate in direct water attack on cytidine. The following compounds, at a concentration (3 × 10^{-4} M) in excess of the K_m value of the substrate, do not show significant inhibition in the standard assay with 1.67 × 10^{-4} M cytidine: 3-N-methylcytidine, 4-O-methyluridine, 4-thiouridine, 4-N,N-dimethylcytidine, and 1-(β-D-ribofuranosyl)-4-sulfonyl-2-pyrimidone. The product ammonia shows no significant inhibition at quite high concentrations, an indication that its K_i value is well in excess of 1 M. Inhibition is not observed in ammonia–ammonium chloride buffers at concentrations as high as 1 M at pH 9.2, at which ammonia and its conjugate acid are present in approximately equal concentration.

B. Cytidine Deaminase from Human Liver[9]

Camiener and Smith[10] examined cytidine deaminase activity in various human tissues (liver, kidney, heart, and muscle) and in liver from various species (man, monkey, rabbit, rat, dog, guinea pig, mouse, frog, pigeon, cat, and pig) and found the highest activity in human liver.

Assay Method

Principle. The deamination of cytidine can be followed by measuring the decrease in A_{290nm} where the change in extinction coefficient corresponding to complete conversion to uridine is −2100 for a 1-cm light path at pH 8.0. Because the K_m for 6-azacytidine ($\Delta\epsilon$ = −800 at 305 nm, pH 8.0) is high, observation of the reaction is facilitated by use of cuvettes of shorter path length (0.5 cm). The deamination of 5-azacyti-

[9] D. F. Wentworth and R. Wolfenden, *Biochemistry* **14**, 5099 (1975).
[10] G. W. Camiener and C. G. Smith, *Biochem. Pharmacol.* **14**, 1405 (1965).

dine ($\Delta\epsilon$ = -1300 at pH 8.0) can be followed at 270 nm, but in the presence of Tris·HCl or ethanolamine buffer a nonenzymic reaction takes place which requires correction.[8]

Reagents

0.002 M cytidine
0.02 M Tris·HCl, pH 8.0

Procedure. To 0.5 ml of buffer and 0.1 ml of cytidine, add that volume of water that will give a final volume of 1 ml after addition of enzyme. Start the reaction by adding enzyme.

Definition of Unit. One unit of activity is that amount of enzyme required to deaminate 1 μmole of cytidine per minute at 25° in the standard assay above.

Purification Procedure

Step 1. Preparation of Crude Extract. The enzyme can be prepared from frozen tissue which has been removed at autopsy. A chilled suspension of tissue (100 g) in Tris·HCl buffer (300 ml, 0.013 M, pH 8.0) was homogenized for 2 min in a Waring Blendor. The homogenate was rapidly frozen and thawed, and then cleared of cell debris by centrifugation at 0° for 20 min at 20,000 g in a Sorvall centrifuge.

Step 2. Heat Treatment. In view of the observed heat stability of this enzyme,[11] the supernatant fluid was incubated in 10-ml portions at 75° for 10 min, rapidly cooled, and then cleared by centrifugation at 0° for 60 min at 20,000 g. This heat-treatment procedure was repeated once.

Step 3. Ammonium Sulfate Fractionation. The supernatant fluid was then adjusted to 40% saturation by addition of solid ammonium sulfate (24 g/100 ml of supernatant fluid) with stirring at room temperature, and stirring was continued for 30 min. After removal of the precipitate, the active supernatant fluid was adjusted to 70% saturation by further addition of ammonium sulfate (22.9 g/100 ml of supernatant fluid). The resulting precipitate, recovered by centrifugation, was dissolved in 5 ml of Tris·HCl buffer (0.01 M, pH 8.0) and dialyzed 3 times against 250 ml of the same buffer. This procedure was found to result in a 21-fold increase in the specific activity of the enzyme as compared with the crude extract so that the stock enzyme solution contained 0.2 unit of activity and 8 mg of protein/ml as determined by the procedure of Lowry *et al.*[5]

---

[11] G. W. Camiener, *Biochem. Pharmacol.* **16**, 1681 (1967).

Properties

Stability. The partially purified enzyme may lose some activity upon storage, but activity may be restored by the addition of 2.5 mM dithiothreitol.

Effect of pH. The enzyme exhibits identical rates of deamination of 10^{-4} M cytidine at pH 5, 8, and 10.

Specificity. The enzyme is specific for cytidine derivatives and can deaminate several antineoplastic agents. At pH 8.0, 5-azacytidine (K_m = 5.8 × 10^{-5} M) is deaminated 0.17 as rapidly as cytidine (K_m = 9.2 × 10^{-6} M), whereas 6-azacytidine (K_m = 4.2 × 10^{-3} M) is converted 6.4 times more rapidly than cytidine. The enzyme can also deaminate cytosine arabinoside[11] (K_m = 1.4 × 10^{-4} M). Cyclocytidine (2,2′-anhydro-1-β-D-arabinofuranosylcytosine) is not a substrate. Camiener[12] has tested a number of cytidine analogs at 1 mM levels, pH 8.0, 37°. Thus certain substituents at the 5-position are tolerated by the enzyme as the order of deamination rates is 5-chloro > 5-bromo > 5-H > 5-iodo > 5-methyl, whereas 3-methylcytidine is not a substrate. Cytosine 2′-deoxyriboside is a substrate, but epimerization at the 3′-position of cytosine riboside or cytosine arabinoside results in the loss of substrate activity. Cytosine is also not a substrate.

Time-dependent Inhibition. In contrast to the rapid inhibition of bacterial cytidine deaminase by 3,4,5,6-tetrahydrouridine, the onset of inhibition of the enzyme from human liver is relatively slow and can be observed with an ordinary recording spectrophotometer. Inhibition is reversible, and the ratio (k^{off}/k^{on}) of rate constants for binding (k^{on} = 2.4 × 10^4 M^{-1} sec^{-1}) and release (k^{off} = 5.6 × 10^{-4} sec^{-1}) is in reasonable agreement with a K_i value (2.9 × 10^{-8} M) measured separately under steady-state conditions. The enzyme from HeLa cells responds similarly to tetrahydrouridine with k^{off} = 1.1 × 10^3 sec^{-1} and k^{on} = 2.3 × 10^4 M^{-1} sec^{-1}. Final steady-state rates indicate K_i = 4.0 × 10^{-8} M in accordance with the ratio (k^{off}/k^{on}).

Effect of D₂O. Substitution of deuterium oxide for water leads to no significant change in the rate of deamination of 10^{-4} M cytidine or 10^{-3} M 6-azacytidine at pH = pD = 8.0, and no isotope effect is observed on the rate of binding or release of tetrahydrouridine.

---

[12] G. W. Camiener, *Biochem. Pharmacol.* **16,** 1691 (1967).

[53] Cytidine Deaminase from Leukemic Mouse Spleen

By Ivan K. Rothman, V. G. Malathi, and Robert Silber

$$\text{Cytidine} + H_2O \rightarrow \text{uridine} + NH_3$$

Cytidine deaminase is widely distributed among mammalian tissues.[1] This enzyme is responsible for the inactivation of cytosine arabinoside, one of the most useful chemotherapeutic agents available for the treatment of acute myeloblastic leukemia. In the mouse spleen, a marked increase in specific activity of cytidine deaminase occurs following infection with Friend leukemia virus.[2] This enzyme is not detected in normal mouse erythrocytes, but appears during periods of accelerated erythrocyte production, and thus can be considered a marker for "stress" erythropoiesis in the mouse.[3,4] Elevated levels of cytidine deaminase are also found in regenerating mouse liver,[5] in human leukemic cells following treatment of the patient with cytosine arabinoside,[6] and in HeLa cells cultured in the presence of cytosine arabinoside.[7] Described below are the purification and properties of this enzyme from leukemic mouse spleen.[8]

Assay Method

Principle. The activity of cytidine deaminase is assayed by measuring the amount of labeled uridine formed by the deamination of 2-^{14}C-labeled cytidine.

Procedure. The assay mixture contains the enzyme preparation, 5 μmoles of Tris·HCl buffer (pH 8.0), and 0.05 μCi of 2-^{14}C-labeled cytidine (23 mmoles/mCi), in a total volume of 40 μl. It is incubated at 37° for 10 min and the reaction terminated by the addition of 20 μl of a mixture of unlabeled marker compounds, cytidine, uridine, and uracil (0.05 M each) in 3 M HCl. Samples to which the acid and reference

[1] G. W. Camiener and C. G. Smith, *Biochem. Pharmacol.* **14**, 1405 (1965).
[2] I. K. Rothman, V. G. Malathi, and R. Silber, *Cancer Res.* **31**, 274 (1971).
[3] I. K. Rothman, E. D. Zanjani, A. S. Gordon, and R. Silber, *J. Clin. Invest.* **49**, 2051 (1970).
[4] D. E. Harrison, V. G. Malathi, and R. Silber, *Blood Cells* **1**, 605 (1975).
[5] I. K. Rothman, R. Silber, K. M. Klein, and F. F. Becker, *Biochim. Biophys. Acta* **228**, 307 (1971).
[6] C. D. Steuart and P. J. Burke, *Nature (London), New Biol.* **233**, 109 (1971).
[7] R. Meyers, V. G. Malathi, R. P. Cox, and R. Silber, *J. Biol. Chem.* **248**, 5909 (1973).
[8] V. G. Malathi and R. Silber, *Biochim. Biophys. Acta* **238**, 377 (1971).

compounds are added prior to the enzyme serve as blanks. The precipitate is removed by centrifugation, and 10-μl aliquots of the supernatant fluid are spotted onto Whatman No. 1 filter paper and subjected to ascending chromatography for 16 hr in *n*-butanol:water:formic acid (77:13:10). The positions of the marker compounds (cytidine, uridine, and uracil) are identified under ultraviolet light, and the spots are cut out. The paper is immersed in 10 ml of scintillation fluid (100 ml of toluene, 4 g of PPO, and 100 mg of POPOP) and the radioactivity determined in a liquid scintillation counter. The combined radioactivity in the uridine and uracil spots is used for calculating the activity, since crude fractions contain nucleoside phosphorylase.

Units. A unit of enzyme activity is defined as the amount of enzyme which deaminates 1 μmole of cytidine in 1 min at 37°. For calculating specific activity (units/mg), protein concentration is measured by the method of Lowry et al.[9]

Purification Procedure

Swiss-Webster female mice weighing 20–25 g are infected with Friend leukemia virus by intraperitoneal injection of a cell-free homogenate prepared from leukemic spleens of DBA/2 mice. Animals are sacrificed 7 days after virus inoculation. Spleens from 4–6 animals (9 g wet weight) are homogenized in 18 ml of 0.01 M Tris·HCl buffer (pH 8.0) and centrifuged at 11,000 g for 10 min at 4°. The following purification procedure is summarized in the table.

PURIFICATION OF CYTIDINE DEAMINASE FROM LEUKEMIC MOUSE SPLEEN

Step	Volume (ml)	Protein (mg/ml)	Total activity (units)	Specific activity (units/mg)
1. Centrifuged homogenate	15.3	27.5	36	0.9
2. Streptomycin fraction	14.6	19.5	34	1.2
3. Ammonium sulfate fraction	37.0	2.6	27	2.8
4. Calcium phosphate gel	36.0	0.92	22	6.3
5. Sephadex G-150	7.8	0.3	19	8.0
6. DEAE fraction	4.0	0.02	7.2	90

[9] O. H. Lowry, N. J. Rosebrough, A. L. Farr, and R. J. Randall, *J. Biol. Chem.* **193**, 265 (1951).

Streptomycin Treatment. Freshly prepared 10% streptomycin sulfate is added to the centrifuged homogenate to a final concentration of 1%. This is stirred for 10 min at 4° and then centrifuged at 11,000 g for 10 min.

Ammonium Sulfate Precipitation. The supernatant fluid is treated with solid ammonium sulfate at 4°, and the protein precipitated between 25–45% saturation is dissolved in 2 volumes of 0.01 M potassium phosphate buffer (pH 7.2).

Calcium Phosphate Gel Treatment. The ammonium sulfate fraction is desalted by passage through a column of Sephadex G-25 (1.75 × 25 cm), and the material eluted in the void volume is diluted to contain 5–6 A units/ml at 280 m. Calcium phosphate gel is added, 11 mg dry weight per milliliter of enzyme solution, and the mixture is stirred for 10 min at 4° and then centrifuged at 11,000 g for 10 min.

Sephadex Gel Filtration. The supernatant fluid is brought to a 50% saturation with solid ammonium sulfate. The precipitate is collected by centrifugation and dissolved in a minimal volume of 0.01 M potassium phosphate buffer (pH 7.2). This solution is passed through a column of Sephadex G-150 (1 × 45 cm) at room temperature and equilibrated with 0.01 M potassium phosphate buffer (pH 7.2) containing 0.005 M dithiothreitol. Proteins are eluted with 36 ml of the same buffer. Two-milliliter fractions are collected, and those containing the enzyme activity are pooled.

Chromatography on DEAE-Cellulose. The above fraction is adsorbed at room temperature onto a DEAE-cellulose column (1 × 20 cm) that has been equilibrated with 0.01 M potassium phosphate buffer (pH 7.2) containing 0.005 M dithiothreitol. The column is washed with 12 ml of the same buffer, and elution is started with 8 ml of 0.025 M buffer containing dithiothreitol. This is followed by a linear gradient of potassium phosphate buffer (pH 7.2), from 0.02–0.2 M, containing 0.005 M dithiothreitol. The enzyme is eluted at a buffer concentration of 0.08–0.10 M. Four-milliliter fractions are collected, and those containing the enzyme activity are pooled. A 100-fold purification of the enzyme from leukemic spleen is attained by these procedures.

For removal of dithiothreitol, the enzyme preparation is concentrated with Aquacide I (Calbiochem). The material is then passed through a Sephadex G-25 column equilibrated with 0.01 M potassium phosphate buffer (pH 7.2) and eluted with the same buffer. This operation occasionally results in inactivation of the enzyme.

Properties

Stability. The crude enzyme is stable for 4 weeks at $-20°$, but repeated freezing and thawing destroy the activity. The enzyme becomes very labile during the later stages of purification, but activity can be restored by the addition of the thiol donors GSH or dithiothreitol. The final preparation (DEAE fraction) is very unstable with up to 50% of the activity lost on storage at room temperature or 4° for 24 hr.

Product Specificity. With the crude enzyme preparations, the main radioactive product formed is uridine. A small amount of uracil (usually less than 10% of the uridine) is also detected because of the presence of nucleoside phosphorylase. The purified preparation is mostly devoid of the latter activity.

Substrate Specificity. The enzyme can deaminate a variety of cytosine nucleosides. The relative rate of deamination as compared to cytidine is as follows: 2-deoxycytidine, 47%; cytosine arabinoside, 35%; 5-methyldeoxycytidine, 60%; 5-iodocytidine, 200%; 5-iododeoxycytidine, 60%; and 5-bromodeoxycytidine, 172%. No activity is observed with cytosine, 5-methylcytosine, 6-methylaminopurine, CMP, dCMP, CDP, dCDP, CTP, or dCTP.

Other Properties. The reaction is linear for 45 min. The purified enzyme is very sensitive to inhibition by SH reagents; p-hydroxymercuribenzoate, N-ethylmaleimide, Hg^{2+}, and Cu^{2+}. Activation of the enzyme occurs with EDTA, GSH, and dithiothreitol. Slight activation is noted with Mn^{2+}. Pyrimidine nucleotides (UDP, dUDP, UTP, dUTP, CDP, dCDP, CTP, dCTP, and dTTP) in concentrations of up to $3.0 \times 10^{-3} M$ do not alter the enzyme activity. The enzyme in crude homogenates is relatively heat stable, with no loss of activity when heated at 50° for 15 min. The activity is lost completely after heating at 75° for 15 min. The molecular weight of the enzyme is 74,000, as estimated by filtration through a Sephadex G-150 column.

Comparison to Normal Spleen Enzyme. When cytidine deaminase is purified from normal mouse spleen through step 3 of the purification procedure, a comparison of its properties with those of the enzyme from leukemic spleen at the same stage of purification reveals the following: The specific activity of the enzyme from leukemic spleen is much higher than that from normal spleen (400-fold higher in the crude extract). The K_m for cytidine is identical for the two enzymes (63 μM), but the K_m for deoxycytidine is about 4 times greater for the leukemic spleen enzyme than for the normal spleen enzyme (340 μM vs 79 μM). The enzyme from normal spleen is less sensitive to stimulation by dithiothreitol or inhibition by p-hydroxymercuribenzoate. In the presence of dithiothreitol, both enzymes have a broad pH optimum ranging from approximately pH 5–8. The absence of dithiothreitol has no effect on the pH response curve of the normal spleen enzyme, but the leukemic spleen enzyme shows a loss of activity most striking in the alkaline range with a pH optimum of 7.2; in the acid pH range much higher activity is noted with citrate than with acetate buffer with an optimum at 5.5. These differences in properties between the normal and leukemic spleen enzymes could be due to other proteins in the reaction mixture since the specific activity of the normal spleen enzyme is very much lower than that of the leukemic spleen enzyme.

Cytidine Deaminase from Other Sources. High levels of cytidine deaminase are found in human polymorphonuclear leukocytes.[10] A 700-fold purification of this human granulocyte enzyme has been reported.[11] Genetic polymorphism of human leukocyte cytidine deaminase has recently been demonstrated.[12] Cytidine deaminase has also been partially purified from mouse kidney[13,14] and sheep liver.[15]

[10] R. Silber, *Blood* **29**, 896 (1967).
[11] B. A. Chabner, D. G. Johns, C. N. Coleman, J. C. Drake, and W. H. Evans, *J. Clin. Invest.* **53**, 922 (1974).
[12] Y. Teng, J. E. Anderson, and E. R. Giblett, *Am. J. Hum. Genet.* **27**, 492 (1975).
[13] W. A. Creasey, *J. Biol. Chem.* **238**, 1772 (1963).
[14] R. Tomchick, L. D. Saslaw, and V. S. Waravdekar, *J. Biol. Chem.* **243**, 2534 (1968).
[15] G. B. Wisdom and B. A. Orsi, *Eur. J. Biochem.* **7**, 223 (1969).

[54] Deoxycytidylate Deaminase from T2-Infected *Escherichia coli*

By Gladys F. Maley

$$dCMP + H_2O \rightarrow dUMP + NH_3$$

Deoxycytidylate deaminase (EC 3.5.4.12) is not detectable in *Escherichia coli* until after infection with T-even bacteriophage.[1] This enzyme provides the infected cell with a new pathway for the synthesis of dUMP, a nucleotide required for the synthesis of dTMP. The phage-induced enzyme, as in the case of a similar enzyme in animal tissues,[2] is subject to feedback regulation by the end-products of its metabolic

[1] K. Keck, H. R. Mahler, and D. Fraser, *Arch. Biochem. Biophys.* **86**, 85 (1960).
[2] F. Maley and G. F. Maley, *Adv. Enzyme Regul.* **8**, 55 (1970).

pathway, dCTP (or HM-dCTP) and dTTP.[3-5] The methodology reported here relies on the excellent cooperation provided by the New England Enzyme Center, Boston, Massachusetts. Although they employ techniques which are not normally available, the procedure can be adapted readily for use in the average laboratory.

Assay Method

The most convenient assay for this enzyme is the spectrophotometric procedure described previously in this series.[6] Although other wavelengths can be used, the decrease in absorbancy at 290 nm is both simple and reliable.

Reagents

Tris·HCl buffer, pH 8.0, 20 mM, containing: dCMP, 1.0 mM; MgCl$_2$, 2.0 mM; and dCTP, 8.0 μM
2-Mercaptoethanol, 0.1 M freshly prepared

Procedure. In practice, 0.5 ml of the substrate solution, 0.05 ml of the 2-mercaptoethanol solution, and 0.45 ml of water are equilibrated in 1.0- or 1.2-ml silica cuvettes of 10-mm light path at 30° or 37°. Enzyme is added in a volume of 50 μl, the solution is mixed, and the decrease in absorbancy at 290 nm is recorded automatically. A unit is defined as the deamination of 1 μmole of dCMP per minute at 37°. Protein is determined by the method of Lowry *et al.*,[7] and specific activity is defined as units of dCMP deaminase per milligram of protein.

Purification Procedure

Reagents

Culture medium[8] containing (in grams per liter) K$_2$HPO$_4$ (7), KH$_2$PO$_4$ (3), (NH$_4$)$_2$SO$_4$ (1), MgSO$_4$ (0.1), sodium citrate (0.5), and glucose (2) is sterilized separately.
Bacteriophage, 4–6 × 10^{12}/ml[9]

[3] F. Maley and G. F. Maley, *J. Biol. Chem.* **240**, PC3226 (1965).
[4] W. H. Fleming and M. J. Bessman, *J. Biol. Chem.* **242**, 363 (1967).
[5] G. F. Maley, D. U. Guarino, and F. Maley, *J. Biol. Chem.* **242**, 3517 (1967).
[6] F. Maley, this series, Vol. 12, p. 171.
[7] O. H. Lowry, N. J. Rosebrough, A. L. Farr, and R. J. Randall, *J. Biol. Chem.* **193**, 265 (1951).
[8] B. D. Davis and E. S. Mingioli, *J. Bacteriol.* **60**, 17 (1950).
[9] G. F. Maley, D. U. Guarino, and F. Maley, *J. Biol. Chem.* **247**, 931 (1972).

Deoxycytidine-5'-monophosphate,   deoxycytidine-5'-triphosphate,
from Calbiochem
2-Mercaptoethanol, from Eastman Organic Chemicals
Ammonium sulfate (enzyme grade)
Streptomycin sulfate, from ICN
DEAE-cellulose and phosphocellulose, from Schleicher and Schuell
Inc.
Sephadex G-200, from Pharmacia

Growth of Bacteria. Escherichia coli B are grown at 37° in an aerated tank containing 530 liters of Davis minimal medium with agitation. The fermentor is inoculated with 1–2% inoculum. When the culture reaches an A_{650} of 0.8, purified T2r+ bacteriophage are added to a multiplicity of infection of 5. After 9 min of additional aeration, the infected culture is transferred to two 600-liter tanks which contain a total of 530 pounds of chipped ice. The cells are harvested from the culture in a refrigerated Sharples Md 16 centrifuge and frozen. The total yield of infected cells is approximately 500 g (net weight), which can be used immediately for the purification of dCMP deaminase or kept frozen at −20° without appreciable effect on the yield of enzyme.

Preparation of Broken-Cell Suspension. The frozen cell paste is thawed overnight at 4° and then suspended in 2 volumes of 10 mM potassium phosphate buffer, pH 7.5, containing 2 mM MgCl$_2$. The suspension is passed through a double layer of cheesecloth, and the cells are broken by three passages through a Manton–Gaulin laboratory homogenizer. The broken cell suspension is used immediately or stored at −20°.

All of the preceding steps are performed at the New England Enzyme Center. The remaining steps are those employed in our laboratory.

Step 1. Crude Extract. The crude broken-cell suspension (750 ml) is thawed overnight at 4°, brought to 1750 ml with 20 mM 2-mercaptoethanol, and sonicated for 2 min at maximal intensity with the standard probe of the Biosonik IV (Bronwill Corp.). The sonic extract is centrifuged at 16,300 g for 10 min, and the pellet is discarded.

Step 2. Streptomycin Precipitation and Autolysis. A 5% solution of streptomycin sulfate (300 ml) is added slowly to the supernatant fluid with continuous stirring. The resultant mixture (1800 ml) is stirred for an additional 10 min and then centrifuged at 16,300 g for 15 min. After decanting, the precipitate is homogenized in the Sorvall Omnimixer for 1 min with 900 ml of buffer containing 0.1 M potassium phosphate (pH

7.1)–3 mM MgCl$_2$–40 μM dCTP–20 mM 2-mercaptoethanol. Chloroform (0.9 ml) is added, and the suspension is incubated at 37° for 3 hr and then at 25° for 12–14 hr. The suspension is chilled in an ice bath and diluted with an equal volume of cold 20 mM 2-mercaptoethanol. The cold mixture is centrifuged at 16,300 g for 15 min to remove a small precipitate which forms during autolysis.

Step 3. Ammonium Sulfate Precipitation. Solid ammonium sulfate is slowly added to the continuously stirred supernatant fluid to a final saturation of 65%. After stirring for an additional 10 min, the precipitate is collected by centrifugation at 16,300 g for 10 min, dissolved in about 200 ml of 10 mM potassium phosphate, pH 7.5, and 20 mM 2-mercaptoethanol, and dialyzed against two 6-liter changes of the same buffer overnight. The dialysis tubing that is used was boiled in 0.1 N NaOH and 0.1 mM EDTA for 30 min, then rinsed in glass-distilled water, boiled in glass-distilled water for 15 min, and finally rinsed and stored in glass-distilled water at 4°.

Step 4. DEAE-Cellulose Chromatography. The dialysate (275 ml) is sonicated for 2 min and applied to a column of DEAE-cellulose (7.5 × 15 cm) equilibrated with 10 mM potassium phosphate, pH 7.5, and 20 mM 2-mercaptoethanol. The column is washed successively with 300 ml of 10 mM, 200 ml of 50 mM, and 500 ml of 0.1 M potassium phosphate, pH 7.5. The elution buffers all contain 2 mM MgCl$_2$ and 20 mM 2-mercaptoethanol. Fractions of about 15 ml are collected, and aliquots are assayed for deaminase activity spectrophotometrically. The fractions containing the peak activity are pooled, and the protein is precipitated by addition of solid ammonium sulfate to 60% saturation. After centrifugation the precipitate is dissolved in a minimal volume of buffer containing 10 mM potassium phosphate (pH 7.1), 2 mM MgCl$_2$, and 50 μM dCTP. The final solution (10–15 ml) is stored at −20°. For the next step, five of the stored fractions are combined and dialyzed against two 5-liter changes of 10 mM potassium phosphate, pH 7.1, and 20 mM 2-mercaptoethanol.

Step 5. Phosphocellulose Chromatography. The dialysate (150 ml) is sonicated for 2 min with the Biosonik IV standard probe at full intensity and passed through a column of phosphocellulose (3.4 × 20 cm) equilibrated with 10 mM potassium phosphate, pH 7.1, and 20 mM 2-mercaptoethanol. The column is washed successively with 500 ml each of solutions containing 10 mM, 50 mM, and 0.1 M potassium phosphate, pH 7.1, and 20 mM 2-mercaptoethanol. Fractions of 10 ml are collected at intervals of 2–3 min, and the deaminase is eluted on addition of the 0.1 M phosphate buffer. Protein in the pooled peak fractions is precipitated by addition of solid ammonium sulfate to 60% saturation. The centrifuged precipitate is dissolved in a minimal volume of buffer containing 50 mM potassium phosphate (pH 7.1), 1 mM MgCl$_2$, 20 μM dCTP, and 20 mM 2-mercaptoethanol.

Step 6. Sephadex G-200 Chromatography. The enzyme solution (10 ml) is passed through a column of Sephadex G-200 (75–120 μm, 2.5 × 90 cm) equilibrated with the buffer used to dissolve the deaminase. Protein is eluted with the same buffer at a flow rate of 25 ml/hr. Fractions of 5 ml are collected and aliquots assayed for deaminase activity. The deaminase is eluted from the column immediately after the void volume and is coincident with the major protein peak. The peak enzyme fractions are pooled, precipitated with solid ammonium sulfate to 60% saturation, and dissolved in 10 mM potassium phosphate, pH 7.1, and 20 mM 2-mercaptoethanol.

Step 7. Isoelectric Precipitation. The enzyme solution (10 ml) is dialyzed overnight against two 2-liter changes of a buffer containing 10 mM potassium phosphate, pH 7.8, and 20 mM 2-mercaptoethanol. During the course of the dialysis, the enzyme precipitates and is collected by centrifugation. The precipitate is washed three times with 5–10 ml portions of the same buffer. The final precipitate is dissolved in a solution of 0.2 M potassium phosphate, pH 7.1, and 0.1 M 2-mercaptoethanol. The maximum solubility is about 5–6 mg of enzyme per milliliter.

The results of the purification are summarized in the table. The method is highly reproducible, and 25–30 mg of homogeneous bacteriophage T2-induced dCMP deaminase can be obtained from 1.5–2 kg of infected cells. The overall recovery is 30–45%. The procedure is equally

PURIFICATION OF DEOXYCYTIDYLATE DEAMINASE

Purification step	Volume (ml)	Protein (g)	Activity units	Specific activity (units/mg)	Yield (%)
1. Crude extract	1700	35.0	4,868	0.14	100
2. Streptomycin autolysate	1740	6.9	4,720	0.68	97
3. Ammonium sulfate	275	3.22	4,452	1.38	91
4. DEAE-cellulose	245	0.285	4,020	14.1	83
5. Phosphocellulose[a]	213	0.207	19,715	95.2	81
6. Sephadex G-200	100	0.115	19,000	165.2	78
7. Isoelectric precipitation	5	0.0255	10,965	430.0	45

[a] Combined five preparations from step 4; see *Purification Procedure.*

effective for isolating the deaminase from *E. coli* infected with either T4 or T6 bacteriophase. Polyacrylamide gel electrophoresis and sedimentation velocity studies indicate that this preparation is homogeneous.[9,10]

Properties

Size and Subunit Structure. The native enzyme has a molecular weight of 124,000 and is composed of six identical subunits.

Composition. A detailed amino acid analysis shows no unusual quantities of amino acids in the deaminase except for methionine and cysteine.[9] Each subunit contains two residues of methionine, one of which occupies the amino terminal end, and eight to nine residues of half-cysteine, all in the reduced form. The carboxy terminus of each subunit is occupied by glutamic acid.

Stability. The pure dCMP deaminase is stable indefinitely when frozen in the absence of 2-mercaptoethanol. When kept at 0–4° and neutral pH in the presence of 0.1 M 2-mercaptoethanol, the enzyme is stable for several months.

Substrate Specificity. Deoxycytidylate deaminase from T2-infected *E. coli* also catalyzes the deamination of 5-MedCMP and the halogenated derivatives of dCMP but appears to be restricted in its ability to deaminate substrates with a large group in the 5-position, such as an iodo- or hydroxymethyl. With respect to the activator ligand, both dCTP and 5-hydroxymethyl dCTP (HM-dCTP) are effective. The true substrate of the enzyme is probably a dCMP-Mg complex, since the reaction is completely dependent on added divalent cation. Mn^{2+} and Ca^{2+} are less effective than Mg^{2+}.[5]

Kinetic Parameters. At the pH optimum, 8.3, in the presence of dCTP (or HMdCTP), the apparent K_m values of the enzyme are 1×10^{-4} M for dCMP and 5-MedCMP and 1×10^{-2} M for CMP.[9] Decreasing the pH to 6.0 lowers the apparent K_m for each of these substrates. In the absence of dCTP the pH optimum is 5.3.

Influence of Regulatory Ligands. Deoxycytidylate deaminase is stabilized by the presence of thiols, which facilitates the measurement of the cooperative response of the enzyme to the substrate, dCMP. The sigmoidal-velocity rates obtained in the absence of dCTP (or HMdCTP) are converted to normal hyperbolic Michaelis–Menton kinetics in their presence. This effect can be completely reversed to sigmoidal kinetics

[10] G. F. Maley, R. MacColl, and F. Maley, *J. Biol. Chem.* **247,** 940 (1972).

by increasing the amount of the inhibitor ligand, dTTP. The low concentrations of these ligands required to produce the observed responses emphasize the importance of the dCTP–dTTP interplay in regulating the activity of the deaminase. As expected, dCTP, and to a lesser extent dTTP, protects the enzyme from inactivation by heat, protein denaturants, and proteolytic enzymes.

[55] dCTP Deaminase from *Salmonella typhimurium*[1]

By Jan Neuhard

$$dCTP + H_2O \rightarrow dUTP + NH_3$$

Assay Methods

Spectrophotometric Assay[2]

Principle. The assay employed for following the purification of dCTP deaminase is based on the difference in molar extinction coefficients between dCTP and dUTP. At 290 nm and pH 2 this difference is 10.3×10^3. The sensitivity of the assay is limited by the absorbance of the substrate dCTP and by the fact that the assay is based on measurements of *decreases* in absorbancies.

Reagents

Standard buffer: potassium phosphate, 50 mM, pH 6.8, which is 2 mM with respect to β-mercaptoethanol and 2 mM with respect to ethylenediaminetetraacetic acid (EDTA)

Substrate: dCTP, 10 mM, which is 50 mM with respect to $MgCl_2$

Procedure. A mixture of 0.35 ml of standard buffer and 0.1 ml of enzyme is preincubated at 37° for 3 min. The reaction is started by the addition of 0.05 ml substrate. At 3, 6, 9, and 12 min, 0.1-ml samples are withdrawn from the assay mixture and added to 0.9 ml of ice-cold 0.5 M perchloric acid. After centrifugation the absorbancy of the supernatant fluid is measured at 290 nm. Assays on crude extracts are very unreliable. It is essential to remove nucleic acids from such extracts with

[1] dCTP aminohydrolase, EC 3.5.4.13.
[2] C. F. Beck, A. R. Eisenhardt, and J. Neuhard, *J. Biol. Chem.* **250**, 609 (1975).

streptomycin sulfate (see *Purification Procedure*) and to dialyze before reproducible activity measurements can be obtained.

Radioactive Assay[2]

Principle. In this assay the dCTP deaminase activity is measured as the amount of [³H]dUTP formed from [³H]dCTP with time. Since the assay is based on a chromatographic separation of dCTP and dUTP, it is essential that the enzyme preparation be free of enzymes that degrade these compounds. Thus, this assay can only be used with fairly purified preparations of dCTP deaminase. The advantage of this assay over the spectrophotometric assay is that it employs much less enzyme due to its larger sensitivity and the smaller assay volume.

Reagents

Buffer: potassium phosphate, 50 mM, pH 6.8, made 2 mM with respect to β-mercaptoethanol

Substrate: [³H]dCTP, 10 mM, specific activity 0.5 μCi/μmole, made 50 mM with respect to MgCl$_2$

Procedure. Buffer and enzyme in a total volume of 0.045 ml are mixed and incubated for 2 min at 37°. The reaction is started by the addition of 0.005 ml substrate. At 3-min intervals 0.01-ml samples of the reaction mixture are withdrawn and applied to the start spots of polyethyleneimine-impregnated cellulose plates for thin-layer chromatography (PEI plates).[3] The spots are dried rapidly with hot air. Prior to the application, 10 nmoles of dCTP and dUTP, and 200 nmoles of EDTA, are applied to each start spot. After termination of the assay chromatograms are washed 10 min in absolute methanol, dried, and developed step by step in one dimension in: (1) absolute methanol, to the start; (2) ammonium sulfate, 0.5 M, to 3.5 cm above the start; and (3) ammonium sulfate, 0.7 M, to 12 cm above the start. After drying, the spots corresponding to dCTP and dUTP are cut out, placed in counting vials, and the nucleotides eluted by the addition of 0.25 ml of ammonia, 2 M. After 20 min, 8 ml of a toluene-based scintillation liquid containing 10% Bio-Solv (BBS 3) are added, and the samples are counted in a scintillation spectrometer. From the percentage conversion of dCTP to dUTP with time, initial velocities are calculated.

[3] PEI plates were prepared according to K. Randerath and E. Randerath, *Anal. Biochem.* **13**, 575 (1965). However, commercial PEI plates may be used with similar results.

Units and Specific Activities

One unit is defined as the amount of enzyme that converts 1 nmole of substrate to product per minute at 37°. Specific activity is units per milligram of protein. Protein is determined by the method of Lowry *et al.*[4]

Purification Procedure[2]

Growth of Cells. Salmonella typhimurium LT2 is grown at 37° with vigorous aeration in a fortified glucose minimal medium.[2] In late exponential phase the culture is chilled by the addition of ice, and the cells are harvested by centrifugation. The cell paste is stored at −60°. All the following steps are performed at 4° unless otherwise stated.

Step 1. Extraction. The frozen cell paste is suspended in 2 volumes of standard buffer. The cells are broken in a French press and the cell debries removed by centrifugation in a refrigerated Sorvall centrifuge at 43,000 g for 90 min. The pH of the supernatant fluid is adjusted to 6.8 by the addition of KOH, 2 *M*.

Step 2. Streptomycin Precipitation. To each milliliter of the supernatant fluid 0.25 ml of a 15% solution of streptomycin sulfate in standard buffer is added slowly with stirring. The stirring is continued overnight, followed by centrifugation for 60 min at 43,000 g.

Step 3. Ammonium Sulfate Fractionation. The streptomycin supernatant fluid is treated with solid ammonium sulfate to 0.35 saturation (0.198 g/ml). After 30 min the precipitate is removed by centrifugation and the pH of the supernatant fluid adjusted to 6.8 by the addition of KOH, 2 *M*. The enzyme is precipitated from the neutralized supernatant fluid by the further addition of solid ammonium sulfate to 0.47 saturation (0.074 g/ml). After centrifugation the precipitate is dissolved in standard buffer (one-fifth the volume of the original extract) and passed through a Sephadex G-25 column (3 × 50 cm) equilibrated with 0.05 *M* potassium phosphate (pH 6.5), 2 m*M* β-mercaptoethanol, 2 m*M* EDTA, 20% ethylene glycol, and 0.05 *M* NaCl.

Step 4. DEAE-Sephadex Chromatography. The desalted solution from the previous step is applied to a DEAE-Sephadex A-50 column (5 × 50 cm) equilibrated with the same buffer as used for the G-25 column,

[4] O. H. Lowry, N. J. Rosebrough, A. L. Farr, and R. J. Randall, *J. Biol. Chem.* **193**, 265 (1951).

and the column is washed with 900 ml of the application buffer. dCTP deaminase is eluted from the column with a linear gradient in NaCl from 0.05–0.35 M, both limiting solutions containing 0.05 M potassium phosphate (pH 6.5), 2 mM β-mercaptoethanol, 2 mM EDTA, and 20% ethylene glycol. Total gradient volume is 6 liters, and the flow rate is 25 ml/hr.[5] The fractions containing the enzyme are pooled. After adjustment of the pH to 6.8, the pooled fraction is concentrated 5-fold by pressure filtration (Amicon). Ethylene glycol is removed from the filtrate by passage through a Sephadex G-25 column (3 × 50 cm) equilibrated with standard buffer, and the enzyme is recovered from the eluate by ammonium sulfate precipitation (0.60 saturation). The precipitate is dissolved in 2 ml of standard buffer.

Step 5. Sephadex G-100 Chromatography. The concentrated DEAE fraction is applied to a Sephadex G-100 column (2.5 × 95 cm) equilibrated with 0.05 M potassium phosphate (pH 7.4), 2 mM β-mercaptoethanol, 2 mM EDTA, and 20% ethylene glycol. The enzyme is eluted from the column with the same buffer (10 ml/hr). The fractions containing the enzyme are pooled and concentrated 10-fold by pressure filtration.

Step 6. Heat Treatment. The concentrated G-100 fraction is heated at 60° for 5 min followed by rapid chilling. Denatured protein is removed by centrifugation. The supernatant fluid may be concentrated further by pressure filtration.

The purification procedure is summarized in the table.

PURIFICATION PROCEDURE FOR dCTP DEAMINASE FROM *Salmonella typhimurium*[a]

Fraction	Total units	Total protein (mg)	Specific activity (units/mg)	Recovery (%)
Streptomycin fraction	13,300	6700	2	—
Ammonium sulfate	11,500	1890	6	86
DEAE-Sephadex (concentrated)	3,400	61	56	26
Sephadex G-100 (concentrated)	1,410	11	128	11
Heat treatment	1,470	1.5	980	11

[a] Data given are from a purification starting with 100 g packed cells of S. *typhimurium* KP-1074.[2]

[5] Essentially identical fractionations are obtained when DEAE-cellulose DE-52 columns are used instead of DEAE-Sephadex A-50 columns.

Properties

Purity. The dCTP deaminase preparation obtained by this procedure is about 20% pure as judged by analytical polyacrylamide gel electrophoresis.

Stability. In the presence of 20% ethylene glycol purified dCTP deaminase is stable for several weeks at 0°, and it may be stored at −50° for at least 2 months without loss of activity.

Substrate Specificity. The enzyme is highly specific for dCTP. None of the following compounds will serve as substrates: dCDP, dCMP, 2'deoxycytidine, CTP, CDP, CMP, cytidine, or cytosine.

pH Optimum. With saturating dCTP concentrations dCTP deaminase shows a pH optimum of 6.8. However, the pH optimum decreases with decreasing dCTP concentration.[2]

Metal Ion Requirement. The enzyme has an absolute requirement for divalent metal ions. Mg^{2+}, Mn^{2+}, and Co^{2+} are equally effective. Kinetic data suggest that a metal–dCTP complex is the true substrate for the enzyme.[2]

Inhibitors. dCTP deaminase is inhibited by *p*-chloromercuribenzoate, 12 μM giving 50% inhibition. The inhibition is completely reversed by β-

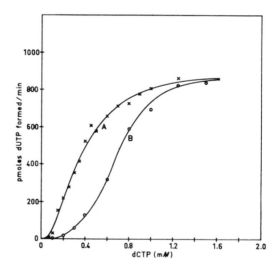

Fig. 1. Kinetics of dCTP saturation of dCTP deaminase in the absence (A) and in the presence (B) of 0.4 mM dTTP. In assays containing dTTP, the Mg^{2+} concentration was increased by 0.4 mM. From Beck *et al.*[2]

mercaptoethanol. Of 14 naturally occurring nucleotides tested, only dTTP and dUTP showed inhibition. With a substrate concentration of 0.2 mM, dTTP and dUTP at 0.2 mM inhibit the enzyme 94% and 83%, respectively (see next paragraph).

Kinetics. The enzyme shows a sigmoid saturation curve with respect to the substrate (curve A on Fig. 1); the Hill coefficient n for dCTP is 1.7. In the presence of dTTP the sigmoidicity of the substrate saturation curves increases (curve B on Fig. 1). With 0.2 mM and 0.4 mM dTTP the n values are 2.5 and 3.0, respectively.

Molecular Weight. The molecular weight of dCTP deaminase as determined by gel filtration on Sephadex G-200 is about 82,000.

Acknowledgment

This work was supported by a grant from the Danish Natural Science Research Council.

[56] Uridine Phosphorylase from Rat Liver

By E. W. YAMADA

$$\text{Uridine} + P_i \rightleftharpoons \text{uracil} + \text{ribose-1-P} \qquad (1)$$

Uridine phosphorylase, because of its role in the degradation of pyrimidine nucleosides as well as in the "salvage" pathway for nucleic acid synthesis,[1] occupies an important amphibolic position in metabolism. Separation of this enzyme from thymidine phosphorylase was achieved first for *Escherichia coli*[2] and later for mammalian tissues.[3,4] The enzyme resides mainly in the cytosol fraction of rat liver.[4]

In the purification scheme described here, major difficulties arising from the instability of the enzyme[5] and irreversible inactivation accompanied by new electrophoretic species are resolved.[6,7]

[1] A. Kornberg, *in* "The Chemical Basis of Heredity" (W. D. McElroy and B. Glass, eds.), p. 579. Johns Hopkins Press, Baltimore, Maryland, 1957.
[2] W. E. Razzell and H. G. Khorana, *Biochim. Biophys. Acta* **28**, 562 (1958).
[3] T. A. Krenitsky, M. Barclay, and J. A. Jacquez, *J. Biol. Chem.* **239**, 805 (1964).
[4] E. W. Yamada, *J. Biol. Chem.* **243**, 1649 (1968).
[5] P. Reichard and O. Sköld, this series, Vol. 6 [VII].
[6] R. Bose and E. W. Yamada, *Biochemistry* **13**, 2051 (1974).
[7] A. Kraut and E. W. Yamada, *J. Biol. Chem.* **246**, 2021 (1971).

Assay Method

Principle. Other assay methods of arsenolysis and detection of uracil or ribose formation[5,8] suffer from the disadvantages that activity is nonlinear with time and the precision of assay is low.[8] The most convenient and reliable method entails measurement of the rate of phosphorolysis of nucleosides by spectrophotometry[9] with modifications for enzyme samples of high[4] or low[10] protein density. The basis of the method is the difference in absorption spectra in the alkaline pH range[11] in ultraviolet (UV) light between pyrimidines and their respective nucleosides. The assay is very sensitive to extraneous contaminants; glassware must be cleaned scrupulously, preferably in P_i-free detergent, dichromate solutions above all are to be avoided, and crude enzyme fractions should be dialyzed to remove endogenous UV-absorbing material. In addition, the enzyme is sensitive to oxidation and metal ions so that suspension media should contain β-mercaptoethanol and EDTA wherever possible.

Reagents

Potassium phosphate buffer, 1.0 M, pH 7.4
Potassium phosphate buffer, 0.05 M, pH 7.0
β-Mercaptoethanol, 0.15 M (105 μl diluted to 10 ml)
Uridine, deoxyuridine, or thymidine, 0.1 M, pH 7.0
Perchloric acid, 2.12 M (96.5 ml of 72% diluted to 500 ml)
NaOH, 10 M (9.6 N by titration)

Procedure. The standard assay mixture contains in a final volume of 1.5 ml and at a pH of 7.4: 150 μl P_i buffer (pH 7.4), 50 μl β-mercaptoethanol, 0.5 ml P_i buffer (pH 7.0), plus enzyme suspended in this buffer and 50 μl nucleoside to initiate the reaction. P_i buffer, pH 7.0, plus enzyme aliquot, together, are kept at 0.5 ml to maintain the P_i concentration and a pH of 7.4. This is of particular concern when enzyme aliquots vary widely as during column chromatography.

Tubes are stored in ice prior to incubation at 37° for 5–10 min; the reaction is linear for 30 min and with increasing protein concentration up to an optical density (OD) increase of 0.390. Controls for each test are incubated or can be left in ice without incubation, in either case, without

[8] W. E. Razzell, this series, Vol. 12 [16].
[9] E. S. Canellakis, *J. Biol. Chem.* **227**, 701 (1957).
[10] A. Kraut, R. Bose, and E. W. Yamada, unpublished observations.
[11] G. H. Beaven, E. R. Holiday, and E. A. Johnson, *in* "The Nucleic Acids" (E. Chargaff and J. N. Davidson, eds.), Vol. 1, p. 493. Academic Press, New York, 1955.

nucleoside. The reaction is stopped with 450 μl perchloric acid; nucleoside is then added to control tubes. After 5 min in ice the contents are transferred to plastic tubes and centrifuged in rotor SM-24 at 16,000 g for 15 min in a Sorvall RC-2B at 2°. One-milliliter aliquots of the supernatant fluids are then added to 70 μl NaOH to give a final pH of 12. Each test solution is read against its control at either 290 nm (uracil) or 295 nm (thymine) in cuvettes of 1-cm light path. Duplicate samples agree within 0.020 OD unit or better.

Calibration graphs for standard solutions of the bases or nucleosides, treated as for enzyme assays but with enzyme omitted, are prepared. From these it is determined that for every μmole of uracil formed from uridine or deoxyuridine the optical density increases 2.11 and 2.26, respectively, and for every μmole of thymine formed from thymidine the optical density increases 2.06. These factors are used to determine directly the μmoles of base formed per tube.

Assays Under Other Conditions. When the enzyme protein contributes less than 0.01–0.02 OD unit the above assay procedure can be simplified.[10] Five minutes after the addition of perchloric acid, 140 μl NaOH are added with mixing. Precipitates form rapidly; supernatant fluids can be decanted immediately and read.

Under some conditions to maintain P_i concentrations as well as the pH of the assay mixture, P_i buffer, pH 7.4, is replaced by 30 μl 1 M P_i buffer, pH 6.85, and 0.5 ml P_i buffer, pH 7.0, plus enzyme is replaced by 1.2 ml enzyme plus 0.02 M P_i buffer, pH 8.0. This is necessary when enzyme fractions are suspended in the latter buffer as during column chromatography.

Reverse Direction Assay. The decrease in absorbance due to the formation of nucleoside from base is measured.[6,12] The method is as described for the standard assay except that P_i buffers are replaced by glycylglycine, acetate, or Tris, depending upon the final pH desired; to each tube 100 μl of 0.01 M uracil (or thymine), pH 7.0, and 100 μl of 0.042 M ribose-1-P (dicyclohexylammonium salt) or deoxyribose-1-P (cyclohexylammonium salt), pH 7.0, are added. The reaction is linear for 70 min at 37° up to an optical density change of 0.350 for uridine synthesis at pH 8.2.[7] The cost of the pentose sugars, the relative insolubility of the bases, as well as the requirement for P_i-free solutions preclude the use of this assay on a routine basis.

Definition of Unit and Specific Activity. A unit of uridine phosphorylase is defined as the amount of the enzyme catalyzing the formation

---

[12] L. M. Paege and F. Schlenk, *Arch. Biochem. Biophys.* **40**, 42 (1952).

of 1 μmole of uracil (or thymine) per minute at 37°. Specific activity is expressed as enzyme units per milligram protein determined by the Folin reagent[13] or in fractions separated by column chromatography by the UV absorption method of Warburg.[13] Crystallized bovine serum albumin is used as standard with the Folin reagent.

Purification Procedure

Fifty male rats of between 250–350 g are decapitated in lots of 6–8; the livers are perfused *in situ* with ice-cold 0.9% NaCl, removed, and weighed. Alternately, frozen livers (50–100) weighing between 5–7 g are obtained from a commercial source; the livers are rinsed in ice-cold sucrose solution (0.25 M sucrose–5 mM β-mercaptoethanol) before use. All subsequent steps are carried out at 2° or in an ice bath. Routinely, all dialyzing tubing is washed with 1% EDTA, pH 7.0, and then with dialysis buffer. Deionized, glass-redistilled water and reagents of highest purity are used throughout. The procedures are summarized in Table I.

Step 1. Homogenate. Minced livers are homogenized in 3 volumes of sucrose solution in a Potter–Elvejhem apparatus equipped with Teflon pestle; homogenates are filtered through four layers of cotton gauze.

TABLE I
PURIFICATION DATA FOR LIVER URIDINE PHOSPHORYLASE

Purification steps	Volume (ml)	Total units	Total protein (mg)	Yield (%)	Specific activity ($\times 10^3$)
1. Homogenate	3673	119.0	94,197	100	1.27
2. Cytoplasm	3735	107.1	48,832	90.0	2.00
3. 160,000 g supernatant	3357	92.8	23,698	78.0	3.83
4. Ammonium sulfate	172	74.5	11,988	62.6	6.20
5. Heat	170	60.8	8,262	51.1	7.33
6. DEAE-Sephadex (pH 8.0) No. 1	49	52.5	1,568	44.1	33.50
7. DEAE-Sephadex (pH 8.0) No. 2	23	33.6	488	28.2[a]	68.30
8. DEAE-Sephadex (pH 7.2)	6	20.4	193	17.2[a]	105.00
9. Sephadex G-200 (pH 8.0)	3	13.9	26	11.7	533.33
10. Hydroxyapatite (pH 7.3)	4	9.63	4	8.1	2400.00

[a] Low recoveries at these stages are due to incomplete precipitation of pooled column fractions by ammonium sulfate. The recovery of activity from the columns (steps 6–10) is 80–90%.

[13] E. Layne, this series, Vol. 3 [73].

Step 2. Cytoplasm. Nuclei and unbroken cells are sedimented at 900 g for 20 min in a Sorvall RC-2B centrifuge. The cytoplasmic fraction is removed; the sediment is rehomogenized with 2 times its weight of sucrose solution and recentrifuged. The supernatant fluid is combined with the cytoplasmic fraction and dialyzed in 100 volumes of buffer A [0.05 M potassium P_i buffer (pH 7.0)–10 mM β-mercaptoethanol–1 mM EDTA (pH 7.0)] in a continuous-flow apparatus.

Step 3. 160,000 g Supernatant. The dialyzed cytoplasmic fraction is centrifuged at 160,000 g for 1½ hr in a Spinco ultracentrifuge. At this point, uridine phosphorylase can be separated from thymidine phosphorylase in 60 ml of supernatant fluid from four livers by DEAE-Sephadex chromatography[4] (step 6); however, for bulk purifications, a more concentrated enzyme extract is preferable.

Step 4. Ammonium Sulfate. Solid ammonium sulfate (21 g/100 ml) is dissolved in the supernatant fluid of step 3 and, after stirring for an additional 20 min, the precipitate is removed by centrifugation at 16,000 g for 20 min. Additional ammonium sulfate (14 g/100 ml original volume) is added to the supernatant fluid. After centrifugation as before the precipitate (21–35% ammonium sulfate residue) is resuspended in buffer A.

Step 5. Heat. The fraction from step 4 is placed in stainless-steel centrifuge tubes and heated, with gentle stirring, for 3 min at 50°. The tubes are immersed quickly in ice, stirred for an additional min, and then centrifuged at 16,000 g for 15 min. The precipitate is washed with a few ml of buffer A and recentrifuged. The residue is discarded; supernatant fluids and washings are combined and saved.

Step 6. First DEAE-Sephadex Chromatography (pH 8.0). A column of DEAE-Sephadex A-50 (2.5 × 90 cm, bed volume, 400 ml) is equilibrated with buffer B [0.02 M potassium P_i buffer (pH 8.0)–10 mM β-mercaptoethanol–1 mM EDTA (pH 8.0)]. The fraction from step 5 is dialyzed in 100 volumes of buffer B for 1½ hr, diluted with buffer B to give a protein concentration of just less than 35 mg/ml, and applied to the column, followed by 1½ bed volumes of buffer B. Twelve-milliliter fractions are collected at a flow rate of 0.4 ml/min. Washing of the column with buffer B is continued until the optical density of effluents is reduced at 260 and 280 nm to low constant levels. Elution with a linear gradient of KCl is started by allowing 0.4 M KCl in 400 ml of buffer B to flow into a mixing chamber containing 400 ml of buffer B and thence into the column. Complete separation of uridine phosphorylase from thymidine phosphorylase is achieved by this step. Fractions containing enzyme activity are pooled; the enzyme is precipitated with ammonium sulfate (56 g/100 ml) and resuspended in buffer A.

Step 7. Second DEAE-Sephadex Chromatography (pH 8.0). The fraction from step 6 is dialyzed in buffer B as described in step 6 and then step 6 is repeated; the KCl gradient can be reduced to be linear from 0–0.3 *M*.

Step 8. Third DEAE-Sephadex Chromatography (pH 7.2). The enzyme preparation is dialyzed in buffer B adjusted to pH 7.2 and applied to a DEAE-Sephadex A-50 column (2.5 × 45 cm, bed volume, 240 ml) by a procedure identical with that of step 6 with the exception that the KCl gradient is now linear from 0–0.15 *M*.

Step 9. Gel Filtration (pH 8.0). A Sephadex G-200 column (2.5 × 90 cm, bed volume, 421 ml) is equilibrated with buffer B. The enzyme fraction from step 8 is concentrated to 4.8 ml in dialyzing tubing, tied at one end, and immersed in a 500-ml beaker containing Carbowax. After every 45 min the wet Carbowax is replaced by dry over 4–6 hr. The surface of the tubing is washed with buffer B and the sample is applied to the column and eluted with buffer B; 8-ml fractions are collected at a flow rate of 32 ml/hr. The fractions of highest activity are pooled, concentrated with Carbowax for 6 hr, and then dialyzed for 2 hr in 2 liters of buffer C [0.02 *M* potassium P$_i$ buffer (pH 7.3)–10 m*M* β-mercaptoethanol].

Step 10. Hydroxyapatite Chromatography (pH 7.3). A column of hydroxyapatite (0.9 × 60 cm, bed volume, 50 ml) is equilibrated with buffer C. Enzyme fraction from step 9 (2.5 ml, 21.7 mg protein) is applied followed by buffer C until absorption at 280 nm reaches low constant readings. The column is then eluted with a linear gradient of P$_i$ by allowing 300 ml of buffer C (P$_i$ adjusted to 0.15 *M*) to flow into a mixing chamber containing 300 ml of buffer C. Combined enzyme fractions are concentrated with Carbowax, dialyzed for 3 hr against buffer A, clarified by centrifugation at 11,000 *g* for 20 min, and stored frozen at −40°−−70°. Purification achieved is 1900-fold (Table I).

Properties

Gel Electrophoresis and Stability. Purified uridine phosphorylase migrates as a single, sharp band in 7% acrylamide with relative electrophoretic mobilities (R_m) of 0.33–0.36 at pH 8.3 or 0.28–0.31 at pH 7.0.[6] Faint satellite bands (enzyme aggregates) are discernible after storage for just over a week at −40° in buffer A; however, 90% of activity remains after storage for 6 weeks. All activity is lost after 4 weeks in Tris buffer [0.05 M Tris·HCl (pH 7.0)–10 mM β-mercaptoethanol] at $-40°$ at protein concentrations of 0.2–2.3 mg/ml; 25% of activity can be restored by incubation for 90 min at 25° in 0.05 M potassium P_i buffer (pH 7.0)–0.075 M β-mercaptoethanol. The reactivated enzyme contains an additional, more rapidly migrating electrophoretic band which does not stain for enzyme activity.[6]

Substrate Specificity. Nucleoside-cleaving activity is negligible in the absence of P_i. Constant ratios of 1.4 (uridine to deoxyuridine) and 10.0 (uridine to thymidine) at pH 7.4 are obtained after step 7 to step 10.

Synthesis of Nucleosides. The synthesis of uridine is inhibited 50% at pH 8.2 by 1.2 mM uracil, and that of deoxyuridine at pH 6.5, by 65%.[7]

pH Optima. The phosphorolysis of uridine, deoxyuridine, and thymidine proceeds with maximal velocity at pH 8.2, 6.5, and 5.7, respectively.[6,7] The pH optimum for the synthesis of uridine is 8.5.[7]

Molecular Weight and Subunit Composition. The molecular weight determined by Sephadex G-200 chromatography is 110,000[7] and by acrylamide gel electrophoresis,[14] 102,500.[6] The protein dissociates into four subunits of molecular weight of 26,000 in 0.5% sodium dodecyl sulfate–4 M urea–0.5% β-mercaptoethanol.[6]

Amino Acid Analysis. All but the tryptophan and tyrosine residues have been determined.[10] The molar ratios resemble those of purine nucleoside phosphorylase of *Bacillus cereus.*[15] Notable are the high half-cystine residues (24/mole) and histidine residues (12/mole).[6]

Inhibitors. Deoxyglucosylthymine inhibits the phosphorolysis of uridine and deoxyuridine and the synthesis of uridine by 50% at concentrations of 0.10 mM, 0.018 mM, and 0.14 mM, respectively. Phosphorolysis of deoxyuridine or thymidine by thymidine phosphorylase of rat liver is not inhibited by 0.19 mM deoxyglucosylthymine. This inhibitor can be used to distinguish the two types of phosphorylase.

Uridine phosphorylase is inhibited, in decreasing order, by p-mercuriphenylsulfonate, p-mercuribenzoate, 5,5′-dithiobis(2-nitrobenzoic acid), o-iodosobenzoate, N-ethylmaleimide, iodoacetamide, and iodoacetic acid in concentrations giving 50% inhibition ranging from 4.0 × 10^{-7} to 5.0 × 10^{-3} M.[7] Neither uridine nor P_i protects the enzyme against inhibition by the mercurials, but both protect substantially against the other sulfhydryl reagents. Uridine affords more protection than P_i, the

[14] J. L. Hedrick and A. J. Smith, *Arch. Biochem. Biophys.* **126,** 155 (1968).
[15] H. L. Engelbrecht and H. L. Sadoff, *J. Biol. Chem.* **244,** 6228 (1969).

preferential protection being most apparent with o-iodosobenzoate which reacts with vicinial sulfhydryl groups. Results with deoxyuridine are similar to those with uridine, at the pH optimum of each.[7]

Equilibrium Constants. K values for the phosphorolysis of uridine are 0.031 at pH 7.4 and 0.078 at pH 8.2.[6]

Kinetic Studies. Initial velocity and product inhibition studies show that the reaction proceeds by an ordered bi bi mechanism[16] in which P_i binds before uridine and ribose-1-P is released after uracil.[7] The kinetic constants are listed in Table II.[17]

Pentosyl Transfer. Ribosyl transfer from uridine to uracil, under steady-state conditions, is maximal when 0.4 mM P_i, 1.0 mM uridine, and 0.9 mM [2-^{14}C]uracil are present at pH 8.2 and is a linear function of time and enzyme concentration. P_i acts as a competitive inhibitor (K_i = 0.083 mM) as predicted by the rate equation[6] for the uracil-uridine exchange. Ribosyl transfer occurs in the absence of P_i only when exogenous ribose-1-P is present; maximal rates obtain with 0.1 mM ribose-1-P, 1.99 mM uracil, and 1.33 mM [U-^{3}H]uridine. In concentrations in excess of 0.6 mM, ribose-1-P inhibits transferase activity;

TABLE II

KINETIC PARAMETERS OF SUBSTRATES FOR URIDINE PHOSPHORYLASE[a]

Parameter	Phosphorolysis (mM)	Nucleoside synthesis (mM)
I. pH 8.2		
Phosphate, K_a	0.349	
Phosphate, K_{ia}	3.215	
Uridine, K_b	0.240	
Ribose-1-P, K_q		0.071
Ribose-1-P, K_{iq}		0.085
Uracil, K_p		0.286
II. pH 6.5		
Phosphate, K_a	0.550	
Phosphate, K_{ia}	2.340	
Deoxyuridine, $K_b{}'$	0.300	

[a] Determined by the computer program described by Cleland.[17]

[16] W. W. Cleland, *Biochim. Biophys. Acta* **67**, 105 (1963).
[17] W. W. Cleland, *Adv. Enzymol.* **29**, 1 (1967).

complete inhibition occurs at 16 mM or greater indicating a sequential mechanism without randomness.[18] Pentosyl transfer occurs by the normal uridine phosphorylase mechanism and can be classified as "indirect."[19] Thymidine phosphorylase, on the other hand, has an additional transferase activity[6,20,21] which has been designated as "direct."[19]

Catalysis Model. Kinetic studies relating changes in V_m or K_m with pH as well as studies with o-iodosobenzoate inhibition allow the model of Koshland[22] for uridine phosphorylase to be extended.[6,7] The active site of the enzyme has three sulfhydryl groups; at pH 8.0, uridine binds to two, one of which has an enzyme-substrate dissociation constant (pK_{es}) of 8.0, while P_i binds to the third. A histidyl residue in the unprotonated form forms a hydrogen bond with the C'-2-OH of the pentose moiety of uridine and one with P_i (pK_{es} = 7.0). The three-point attachment of uridine results in the withdrawal of electrons from N-1 of the β-N-glycosidic bond and orients this bond for "backside"[22] nucleophilic attack by P_i. There is facilitated proton transfer[23] and replacement of the more weakly nucleophilic N-1 of uridine by P_i. The model for deoxyuridine is similar except that at the pH optimum of 6.5 the protonated form of the histidyl residue forms a salt bridge with P_i (pK_{es} = 6.5), there is no interaction between the histidyl residue and the deoxyribose moiety, neither sulfhydryl group is ionized, and both bind to the uracil moiety of deoxyuridine by hydrogen bonds.

Tissue Distribution. Rat tissues which contain uridine phosphorylase include small intestine, heart, kidney, spleen, liver, brain, and testes, in decreasing order of specific activity which ranges from 16.5–0.67 nmoles of uracil formed per minute per milligram protein.[10]

[18] W. W. Cleland, *in* "The Enzymes" (P. D. Boyer, ed.), 3rd ed., Vol. 2, p. 1. Academic Press, New York, 1970.
[19] T. A. Krenitsky, *J. Biol. Chem.* **243,** 2871 (1968).
[20] R. C. Gallo and T. R. Breitman, *J. Biol. Chem.* **243,** 4936 (1968).
[21] B. K. Kim, S. Cha, and R. E. Parks, Jr., *J. Biol. Chem.* **243,** 1771 (1968).
[22] D. E. Koshland, Jr., *in* "The Enzymes" (P. D. Boyer, H. Lardy, and K. Myrbäck, eds.), 2nd ed., Vol. 1, p. 305. Academic Press, New York, 1959.
[23] J. H. Wang, *Science* **161,** 328 (1968).

[57] Pyrimidine Nucleoside Phosphorylase from Haemophilus influenzae[1,2]

By J. J. Scocca

Uridine + P_i ⇌ uracil + D-ribose-1-phosphate
Thymidine + P_i ⇌ thymine + 2-deoxy-D-ribose-1-phosphate

A single pyrimidine nucleoside phosphorylase comprises both uridine phosphorylase (uridine: orthophosphate ribosyltransferase, EC 2.4.2.3) and thymidine phosphorylase (thymidine: orthophosphate deoxyribosyltransferase, EC 2.4.2.4) in *Haemophilus influenzae*. The enzyme has been purified approximately 400-fold by the procedure described here; the pattern of substrate specificity was the same for crude extracts and for the most purified fractions. It appears that a single enzyme is responsible for cleaving the glycosidic bond of 2,6-diketopyrimidine nucleosides to the corresponding base and pentose-phosphate moieties.[2]

Assay Method

Principle. The enzyme is conveniently assayed by spectrophotometric measurement of the free pyrimidine base produced.[3] The routine assay employed uridine as substrate.

Procedure. Reaction mixtures (0.35 ml) contained 0.14 M potassium phosphate buffer (pH 8.2), 5 mM dithiothreitol, 12 mM uridine, and enzyme. After incubation for 20 min at 37°, the reaction was stopped by addition of 3 ml of 0.01 N NaOH. The absorbance was determined and corrected for a blank which had been stopped at zero time. The assay for cleavage for deoxynucleosides was done similarily, except that the pH was 7.4 and dithiothreitol was omitted. Modification of the assay for initial rate measurements has been described.[2]

Definition of Units. The wavelengths and extinction coefficients used in the assay of the enzyme are presented in Table I. *One unit* of enzyme produced 1 nmole of base per minute under the conditions of the routine assay; the assay was linear over the range of 1–5 units.

[1] Supported by NIH Grant GM 21890 and by a Research Career Development Award (GM 70663).
[2] J. J. Scocca, *J. Biol. Chem.* **246**, 6606 (1971).
[3] W. E. Razzell and H. G. Khorana, *Biochim. Biophys. Acta* **28**, 562 (1958).

TABLE I
WAVELENGTH AND EXTINCTION COEFFICIENTS FOR THE ASSAY

Base produced	Wavelength (nm)	$\epsilon_{mM}^{1\ cm}$
Uracil[a]	290	5.7
Thymine[a]	300	3.7
5-Bromouracil[b]	312	2.2

[a] Razzell and Khorana.[3]
[b] Scocca.[2]

Purification of the Enzyme

Growth of Cells. H. influenzae Rd stocks are maintained by adding sterile glycerol to exponentially growing cultures to a final concentration of 15% (v/v) and storing aliquots at −70°. For enzyme preparation, bacteria were grown at 37° with rotary shaking overnight in 3.7% Brain Heart Infusion (or 2.5% Heart Infusion) supplemented with 1 mg/liter of hemin and 0.4 mg/liter of NAD. Cells were harvested by centrifugation at 5000 g for 20 min at 4° and were washed with 0.1 culture volume of cold 0.15 M NaCl. Between 3–4 g of wet cell paste were obtained from each liter of culture.

All centrifugations were at 10,000 g for 20 min, and all operations were carried out at 0–4°. All solutions used in the purification contained 2 mM dithiothreitol.

Preparation of Crude Extract. To 40 g of packed H. influenzae cells, 80 g of alumina A 305 (Sigma) were added, and the mass was triturated in a chilled mortar until a homogeneous tacky paste was obtained. The homogenate was extracted with 160 ml of 0.02 M potassium phosphate buffer, pH 7.4, and centrifuged. The pellet was reextracted with 80 ml of the same buffer and centrifuged. The pooled supernatant solutions constituted fraction 1, the crude extract.

Precipitation with Streptomycin Sulfate. Solutions of streptomycin sulfate (10% in H₂O) were prepared just before use. Fraction 1 (230 ml) was stirred while 0.2 volume of streptomycin sulfate solution was added drop by drop. The suspension was stirred for an additional 30 min; the precipitate was centrifuged down and discarded. The supernatant solution was fraction 2.

Treatment with Acid and Adsorption to Calcium Phosphate Gel. Sodium acetate buffer (1 M), pH 4.8, was added drop by drop with stirring to fraction 2 to a final concentration of 0.05 M. Stirring was continued for 20 min; the suspension was centrifuged and the precipitate was discarded. Calcium phosphate gel[4] (50 mg of solids per milliliter) was added to the supernatant solution (0.15 ml of gel per milliliter); the suspension was stirred for 10 min and centrifuged, and the packed gel was discarded. Uridine phosphorylase was adsorbed by adding 0.2 ml of gel per milliliter of supernatant solution, stirring for 10 min, and centrifuging as before. The packed gel was washed twice with 50-ml portions of cold H_2O and collected by centrifugation. Enzymic activity was eluted by suspending the gel in 50 ml of 0.1 M potassium phosphate buffer, pH 7.4, stirring for 10 min, and centrifuging. The packed gel was reextracted with 25 ml of the same buffer, stirred, and centrifuged as before. The combined supernatant solutions constituted fraction 3.

Fractionation by Ammonium Sulfate. A solution of 1 mM EDTA–2 mM dithiothreitol saturated with $(NH_4)_2SO_4$ at 25° was used. To 75 ml of fraction 3, 49.5 ml of ammonium sulfate solution were added slowly with continuous stirring. Stirring was continued for 20 min; the precipitate was collected by centrifugation and discarded. To the supernatant solution, 43.0 ml of ammonium sulfate solution were added; the mixture was stirred for 20 min and centrifuged. The precipitate (fraction 4) was dissolved in 10 ml of 0.02 M potassium phosphate buffer, pH 7.4, containing 2 mM dithiothreitol and 20% (v/v) glycerol. In all subsequent operations, the buffers used contained both dithiothreitol and glycerol at these concentrations.

Chromatography on DEAE-Cellulose. A column (1.4 × 25 cm) of DEAE-cellulose was equilibrated with 0.02 M Tris-Cl, pH 7.4 (buffer 1). Fraction 4, previously dialyzed for 18 hr against 15 volumes of this buffer, was applied to the column at a flow rate of 0.02 ml/min. The column was washed with 50 ml of buffer 1 and eluted with a linear gradient established between 400 ml of 0.05 M KCl in buffer 1 and 400 ml of 0.4 M KCl in the same buffer. The enzyme emerged after 200 ml of the eluant had passed through the column. The active fractions were pooled to give fraction 5.

Chromatography on Hydroxylapatite. Fraction 5 was dialyzed for 18 hr against 0.01 M potassium phosphate buffer, pH 6.8, and applied to a column of hydroxylapatite (1 × 5 cm) which had been equilibrated with the same buffer. The column was washed with buffer at a flow rate of 0.2 ml/min. The enzyme passed through the column essentially unretarded and was collected in 3-ml fractions. These were pooled, and the

[4] D. Keilin and E. F. Hartree, *Proc. R. Soc. London, Ser. B* **124**, 397 (1938).

enzyme was concentrated by placing the solution in a dialysis bag, which was then immersed in dry Sephadex G-200 for 48 hr. The concentrated material was designated fraction 6.

The results of this purification procedure are summarized in Table II. Results of polyacrylamide gel electrophoresis of fraction 6 suggested that the enzyme constituted about 20% of the protein in this fraction; no interfering activities were found.

Properties of the Enzyme

Stability. Pyrimidine nucleoside phosphorylase from *H. influenzae* was found to be quite labile during the course of purification. The use of dithiothreitol and glycerol was necessary to obtain acceptable recoveries in the various steps; in addition, repeated cycles of freezing and thawing inactivated the enzyme if phosphate was absent. When these precautions were taken, fractions 4–6 were stable for 6 months when stored at $-20°$.

Effect of pH. The initial rates of phosphorolysis of ribonucleosides and of uracil arabinoside were maximal at pH 7.4. The optimum pH found with deoxynucleoside substrates was slightly more acidic, with a maximum at pH 6.9. At pH 7.4, deoxynucleoside cleavage proceeded at 75% of the rate observed at pH 6.9.

Substrate Specificity. Table III presents the results of a study of the abilities of various nucleosides to serve as substrates. The enzyme cleaved all the 4-keto pyrimidine nucleosides tested, regardless of the nature of the pentose moiety. Purine- and 4-amino pyrimidine-nucleosides were not split. Assays of the crude extract and of less purified

TABLE II
PURIFICATION OF PYRIMIDINE NUCLEOSIDE PHOSPHORYLASE

Fraction	Volume (ml)	Protein (mg/ml)	Activity (units/ml)[a]	Recovery (%)	Specific activity (units/mg)
1. Extract	230	6.4	410	100	70
2. Streptomycin	250	4.5	390	103	90
3. Calcium phosphate gel	75	2.3	1200	96	550
4. Ammonium sulfate	10	3.0	6100	65	2,030
5. DEAE-cellulose	37	0.08	900	35	12,200
6. Hydroxyapatite	16	0.04	1020	18	25,500

[a] Determined with uridine as substrate.

TABLE III
SUBSTRATE SPECIFICITY

Nucleoside	Concentration (mM)	Activity (units/ml)
Uridine	8.3	100
5-Bromouridine	5.6	210
5-Methyluridine	5.6	190
Uracil arabinoside	1.2	100
Deoxyuridine	10.0	270
Thymidine	8.0	265
5-Bromodeoxyuridine	12.0	1600
Adenosine, guanosine, or cytidine	6.0	<3[a]

[a] Determined by measuring the ribose rendered nonadsorbable to Norit A; all other values determined spectrophotometrically.

fractions showed that the ratio of uridine phosphorylase activity to thymidine phosphorylase activity had remained constant at a value of approximately 3.8 throughout the purification. The effects of varying the substituent at position 5 of the pyrimidine ring were dissimilar for ribo- and deoxyribonucleosides; uridine was the best ribonucleoside substrate tested whereas 5-bromodeoxyuridine was the deoxynucleoside most readily cleaved. This pattern of substrate specificity did not change during purification. It is of interest that the presence of ribonucleoside substrates protected both thymidine and uridine phosphorylase activities against heat, while deoxynucleoside substrates were ineffective.

Kinetics. Table IV summarizes the K_m and relative V_{max} values obtained for the nucleoside substrates; all these values were determined

TABLE IV
SUMMARY OF KINETIC PARAMETERS OF PYRIMIDINE NUCLEOSIDE
PHOSPHORYLASE WITH VARIOUS NUCLEOSIDE SUBSTRATES

Nucleoside	K_m (mM)	Relative V_{max}
Uridine	0.24	1.0
5-Methyluridine	0.07	0.27
5-Bromouridine	0.03	0.40
Deoxyuridine	0.13	0.12
Thymidine	0.11	0.21
5-Bromodeoxyuridine	0.10	0.74

at pH 7.4. Fraction 6 was used as enzyme; the measured V_{max} with uridine as substrate was 63.5 μmoles/min/mg of protein. The K_m for P_i, determined with uridine (1.2 mM) as substrate, was 0.38 mM. Uridine concentrations above 15 mM were somewhat inhibitory; no other nucleoside substrate tested had this effect.

The conclusion that a single pyrimidine nucleoside phosphorylase cleaves both uridine and thymidine in *H. influenzae* rests on several lines of evidence. First, the two activities maintained a constant ratio over a 400-fold purification and were not resolved by analytical poly-acrylamide gel electrophoresis. Second, the activities showed identical kinetics of heat inactivation and were identically protected against heating by several nucleoside substrates. Third, kinetic studies with ribo- and deoxynucleoside substrate pairs gave results which conformed quantitatively with expectation for a single enzyme comprising both activities.[2]

The pattern of substrate specificity observed is quite similar to that of the uridine phosphorylase purified from several mammalian sources, which were observed to retain significant activity toward deoxynucleo-sides.[5-8] *Bacillus stearothermophilus* was found to possess a single pyrimidine nucleoside phosphorylase which was 1.5 times as active with thymidine as with uridine;[9] this ratio for the *H. influenzae* enzyme was 0.25. The enzyme described here shows a striking preference for 5-bromodeoxyuridine as a substrate; the initial rate observed with this substrate was 5 times that found with any other deoxynucleoside.

[5] T. A. Krenitsky, M. Barclay, and J. A. Jacquez, *J. Biol. Chem.* **239**, 805 (1964).
[6] E. W. Yamada, *J. Biol. Chem.* **243**, 1649 (1968).
[7] A. Kraut and E. W. Yamada, *J. Biol. Chem.* **246**, 2021 (1971).
[8] H. Pontis, G. Degerstedt, and P. Reichard, *Biochim. Biophys. Acta* **51**, 139 (1961).
[9] P. P. Saunders, B. A. Wilson, and G. F. Saunders, *J. Biol. Chem.* **244**, 3691 (1969).

[58] Thymidine Phosphorylase from *Salmonella typhimurium*

By PATRICIA A. HOFFEE and JAMES BLANK

Thymidine + P_i ⇌ thymine + 2-deoxy-D-ribose-1-phosphate

In *Escherichia coli* and *Salmonella typhimurium*, thymidine phospho-rylase plays an important role in the metabolism of thymine auxotrophs and is necessary for the conversion of exogenous thymine to thymidine. In low-thymine-requiring strains, thymidine phosphorylase becomes phenotypically constitutive in strains containing a mutation in deoxyribose-5-phosphate aldolase.[1-3] Thymidine phosphorylase is regulated along with deoxyribose-5-phosphate aldolase, phosphodeoxyribomutase, and purine nucleoside phosphorylase as part of the *deo* regulon.[2-9]

Assay Method[10]

Principle. Enzyme activity is measured spectrophotometrically by measuring the formation of thymine from thymidine in the presence of sodium arsenate.

Reagents

Tris (hydroxymethyl) aminomethane-HCl buffer 0.2 M, pH 7.4, with 0.1 M sodium arsenate
Thymidine, 4 mg dissolved in 1 ml of the Tris·HCl buffer
NaOH 0.3 N

Procedure. Add 0.2 ml of the thymidine–buffer solution to 3 tubes marked 0, 5, and 10 min. Add enzyme solution at 0°. To the 0 time tube add 1 ml of NaOH and incubate the 5- and 10-min tubes at 37°. Stop the reaction at the specified time by adding 1 ml of 0.3 N NaOH. The optical density of each sample is then determined in a spectrophotometer set at 300 nm. The increase in optical density is plotted against time.

Definition of Units. One unit of activity is defined as the number of μmoles of thymine produced per minute at 37°. An extinction coefficient of 4.04 is used. Protein is determined by the method of Lowry *et al.*[11] before ammonium sulfate fractionation (see *Purification*) and by the method of Bücher[12] after ammonium sulfate fractionation.

[1] W. E. Razzell and G. G. Khorana, *Biochim. Biophys. Acta* **28**, 562 (1958).
[2] T. R. Breitman and R. M. Bradford, *Biochim. Biophys. Acta* **138**, 217 (1967).
[3] P. Hoffee, *J. Bacteriol.* **95**, 449 (1968).
[4] A. Munch-Peterson. *Eur. J. Biochem.* **6**, 432 (1968).
[5] M. S. Lomax and G. R. Greenberg. *J. Bacteriol.* **96**, 501 (1968).
[6] P. Hoffee and B. C. Robertson, *J. Bacteriol.* **97**, 1386 (1969).
[7] B. C. Robertson, P. Jargiello, J. Blank, and P. Hoffee, *J. Bacteriol.* **102**, 628 (1970).
[8] S. Ahmad and R. H. Pritchard, *Mol. Gen. Genet.* **104**, 351 (1969).
[9] J. Blank and P. Hoffee, *Mol. Gen. Genet.* **116**, 291 (1972).
[10] M. Friedkin and D. Roberts, *J. Biol. Chem.* **207**, 245 (1954).
[11] O. H. Lowry, N. J. Rosebrough, A. L. Farr, and R. J. Randall, *J. Biol. Chem.* **193**, 265 (1951).
[12] T. Bücher, *Biochim. Biophys. Acta* **1**, 292 (1947).

Purification of Enzyme[13]

Bacterial Strain. Best results are obtained with a strain of *S. typhimurium* which is constitutive for thymidine phosphorylase and defective in purine nucleoside phosphorylase. Although the following procedure can be used with wild-type strains (inducible for the enzyme and producing an active purine nucleoside phosphorylase), the final enzyme solution still contains trace amounts of purine nucleoside phosphorylase. A suitable strain (*deo*R^c, *deo*D$^-$) can be obtained by published procedures[7,9] or directly from the author.

Growth of Cells

WILD-TYPE *Salmonella typhimurium* (ATCC 15277). Inoculate 2-liter flasks containing 1-liter of Casamino acid medium, buffered at pH 7.0 with phosphate, with 25 ml of an overnight culture grown on the same medium. Incubate the flasks with shaking at 37° for 4 hr or until cells are in mid-log phase. Add sterile 2-deoxyribose to the cells to a final concentration of 0.2%. Continue incubation of the cells at 37° with aeration for an additional 90 min and then rapidly chill and harvest. Wash the cell paste in 1 m*M* EDTA and store frozen at −10°.

Salmonella typhimurium STRAINS CONSTITUTIVE FOR THYMIDINE PHOSPHORYLASE (*deo*R^c). Inoculate 2-liter flasks containing 1-liter of Casamino acid medium, buffered at pH 7.0 with phosphate and containing 10 µg/ml of thymine, with 25 ml of an overnight culture grown on the same medium. Incubate the flasks with shaking at 37° overnight and harvest in the morning. Wash the cell paste as above and store frozen at −10°. The use of the constitutive strains is much more convenient since it does not require induction with 2-deoxyribose, and the amount of enzyme obtained is about 1.5–2 times the amount obtained from the same number of wild-type induced cells.

Preparation of Cell Extract. Resuspend 24 g of frozen *S. typhimurium* in 150 ml of 0.04 *M* triethanolamine buffer, pH 7.6, 1 m*M* in EDTA. Disrupt the cells in a Branson Sonifier keeping the mixture below 10° during oscillation with a NaCl–ice bath. Centrifuge the disrupted suspension for 45 min at 27,000 *g*. Assay the supernatant fluid for activity and protein. Perform all subsequent procedures at 4°.

Protamine Sulfate Step. Make a 2% solution of protamine sulfate (Eli Lilly). To each ml of extract add 0.05 ml of protamine sulfate, and after 10 min remove the precipitate by centrifugation. The enzyme activity

[13] Procedure based on work previously published: J. Blank and P. Hoffee, *Arch. Biochem. Biophys.* **168**, 259 (1975).

should remain in the supernatant fluid. Add 0.2 ml of the 2% solution of protamine sulfate to each ml of the supernatant fluid. After 10 min collect the precipitate by centrifugation and dissolve it in 100 ml of 0.05 M potassium phosphate buffer, pH 7.5, 10 mM in 2-mercaptoethanol. Centrifuge the suspension and reserve the supernatant fluid for assay and further purification.

Ammonium Sulfate Fractionation. Bring the phosphate eluate to 40% ammonium sulfate saturation by the addition of 231 mg/ml of solid ammonium sulfate. After 30 min centrifuge the suspension and discard the precipitate. Bring the supernatant fluid to 60% ammonium sulfate saturation by the addition of 125 mg/ml of solid ammonium sulfate. After 30 min centrifuge the suspension. Dissolve the precipitate in 0.05 M potassium phosphate buffer, pH 7.5, and dialyze the solution overnight against 2000 volumes of 0.01 M potassium phosphate buffer, pH 7.5, 10 mM in 2-mercaptoethanol.

DEAE-Cellulose Chromatography. Apply the dialyzed ammonium sulfate fraction to a DEAE-cellulose column (2.5 × 27 cm) which has been equilibrated with 0.01 M potassium phosphate buffer, pH 7.5, 10 mM in 2-mercaptoethanol. Wash the column with about 200 ml of the equilibrating buffer or until no more 280 nm absorbing material is eluted. Thymidine phosphorylase is then eluted with a linear gradient of potassium phosphate buffer, from 0.01 M, pH 7.5 to 0.2 M, pH 8.0, 10 mM in 2-mercaptoethanol. Combine fractions containing enzyme activity and add sucrose to a final concentration of 15% (w/v). Bring the solution to 90% ammonium sulfate saturation by the slow addition of 630 mg/ml of solid ammonium sulfate. After 30 min centrifuge the suspension and dissolve the precipitate in 0.01 M potassium phosphate buffer, pH 7.5, 10 mM in 2-mercaptoethanol. Dialyze the solution overnight against 1000 ml of the same buffer.

Sephadex G-150 Chromatography. Apply the dialyzed DEAE-cellulose fraction to a Sephadex G-150 column (2.5 × 70 cm) which has been equilibrated with 0.01 M potassium phosphate buffer, pH 7.5, 10 mM in 2-mercaptoethanol. Thymidine phosphorylase is eluted from the column by washing with the equilibration buffer. Fractions containing activity are pooled.

Hydroxyapatite Chromatography. Apply the Sephadex G-150 fraction to a column (2.5 × 10 cm) of hydroxyapatite (Hypatite C, Clarkson Chemical Company, Inc., Williamsport, Pa.) which has been equilibrated with 0.01 M potassium phosphate buffer, pH 7.5, 10 mM in 2-mercaptoethanol. Thymidine phosphorylase does not adsorb to the

PURIFICATION PROCEDURE FOR THYMIDINE PHOSPHORYLASE FROM *Salmonella typhimurium*[a]

Fraction	Volume (ml)	Total units	Specific activity[b]	Purification (fold)	Recovery (%)
1. Crude extract	133	5800	1.8	1	100
2. Protamine sulfate step	106	5500	5.8	3	94
3. Ammonium sulfate (40–60% sat.)	9	4300	17.4	10	74
4. DEAE-cellulose	16	2500	45.5	25	43
5. Sephadex G-150	32	1910	200.0	111	33
6. Hydroxyapatite	2	1180	465.0	258	20

[a] Data given for constitutive mutant JB-3041. Constitutive strains are available from the author or can be selected by procedures described in Blank and Hoffee.[9]

[b] Specific activity is expressed in units/mg protein. One unit is defined as the number of μmoles of thymine produced per minute at 37°.

hydroxyapatite under these conditions and is eluted with the same buffer. Pool the fractions containing enzyme activity and add sucrose to a final concentration of 20%. Bring the solution to 90% ammonium sulfate saturation by the addition of 630 mg/ml of solid ammonium sulfate. After 30 min, centrifuge and dissolve the precipitate in 0.01 *M* potassium phosphate buffer, pH 7.5, 10 m*M* in 2-mercaptoethanol and 20% (w/v) in sucrose. This fraction is homogenous by polyacrylamide gel electrophoresis. The purification procedure is summarized in the table.

Properties[13]

Molecular Weight. Thymidine phosphorylase purified from *S. typhimurium* has a molecular weight of 100,000 ± 10% and is composed of two subunits of molecular weight 47,000.

Substrate Specificity. Purified thymidine phosphorylase can catalyze the conversion of thymidine and deoxyuridine to their respective bases and deoxyribose-1-phosphate. Both substrates show equal activity. Bromodeoxyuridine and iododeoxyuridine react at 40% the rate of thymidine. No activity is seen with the following nucleosides: deoxycytidine, deoxyadenosine, deoxyguanosine, or uridine. The Michaelis constants for thymidine and deoxyuridine are 2.1 m*M* and 8.0 m*M*, respectively, and for phosphate and arsenate they are 2.3 m*M* and 1.3 *M*, respectively.

Stability. During the latter states of purification thymidine phosphorylase is quite labile. The use of 2-mercaptoethanol and sucrose during precipitation procedures with ammonium sulfate will stabilize the enzyme. Purified thymidine phosphorylase can be stored in 10 mM potassium phosphate, pH 7.5, 10 mM in 2-mercaptoethanol and containing 20% sucrose for 3 months at 4° without loss of enzyme activity.

Effect of pH. Optimum activity is found at pH 7.5–8.0 with a rapid decrease in activity both above and below this range.

[59] Thymidine Phosphorylase from *Escherichia coli*

By MARIANNE SCHWARTZ

Thymidine + phosphate ↔ thymine + deoxyribose-1-phosphate

Assay of Enzymic Activity

Principle. The assay used for following the purification of thymidine phosphorylase is based on the difference in molar extinction coefficient between thymidine and thymine at alkaline pH.[1] At 300 nm this difference is 3.4×10^3. For more accurate activity determinations the diphenylamine assay is employed.[2]

Both assays can be used on crude extracts.

Reagents
Thymidine 0.1 M
K-phosphate, 0.1 M, pH 7.1
Buffer: Tris·HCl buffer, 10 mM, pH 7.3, is made 2 mM with respect to EDTA
NaOH, 0.5 N

Procedure. Buffer and enzyme in a total volume of 0.850 ml are mixed at 0°. Then 0.1 ml K-phosphate is added and the mixture transferred to 37° and incubated for 2 min. The reaction is started by addition of 0.05 ml thymidine.

At 5-min time intervals 0.300-ml samples are transferred to 0.700-ml NaOH. The amount of thymine formed is measured at 300 nm. If the

[1] R. D. Hotchkiss, *J. Biol. Chem.* **175**, 315 (1948).
[2] K. Burton, *Biochem. J.* **62**, 315 (1956).

diphenylamine method is employed the reaction is stopped by transferring the samples to tubes containing 0.150 ml of 30% perchloric acid.

Definition of Unit and Specific Activity. One unit is defined as the amount of enzyme that cleaves 1 μmole of thymidine in 1 min at 37°. Specific activity is units per milligram of protein. Protein is determined by the method of Lowry et al.[3]

Purification Procedure

Bacterial Strains and Growth of Cells. E. coli K-12 ($thyA^-$, $deoC^-$, met^-) is used for the purification described below, taking advantage of the high specific activity obtained following thymine starvation. Cells are grown in a glucose minimal medium with addition of 0.016 mM thymine and 0.05 mg/ml methionine. At an absorbance at 450 nm of about 1.0, 2 liters of culture are diluted into 9 liters of prewarmed medium without thymine. The resulting thymine concentration should be less than 3 μM. After 8 hr of thymine starvation the absorbance at 450 nm is about 2.0. The cells are then harvested by centrifugation. The cells are washed in 0.9% NaCl and stored at −60°.

E. coli Strains Constitutive for Thymidine Phosphorylase ($deoR^c$). By employing these strains thymine starvation can be avoided. Two-hundred-milliliter overnight cultures in glycerol minimal medium are used as inoculum for 5-liter flasks each containing 1.8 liters of prewarmed glycerol minimal medium. The flasks are incubated with shaking at 37° overnight. The cells are harvested as described above. The use of $deoR^c$ strains results in the same specific activity in crude extracts as thymine starvation, i.e., about 100 times the wild-type level. The purification procedure is identical for both strains.

Step 1. Disruption of Cells. The cell pellet is thawed and suspended in 2.5 volumes of 0.01 M Tris·HCl, 2 mM EDTA, pH 7.9. The cells are broken by sonic treatment or by using a French press. The cell debris is removed by centrifugation in a refrigerated Sorvall centrifuge at 40,000 g for 60 min.

Step 2. Streptomycin Sulfate Precipitation. To each milliliter of the supernatant fluid 0.23 ml of 10% streptomycin sulfate is added over a period of 30 min to a final concentration of 1.9%. After another 30 min the precipitate is removed by centrifugation, and the pH of the supernatant fluid is adjusted to 7.8 by addition of 1.0 M Tris·HCl, pH 10.0.

[3] O. H. Lowry, N. J. Rosenbrough, and R. J. Randall, *J. Biol. Chem.* **193**, 265 (1961).

Step 3. Ammonium Sulfate Fractionation. The streptomycin sulfate supernatant fluid is treated with solid ammonium sulfate (260 mg/ml) to 0.43 saturation. After standing 30 min at 0°, the suspension is centrifuged and the pellet discarded. Solid ammonium sulfate (205 mg/ml) is further added to 0.73 saturation to precipitate the enzyme. After centrifugation the precipitate is dissolved in 0.1 M Tris·HCl–2 mM EDTA (pH 7.6) to a total volume equal to one-tenth of the original extract.

Step 4. Sephadex G-100 Chromatography. The dissolved precipitate from ammonium sulfate fractionation is then directly applied on a G-100 Sephadex column (2.5 × 90 cm), equilibrated with 10 mM Tris-succinate–50 mM NaCl–1 mM EDTA (pH 6.8). Flow rate is maintained at 18 ml/hr. Fractions of 3 ml are collected, and those containing the enzyme are pooled.

Step 5. DEAE-Sephadex Chromatography.[4] The pooled fractions from step 4 are applied directly on a 2.5 × 40 cm DEAE-Sephadex column, equilibrated with the same buffer as the G-100 column. The column is eluted with a linear gradient from 50 mM NaCl to 300 mM NaCl in 10 mM Tris-succinate–1 mM EDTA (pH 6.8). The total gradient volume is 2.5 liters. Flow rate is maintained at 20 ml/hr. The fractions containing the enzyme are pooled. The enzyme is eluted at a concentration of about 150 mM NaCl.

Step 6. Hydroxyapatite Chromatography. The pooled fractions are concentrated 5- to 10-fold by pressure filtration (Amicon PQM 10 filter) and dialyzed overnight against 20 mM K-phosphate, pH 6.8. The dialyzed solution is applied on a 1.5 × 30 cm hydroxyapatite column, equilibrated with the above buffer. Thymidine phosphorylase is eluted with a linear gradient of K-phosphate buffer from 20–150 mM, pH 6.8. Total gradient volume is 600 ml. Flow rate is maintained at 10 ml/hr. Fractions containing the enzyme are collected. The enzyme is eluted at a phosphate concentration of approximately 100 mM.

The purification procedure is summarized in the table.[5]

Properties

Specificity. Thymidine phosphorylase is specific for pyrimidine deoxyribonucleosides. Purine deoxyribonucleosides and ribonucleosides as

[4] DEAE-cellulose DE-52 gives identical elution profiles.
[5] A. Munch-Petersen, P. Nygaard, K. Hammer-Jespersen, and N. Fiil, *Eur. J. Biochem.* **27**, 208 (1972).

PURIFICATION PROCEDURE FOR THYMIDINE PHOSPHORYLASE FROM *Escherichia coli*[a]

Fraction	Volume (ml)	Specific activity (units/mg)	Recovery (%)
Sonic extract	65	5.22	100
Streptomycin supernatant	77	3.85	73
Ammonium sulfate precipitate	13	7.02	68
G-100 eluate	72	11.95	51
DEAE-sephadex (not concentrated)	250	28.7	38
Hydroxyapatite	87	122	36

[a] Data given are from a purification starting with 20 g of cells of *E. coli* So 103 (*thyA⁻, deoC⁻, met⁻*). Strains constitutive for thymidine phosphorylase (*deoRᶜ*) are available from this laboratory or can be selected by the procedure described by Munch-Petersen *et al.*[5]

well as deoxycytidine are not cleaved by the enzyme. Arsenate can replace phosphate.

Stability. The enzyme is stable for several months when kept in 10 mM Tris·HCl, pH 7.3, at −20°. At 4° the activity is gradually lost. The stability is dependent on the protein concentration.

Effect of pH. The enzyme shows a rather broad pH dependency, with maximum activity at pH 6.3.

Kinetic Constants. The enzyme shows Michaelis–Menten kinetics. Initial velocity studies indicate an ordered sequential reaction mechanism.[6] Phosphate is the first substrate to bind to and deoxyribose the last product to dissociate from the enzyme. K_m: phosphate 0.89 mM; thymidine 0.38 mM. The enzyme is competitively inhibited by ribose-1-phosphate.

Molecular Weight. The enzyme has a molecular weight of 90,000[7] as determined by G-200 gel filtration. The enzyme seems to consist of two identical subunits showed by SDS polyacrylamide electrophoresis after intensive boiling with SDS.[8]

---

[6] M. Schwartz, *Eur. J. Biochem.* **21**, 191 (1971).
[7] This is a correction of the earlier published value.[6] M. Schwartz, unpublished results.
[8] B. Svenningsen, unpublished results.

[60] Nucleoside Deoxyribosyltransferase from *Lactobacillus helveticus*

By R. CARDINAUD

Purine$_1$ (pyrimidine$_1$) deoxyriboside + purine$_2$ (pyrimidine$_2$)
$\rightleftarrows$ purine$_2$ (pyrimidine$_2$) deoxyriboside + purine$_1$ (pyrimidine$_1$)

Assay of Enzyme Activity

Principle. Nucleoside deoxyribosyltransferase (originally named trans-*N*-glycosidase[13], then trans-*N*-deoxyribosylase) catalyzes three types of transfer of the deoxyribosyl moiety according to the nature of the donor and acceptor bases:

$$dRib\text{-}Pur + Pur' \rightleftarrows dRib\text{-}Pur' + Pur \quad (1a)$$

$$dRib\text{-}Pur + Pyr \rightleftarrows dRib\text{-}Pyr + Pur \quad (1b)$$

$$dRib\text{-}Pyr + Pyr' \rightleftarrows dRib\text{-}Pyr' + Pyr \quad (1c)$$

Abbreviations: Pur: a purine base; Pyr: a pyrimidine base

It is now well established[1] that *Lactobacillus helveticus* extracts contain two different enzymes: nucleoside deoxyribosyltransferase-I or purine nucleoside: purine deoxyribosyltransferase strictly specific for transfer between purine bases, and nucleoside deoxyribosyltransferase-II or purine (pyrimidine) nucleoside:purine (pyrimidine) deoxyribosyltransferase which catalyzes the transfer of the deoxyribosyl moiety between purines or pyrimidines as well as from a purine to a pyrimidine. Since the amount of these two types of enzyme varies significantly from strain to strain, it is desirable to perform at least the assay reactions (1a) and (1b). Both enzyme activities can be measured spectrophotometrically although other methods have also been used.[2-5]

[1] J. Holguin and R. Cardinaud, *Eur. J. Biochem.* **54**, 505 (1975).
[2] E. Hoff-Jorgensen, *Biochem. J.* **50**, 400 (1951).
[3] W. S. McNutt, this series, Vol. 2 [67].
[4] W. S. Beck and M. Levin, *J. Biol. Chem.* **238**, 702 (1963).
[5] W. Uerkwitz, *Eur. J. Biochem.* **23**, 387 (1971).

Spectrophotometric Assay

A. *Coupled Enzyme System with Xanthine Oxidase*[6,7]

Hypoxanthine formed (using deoxyinosine as donor) is rapidly oxidized to uric acid by xanthine oxidase. The activity is measured at 40°. The change in absorbance is followed at 290 nm where most common bases and nucleosides have low extinction coefficients.

Reagents
Deoxyinosine, 1 mM
Adenine, 1 mM
PO$_4$ buffer, 0.3 M, pH 6.0
Xanthine oxidase (10 mg/ml suspension in 3.2 M ammonium sulfate from Boehringer Mannheim diluted to 1:10 with ice-cold 2 M ammonium sulfate solution and kept at 0° until use)

Procedure. One milliliter of PO$_4$ buffer, 0.8 ml of water, and 0.5 ml of deoxyinosine are placed in a quartz cell of 1-cm path length; 0.1 ml of xanthine oxidase is added. The reaction mixture is checked to see that the slight amount of hypoxanthine occasionally present in deoxyinosine is completely oxidized before adding 0.5 ml of adenine. The change in absorbance due to the slow oxidation of adenine[8] is recorded. The enzyme (0.1 ml) is added, and the absorbance change observed is corrected for the oxidation of adenine. It is important to check that the rate of oxidation of hypoxanthine is not limiting and that no deoxyinosine hydrolase, adenine deaminase, or nucleoside phosphorylase are present.

Definition of Unit and Specific Activity. A unit of enzyme activity is defined as the quantity of enzyme necessary to produce 1 μmole of hypoxanthine in 1 min under the conditions defined above. One activity unit corresponds to an absorbance change of 4.07/min (ϵ for uric acid at 290 nm is 12,200 M^{-1} cm^{-1}) for a final solution of 3 ml. Specific activity is defined as units per milligram of protein. Protein is determined by the Folin method as described in Leggett Bailey[9] using crystalline serum bovine albumin as a standard.

[6] H. M. Kalckar, *J. Biol. Chem.* **167**, 429 (1947).
[7] C. Danzin and R. Cardinaud, *Eur. J. Biochem.* **48**, 255 (1974).
[8] H. Klenow, *Biochem. J.* **50**, 404 (1952).
[9] J. Leggett Bailey, "Techniques in Protein Chemistry," 2nd ed., p. 340. Elsevier, Amsterdam, 1967.

B. Direct Spectrophotometric Determination.[10]

All assays are carried out at 40° in 0.1 M phosphate buffer, pH 6.0, in a quartz cell of 1-cm path length (final volume 3 ml). The deoxyribonucleoside and the base (1 mM solution, 0.166 mM final concentrations) are mixed with a sufficient volume of buffer. The enzyme (0.1 ml) is added to start the reaction and the absorbance change followed with time. A suitable wavelength is selected according to the nature of the substrate couple: deoxycytidine-adenine and deoxyadenosine-cytosine, $|\Delta\epsilon_{280}|$ = 3864 M^{-1} cm^{-1}; deoxyguanosine-cytosine, $\Delta\epsilon_{290}$ = 2178 M^{-1} cm^{-1}; deoxyinosine-cytosine, $\Delta\epsilon_{280}$ = 2526 M^{-1} cm^{-1}; thymidine-cytosine, $\Delta\epsilon_{240}$ = 2248 M^{-1} cm^{-1}. It is important to test for the presence of the corresponding hydrolases and deaminases.

Purification of Enzyme

Growth of Bacteria [*Lactobacillus helveticus CNRZ 66 (NCDO 30) or CNRZ 303 Strains from CNRZ, Jouy en Josas*]. The bacteria are grown in a slightly modified MRS medium:[11] peptone, 20 g/liter; $K_2H PO_4$, 2 g/liter; $Na_2SO_4 \cdot 10 H_2O$, 8.2 g/liter; ammonium citrate, 2 g/liter; $MgSO_4 \cdot 7 H_2O$, 0.2 g/liter; $MnSO_4 \cdot H_2O$, 0.038 g/liter. The pH is adjusted to 6.7 with a 10 N NaOH solution. The medium is sterilized at 120° for 20 min, and a solution containing glucose (1 g/ml) and yeast extract (0.5 g/ml), sterilized at 110° for 10 min, is added to a final concentration of 20 g/liter for glucose and 5 g/liter for yeast extract. The medium is inoculated with 1/10 volume of a 14–16 hr culture in the same medium. The culture is incubated at 40° for about 20 hr with gentle stirring of the broth. To limit the inhibitory effect of the lactic acid liberated, a constant pH value of 6.5 is maintained by adding 5 N NaOH with a pH-stat. The bacteria are collected by centrifugation at 20,000 g and washed in a 0.05 M citrate buffer, pH 5.9, centrifuged, and resuspended in a 0.1 M phosphate buffer, pH 6.0 (1 ml/g bacteria).

Step 1. Preparation of Crude Extract. The bacteria are disrupted either by sonication in a Biosonik III sonifier or through a press at 500 kg/cm², with the temperature kept below 12°, and then centrifuged at 27,000 g (0–4°). The cell debris is resuspended in 0.1 M phosphate buffer, pH 6.0, and centrifuged under the same conditions. The two supernatant fluids are pooled and assayed for transfer activities. Henceforward all operations are carried out at 4° unless otherwise specified.

---

[10] C. Danzin and R. Cardinaud, *Eur. J. Biochem.* **62**, 365 (1976).
[11] J. C. de Man, M. Rogosa, and M. E. Sharpe, *J. Appl. Bacteriol.* **23**, 130 (1960).

Step 2. Ammonium Sulfate Precipitation. The pooled supernatant fluids are fractionated by the addition of solid ammonium sulfate. The material which precipitates between 30–70% saturation is collected by centrifugation at 20,000 g for 30 min, dissolved in a 0.1 M phosphate buffer, pH 6.0 (100 ml/g protein), and dialyzed against the same buffer.

Step 3. Protamine Sulfate Precipitation. A 1% protamine sulfate solution is added slowly with continuous stirring until the nucleic acid peak at 260 nm has virtually disappeared. After centrifugation at 20,000 g (30 min) to remove the precipitate, the supernatant fluid is dialyzed against the 0.1 M phosphate buffer (pH 6.0).

Step 4. Heat Treatment. The solution is heated at 60° for 5 min, and the precipitate is spun down at 20,000 g (30 min) and discarded. When the volume of solution is large this step can be conveniently carried out in coiled glass tubing immersed in a water bath. The equivalent of a 5-min heating is obtained by adjusting the diameter and length of the tubing and the flow rate.

Step 5. Affinity Chromatography. This stage separates and purifies the two nucleoside deoxyribosyltransferases.

A. SEPARATION OF NUCLEOSIDE DEOXYRIBOSYLTRANSFERASE-I. The specific adsorbent is Sepharose IVB substituted with m-phenylene diamine onto which is coupled a diazonium salt of 6-(p-amino benzylamino-) purine.[1] This column retains both enzymes. The column is washed with the 0.1 M phosphate buffer, pH 6.0, and nucleoside deoxyribosyltransferase-I is selectively eluted with a 10 mM deoxyinosine solution in the same buffer. The eluate is concentrated by ammonium sulfate precipitation at 70% saturation. The precipitate is dissolved in the phosphate buffer, dialyzed against the same buffer, and centrifuged at 27,000 g for 30 min. The supernatant fluid is stored at 2–4°.

B. SEPARATION OF NUCLEOSIDE DEOXYRIBOSYLTRANSFERASE-II. The specific adsorbent is Sepharose IVB substituted with m-phenylene diamine onto which is coupled a diazonium salt of 5-(p-amino phenyl-n-propyl) uracil.[1] After elution of nucleoside deoxyribosyltransferase-I, adsorbent-I is eluted with a 2 M urea solution. The eluate is dialyzed against 0.1 M phosphate (pH 6.0) and passed through a column of adsorbent-II. This adsorbent selectively retains nucleoside deoxyribosyltransferase-II. After washing the column with the phosphate buffer, nucleoside deoxyribosyltransferase-II is eluted with 10 mM adenine. The protein is concentrated by ammonium sulfate precipitation at 70% saturation. The precipitate is spun down at 27,000 g (30 min), dissolved in the 0.1 M phosphate buffer, and the solution exhaustively dialyzed against the same buffer. After centrifugation (27,000 g, 30 min) the solution is stored at 2–4°.

A summary of the purification is given in Table I.

Properties

Physical Properties. The apparent molecular weight of nucleoside deoxyribosyltransferase-I is about 86,000 ± 4,000,[1] whereas a value of 82,000 has been found for a mixture of nucleoside deoxyribosyltransferase-I and -II. This mixture has been crystallized.[5] The optimal stability of the enzyme is around pH 6.5.[12] Crude extracts and pure preparations are stable and can be stored at 0–4° for several months without significant loss of activity. For longer periods they can be frozen, but repeated freezing and thawing result in a rapid decrease of the activity.

Substrate Specificity. The two distinct nucleoside deoxyribosyltransferases differ at least by their specificity since nucleoside deoxyribosyltransferase-I is specific for the transferase of deoxyribosyl between purine bases whereas nucleoside deoxyribosyltransferase-II catalyzes the transfer between purines or pyrimidines. A great number of purine analogues have been found to be acceptors.[4,12–15] Semiempirical rules have been devised[15] to predict if a purine base is a competent acceptor: (1) A tautomeric proton must be present on the imidazole ring. The "usual" shift is between position 9 and 7. (2) The position of the tautomeric proton governs the site of substitution. (3) For steric reasons no substituent is allowed on position 8. A wide variety of pyrimidine analogues are also acceptors,[12–14] and similar rules are available:[16] (1) A lactim–lactam type tautomerism must be possible between the nitrogen atom N-1 and a suitable group on position 2 (or 6). (2) No other substituent can be present on position 6 (or 2) for steric reasons. (3) The aromatic properties of the ring must be preserved. On the other hand, both enzymes appear to be strictly specific for the deoxyribosyl moiety, and only 2'deoxy-β-D-ribofuranosides are donors. However 5'-fluoro-5' deoxythymidine gave a positive reaction with adenine in the presence of nucleoside deoxyribosyltransferase-II.[17]

[12] A. H. Roush and R. F. Betz, *J. Biol. Chem.* **233**, 261 (1958).
[13] W. S. McNutt, *Biochem. J.* **50**, 384 (1952).
[14] K. Baranski, T. J. Bardos, A. Bloch, and T. I'. Kalman, *Biochem. Pharmacol.* **18**, 347 (1969).
[15] J. Holguin and R. Cardinaud, *Eur. J. Biochem.* **54**, 515 (1975).
[16] J. Holguin and R. Cardinaud, unpublished.
[17] R. Cardinaud, unpublished observation.

TABLE I

PURIFICATION OF NUCLEOSIDE DEOXYRIBOSYLTRANSFERASE-I AND -II FROM *Lactobacillus helveticus*

Fractions	Total protein (mg)	Activity (unit)		Recovery (%)		Purification (fold)		Specific activity (units/mg)		
		dIno → Ade[b]	dCyd → Ade	dIno → Ade	dCyd → Ade	dIno → Ade	dCyd → Ade	dIno → Ade	dCyd → Ade	dThd → Cyt
Initial extract	7750	387.50	1550	—	100	1	1	0.050	0.20	0.033
Ammonium sulfate	3915	371.9	1504	—	97	1.50	1.92	0.095	0.384	0.061
Protamine sulfate	3396	434.7[a]	1450	100	93.5	2.56	2.13	0.128	0.427	0.052
Heat treatment	2360	361.1	1410	83	91.0	3.06	2.99	0.153	0.598	0.082
Eluate of adsorbent I	10.72	322.0	—	74	—	600.0	—	30.0	—	—
Supernatant after centrifugation	5.01	308.1	—	71	—	1200.0	—	60.0	—	—
Eluate of adsorbent II	28.1	—	1210	—	78.0	—	215.0	—	43.0	7.75

[a] A high value occasionally observed at this stage indicates the removal of an inhibitor.
[b] dIno → Ade denotes a transfer in which deoxyinosine and adenine are the substrates.

Xanthine is a substrate for both enzymes. Using thymidine as the donor a crude extract of nucleoside deoxyribosyltransferase catalyzes the formation of two products: 9-deoxyribosylxanthine (20%) and 7-deoxyribosylxanthine (80%).[18] This observation is correlated to the fact that xanthine has its tautomeric proton mainly on N-7.[19]

Effect of Ionic Conditions. The optimum activity is near pH 5.8.[13] In a 0.02 M phosphate buffer the pH optimum of pure nucleoside deoxyribosyltransferase-I is 5.75.[20] No cations have been found to play any role in the catalytic activity, and SH reagents do not hinder the transfer reaction. Tris has a specific effect since it completely inhibits the nucleoside deoxyribosyltransferase-II reaction at concentration in the range of 0.1 M, pH 6.9,[12] and also the nucleoside deoxyribosyltransferase-I reaction[1] but at much higher concentrations (62% inhibition at 0.8 M, pH 5.9).

Kinetic Parameters. According to the criteria given by Cleland[21] the various purine-to-purine transfers catalyzed by nucleoside deoxyribosyltransferase-I follow a ping-pong bi-bi mechanism[7] in which the first product is released leaving the deoxyribosyl enzyme before the second substrate combines with the modified enzyme. A number of kinetic constants for some transfer reactions have been determined (Table II). Adenine (second substrate) is a competitive inhibitor of the first substrate (deoxyribonucleosides). The mechanism of nucleoside deoxyribosyltransferase-II action is also ping-pong bi-bi[10] where adenine is a competitive inhibitor of the first substrate. Some kinetic constants are given in Table III.

Distribution. The enzyme was first discovered by McNutt[13] in *L. helveticus, L. delbrueckii,* and *Thermobacterium acidophilus.* Subsequently deoxyribosyltransferases have been found in all species of the subgenus *Thermobacterium* of *Lactobacilli.* An independent direct deoxyribosyl transfer activity has not been detected so far in any other organism, but in some cases such a reaction may be catalyzed either by a purine nucleoside phosphorylase or a thymidine phosphorylase. However, this question is usually considered as not completely resolved.[22] It is interesting to note that *Lactobacilli* lacking the phosphorylases have a

[18] J. Holguin-Hueso and R. Cardinaud, *FEBS Lett.* **20**, 171 (1972).

[19] D. Lichtenberger, F. Bergmann, and Z. Neiman, *J. Chem. Soc. C,* p. 1676 (1971).

[20] J. Holguin, Doctorate Dissertation, Paris-Orsay (1974).

[21] W. W. Cleland, *Biochim. Biophys. Acta* **67**, 104, 173, and 188 (1963).

[22] R. E. Parks and R. P. Agarwal, *in* "The Enzymes" (P. D. Boyer, ed.), 3rd ed., Vol. 7, p. 483. Academic Press, New York, 1972.

TABLE II

KINETIC CONSTANTS FOR NUCLEOSIDE DEOXYRIBOSYLTRANSFERASE-I[a]

Substrate	Michaelis constant (mM)	Inhibition constant (mM)	Maximum velocity (μmole min^{-1} μg^{-1})	Equilibrium constant K
Forward reaction:				
Deoxyinosine (dIno)	$K_{\text{dIno}} = 0.35$	$(K_i)_{\text{dIno}} = 0.34$	$V_{\text{(dIno}\rightarrow\text{Ade)}} = 0.42$	2.34
Adenine (Ade)	$K_{\text{Ade}} = 0.041$	$(K_j)_{\text{Ade}} = 0.039$ $K_1 = 0.12^b$		
Reverse reaction:				
Hypoxanthine (Hyp)	$K_{\text{Hyp}} = 0.086$	$(K_i)_{\text{Hyp}} = 0.079$	$V_{\text{(dAdo}\rightarrow\text{Hyp)}} = 0.46$	0.42
Deoxyadenosine (dAdo)	$K_{\text{dAdo}} = 0.45$	$(K_i)_{\text{dAdo}} = 0.43$		

[a] From Ref. 7.
[b] Dead-end inhibition.

TABLE III

KINETIC CONSTANTS FOR THE TRANSFERS (dCyd → Ade) AND (dAdo → Cyt) CATALYZED BY NUCLEOSIDE DEOXYRIBOSYLTRANSFERASE-II[a]

Substrate	Michaelis constant (mM)	Inhibition constant (mM)	Maximum velocity (μmole min^{-1} μg^{-1})	Equilibrium constant K
Forward reaction:				
Deoxycytidine	$K_{dCyd} = 0.090$	$(K_i)_{dCyd} = 0.095$		
Adenine	$K_{Ade} = 0.019$	$(K_i)_{Ade} = 0.021$	$V_{(dCyd \to Ade)} = 0.085$	8.15
		$(K_i)_{Ade} = 0.41^{b}$		
Reverse reaction:				
Cytosine	$K_{Cyt} = 0.22$	$(K_i)_{Cyt} = 0.17$		
Deoxyadenosine	$K_{dAdo} = 0.12$	$(K_i)_{dAdo} = 0.092$	$V_{(dAdo \to Cyt)} = 0.113$	0.115

a From Ref. 10.
b Dead-end inhibition.

direct transfer system instead, which may play a similar role in a salvage mechanism although the reaction is limited to the deoxyribosyl transfer. The general characteristics of the enzymes found in *Thermobacteria* other than *L. helveticus* are very similar. In spite of the fact that none has been studied in the pure form it is easily recognized that adenine is a competitive inhibitor of the donor[4] and the same broad specificity with regard to the acceptor bases[4,23-25] is observed on the enzymes from different sources.

[23] M. Kanda and Y. Takagi, *J. Biochem.* (*Tokyo*) **46**, 725 (1959).
[24] A. Minghetti, *Ital. J. Biochem.* **8**, 224 (1954).
[25] J. C. Marsh and M. E. King, *Biochem. Pharmacol.* **2**, 146 (1959).

Section VII
Purine Metabolizing Enzymes

A. Kinases
Articles 61 through 64

B. Deaminases
Articles 65 through 69

C. Phosphorylases
Articles 70 through 73

D. Phosphoribosyltransferases
Articles 74 through 78

E. General Methods
Article 79

[61] Purification and Characterization of Adenylate Kinase from Rat Liver

By WAYNE E. CRISS and TAPAS K. PRADHAN

$$\text{ATP} + \text{AMP} \rightleftharpoons 2\,\text{ADP}$$

Adenylate kinase (ATP:AMP phosphotransferase, EC 2.7.4.3), which catalyzes the above reaction, is a central component in the adenylate energy system (ATP, ADP, AMP, and adenylate kinase) of all biological cells. The adenylate energy system plays a very important role in the regulation of various metabolic pathways such as glycolysis, gluconeogenesis, lipolysis, lipogenesis, glycogenesis, glycogenolysis, and oxidative phosphorylation.[1-3] Adenylate kinase has been highly purified from Baker's yeast,[4] rabbit muscle,[5] bovine liver mitochondria,[6] human tissues,[7] and a AgNO$_3$-treated extract of swine liver.[8] We have previously reported that the major liver form of adenylate kinase has a subcellular location in the outer mitochondrial compartment,[9] shows activity related to tissue respiration,[10] is very low in fetal liver and hepatomas[11] and responds to hormonal and dietary regulation.[12] This enzymic reaction is unusual in that it has a K_{eq} near 1.

Assay Method

Principle. ATP, ADP, and AMP are separated on Whatman 3 MM chromatography paper by high-voltage electric current, according to their respective R_f values. AMP which has the lowest R_f value is separated near the base line, while ATP having the highest R_f value

[1] D. E. Atkinson, *Annu. Rev. Biochem.* **35**, 85 (1966).
[2] D. E. Atkinson, *Annu. Rev. Microbiol.* **23**, 47 (1969).
[3] H. A. Krebs, *Proc. Soc. London, Ser. B* **159**, 545 (1964).
[4] S. Su and P. J. Russell, *Biochim. Biophys. Acta* **132**, 370 (1967).
[5] L. Noda, *J. Biol. Chem.* **232**, 237 (1958).
[6] F. S. Markland and C. L. Wadkins, *J. Biol. Chem.* **241**, 4124 (1966).
[7] E. Thuma, R. H. Schirmer, and I. Schirmer, *Biochim. Biophys. Acta* **268**, 81 (1972).
[8] M. Chiga and W. E. Plaut, *J. Biol. Chem.* **235**, 3260 (1960).
[9] W. E. Criss, *J. Biol. Chem.* **245**, 6352 (1970).
[10] W. E. Criss, *Arch. Biochem. Biophys.* **144**, 138 (1971).
[11] R. Filler and W. E. Criss, *Biochem. J.* **122**, 553 (1971).
[12] W. E. Criss, G. Litwack, H. P. Morris, and S. Weinhouse, *Cancer Res.* **30**, 370 (1970).

METHODS IN ENZYMOLOGY, VOL. LI

separates on the farthest zone from the base line. ADP migrates between ATP and AMP.

Reagents

Triethanolamine-HCl buffer, 0.05 M, pH 7.0
$MgCl_2$, 0.015 M, pH 7.0
ATP, 0.015 M, pH 7.0
AMP, 0.0045 M, pH 7.0
Perchloric acid, 1.5 N
Citrate buffer, 0.05 M, pH 3.3
Ammoniumbicarbonate, 0.75 M.

Procedure. The assay system totals 1 ml in volume and contains 0.45 ml of buffer, 0.1 ml of $MgCl_2$, 0.1 ml of ATP, and 0.25 ml of AMP. The system is preincubated at 37° for 15 min, and the enzymic reaction is started by adding 100 μl enzyme. The reaction is incubated for 15 min. The reaction is stopped by adding 1 ml of cold perchloric acid, mixing, and immediately placing the reaction tube in ice. Then 100 μl of the reaction mixture are spotted on Whatman 3 MM filter paper. High-voltage electrophoresis is next carried out in citrate buffer, pH 3.3, using a Savant flat plate system. Electrophoresis is at 1500 V for 3 hr at 2°. After the separation, the paper is dried at 80° for 1 hr, and ADP spots are detected by an ultraviolet light lamp, cut into small pieces, and incubated in 3 ml of ammonium bicarbonate solution at 37° overnight. The eluant from the spots is diluted 1:10 with ammonium bicarbonate solution and assayed spectrophotometrically at 259 nm using a Beckmann Acta II spectrophotometer.

Definition of Unit and Specific Activity. One unit is defined as the amount of enzyme activity expressed in μmoles of adenine nucleotide produced per milliliter of assay medium per minute at 37°. Specific activity is expressed in terms of units per milligram of protein. Protein is determined by the method of Lowry et al.[13]

Purification

Step 1. Preparation from Liver Tissue. Rats are decapitated by guillotine and exsanguinated. The liver is immediately removed and placed in cold 12.5 mM sucrose solution containing 1 mM cysteine. All procedures used for the purification are performed at 4°. The cold liver is blotted, weighed, and homogenized with 2 volumes of cold 12.5 mM

---

[13] O. H. Lowry, N. J. Rosebrough, A. L. Farr, and R. J. Randall, *J. Biol. Chem.* **193**, 265 (1951).

sucrose solution containing 1 mM cysteine in a Waring Blendor for 2 min. The homogenate is centrifuged in a Spinco model L ultracentrifuge at 30,000 rpm for 2 hr. This supernatant fraction is used as the cytosol (pH 6.8).

Step 2. pH Fractionation. The pH of the cytosol is lowered from 6.8 to 3.0 by adding 3 N HCl very rapidly (in less than 1 min). It is maintained at pH 3.0 by stirring for 10 min. Then 1 N NaOH solution is added very slowly (5–10 min) until the pH is restored to 6.8. The solution is continually stirred for 20 min and then centrifuged at 16,000 g for 30 min. The supernatant fraction is concentrated by Amicon ultrafiltration using a UM-10 membrane.

Step 3. Sephadex G-75 Chromatography. The concentrated supernatant fraction is passed through a 5.6 × 45 cm column, containing Sephadex G-75, which is previously equilibrated with 5 mM sodium phosphate buffer containing 1 mM β-mercaptoethanol at pH 7.2. The enzyme is eluted from the column with the same buffer. Fractions of 8 ml each are collected and assayed for enzymic activity. Fractions containing approximately 90% of the enzymic activity are combined and concentrated by ultrafiltration using a UM-10 membrane.

Step 4. Isoelectrofocusing. This procedure is carried out according to the method described in the LKB Instruments Instruction Manual with the following modifications. Phosphoric acid is used as the anode solution at the bottom of the column and ethylenediamine as the cathode solution at the top of the column. Thirty–forty milligrams of protein in 5–6 ml of cytosol are placed on the 440-ml capacity column. Ampholytes having a pH range of 3–10 are used. The current is turned on and maintained at 10 mA for 16 hr. The current is then maintained at 2 mA for 48 hr. Next, 150 fractions of 3 ml each are collected from the bottom of the column and assayed for enzymic activity. Fractions containing approximately 90% enzymic activity are combined and concentrated by ultracentrifugation using a UM-10 membrane.

Step 5. Rechromatography of Sephadex G-75. The concentrated electrofocused fraction is ultradialyzed with 30 volumes of 5 mM sodium phosphate buffer at pH 7.2 in the ultrafiltration cell using a PM-10 membrane. The washed concentrate is layered on a Sephadex G-75 column, 3 × 77 cm, which is previously equilibrated with 5 mM sodium phosphate buffer at pH 7.2. The enzyme is eluted with the same buffer. Then 3-ml fractions are collected and assayed for enzymic activity. Fractions containing approximately 90% of the enzymic activity are combined and concentrated by ultrafiltration using a UM-10 membrane.

A summary of the purification scheme is given in Table I.

TABLE I
PURIFICATION OF LIVER ADENYLATE KINASE

Fraction	Volume (ml)	Units (μmoles/min)	Protein (mg)	Specific activity (unit/mg)	Recovery (%)	Purification (-fold)
Cytosol	300	17,700	16,500	1.07	98.3	2.4
pH Fractionation	280	10,720	2,800	3.82	59.6	8.5
Sephadex G-75	114	9,010	682	13.2	50.1	29.3
Isoelectrofocusing	32	5,280	8.3	636.1	29.3	1413.5
Rechromatography of Sephadex G-75	22	5,300	7.5	1000	29.4	1570.4

Properties

Stability. The purified liver mitochondrial adenylate kinase preparations were stable up to about 6 months when frozen, but steadily lost activity when repeatedly thawed and frozen. The preparations lost less than 10% activity in 1 month when stored in 5 mM sodium phosphate buffer containing 1 mM β-mercaptoethanol at pH 7.0 at 4°.

Enzyme Homogeneity. Polyacrylamide gel electrophoresis using a 2% spacer gel and a 7.5% running gel (according to the procedure of Dunker and Rueckert[14]) revealed one protein band. This band contained all of the adenylate kinase activity.

Molecular Weight Determination with Sephadex G-100. These studies were performed at 4° as described by Andrews.[15] The molecular weight standards which were employed included chicken heart lactate dehydrogenase (molecular weight 135,000), human hemoglobin (molecular weight 64,500), pepsin (molecular weight 37,000) and horse heart cytochrome c (molecular weight 12,400). A 0.3-ml aliquot of enzyme was layered on the 48 × 1 cm calibrated column of Sephadex G-100 which was equilibrated with 5 mM sodium phosphate buffer, pH 7.2, containing 1 mM β-mercaptoethanol. The same buffer was used to elute the enzyme from the column. From a plot of elution volume versus log molecular weight of the standards, the molecular weight of the enzyme was determined to be about 46,000.

Molecular Weight Determination by Analytical Ultracentrifugation. Ultracentrifugation analysis was carried out in a Spinco Model E ultracentrifuge with a 3-channel Yphantis cell. Equilibrium analysis was performed by the method of Yphantis.[16] Calculation of the molecular weight was made by computer. The molecular weight of the native enzyme was calculated to be 43,000.

Determination of Subunits. The determination of subunits was accomplished by SDS–polyacrylamide disc gel electrophoresis. The liver enzyme was observed to be a trimer containing two subunits of similar size and one slightly smaller subunit.[17] The two larger subunits each had a molecular weight near 13,000, while the third subunit was near 11,000.

Amino Acid Studies. Amino acid analysis of the native liver mitochondrial adenylate kinase was performed on a Beckmann 120C amino

[14] A. K. Dunker and R. R. Rueckert, *J. Biol. Chem.* **244**, 5074 (1969).
[15] P. Andrews, *Biochem. J.* **96**, 595 (1965).
[16] D. A. Yphantis, *Biochemistry* **3**, 297 (1964).
[17] W. E. Criss, T. K. Pradhan, and H. P. Morris, *Cancer Res.* **34**, 3062 (1974).

acid analyzer. Based on a subunit molecular weight of 13,000, all determined amino acid values were rounded off to the closest single integer. There were 1 methionine, 2 tyrosine, 3 phenylalanine, 4 each of histidine, arginine, proline, and isoleucine residues, 5 each of threonine, serine, and valine residues, 8 each of lysine, glycine, leucine, and aspartic acid residues, 9 glutamic acid, and 11 alanine residues.

Table II describes a number of physical properties which are characteristic for the liver enzyme.

Substrate Specificity. Liver mitochondrial adenylate kinase was specific for ATP, dATP, and dGTP as a phosphoryl donor. The V_{max} values were 16,340, 40,850, and 25,510 moles of diphosphate formed min^{-1} $mole^{-1}$ of enzyme, respectively. The enzyme did not use GTP, ITP, UTP, or CTP as triphosphate donors. 3':5'-cyclic AMP, dAMP, 2'-AMP, 3'-AMP, TMP, CMP, UMP, GMP, IMP, and dGMP did not serve as substrate phosphate acceptors.

Kinetic Regulation. Table III summarizes several kinetic properties of liver adenylate kinase.[18,19]

Effect of Citric Acid Cycle Intermediates. Most citric acid cycle intermediates, except succinate and oxaloacetate, activated the enzyme, as shown in Table IV.

Effect of Nucleotides. Only the nucleoside 5'-diphosphates stimulated the enzyme, e.g., GDP (+40%), TDP (+18%), CDP (+15%), and UDP (+12%). Triphosphates such as GTP, CTP, UTP, TTP, 2'-dGTP, 2'-dCTP, and 2'-dATP and monophosphates such as GMP, CMP, UMP,

TABLE II
PHYSICAL PROPERTIES OF LIVER ADENYLATE KINASE

Characteristics	Values
Diffusion coefficient	4.8 (10^{-7})
Partial specific volume	0.74
Frictional ratio	1.1
Axial ratio	4.0
Svedberg values	1.23, 3.52
Isoelectric point	8.0
Molecular weight	43,000
Subunits	3

[18] T. K. Pradhan and W. E. Criss, *Eur. J. Biochem.* **43**, 541 (1974).
[19] T. K. Pradhan, W. E. Criss, and H. P. Morris, *Cancer Res.* **34**, 3058 (1974).

TABLE III
KINETIC STUDIES ON LIVER ADENYLATE KINASE[a]

Characteristics	Values
K_m (ATP) (+ citrate)	7.0
K_m (ADP) (+ citrate)	16.8
K_m (AMP) (+ citrate)	1.9
Hill slope (ATP) (+ citrate)	2.0
Hill slope (ADP) (+ citrate)	2.3
Hill slope (AMP) (+ citrate)	0.5
$K_a{}^b$ (+ citrate)	0.09
$K_i{}^c$ (phMPS[d]) (+ citrate)	30.0
K_i (pCMB[e]) (+ citrate)	0.62
K_i (SDS[f]) (+ citrate)	0.7
Palmitic acid and stearic acid	No effect
Oleic acid	Inhibition
Ethylmercurithiosalicylate	No effect
Dithiothreitol	· No effect

[a] Treatment included preincubation of nucleotides and additives in 50 mM triethanolan-ine-HCl buffer at pH 7.0 for 15 min at 37°, after which they were assayed. The final concentration of citrate in the assay system was 10 mM. K_a and K_i values are in mM.
[b] Activation constant.
[c] Inhibitor constant.
[d] p-Hydroxymercuriphenylsulfonate.
[e] p-Chloromercuribenzoate.
[f] Sodium dodecyl sulfate.

TABLE IV
EFFECT OF CITRIC ACID CYCLE INTERMEDIATES ON THE LIVER ADENYLATE KINASE[a]

Intermediates	Stimulation
Citrate	$+166 \pm 9$
Isocitrate	$+34 \pm 6$
Malate	$+40 \pm 5$
Fumarate	$+30 \pm 6$
α-Ketoglutarate	$+17 \pm 4$
cis-Aconitate	$+28 \pm 4$
Succinate	0
Oxaloacetate	0

[a] Treatment included preincubation of nucleotides and intermediates in 50 mM trietha-nolamine-HCl buffer at pH 7.0 for 15 min at 37°C, after which they were assayed. Final concentration of each intermediate was 50 mM. Stimulation is expressed as mean $\pm$SE (%) of control activity.

TABLE V
EFFECT OF PROSTAGLANDINS ON THE ACTIVITY OF LIVER ADENYLATE KINASE[a]

Prostaglandins	Stimulation	K_m
A_2	$+149 \pm 10$	24
E_1	$+129 \pm 8$	25
E_2	$+27 \pm 4$	16
$F_2\alpha$	$+148 \pm 9$	35

[a] Treatment included preincubation of nucleotides and prostaglandins in 50 mM triethanolamine-HCl buffer at pH 7.0 for 15 min at 37°, after which they were assayed. Prostaglandins were dissolved in 95% ethanol and added in 10-μl aliquots. Final concentration of prostaglandin was 10^{-4} M. Stimulation is expressed as mean $\pm$SE (%) of control activity. K_m values are in μM.

TMP, IMP, 2'-dAMP, and 3':5'-cAMP had no effect on the activity of the enzyme.

Effect of Prostaglandins. Liver and adenylate kinase were stimulated by prostaglandins A_2, E_1, E_2, and $F_2\alpha$, and the apparent Michaelis constant values ranged from 16 μM to 35 μM, as described in Table V.[20]

Concluding Remarks

Since liver adenylate kinase is located in the outer mitochondrial compartment and is stimulated by carboxylic acids (citric acid cycle intermediates), it could play a functional role in the regulation of mitochondrial ATP production. Transportation of dicarboxylic and tricarboxylic acids and the translocation of the adenine nucleotides into the mitochondrial matrix are dependent upon the external and internal mitochondrial cations and hydrogen ion concentrations. The inner membrane of the mitochondrion is apparently impermeable to most ions. Mitochondria have exchange diffusion mechanisms for the entry of phosphate and an energy-dependent transportation system for the transport of many multivalent ions, such as the mitochondrial substrates. It seems that liver mitochondrial adenylate kinase, which is localized in the outer compartment, which has been observed to undergo mitochondrial flux, which is stimulated by citrate, and which has a substrate (ATP) that chelates cations 10 times more readily than a product (ADP), may be involved not only in directly altering the phosphorylated adenine nucleotide levels, while leaving the concentration of PO_4^{-3} unchanged,

[20] T. K. Pradhan and W. E. Criss, *Oncology* **33**, 15 (1976).

but also in altering the surface charge on the inner mitochondrial membrane via the chelation–nonchelation of cations. Hence, liver mitochondrial adenylate kinase could play a homeostatic role by maintaining an optimum level of phosphorylated adenine nucleotides and the proper inner membrane charge to allow efficient mitochondrial functioning. Therefore, adenylate kinase might be able to provide the liver cell with an excellent homogeneous mechanism to maintain efficiently coupled mitochondria. A more detailed description of this mechanism has already been published.[21]

Acknowledgment

This work was supported by Grants CA-11818 from the National Institutes of Health and F71UF from the American Cancer Society. One of us (Wayne E. Criss) was the recipient of a Research Career Development Award (CA-70187) from the United States Public Health Service.

[21] W. E. Criss, in "Hormones and Cancer" (K. W. McKerns, ed.), p. 169. Academic Press, New York, 1974.

[62] AMP (dAMP) Kinase from Human Erythrocytes

By Kenneth K. Tsuboi

$$AMP + ATP \rightleftarrows 2\ ADP$$

AMP (dAMP) kinase or adenylate kinase (ATP-AMP phosphotransferase, EC 2.7.4.3) exists in the human erythrocyte in multiple molecular form and comprises about 0.005% of the cell protein. Although most of the enzyme is soluble, significant activity is retained with the stroma at low ionic strength. The stroma-associated enzyme is released in 0.3 M KCl and does not differ from the soluble enzyme forms. Genetic studies indicate that two alleles at a single polymorphic locus code for most of the enzyme forms present in the human erythrocyte, with two common phenotypes, AK 1 (homozygote) and AK 2-1 (heterozygote) exhibited at frequencies of about 90 and 10%.[1] AK 1 exhibits a major and several minor secondarily derived electrophoretic forms. AK 2-1 exhibits two major and several minor derived electrophoretic forms. The isolation and properties of the predominant enzyme form are described.

[1] J. A. Bowman, H. Frischer, F. Ajmar, P. E. Carson, and M. K. Gower, Nature (London) **214**, 1156 (1967).

METHODS IN ENZYMOLOGY, VOL. LI

Assay

Principle. Adenylate kinase activity can be determined either in the forward or back reaction by measuring ATP or ADP production or consumption. Rates of ATP or ADP production are most conveniently determined by one-step combined assays at high sensitivity by fluorimetric measurements[2] involving enzyme-coupled NADP reduction or NADH oxidation, respectively. Assays of adenylate kinase activity involving two steps require special consideration due to the extreme stability of this enzyme towards the usual denaturants employed to terminate enzyme reactions (heat, acid precipitants). For routine adenylate kinase assay, the rate of ATP production (back reaction) is followed fluorimetrically in a one-step combined assay, which is described.

Reagents

Tris-chloride buffer, 0.5 M, pH 8.0
ADP, 0.1 M Na salt
NADP, 10 mM
Glucose, 0.2 M
MgCl$_2$, 0.1 M
Hexokinase, 300 units/mg protein (myokinase-free)
Gluc-6-P dehydrogenase, 300–400 units/mg protein (myokinase-free)

Enzyme. Enzyme is diluted for assay in the presence of Triton X-100 (0.2%), EDTA at 0.1 mM, dithiothreitol at 0.5 mM, and excess electrolyte (0.1 M KCl and 10 mM P$_i$ buffer at pH 7.0). The enzyme is particularly unstable in solutions of low ionic strength.

Procedure. Adenylate kinase activity is measured by the rate of ATP produced in reaction mixtures containing in 1.0-ml volume: 50 mM Tris-chloride buffer at pH 8.0, 1 mM ADP, 2 mM MgCl$_2$, 0.1 mM NADP, 2 mM glucose, 0.1 unit hexokinase, 0.1 unit gluc-6-P dehydrogenase, and 0.0001–0.001 units of enzyme to be assayed. Reactions are carried out at 25° in a Turner fluorometer equipped with a thermostated sample compartment and attached linear recorder. The reaction leading to NADPH generation is recorded as a linear increase in fluorescence, with the rate determined by the slope of the straight line.

Definition of Unit and Specific Activity. One unit of enzyme activity catalyzes the formation of 1 μmole of ATP per minute at 25° and pH 8.0 under the specified assay conditions. Specific activity is defined as units of enzyme per milligram protein. Protein is determined by hemoglobin

---

[2] P. Greengard, *Nature (London)* **176**, 632 (1956).

analysis[3] in initial crude solutions, by absorbancy at 290 nm in 0.1 N NaOH in partially purified enzyme solutions, and by absorbancy at 280 nm at pH 8.0 in the highly purified enzyme solutions. One mg of protein per milliliter was assumed to give an absorbancy of 1.0 at 290 nm in alkali and 0.56 at 280 nm at pH 8.0.

Purification Procedure[4]

Preliminary Typing. Outdated human blood is screened by starch gel electrophoresis to separate approximately 1 out of 10 blood samples exhibiting the less frequent AK 2-1 phenotype (Fig. 1b) from the predominant AK-1 phenotype (Fig. 1a). The AK-1 phenotype displays a major (least anodal) and two associated minor enzyme forms. Blood samples displaying the AK-1 phenotype are combined. Purification of the major enzyme form to apparent homogeneity from 2 liters of AK-1 red cells is described (see purification protocol in the table).

Chloroform–Methanol Denaturation. To 2 liters of washed red cells are added: 2 liters of water, AMP to 1 mM, and conc. NH$_4$OH to pH 8.7. The mixture is cooled to 2–4°, and 1200 ml of an equal mixture (v/v) of chloroform–methanol (precooled to −20°) are added. The mixture is immersed in an ice bath and stirred at 5-min intervals for 1 hr. After adding 2.8 liters of cold water, the mixture is centrifuged to remove denatured Hb. To the resulting red solution are added 1 N acetic acid to

PURIFICATION PROTOCOL OF ERYTHROCYTE ADENYLATE KINASE

Step	Volume (ml)	Protein (mg)	U.E. (μmole/ min)	Specific activity (U.E./mg protein)	Yield (%)
1. Hemolysate	4000	540,000	90,000	0.17	100
2. CHCL$_3$–MePH conc. (1)	385	50,000	60,000	1.2	67
3. CHCL$_3$–MePH conc. (2)	70	20,800	42,000	2.0	47
4. Sephadex G-100 conc.	60	7,000	37,000	5.3	41
5. DEAE-cell. filtrate	490	3,400	36,000	10.6	40
6. CM-cell. batch	75	140	30,000	210	33
7. CM-cell. chrom. (1)	6.0	20	24,000	1200	27
8. CM-cell. chrom. (2)	2.0	6.6	17,000	2600	19
9. CM-cell. chrom. (3)	1.0	4.1	13,000	3200	14

[3] G. E. Cartwright, *in* "Diagnostic Laboratory Hematology," 3rd. ed., p. 42. Grune & Stratton, New York, 1963.
[4] K. K. Tsuboi and C. H. Chervenka, *J. Biol. Chem.* **250,** 132 (1975).

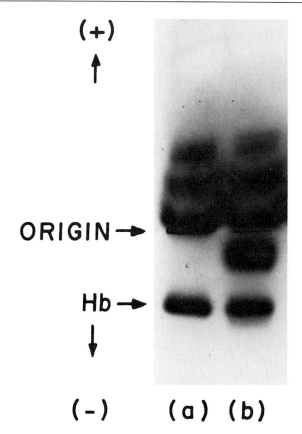

FIG. 1. Electrophoretic patterns in starch gel of the adenylate kinases of the human red cell, demonstrating the predominant AK-1 phenotype (a) and the less frequent AK 2-1 phenotype (b).

pH 7.0, Triton X-100 to 0.002%, and caprylic acid to 0.01%, prior to concentration by rotary evaporation initially at room temperature and then at 35° after removal of organic solvents. The solution is concentrated to contain about 130 mg Hb per ml (step 2).

Second Chloroform–Methanol. The still Hb-rich solution is adjusted to pH 8.7 with conc. NH₄OH, cooled to 2–4°, and mixed with 0.3 volume of an equal mixture (v/v) of chloroform–methanol (precooled to −20°). The mixture is immersed in an ice bath and stirred at 10-min intervals over a 45-min period. After adding an equal volume of cold water, the mixture is centrifuged. To the resulting solution are added acetic acid to pH 7.0, Triton X-100 to 0.002%, caprylic acid to 0.01%;

the solution is concentrated to about 70 ml volume by rotoevaporation under reduced pressure as before (step 3).

Fractionation on Sephadex G-100. The concentrated enzyme solution is filtered, divided into two equal aliquots, and each is successively fractionated on a preparative column of Sephadex G-100 (5 × 100 cm) in buffer medium containing 50 mM KCl, 20 mM P_i at pH 7.0, 0.1 mM EDTA, and 0.002% Triton X-100. The active enzyme fractions from both column runs are combined, concentrated by rotoevaporation to about 40-ml volume, and dialyzed overnight against 2 liters of a solution containing 10 mM P_i at pH 7.0, 0.1 mM EDTA, 0.002% Triton X-100, and 0.1 mM dithiothreitol (step 4).

Treatment with DEAE-Cellulose and Batch Adsorption on CM-Cellulose. The dialyzed enzyme solution is diluted with 9 volumes of a cold solution containing 0.1 mM EDTA plus 0.002% Triton X-100 and mixed immediately with 175 ml of an aqueous 50% DEAE-cellulose suspension (1 ml/40 mg protein) equilibrated at pH 8.5 in Tris-chloride buffer. The resulting mixture is filtered immediately through a coarse scintered-glass funnel into a pressure flask containing 175 ml of a 50% suspension of CM-cellulose (preequilibrated to pH 7.0 with sodium P_i buffer) in 2 mM P_i buffer at pH 7.0, 0.1 mM EDTA, and 0.002% Triton X-100. All operations to this point should be completed as rapidly as possible at temperatures near 0° to avoid excessive enzyme loss induced at low ionic strength. About 10% of the enzyme activity remains unadsorbed on the CM-cellulose and consists primarily of the minor enzyme forms of greater anodal migration in starch gel (see Fig. 1). The CM-cellulose, containing adsorbed enzyme, is transferred to a 125-ml-volume coarse scintered-glass funnel, washed, and eluted by gravity flow in 0.1 M KP_i buffer, pH 7.0, containing 0.1 mM EDTA, 0.2 mM dithiothreitol, and 0.002% Triton X-100, by collecting the initial 75-ml elution volume (steps 5 and 6).

CM-Cellulose Chromatography. The eluted enzyme solution is concentrated to less than 10 ml (UM-10 Amicon filter) and dialyzed 3 hr (1.0-cm-diameter tubing) with stirring against a liter of solution containing 10 mM Tris-chloride (pH 7.5), 0.2 mM EDTA, 0.4 mM dithiothreitol, and 0.004% Triton X-100. The dialyzed solution is diluted with an equal volume of water and immediately transferred to a CM-cellulose column (1.5 × 27 cm). The enzyme is eluted by a linear Tris gradient (pH 7.5) extending from 5–30 mM (total 250 ml volume) in the presence of 0.2 mM dithiothreitol, 0.1 mM EDTA, and 0.002% Triton X-100. Enzyme fractions are collected in tubes containing preadded salt (KCl to final 0.1 M) for stabilization. Fractions comprising the major enzyme peak (minor enzyme peak also present) are pooled, concentrated, and dialyzed as before preparatory to rechromatography (step 7).

CM-Cellulose Chromatography. The dialyzed enzyme solution is diluted with an equal volume of water and rechromatographed as before on CM-cellulose. Enzyme activity is eluted in a narrow and almost coincident band of protein, indicating probable near-homogeneity. The active fractions are again combined, concentrated, and dialyzed for final chromatography (step 8).

CM-Cellulose Chromatography to Constant Specific Activity. Chromatography on CM-cellulose is repeated a third time under identical conditions. The chromatography results in only little further increase in specific activity and yields a single apparent homogeneous enzyme–protein peak. The combined enzyme fractions are concentrated by vacuum dialysis, have a specific activity exceeding 3000, and are stored as the final product (step 9).

Overall yield is usually between 10–20% with most of the loss attributable to instability associated with fractionation sequences requiring low ionic strengths.

Properties

Stability. Enzyme preparations purified to homogeneity maintain specific activity at 3000 for only a few days, dropping to around 1000 with storage at near 0°. Purified enzyme solutions are maximally stable to dilution and heat only in the combined presence of dithiothreitol (2 mM), Triton X-100 (0.02%), and salt (0.2 M KCl). Under these conditions little inactivation is found after 10-min heating at 90° (pH 6–9), even in the presence of 5 M guanidine hydrochloride.

Purity. The enzyme appears homogeneous by acrylamide gel electrophoresis, sedimentation analysis, and by specific activity analysis across elution peaks following gel filtration and CM-cellulose chromatography.

Electrophoretic Behavior and pI. The enzyme migrates as a weak anion at pH 7.0 in starch gel[5] (see Fig. 1), while demonstrating strong basic properties with ion-exchange celluloses and a pI of 9.0 by isoelectric focusing. Anomalous electrophoretic behavior in starch gel is thereby indicated.

Sedimentation and Molecular Weight. A sedimentation coefficient of 2.0–2.1 S, partial specific volume of 0.722 ± 0.004 cm³/g, and molecular

---

[5] R. A. Fildes and H. Harris, *Nature (London)* **209**, 261 (1966).

weight of 21,300 are obtained by sedimentation analyses of the native enzyme. Sedimentation analysis and electrophoresis of the enzyme in the presence of strong denaturants show no evidence of subunit structure and yield molecular weight estimates of around 23,000.

Ultraviolet Absorbance. The enzyme demonstrates a broad absorbance maximum at 280 nm and 280:290 nm absorbance ratio of 1.84 at neutral pH. The absorbance coefficient is $E_{1cm}^{1\%} = 5.6$ at 280 nm at neutral pH, which is similar to the enzyme prepared from rabbit muscle.[6]

Catalytic Properties. Relative catalytic activity when tested using various nucleoside monophosphate and nucleoside triphosphate pairs shows: AMP + ATP (100%), AMP + dATP (52%), dAMP + ATP (10%), AMP + CTP (2.6%), CMP + ATP (1.1%), AMP + dGTP, GTP, UTP, and ITP (0.0%), and ATP + dGMP, GMP, UMP, and IMP (0.0%).

K_m values are 9 and 8 × 10^{-5} M for ATP and AMP, respectively, when tested at magnesium concentrations maintained at 2:1 ratio relative to ATP. A K_m value of 1.1 × 10^{-4} M is shown for ADP at constant 1.0 mM magnesium (at ADP concentrations tested at 0.01–0.05 mM). All other combinations of magnesium to ADP concentrations yield larger apparent K_m values.

Immunological Properties. Repeated injections of the enzyme in complete Freund's adjuvant into rabbit footpads at 3-week intervals yield no detectable antibody titer. The apparent lack of antigenicity is suggestive of a high degree of structural similarity between the human and rabbit enzymes.

---

[6] L. Noda and S. A. Kuby, *J. Biol. Chem.* **226**, 541 (1957).

[63] Guanylate Kinase from *Escherichia coli* B

By Max P. Oeschger

$$\text{(d)GMP} + \text{(d)ATP} \rightarrow \text{(d)GDP} + \text{(d)ADP} \tag{1}$$

Guanylate kinase from *E. coli* actively phosphorylates guanine and 2-aminopurine nucleotides but shows only slight activity with other purine nucleotides and no activity with pyrimidine nucleotides.[1-3] The enzyme

---

[1] M. J. Bessman and M. J. Van Bibber, *Biochem. Biophys. Res. Commun.* **1**, 101 (1959).
[2] M. P. Oeschger and M. J. Bessman, *J. Biol. Chem.* **241**, 5452 (1966).
[3] E. G. Rogan and M. J. Bessman, *J. Bacteriol.* **103**, 622 (1970).

will phosphorylate both ribo- and deoxyribonucleotide substrates although the K_m value for dGMP is 5 times that of the K_m value for GMP.[2] The activity of the enzyme is markedly stimulated by K^+ and NH_4^+, but the mechanism of activation by the two ions appears to be quite different.[2]

Assay Methods

Spectrophotometric Assay

The spectrophotometric assay is the more convenient of the two assays presented but is subject to high blank values with crude extracts. Ninety percent of the interfering activities can be removed from crude extracts by treatment with acetone permitting use of the spectrophotometric assay at all stages of the purification.

Principle. This method is based on the following sequence of reactions:

$$(d)GMP + (d)ATP \xrightarrow{\text{guanylate kinase}} (d)GDP + (d)ADP \qquad (2)$$

$$2\text{ phosphoenolpyruvate} + (d)ADP + (d)GDP \xrightarrow{\text{pyruvate kinase}}$$

$$2\text{ pyruvate} + (d)ATP + (d)GTP \qquad (3)$$

$$2\text{ pyruvate} + 2\text{ NADH} + 2\text{ H}^+ \xrightarrow{\text{lactate dehydrogenase}} 2\text{ lactate} + 2\text{ NAD}^+ \qquad (4)$$

With pyruvate kinase and lactate dehydrogenase present in excess, the rate of guanylate phosphorylation is proportional to the rate of NADH oxidation, which is measured by the absorbance decrease at 340 nm.[4,5]

Reagents. The composition of the assay mixture is shown in Table I.

Procedure. The constituents of the assay (Table I) are mixed in a cuvette (1-cm path length), and the oxidation of NADH is followed spectrophotometrically at 340 nm. With impure enzyme preparations, a background rate of NADH oxidation is obtained for 2–3 min without (d)GMP in the reaction mixture. (Deoxy)guanosine monophosphate is then added to initiate the guanylate kinase catalyzed reaction. The background rate, which is due to ATPase and DPNH oxidase activities, is subtracted from the rate obtained after the addition of (d)GMP.

Treatment of Crude Extract. The spectrophotometric assay can be used with crude extracts after removing proteins which give high

[4] A. Kornberg and W. E. Pricer, *J. Biol. Chem.* **193**, 481 (1951).
[5] I. Liberman, A. Kornberg, and E. S. Simms, *J. Biol. Chem.* **215**, 429 (1955).

TABLE I
SPECTROPHOTOMETRIC ASSAY MIXTURE

Component	Volume (ml)	Final concentration (mM)
H_2O	0.39	—
Tris·HCl, 1 M (pH 8 at 50 mM)	0.10	100
KCl, 2 M	0.12	240
MgCl$_2$, 0.1 M	0.20	20
ATP, 0.1 M (pH 7)	0.05	5
Phosphoenolpyruvate, 50 mM	0.02	1
NADH, 25 mM	0.04	1
Lactic dehydrogenase[a]	0.03	—
Guanylate kinase[b]	0.02	—
dGMP 0.1 M	0.03	3
Vf	1.00	

[a] Lactic dehydrogenase is Type I from Sigma which contains pyruvate kinase. Pyruvate kinase is essential for the assay. The enzyme is prepared by dilution 1:5 with and subsequent dialysis against 0.05 M Tris·HCl (pH 8 at 0.05 M) or 0.1 M KPO$_4$ buffer (pH 7.5).

[b] Highly purified samples of enzyme may be diluted in the solution of lactic acid dehydrogenase to maintain full activity.

endogenous rates of NADH oxidation in the absence of substrate. A convenient way to achieve this fractionation is to slowly add an equal volume of acetone (at $-10°$) to a small volume of crude extract (approximately 10 mg of protein per milliliter) at $0°$.[6] The precipitate is removed by centrifugation and the acetone evaporated from the supernatant fluid by further centrifugation at $10°$ in an uncovered rotor. This treatment removes about 90% of the ATPase and DPNH oxidase activities from the crude extract while leaving the guanylate kinase activity unchanged.[2]

Definition of Unit and Specific Activity. One unit of guanylate kinase activity is defined as the amount of enzyme which catalyzes the phosphorylation of 1.0 μmole of (d)GMP per minute at $37°$.

The specific activity of a preparation is defined as the units of guanylate kinase activity present per milligram of protein.

[6] The temperature may rise to $10°$ during the addition without affecting the guanylate kinase activity.

Radiochemical Assay

Principle. In crude extracts the radiochemical assay gives the most accurate measure of enzyme activity. This method measures the formation of a phosphomonoesterase-insensitive material (nucleoside di- or triphosphate) from a phosphomonoesterase-sensitive material (nucleoside monophosphate).[7]

Reagents. Composition of the assay mixture is shown in Table II.

Procedure. The constituents of the assay (Table II) are mixed in a 13 × 100 mm tube and incubated at 37° for 20 min.[8] The reaction is

TABLE II
RADIOCHEMICAL ASSAY MIXTURE

Component	Volume (ml)	Final concentration (mM)
First incubation[a]		
H_2O	0.045	—
Tris·HCl, 1 M (pH 8 at 50 mM)	0.025	100
KCl, 2 M	0.03	240
MgCl₂, 0.1 M	0.05	20
ATP, 0.125 M (pH 7)	0.01	5
Phosphoenolpyruvate, 50 mM	0.02	4
Lactic dehydrogenase[b]	0.03	—
[32P]dGMP, 12.5 mM (sp act ≥ 0.1 mCi/mM)	0.02	1
Guanylate kinase	0.02	—
Vf	0.25	
Second incubation[c]		
H_2O	0.5	
Potassium acetate, 1 M (pH 5.1)	0.1	
Semen phosphomonoesterase	0.02	
Norit adsorption[d]		
H_3PO_4, 5 M	0.1	
Norit 20% (v/v) in H_2O	0.1	

[a] Reaction set up on ice, incubated for 20 min at 37°, stopped by heating to 100° for 2 min.

[b] Lactic dehydrogenase is Type I for Sigma which contains pyruvate kinase. The enzyme is prepared by dilution 1:5 with and subsequent dialysis against 0.1 M KPO₄ buffer, pH 7.5.

[c] 30 min at 37°, stopped by placing tubes in ice-water bath.

[d] Allowed to adsorb in ice-water bath for 5 min, Norit collected by filtration on RA 934 filters.

[7] I. R. Lehman, M. J. Bessman, E. S. Simms, and A. Kornberg, *J. Biol. Chem.* **233,** 163 (1958).

[8] The tubes are topped with glass marbles to prevent evaporation in this step.

terminated by heating the tube at 100° for 2 min.[8] The unreacted substrate is degraded following acidification of the reaction mixture with potassium acetate (pH 5.1) by incubation with a crude preparation of human semen phosphomonoesterase at 37° for 20 min.[9] After treatment with semen phosphomonoesterase the labeled product is adsorbed onto Norit, collected by filtration on a glass fiber disc, washed, dried, and counted.

Purification Procedure[2]

The results of a typical purification procedure are summarized in Table III. After the third step (ammonium sulfate fractionation) the enzyme is relatively free of most noxious contaminants and can be used for the preparation of phosphorylated nucleotides. Unless otherwise stated all operations are carried out between 0°–4°. A buffer of 0.1 M potassium phosphate, pH 7.5, containing 1 mM EDTA is used. For large-scale work centrifugations are performed in a Lourdes continuous-flow refrigerated centrifuge at 25,000 g.

Step 1. Crude Extract. First 500 g of washed (0.5% NaCl, 0.5% KCl), packed *E. coli* B cells (stationary phase, grown aerobically at 37° in a medium composed of 1.1% K_2HPO_4, 0.85% KH_2PO_4, 0.6% Difco yeast extract, and 1% dextrose) are suspended in 5 liters of buffer with the aid of a Waring Blendor and then disrupted by sonic oscillation (a Branson sonifier tuned to maximum output fitted with a continuous-flow

TABLE III
PURIFICATION OF GUANYLATE KINASE FROM *E. coli*[a]

Step and fraction[b]	Total volume (ml)	Enzyme activity (units/ml)	Protein (mg/ml)	Specific activity (units/mg protein)	Relative purity	Recovery (%)
1. Crude (I)	23,000	0.22	8	0.027	1	100
2. Acetone (II)	3,550	1.0	3	0.33	12	70
3. $(NH_4)_2SO_4$ (III)	124	24.8	26	0.95	35	61
4. K_2HPO_4 (IV)	14	200	99	2.0	74	56
5. Sephadex (V)	330	6.4	0.43	15	556	42
6. DEAE-cellulose (VI)	160	8.4	0.22	38.2	1415	27
7. DEAE-cellulose (VII)	214	4.3	0.023	187	6926	18

[a] Activity was measured with the spectrophotometric assay using dGMP as the substrate.
[b] Steps 1, 2, 3, 5, and 6 were carried out on a smaller scale, and the fractions were stored until sufficient quantities were accumulated.

[9] J. Wittenberg and A. Kornberg, *J. Biol. Chem.* **202**, 431 (1953).

attachment employing a flow rate of 18–22 ml/min). The supernatant fluid is collected by centrifugation[10] and is either stored frozen where it is stable or at 4° where it loses less than 10% of its activity per week.

Step 2. Acetone Fractionation. To each liter of fraction 1 (protein concentration 7–8 mg/ml) 1.5 liters of acetone are added over a 10-min period with constant, rapid stirring.[6] Immediately after the addition of the acetone the precipitate is removed by centrifugation at −3° using a flow rate of 300 ml/min.[10] This supernatant fluid may be stored at −15° for at least 1 week without the loss of activity. To each 1.5 liters of the supernatant fluid, at −12°, 0.8 liter of acetone at 0° is added with constant, rapid stirring over a period of 10 min. During the addition the temperature is maintained at −10°. The precipitate is collected immediately by centrifugation at −15° using a flow rate of 250 ml/min[10] and dissolved in buffer with the aid of a probe-type sonic oscillator. A ratio of 2.5 ml of buffer to 10^4 units of fraction 1 is used. Fraction 2 may be stored at −15° for at least 1 month with no loss of activity.

Step 3. Ammonium Sulfate Fractionation. To 1.6 liters of fraction 2 (adjusted to 550 units/ml), 1.04 liters of saturated ammonium sulfate solution containing 1 mM EDTA are added with constant stirring over a period of 15 min. The suspension is allowed to equilibrate for 1 hr with constant stirring, and the precipitate is removed by centrifugation for 20 min at 16,000 g. To the supernatant fluid are added 925 ml of saturated ammonium sulfate solution, containing 1 mM EDTA, over a period of 10 min. The suspension is allowed to equilibrate for 1 hr with constant stirring, and the precipitate is then collected by centrifugation with a flow rate of 50 ml/min[10] and dissolved in buffer (a ratio of 6 ml of buffer to 10^5 units of fraction 2). This fraction is stable for at least 6 months when stored at −15°.

Step 4. Potassium Phosphate Fractionation. To 125 ml of fraction 3 (adjusted to 15,000 units/ml) 21 ml of 5 M K$_2$HPO$_4$ containing 1 mM EDTA are added drop by drop with constant stirring over a period of 5 min. The suspension is allowed to equilibrate for 10 min and the precipitate removed by centrifugation at 16,000 g for 20 min. To the supernatant fluid an additional 7.7 ml of the phosphate solution are added as above. After an additional 10 min for equilibration, the precipitate is collected by centrifugation for 20 min at 16,000 g and dissolved in 5 ml of distilled water. To this solution (13 ml) 1 ml of 1 M potassium phosphate buffer (pH 7.5 at 0.1 M) containing 10 mM EDTA is added. This fraction is stable at −15° for at least 6 months.

---

[10] For smaller volumes, batchwise centrifugation at 15,000 g for 20 min is used.

Step 5. Sephadex G-100 Fraction. A column of Sephadex G-100 (33.3 cm² × 100 cm) is equilibrated with 0.1 M potassium phosphate buffer, pH 7.5, containing 1 mM EDTA and 0.1 M potassium chloride. Then 5 ml of fraction 4 are layered on the top of the column.[11] The sample is run into the gel, and the top of the column is layered with buffer and elution begun. The flow rate is set at 60 ml/hr, and 10-ml fractions are collected. The enzymic activity is quantitatively recovered in a peak between 0.39–0.46 column bed volumes. The peak fractions (with specific activities greater than 5000 units/mg protein) are pooled (fraction 5, Table III). The active fractions are conveniently identified because of the coincidental elution of a flavoprotein which is visible as a discrete yellow band migrating down the column. The pooled Sephadex fraction is stable for 1 week when stored at 4°.

Step 6. First Chromatography on DEAE-Cellulose. Fraction 5 (200 ml containing 85 mg of protein) is transferred by dialysis to 0.02 M potassium phosphate buffer, pH 7.0, containing 1 mM EDTA and applied to a column of DEAE-cellulose (0.1 cm² × 100 cm) equilibrated with the same buffer. After washing the column with 15 ml of the phosphate solution, an exponential gradient is applied with 0 M and 0.2 M potassium chloride as limiting concentrations. The total volume of the gradient is 350 ml; 0.02 M potassium phosphate buffer, pH 7.0, and 1 mM EDTA are present throughout. The flow rate is set to approximately 7 ml/hr, and 3.5-ml fractions are collected. Of the activity applied, 80% is eluted in a single broad peak between 3.5–20 bed volumes (0.015–0.08 M potassium chloride).

The peak fractions containing enzyme of specific activity above 20,000 units/mg of protein (64% of that applied to the column) are pooled (fraction 6, Table III). The flavoprotein which followed the enzymic activity in the previous column fractions is separated in this step. Analysis of the pooled fractions by disc electrophoresis reveals two major bands of protein. This fraction may be stored indefinitely at 4° without loss of activity.

Step 7. Second Chromatography on DEAE-Cellulose. A column (0.1 cm² × 100 cm) is prepared with a washed DEAE-cellulose suspension and equilibrated with 0.02 M potassium phosphate buffer, pH 8.3, containing 1 mM EDTA. Fraction 6 (135 ml containing 30 mg of protein) is equilibrated with the same buffer by dialysis and applied to the

---

[11] A convenient method for the even application of the sample to large diameter columns is to freeze it as a disc (2 mm thick, slightly smaller in diameter than the column i.d.) supported by a nichrome wire frame. The frozen disc is suspended just above the gel at the top of the column and allowed to melt. This method allows an even application of material without disturbing the surface of the gel.

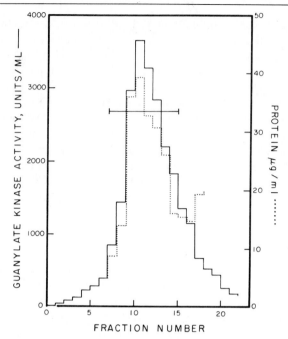

FIG. 1. Chromatography of fraction 6 on DEAE-cellulose at pH 8.3. Activity was measured with the spectrophotometric assay using dGMP as substrate. The fractions indicated by the bar were pooled and constitute fraction 7. From M. P. Oeschger and M. J. Bessman, *J. Biol. Chem.* **241**, 5452 (1966).

adsorbent at a rate of 12 ml/hr. The column is washed with 40 ml of the phosphate solution, and a linear gradient is applied with 0 M and 0.2 M potassium chloride as limiting concentrations. The total volume of the gradient is 1.5 liters, and 0.02 M potassium phosphate buffer, pH 8.3, and 1 mM EDTA are present throughout. The flow rate is set at 12 ml/ hr, and 22-ml fractions are collected. Of the activity applied to the column, 90% is eluted in a single peak between 2–60 bed volumes (0.003–0.08 M potassium chloride) (Fig. 1). The peak fractions (67% of the total activity) are pooled. As can be seen in Fig. 1 these fractions contain enzyme of constant specific activity (116,000 units/mg of protein) and reveal only one band when analyzed with disc electrophoresis. This fraction is stable indefinitely when stored at 4° in the eluting buffer or when dialyzed into 0.05 M Tris-chloride buffer at pH 8.1.

Properties[2]

Stability. The activity of guanylate kinase is stable at −20° at all stages of the purification. The activity is unstable in dilute solution (less than 5 μg protein per milliliter) but may be stabilized by the addition of bovine serum albumin or other suitable protein. Highly purified guanylate kinase is stabilized in the assay solutions by the presence of lactic dehydrogenase and pyruvate kinase.

Substrate Specificity. The activity of guanylate kinase is specific for purine nucleotides. It is highest with guanine, good with 2-aminopurine (one-tenth that of guanine), very limited with inosine (1% that of guanine), and barely detectable with adenine as the purine bases. The enzyme is active with ribo- or 2'-deoxyribonucleotides. The K_m value for dGMP when using NH_4^+ as the activating ion is 0.3 mM.

Cationic Requirements. Guanylate kinase requires Mg^{2+} for activity. Mn^{2+} will substitute for Mg^{2+} and to good measure so will Co^{2+}. The K_m values for the three ions are 0.75 mM, 1.0 mM, and 1.25 mM, respectively. Guanylate kinase also requires either the monovalent ion K^+ or NH_4^+ for activity. The apparent mechanism for activation by the two ions is quite different, for while both K^+ and NH_4^+ enhance the V_{max}, K^+ reduces the K_m value for dGMP 32-fold while NH_4^+ does not alter the substrate K_m value (Figs. 2 and 3). The K_m values for K^+ range between 10.8–347 mM while the K_m value for NH_4^+ remains fixed at 39 mM.

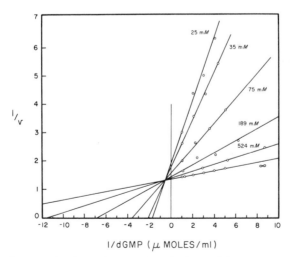

Fig. 2. Influence of K^+ on the K_m for dGMP. Activity was measured with the spectrophotometric assay at five concentrations of K^+. The values for the line marked ∞ were obtained by the extrapolation to infinite K^+ concentration of the observed rates with different concentrations of dGMP. The source of the enzyme was fraction 7. From M. P. Oeschger and M. J. Bessman, *J. Biol. Chem.* **241**, 5452 (1966).

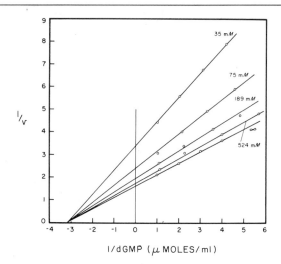

$1/dGMP$ (μ MOLES/ml)

FIG. 3. Influence of NH_4^+ on the K_m for dGMP. Activity was measured with the spectrophometric assay at four concentrations of NH_4^+. The values for the line marked ∞ were obtained by the extrapolation to infinite NH^+ concentration of the observed rates with different concentrations of dGMP. From M. P. Oeschger and M. J. Bessman, *J. Biol. Chem.* **241**, 5452 (1966).

pH Optimum. Guanylate kinase is maximally active in Tris-chloride buffer between pH 7.3 and pH 8.2. The enzyme shows 70% of its maximal activity at pH 6.5 and pH 9.0.

Molecular Properties. Purified guanylate kinase migrates as a single sharp band during electrophoresis on 7.5% polyacrylamide gels. At equilibrium the purified enzyme distributes in a centrifugal field so that a plot of the log of the concentration versus the square of the radius of rotation gives a straight line. The slope of the line suggests a molecular weight for the native enzyme of 88,000. Fluorescence studies indicate that the enzyme may not contain tryptophan, for only a single emission maximum at 3.2 μm is observed. These results indicate that the guanylate kinase activity is associated with a single protein, native molecular weight 88,000, which does not contain tryptophan.

[64] Guanylate Kinases[1] from Human Erythrocytes, Hog Brain, and Rat Liver

By K. C. AGARWAL, R. P. MIECH, and R. E. PARKS, JR.

$$ATP + GMP \text{ (dGMP)} \rightleftharpoons ADP + GDP \text{ (dGDP)}$$

Assay Method[2]

Principle. A convenient assay involves the measurement of both ADP and GDP formed from ATP and GMP by coupling the above reaction to the pyruvate kinase and lactate dehydrogenase reactions. The decrease in absorbancy at 340 nm associated with NADH oxidation is measured with a spectrophotometer using a blue filter.[3] Alternative enzyme assays have been described.[4-7]

$$ATP + GMP \rightleftharpoons ADP + GDP$$

$$\frac{ADP}{GDP} + 2 \, PEP \rightleftharpoons \frac{ATP}{GTP} + 2 \, \text{pyruvate}$$

$$2 \, \text{Pyruvate} + 2 \, NADH + 2 \, H^+ \rightleftharpoons 2 \, \text{lactate} + 2 \, NAD^+$$

Reagents

Stock solutions	Final concentration (1 ml reaction volume)
1. Tris·HCl buffer, 1 M, pH 7.5	0.1 M
2. KCl, 1 M	0.1 M
3. MgCl$_2$, 0.1 M	0.01 M
4. Sodium phosphoenol pyruvate, 15 mM	1.5 mM
5. Pyruvate kinase, 25 units/ml	2.5 units
6. Lactate dehydrogenase, 33 units/ml	3.3 units

[1] ATP:GMP phosphotransferase, EC 2.7.4.8.

[2] R. P. Miech and R. E. Parks, Jr., *J. Biol. Chem.* **240**, 351 (1965).

[3] Without the use of the blue filter a large error in the measurement of NADH oxidation may occur. For references related to the effects of stray light in spectrophotometry and how they may introduce reproducible artifacts in this coupled enzyme assay, the reader is referred to R. L. Cavalieri and H. Z. Sable, *Anal. Biochem.* **59**, 122 (1974) and R. L. Miller *et al.*, *Biochem. Pharmacol.* (in press).

[4] R. L. Miller, D. L. Adamczyk, T. Spector, K. C. Agarwal, R. P. Miech, and R. E. Parks, Jr., *Biochem. Pharmacol.* **26**, 1573 (1977).

[5] R. J. Buccino, Jr. and J. S. Roth, *Arch. Biochem. Biophys.* **132**, 49 (1969).

[6] M. P. Oeschger and M. J. Bessman, *J. Biol. Chem.* **241**, 5452 (1966).

[7] Y. Sugino, H. Teraoka, and H. Shimono, *J. Biol. Chem.* **241**, 961 (1966).

7. NADH, 1.5 mM 0.15 mM
8. ATP, 40 mM (neutralized with Tris to pH   4.0 mM
7.0)
9. GMP, 1 mM 0.1 mM

Procedure. With the omission of GMP, equal volumes of the above reagents are mixed to form a GMP kinase reaction mixture (GKRM). Pipette 0.8 ml of GKRM into a 1-ml cuvette and record the absorbance at 340 nm. The absorbance reading (~0.930 A_{340}) should be constant with time. The guanylate kinase solution (0.1 ml) is added to the cuvette containing GKRM, and the rate of change of A_{340} with time is recorded at 30°. Any contaminating enzymes that may degrade ATP, PEP, or NADH result in a constant rate of decrease in A_{340} absorbance which represents a background rate. The background rate is usually high in initial crude tissue or cellular extracts which must be diluted sufficiently so that the background rate does not exceed 0.050 A_{340} per minute. After the background rate has become constant, 0.1 ml GMP (1 mM) is added to the cuvette to initiate the guanylate kinase reaction. The constant rate of decrease in A_{340} is recorded and represents the sum of the background rate and guanylate kinase rate. The difference between this rate of change in A_{340} and the background rate represents the rate of guanylate kinase activity expressed as ΔA_{340} per minute. Since 2 μmoles of NADH are oxidized per μmole of GMP converted to GDP, the guanylate kinase activity expressed as ΔA_{340}/min must be divided by 12.4 A_{340}/μmole ml^{-1} to yield the number of μmoles of GMP converted to GDP per minute in the 1-ml reaction volume.

Definition of Enzyme Unit and Specific Activity. One enzyme unit is defined as the amount of enzyme that catalyzes the conversion of 1 μmole of GMP to GDP per minute under the above reaction conditions. Specific activity is expressed as units per milligram of protein. Protein concentrations may be determined conveniently by methods such as that of Lowry *et al.*[8] or by ultraviolet (UV) absorption at 280 nm.[9]

Purification Procedures

All procedures steps are performed at 0–4°. Enzymic activities and protein concentrations are determined after each step of purification.

[8] O. H. Lowry, N. J. Rosebrough, A. L. Farr, and R. J. Randall, *J. Biol. Chem.* **193**, 265 (1951).
[9] O. Warburg and W. Christian, *Biochem. Z.* **310**, 384 (1941).

Hog Brain[2] *or Perfused Rat Liver:*[10] *Step 1. Extraction.* Either frozen tissue that has been thawed or fresh tissue may be used. Trimmed tissue (100 g) is added to about 4 volumes of cold 0.15 M KCl or 0.1 M Tris-chloride buffer, pH 7.5, homogenized in a Waring Blendor at medium speed for 5–10 sec. Prolonged homogenization is to be avoided, since the resulting lipid emulsion interferes with ammonium sulfate fractionation. The homogenate is centrifuged at 20,000 g for 20 min, and the supernatant fluid (fraction 1) is collected.

Step 2. Ammonium Sulfate Fractionation. Solid ammonium sulfate (231.1 g/liter, 40% saturation) is added slowly with stirring to fraction 1. The resulting suspension is centrifuged at 20,000 g for 20 min. The precipitate is discarded, and the volume of the supernatant fluid is measured. More solid ammonium sulfate (123.1 g/liter, 60% saturation) is added slowly, and the mixture is stirred until the ammonium sulfate is completely dissolved. The resulting suspension is permitted to stand overnight without stirring to permit aggregation of the fine precipitate. The suspension is centrifuged at 20,000 g for 20 min, and the supernatant fluid is discarded. The precipitated proteins are dissolved in a minimal amount of 0.005 M Tris-chloride buffer, pH 7.5, and the protein solution is dialyzed overnight against several changes of the same buffer. Precipitated proteins that form during dialysis are removed by centrifugation, and the supernatant fluid (fraction 2) is employed for further purification described in step 3.

Step 3. Calcium Phosphate Gel Adsorption. Calcium phosphate gel is prepared according to the method of Tsuboi and Hudson[11] (also see Chapter 79, Agarwal *et al.*[12]). The final suspension is adjusted with water to a concentration of 15–30 mg of dry solids per milliliter. The calcium phosphate gel suspension is added to fraction 2 of rat liver to produce a gel to protein ratio of 1:20. In the case of hog brain fraction 2, gel to protein is 2:1. After 30 min of gentle stirring the supernatant fluid is collected by centrifugation at 5000 g for 10 min. The gel-pellet is washed twice with 0.005 M Tris-chloride buffer, pH 7.5. The washings are added to the supernatant fluid. Solid ammonium sulfate (445.3 g/liter, 70% saturation) is added slowly to the above clear supernatant fluid and after complete dissolution of the ammonium sulfate, the cloudy suspension is stored at 4° for 12–72 hr. The suspension is centrifuged at 20,000 g for 20 min. The supernatant fluid is discarded, and the precipitate is dissolved in a minimal amount of 0.005 M Tris-chloride buffer, pH 7.5,

[10] K. C. Agarwal and R. E. Parks, Jr., *Biochem. Pharmacol.* **24,** 791 (1975).
[11] K. K. Tsuboi and P. B. Hudson, *J. Biol. Chem.* **224,** 879 (1957).
[12] See R. P. Agarwal, K. C. Agarwal, and R. E. Parks, Jr., this volume [79].

and dialyzed against the same buffer. The dialyzed protein solution (fraction 3) is used for further purification.

Step 4. DEAE-Cellulose Column Chromatography. Fraction 3 is added to a DEAE-cellulose column (chloride form, 2.5 × 20 cm).[13] The column is washed with 0.005 M Tris-chloride buffer, pH 7.5 (approximately 2 bed volumes), until the 10-ml fractions of effluent give a constant low absorbancy at 280 nm. Guanylate kinase is eluted from the column by means of a linear gradient of Tris-chloride buffer, pH 7.5 (0.005–0.11 M over 2 liters). Guanylate kinase activity relatively free of adenylate kinase activity appears in the eluate when the concentration of the buffer reaches approximately 0.04 M. The fractions containing the highest amounts of enzymic activity are pooled, solid ammonium sulfate (445.3 g/liter, 70% saturation) is added, and the resulting suspension is allowed to stand for 48–72 hr. The precipitated proteins are collected by centrifugation and dissolved in a minimal volume of 0.005 M Tris-chloride buffer, pH 7.5 (fraction 4).

Step 5. Molecular Sieving. Fraction 4 is applied to a Bio-gel P-100 or Sephadex G-100 column (2.5 × 50 cm) packed in 0.1 M Tris-chloride buffer, pH 7.5. The enzyme is eluted with the same buffer with collection of 2-ml fractions. The fractions containing guanylate kinase activity are pooled, and solid ammonium sulfate (445.3 g/liter, 70% saturation) is added. The resulting suspension is allowed to stand for 48–72 hr and centrifuged. The precipitated proteins are dissolved in a small volume of 0.1 M Tris-chloride buffer, pH 7.5, and may be stored frozen for periods of a year or more without a significant loss of enzymic activity. A summary of the enzyme purification data from hog brain is given in Table I.

Human Erythrocytes. For the initial stages of purification of human erythrocytic guanylate kinase, the General Procedure[12] has been found satisfactory. In a typical experiment, guanylate kinase is recovered in fraction 8 in yields of about 40% and at a specific activity of about 0.14, i.e., at an overall purification of about 140-fold from hemolysates. For further purification, the ammonium sulfate precipitate of fraction 8 is dissolved in a minimal amount of 0.005 M Tris-chloride buffer, pH 7.5, and dialyzed against the same buffer until free of ammonium sulfate. Thereafter, the guanylate kinase may be purified by DEAE-cellulose chromatography and molecular sieving as described in steps 4 and 5 above. In several experiments overall yields of guanylate kinase from

---

[13] A DEAE-cellulose column (chloride form) of dimensions 2.5 × 20 cm is suitable for processing up to 750 mg of protein present in fraction 3.

TABLE I
PURIFICATION OF GUANYLATE KINASE FROM HOG BRAIN[a]

Purification steps	Total activity (units)	Specific activity (units/mg protein)	Purification (fold)	Recovery (%)
1. Crude homogenate	33	0.017	1	100
2. (NH$_4$)$_2$SO$_4$ fractionation	32	0.061	4	97
3. Ca$_3$(PO$_4$)$_2$-gel adsorption	20.2	0.586	34	61
4. DEAE-cellulose column	6.6	6.5	382	20
5. Sephadex-75 column	5.3	28.0	1647	16

[a] This purification scheme is based on starting with 100 g of hog brain. In practice, starting with 2–3 kg of this tissue is preferable to insure an adequate amount of purified guanylate kinase.

hemolysates were in the order of 15% at a specific activity of about 3.0, about 3000-fold purification from hemolysates.[14]

Step 6. Isoelectric Focusing. Guanylate kinases from human erythrocytes[14] and rat liver.[10] purified through step 5, above, have been subjected successfully to isoelectric focusing. About 10 units of enzymic activity are electrofocused in a 110-ml column (LKB Model 8100-10) in a sucrose gradient with 2% Ampholine (pH 4–6). After 50 hr of electrofocusing at 500 V, successive 1-ml fractions are collected. The guanylate kinase activity, protein concentration,[9] and pH are determined in each fraction, and the fractions containing guanylate kinase activity are pooled separately. A typical isoelectric profile of human erythrocytic guanylate kinase is presented in Fig. 1.

Properties

Stability. The purified guanylate kinases from the above sources have been found stable for several years when stored below 0° in the presence of 70% saturated ammonium sulfate. Studies with hog brain guanylate kinase[2] indicate that the enzyme is stable over a range of pH 5.5–8.5 for at least 15 min at 30°.

In experiments with partially purified erythrocytic guanylate kinase, significant activity was retained for several minutes after precipitation at 3% perchloric acid followed by neutralization with potassium hydroxide. This could yield erroneous data in kinetic experiments. Addition of a combination of formic acid, 0.7 *N*, and 3% perchloric acid gave rapid

[14] K. C. Agarwal and R. E. Parks, Jr., *Mol. Pharmacol.* **8,** 128 (1972).

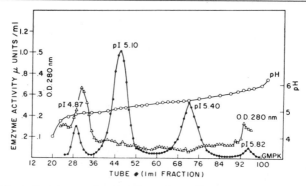

FIG. 1. Electrofocusing profile of erythrocytic purified guanylate kinase. About 20 units (specific activity, 3.3 units/mg of protein) of guanylate kinase are electrofocused in an isoelectrofocusing column (110 ml) containing 2% ampholine (pH 4–6) in a sucrose gradient. After electrofocusing for 50 hr at 500 V, 1-ml fractions are collected. Enzymic activity and pH are determined. Fractions containing guanylate kinase activity are pooled and again subjected to electrofocusing for 60 hr after the establishment of a new sucrose gradient. Enzymic activity, pH, and absorbance at 280 nm are determined in each 1-ml fraction. (Data from Agarwal and Parks.[14])

and complete enzymic inactivation. With purified erythrocytic guanylate kinase isozymes, about 70% of the activity was lost by heating at 50° for 10 min.[15]

Molecular Properties. Agarose gel electrophoresis and isoelectric focusing of the erythrocytic enzyme revealed the presence of four isoelectric variants with pI values of 4.9, 5.1, 5.4, and 5.8.[14] Guanylate kinase purified from rat liver yielded three variants with isoelectric points 4.7, 4.9, and 5.1 with more than 90% of the enzymic activity in the pI 4.9 fraction.[10] Hog brain guanylate kinase gives two bands of enzymic activity on agarose-gel electrophoresis.[15]

Molecular weight values of 18,500 to 24,000 have been reported for guanylate kinases from various tissues.[5,14,16]

Specificity and Kinetic Properties. Guanylate kinases isolated from the above tissues show that the nucleoside monophosphate binding site is highly specific for the guanine moiety. Three nucleoside monophosphates, GMP, dGMP, and 8-azaGMP, serve as active substrates.[2,5,10,17] The K_m values of these nucleoside monophosphates with various guanylate kinases are presented in Table II. Guanosine, AMP, CMP, UMP, XMP, and 6-thioIMP are neither substrates nor inhibitors. 6-ThioGMP

[15] K. C. Agarwal and R. E. Parks, Jr., unpublished results.
[16] R. P. Miech, R. York, and R. E. Parks, Jr., *Mol. Pharmacol.* **5**, 30 (1969).
[17] R. P. Agarwal, E. M. Scholar, K. C. Agarwal, and R. E. Parks, Jr., *Biochem. Pharmacol.* **20**, 1341 (1971).

TABLE II
K_m VALUES FOR GUANYLATE KINASE

Substrates	Hog brain $(10^{-5} M)$	Rat liver[a] $(10^{-5} M)$	Human erythrocytes $(10^{-5} M)$
GMP	0.6	1.0	1.8
dGMP	1.0	2.8	7.4
8-AzaGMP	1.6	7.0	9.1
ATP	12.0	18.0	19.0

[a] Values for rat liver guanylate kinase variant pI 4.9.

functions as substrate, but with a high K_m (2.1 mM) and low V_{max} (3% compared to GMP) with human erythrocytic guanylate kinase.[4] The erythrocytic enzyme catalyzes a very slow reaction with IMP (V_{max}, <1% compared to GMP).[14,17]

Studies with rat liver and erythrocytic guanylate kinases show that the purified enzyme is inhibited by several sulfhydryl reagents such as p-hydroxymercuribenzoate, N^6-ethylmaleimide, and 5,5'-dithiobis (2-nitrobenzoic acid).[5,17] The enzyme is unaffected by mercaptoethanol or dithiothreitol.[5] However, the enzyme inactivated with p-chloromercuribenzoate can be reactivated with dithiothreitol.[17]

Metal Requirements. Guanylate kinases require both mono- and divalent cations.[5,6,18] Since pyruvate kinase also requires these cations, the standard assay described above must be replaced by an alternative method such as an isotope assay. Of divalent cations examined with the hog brain enzyme in the presence of K^+, Mg^{2+} showed the greatest activity and Mn^{2+} about 60% of optimal activity. Ca^{2+} and Sr^{2+} could not replace Mg^{2+}. The optimal ratio of ATP to Mg^{2+} is 1:1. Of the monovalent cations, optimal activity was observed in the presence of Mg^{2+} with K^+ or NH_4^+. Partial activity was seen with Li^+ and Na^+ but no activity occurred with Cs^+. Similar but not identical cation requirements for guanylate kinase from other tissues have been reported.[5,6]

Comments

Guanylate kinase, as with other nucleoside monophosphate kinases, is notable for its relatively high substrate specificity. Among the monophosphate nucleotides tested today, only GMP, dGMP, and the

---

[18] N. Reed-Hackett and R. P. Miech, unpublished results.

analog nucleotide, 8-azaGMP, displayed significant activity. The enzyme also appears highly specific for ADP or ATP. Since the reaction mechanism follows a "random" bi-bi reaction sequence that does not involve the formation of a phosphorylated enzyme intermediate, it may be concluded that the enzyme contains two specific binding sites: one for GMP or GDP and the second for ADP or ATP. The formation of an abortive ternary complex between GMP, ADP, and hog brain guanylate kinase has been demonstrated.[19] The high specificity of guanylate kinase for GMP, dGMP, and ATP has made guanylate kinase a useful biochemical tool for the development of highly specific and sensitive assays, often involving enzyme cycling for biologically important molecules such as 3',5'-cGMP, GMP, and ATP.[20,21] A point of interest is that in most tissues examined, the activity of guanylate kinase is relatively low and, to date, no evidence has been found that indicates that it is subject to allosteric regulation or metabolic induction.

It should be noted that recent advances in the use of affinity chromatography with a substrate, an inhibitor, or a reactive dye coupled to cross-linked agarose beads for enzyme purification have not been applied to date to the purification of guanylate kinase. The use of these techniques could lead to major modifications in the purification scheme described above.

[19] R. P. Miech, Ph.D. Thesis, University of Wisconsin, Madison (1963).
[20] N. D. Goldberg, S. B. Dietz, and A. G. O'Toole, *J. Biol. Chem.* **244**, 4458 (1969).
[21] R. P. Miech and M.-C. Tung., *Biochem. Med.* **4**, 435 (1970).

[65] AMP Deaminase from Rat Skeletal Muscle

By CAROLE J. COFFEE

$$AMP + H_2O \rightleftarrows IMP + NH_3$$

AMP deaminase (EC 3.5.4.6, AMP aminohydrolase) catalyzes the hydrolytic deamination of 5'-adenosine monophosphate as illustrated in the above reaction. The enzyme is widespread in animal tissues, although a considerably higher concentration is found in skeletal muscle than in other tissues including cardiac and smooth muscle.[1] Moreover, the distribution of the enzyme in skeletal muscle varies greatly, with white muscle having significantly higher concentrations than red muscle.[2] Evidence has been presented recently which suggests that different

[1] E. J. Conway and R. Cooke, *Biochem. J.* **33**, 479 (1939).
[2] A. Raggi, S. Ronca-Testoni, and G. Ronca, *Biochim. Biophys. Acta* **178**, 169 (1969).

isozymes of AMP deaminase may be found in red and white muscle fiber types.[3] The enzyme in skeletal muscle is tightly associated with actomyosin[4] and can be dissociated either by heat[5] or by inorganic phosphate.[6] The enzyme has been demonstrated to participate in the purine nucleotide cycle,[7,8] and the reaction catalyzed by AMP deaminase is the major source of ammonia in skeletal muscle.[9,10] Highly purified preparations of the enzyme have been reported from skeletal muscle of rabbit,[11] hen,[12] fish,[13] and rat.[14]

Assay Method

Principle. The enzyme may be assayed either indirectly by monitoring NH_3 production by the microdiffusion technique[15] or directly by the more convenient spectrophotometric method developed by Kalckar.[16] Absorbancy changes resulting from the enzymic hydrolysis of the purines or purine derivatives have been tabulated by Zielke and Seulter.[17] Routinely, the activity of the enzyme is measured by monitoring the change in absorption at either 265 or 285 nm which accompanies the conversion of AMP to IMP. The amount of AMP converted to IMP is calculated using $\Delta\epsilon_{mM}$ values of 8.86 and 0.23 at 265 and 285 nm, respectively.

Stock Reagents

Imidazole-HCl, 0.5 M, pH 6.5
Potassium chloride, 1 M
AMP, 20 mM, pH 6.5

Procedure.[14] An assay mix sufficient for 100 assays is prepared from the above stock reagents as follows: Imidazole-HCl (10 ml), KCl (10 ml),

[3] A. Raggi, C. Bergamini, and G. Ronca, *FEBS Lett.* **58**, 19 (1975).
[4] E. W. Byrnes and C. H. Suelter, *Biochem. Biophys. Res. Commun.* **20**, 422 (1965).
[5] Y. P. Lee, *J. Biol. Chem.* **227**, 987 (1957).
[6] R. D. Currie and H. L. Webster, *Biochim. Biophys. Acta* **64**, 30 (1962).
[7] J. M. Lowenstein and K. Tornheim, *Science* **171**, 397 (1971).
[8] K. Tornheim and J. M. Lowenstein, *J. Biol. Chem.* **247**, 162 (1972).
[9] G. Embden and H. Wassermeyer, *Hoppe-Seyler's Z. Physiol. Chem.* **179**, 226 (1928).
[10] J. K. Parnas, *Biochem. Z.* **206**, 16 (1929).
[11] K. L. Smiley, A. J. Berry, and C. H. Suelter, *J. Biol. Chem.* **242**, 2502 (1967).
[12] H. Henry and O. P. Chilson, *Comp. Biochem. Physiol.* **29**, 301 (1969).
[13] W. Makarewicz, *Comp. Biochem. Physiol.* **29**, 1 (1969).
[14] C. J. Coffee and W. A. Kofke, *J. Biol. Chem.* **250**, 6653 (1975).
[15] G. Schmidt, *Hoppe-Seyler's Z. Physiol. Chem.* **179**, 243 (1928).
[16] H. M. Kalckar, *J. Biol. Chem.* **167**, 461 (1947).
[17] C. L. Zielke and C. H. Suelter, *in* "The Enzymes" (P. D. Boyer, ed.), 3rd ed., Vol. 4, p. 47. Academic Press, New York, 1971.

AMP (10 ml), and H_2O (60 ml) are mixed and equilibrated in a constant-temperature bath at 20°. An aliquot of 0.9 ml of this mixture is pipetted into a quartz cuvette with a 1-cm pathlength. The reaction is initiated by the addition of 0.1 ml of appropriately diluted enzyme solution. The increase in absorbance at 285 nm is recorded as a function of time. Best results are obtained using an expanded-scale recorder with a full-range scale of 0.2 optical density (OD) units. The final concentration of AMP in the standard assay mixture is 2.0 mM.

In order to measure the activity of the enzyme at levels of AMP $\leq$ 0.15 mM, the decrease in absorbance at 265 nm is recorded.

Units. One unit of enzyme activity is the amount of enzyme required to catalyze the deamination of 1 μmole of AMP per minute at 20° under the standard assay conditions described above.

Purification Procedure

Preparation of Crude Extract. Leg and back muscles (1 kg) from rats are excised, sliced into small pieces, and passed through a meat grinder. The ground muscle is mixed with 3.3 volumes (w/v) of 0.1 M potassium phosphate buffer (pH 6.5) containing 0.18 M KCl and 2 mM β-mercapto-ethanol and homogenized in a Waring Blendor for 1 min at high speed. The resulting slurry is stirred at room temperature for 1 hr and then centrifuged at 20,000 g for 15 min. The supernatant fluid is filtered through cheesecloth.

Fractionation on Cellulose-Phosphate. Cellulose-phosphate (What-man P-11) is prewashed successively with 0.5 N KOH, H_2O, 0.5 N HCl, H_2O, 5 mM EDTA, H_2O, and 0.1 M potassium phosphate buffer (pH 6.5) containing 0.18 M KCl and 2 mM β-mercaptoethanol. For each liter of crude extract, 25 ml of cellulose-phosphate are added, and the slurry is stirred at room temperature for 30 min. Approximately 90% of the enzyme activity is absorbed to the cellulose-phosphate under these conditions. The slurry is centrifuged at 10,000 g for 10 min, and the supernatant fluid is discarded. The resin is washed 3 times with 4 volumes of extraction buffer each time and recovered by centrifugation. The slurry is transferred to a 2 × 50 cm glass column and washed with 0.45 M KCl–2 mM β-mercaptoethanol (adjusted to pH 7.0 with K_2HPO_4) until the effluent has an absorbance at 280 nm of $\leq$ 0.01. A linear gradient, consisting of 150 ml of 0.45 M KCl–2 mM β-mercapto-ethanol (pH 8.0) and 150 ml of 1.5 M KCl–2 mM β-mercaptoethanol (pH 8.0), is applied to the column. The enzyme elutes between 100–170 ml.

The yield of enzyme at this step is about 80%, and it is estimated to be approximately 50% pure by the criteria of SDS gel electrophoresis.[14]

Affinity Chromatography on 5'-AMP Sepharose. The preparation and purification of N^6-(6-aminohexyl)-5'-AMP is carried out as described in exquisite detail by Craven *et al.*[18] In order to couple the ligand to Sepharose, 10 mg of N^6-(6-aminohexyl)-5'-AMP are dissolved in 5 ml of $0.1 M$ NaHCO₃ (pH 10) and mixed with 10 g (moist weight) of cyanogen-bromide-activated Sepharose 4B (Sigma) which had been prewashed with 500 ml of ice-cold $0.1 M$ NaHCO₃ (pH 10). The suspension is mixed gently by rotation overnight at 4°. The gel is washed free of excess ligand by successive washings with 500 ml each of $0.1 M$ NaHCO₃ (pH 10), H₂O, $1 M$ KCl, and H₂O. The gel was stored at 4° in 0.02% sodium azide when not in use.

Prior to fractionation of AMP deaminase on the AMP-Sepharose gel, the pool of enzyme obtained from the previous cellulose-phosphate column was dialyzed against $0.02 M$ potassium phosphate buffer (pH 7.5). Dialysis was performed for approximately 24 hr against 3 changes of buffer (20 volumes each). The AMP-Sepharose gel is poured into a glass column and equilibrated with $0.02 M$ potassium phosphate (pH 7.5). The enzyme solution is applied to the column at a flow rate of approximately 10 ml/hr. Following application of the sample, the column is washed with the same buffer until no more protein can be detected in the effluent (A_{280} ≤ 0.01). The column is then eluted with $0.02 M$ potassium phosphate–$0.1 M$ KCl–2 mM β-mercaptoethanol. The elution profile obtained in this step is shown in Fig. 1.[19] Approximately half of the total protein applied to the column is unabsorbed and comes off in the wash. However, all of the enzyme activity remains absorbed and is eluted with $0.1 M$ KCl in 20 mM potassium phosphate (pH 7.5).

Comments on the Purification Procedure. A summary of the purification is shown in the table.[20] The procedure is rapid, easily performed, and results in a high-yield preparation (50–60%) which is homogeneous by the criteria of sedimentation velocity and equilibrium analyses,[14] standard polyacrylamide gel electrophoresis,[21] and SDS polyacrylamide gel electrophoresis.[14] The purification scheme reported here is a simplification of a procedure previously described by Coffee and Kofke.[14] In the

---

[18] D. B. Craven, M. J. Harvey, C. R. Lowe, and P. D. G. Dean, *Eur. J. Biochem.* **41**, 329 (1974).

[19] P. Bohlen, S. Stein, W. Dairman, and S. Udenfriend, *Arch. Biochem. Biophys.* **155**, 213 (1973).

[20] R. F. Itzhaki and D. M. Gill, *Anal. Biochem.* **9**, 401 (1964).

[21] C. J. Coffee and C. Solano, in preparation.

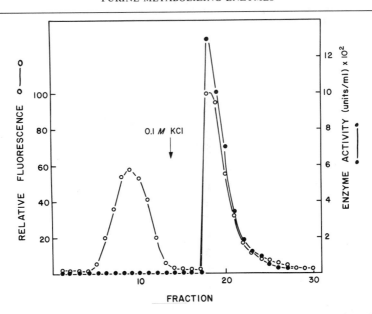

FIG. 1. Fractionation of AMP deaminase on AMP-Sepharose-4B. A solution of AMP deaminase in 20 mM potassium phosphate buffer (pH 7.5) was applied to a 1 × 10 cm column of N^6-(6-aminohexyl)-5'-AMP at a flow rate of 10 ml/hr, and the column was washed with 20 mM potassium phosphate (pH 7.5) until no more protein emerged. Elution was achieved with 0.1 M KCl in 20 mM potassium phosphate (pH 7.5). Fractions of 4 ml each were collected. Aliquots were assayed for protein by the fluorescamine procedure,[19] and for enzyme activity as described in the text.

previous method, purification to homogeneity was achieved by successive chromatographic fractionation on cellulose-phosphate, DEAE-cellulose, and Bio-Gel A-5m. In the simplified procedure described here, the final two chromatographic steps on DEAE-cellulose and Bio-Gel have been replaced by a single step utilizing an affinity column of AMP-Sepharose. The modified procedure has the advantage that it is both more rapid and it results in a higher yield.

PURIFICATION OF AMP-DEAMINASE FROM 1 KG OF RAT SKELETAL MUSCLE

Purification step	Total protein (mg)^a	Total activity (× 10^{-3} units)	Specific activity (units/mg)	Yield (%)
Crude extract	38,700	130	3.3	100
Cellulose-phosphate	116	104	896	80
5'-AMP-Sepharose-4B	47	74	1580	57

^{a} Determined by the microbiuret method.[20]

Properties of the Enzyme

Stability. The enzyme can be stored in 0.1 M potassium phosphate–1.0 M KCl–2 mM β-mercaptoethanol (pH 6.5) at 4° for several weeks without a significant decrease in specific activity.[14] However, rapid loss of enzyme activity is observed at pH values higher than 6.8 or lower than 5.6.[22] Freezing the enzyme at −70° results in inactivation which is associated with the formation of insoluble aggregates.[22,23] The enzyme activity is decreased by a number of chelating agents which is likely a reflection of the fact that the enzyme contains zinc.[24]

Molecular Weight and Subunit Structure. The molecular weight of the homogeneous enzyme has been estimated to be 238,000 by sedimentation equilibrium analysis.[14] A somewhat higher value of 290,000 obtained by sucrose density centrifugation and with a less pure preparation has been reported.[22] In the presence of denaturing agents such as guanidine hydrochloride or sodium dodecyl sulfate, the enzyme dissociates into polypeptides having a weight average molecular weight of approximately 60,000. These data indicate that the enzyme is composed of four subunits which are identical with respect to size. A tetrameric structure of the native enzyme is supported by tryptic peptide mapping, and the maps further indicate that the primary structure of the individual subunits is either identical or very similar.[14]

Amino Acid Composition. Amino acid analysis of the enzyme indicated the presence of 530 amino acids per subunit.[14] The tryptophan content was observed to be 4 residues per subunit. The absorption spectrum of the enzyme is that expected for a tryptophan-containing protein with no ultraviolet-absorbing cofactors. For the native enzyme, the absorbance maximum is at 280 nm ($E_{280\,nm}^{1\%} = 9.8$) and the $A_{280}/A_{260} = 1.85$.[22]

Catalytic Properties. The fundamental enzymology of AMP deaminase has been reviewed earlier by Ziekle and Suelter,[17] and here only selected basic properties are repeated along with information resulting from more recent studies.

Although the enzyme requires zinc[24] and monovalent cations for activity,[22,25,26] no other cofactors have been identified. Potassium is the most effective monovalent cation and the one considered to be physiol-

[22] S. Ronca-Testoni, M. Ranieri, and G. Ronca, *Ital. J. Biochem.* **19**, 262 (1970).
[23] C. Coffee, unpublished observations.
[24] A. Raggi, M. Ranieri, G. Taponeco, S. Ronca-Testoni, G. Ronca, and C. A. Rossi, *FEBS Lett.* **10**, 101 (1970).
[25] G. Ronca, A. Raggi, and S. Ronca-Testoni, *Biochim. Biophys. Acta* **167**, 626 (1968).
[26] C. J. Coffee and C. Sonalo, *J. Biol. Chem.* **252**, 606 (1977).

ogically important. The concentration of K^+ required for maximum activation varies from 25–100 mM depending on the concentration of substrate present.[25,26] As the substrate concentration approaches saturation, the level of K^+ required for maximum activation decreases. The pH optimum of the reaction catalyzed by AMP deaminase is approximately 6.5.[22]

The enzyme presents a sigmoidal substrate-velocity curve when assayed at low levels of potassium ion (5 mM).[25,26] The sigmoidicity disappears at higher concentrations of KCl (100 mM) or in the presence of ADP.[25-27] The concentration of AMP required for half-maximum velocity is approximately 0.4–0.5 mM.[25,26,28] Hill plots indicate a cooperativity coefficient (n_H) of 1.1[27] to 1.5[26] when assayed in the presence of 100 mM KCl, whereas at a KCl concentration of 5 mM, the n_H is 3.2.[26]

Regulatory Properties. The activity of AMP deaminase is affected by a number of biologically important phosphorylated metabolites. Nucleoside triphosphates, creatine phosphate, and P_i, at concentrations comparable to those found in skeletal muscle, all inhibit the enzyme when assayed at physiological concentrations of K^+ (100–150 mM) and AMP (0.1–0.2 mM).[25,26,28] The inhibition by nucleoside triphosphates is sensitive to alterations in pH. For example, the extent of inhibition is significantly higher at pH 7.1 than at pH 6.5.[27] The inhibition observed with nucleoside triphosphates and P_i is reversed by ADP. In addition to this effect, ADP strongly activates the enzyme at low concentrations of KCl. However, at physiological concentrations of KCl, the activation is either very weak[27] or completely absent.[26]

The profile of AMP deaminase activity generated in response to variations in the adenylate energy charge shows that within the physiological range of energy charge values (0.75–0.95),[29] the activity increases linearly with decreasing energy charge. Moreover, this response is insensitive to the total adenylate pool size (from 1–10 mM) and the presence of P_i or creatine phosphate.[26] These observations suggest that although P_i and creatine phosphate exert dramatic effects on the enzyme activity in isolated *in vitro* experiments, they may play an insignificant role in the *in vivo* regulation of skeletal muscle AMP deaminase. Perhaps in an *in vivo* situation where creatine phosphate, ADP, and phosphocreatine kinase are all present, creatine phosphate may play an indirect role in the modulation of AMP deaminase activity by serving as a precursor of ATP. However, it appears that the most important regula-

[27] S. Ronca-Testoni, A. Raggi, and G. Ronca, *Biochim. Biophys. Acta* **198**, 101 (1972).
[28] S. Ronca-Testoni and G. Ronca, *J. Biol. Chem.* **249**, 7723 (1974).
[29] A. G. Chapman, L. Fall, and D. E. Atkinson, *J. Bacteriol.* **108**, 1072 (1971).

tory factor at the direct level of metabolite–enzyme interaction is the relative concentrations of the three adenine nucleotides.

Under conditions similar to those found intracellularly within skeletal muscle (total adenylate pool sizes of 5–8 mM; energy charge ratios of 0.75–0.95), the activity of AMP deaminase is very stringently regulated as evidenced by the fact that under these conditions, the enzyme is working at only 10–15% of its capacity.[26] These observations suggest that in resting muscle or in muscle undergoing moderate activity, the enzyme may be inhibited to a great extent. In fact, maximal activity may be approached only under conditions of extreme stress.

[66] AMP Deaminase from Human Erythrocytes[1]

By GENE R. NATHANS, DONALD CHANG, and THOMAS F. DEUEL

$$AMP + H_2O \rightarrow IMP + NH_3$$

Assay Method

Principle. AMP deaminase (AMP aminohydrolase, EC 3.5.4.6) activity is measured by colorimetric means in an assay adapted from Chaney and Marbach.[2] Ammonia released in the deamination of AMP generates a stable blue color in the catalyzed indophenol reaction. The absorbance is measured at 625 nm and related to a standard curve obtained with ammonium chloride. The method is used when crude extracts are assayed and when assays are performed in the presence of high concentrations of nucleotides other than AMP.

AMP deaminase activity may also be measured by a spectrophotometric assay[3] which records either the decrease in optical density at 265 nm as AMP is hydrolyzed or the increase in optical density at 240 nm as IMP is formed. This latter assay has been used to establish both the hydrolysis of AMP and formation of IMP in order to validate the

[1] This work was supported by Contract EY-76-C-02-0069, awarded to the Franklin McLean Memorial Research Institute (operated by the University of Chicago for the United States Energy Research and Development Administration), and by Grant CA-13980 from the National Institutes of Health. Gene R. Nathans was supported by Training Grant 5-T01-CA-05250 from the National Institutes of Health.

Thomas F. Deuel holds Faculty Research Award No. 133 from the American Cancer Society.

[2] A. L. Chaney and E. P. Marbach, *Clin. Chem. (Winston-Salem, N. C.)* **8**, 130 (1962).
[3] H. M. Kalckar, *J. Biol. Chem.* **169**, 445 (1947).

METHODS IN ENZYMOLOGY, VOL. LI

precursor–product relationship of the catalyzed reaction and is used with purified extracts when the background absorbance is not increased to high levels by nucleotides other than AMP.

Reagents

Solution 1: 50 g phenol and 0.25 g sodium nitroprusside, to 1 liter with deionized water

Solution 2: 25 g sodium hydroxide and 124 ml of 4–6% laboratory grade sodium hypochlorite, to 1 liter with deionized water

Both Solution 1 and 2 may be stored in the refrigerator for up to 2 months if kept in an amber bottle protected from light.

5'-AMP, 0.1 M, neutralized to pH 7.0 with NaOH, stored frozen

5'-ATP, 0.1 M, neutralized to pH 7.0 with NaOH, stored frozen

Imidazole-HCl, 1.0 M, pH 7.0

KCl, 1 M

Procedure. The reaction mixture contains 8 μl of 0.1 M 5'-AMP, 8 μl of 0.1 M 5'-ATP, 16 μl of imidazole-HCl, 40 μl of KCl, and water and enzyme to give a final volume of 0.4 ml. The reaction mixture is preincubated to 37°. The reaction is initiated by the addition of enzyme, and incubation·continues for 10 min. The reaction is terminated by the addition of 0.85 ml of solution 1. After vigorous mixing, 0.55 ml of solution 2 is added and color development obtained by incubation in a water bath at 53° for at least 40 min. The tubes are cooled to room temperature, and the optical density at 625 nm is determined. The absorbance is stable at room temperature for at least 24 hr. Blanks for each series of assays are obtained by adding enzyme to the reaction mixture after the addition of solution 1.

Definition of Unit and Specific Activity. One unit of enzyme activity is that amount which generates 1 μmole of ammonia per minute under stated assay conditions. Specific activity is expressed as units per milligram of protein (method of Lowry *et al.*[4]).

Purification Procedure

All procedures were carried out from 0–5°.

Step 1. Preparation of Human Erythrocytes. Outdated human blood, stored in acid-citrate-dextrose, is obtained from local blood banks and centrifuged at 4500 rpm for 15 min using a Sorvall GS-3 rotor. The

---

[4] O. Lowry, N. J. Rosebrough, A. Z. Tass, and R. J. Randall, *J. Biol. Chem.* **193**, 265 (1951).

plasma and buffy coat were removed. The erythrocytes are washed 3 times in 0.15 M NaCl–5 mM sodium phosphate buffer (pH 8.0) and the top layers of erythrocytes sacrificed to insure complete removal of leukocytes. The washed erythrocytes may be stored frozen without subsequent loss of activity.

Step 2. Preparation of Hemolysates. Washed, packed erythrocytes are melted, frozen, and thawed 3 times. An equal volume of deionized water is added and the hemolysate stirred for 1 hr.

Step 3. DEAE-Cellulose Fractionation. A slurry of DEAE-cellulose (DE-52, Whatman) is prepared as follows: 133 g of DEAE-cellulose are added to 1 liter of 1 M potassium phosphate buffer (pH 9.4), equilibrated for 2 days, washed with 100 M potassium phosphate (pH 6.5) until the pH is below 7.0, washed with 100 mM potassium phosphate (pH 7.0) until a stable pH of 7.0 is achieved, and then washed with deionized water until the conductivity is less than 0.6 mmho. The DEAE-cellulose is stored in deionized water (1 : 1) until used.

An equal volume of red cell hemolysate and DEAE-cellulose slurry are mixed, gently stirred for 30 min, and filtered through a large Büchner funnel using Whatman No. 1 filter paper. The DEAE-cellulose is washed in the Büchner funnel with 20 mM KCl and imidazole-HCl (pH 7.5) until the filtrate is colorless. AMP deaminase is eluted with 500 mM KCl and 20 mM imidazole-HCl (pH 7.0).

Step 4. Concentration with Ammonium Sulfate. The DEAE-cellulose eluate is precipitated by addition of solid ammonium sulfate to 65% saturation, and the precipitate is then resuspended in 0.3 M KCl–3 mM potassium phosphate (pH 7.0)–1 mM 2-mercaptoethanol, using as low a volume as necessary to solubilize the precipitate. The extract, a dark maroon in color, is dialyzed exhaustively with the same buffer. If dialysis is incomplete, the residual ammonium ion will be measured in the subsequent assay of the enzyme. After dialysis, the extract is centrifuged at 18,000 rpm in a Sorvall SS-34 rotor for 15 min and the supernatant fluid retained. This material may be stored at 4–6° for at least 2 months without loss of catalytic activity. Sodium azide, 2 mM, does not interfere with catalytic activity and will prevent bacterial contamination during storage.

Step 5. Phosphocellulose Chromatography. Phosphocellulose (the Brown Company, Berlin, NH) is prepared as follows: 15 g of phosphocellulose are equilibrated for 30 min with 450 ml of 0.5 M NaOH. The orange-brown supernatant fluid is decanted, and the settled phosphocellulose is washed with deionized water in a Büchner funnel until the effluent is reduced to pH 10.0. The phosphocellulose is then resuspended twice within a 45-min period in 450 ml of 0.1% phosphoric acid, washed with deionized water until the pH reaches 5.0, and equilibrated for 2 hr with 5 mM EDTA. The phosphocellulose is washed with 0.3 M NaCl–10 mM imidazole-HCl (pH 6.5)–0.1 mM 2-mercaptoethanol in a Büchner funnel until equilibrium is reached as judged by conductivity and pH measurements. Fresh phosphocellulose was used for each enzyme purification.

Approximately 250 units of enzyme activity per gram of phosphocellulose (4000–5000 units per purification) are combined, stirred for 15 min, and allowed to settle. Over 90% of the enzyme activity should bind to the phosphocellulose under these conditions. The material is poured into a column (2 × 50 cm). The column is washed with starting buffer until the absorbance at 280 nm is less than 0.01. The column is further washed with 0.45 M NaCl, 10 mM imidazole HCl (pH 6.5), and 0.1 mM 2-mercaptoethanol until the absorption at 280 nm is less than 0.01. AMP deaminase is eluted with 0.45 M NaCl, 10 mM imidazole-HCl (pH 6.5), 0.1 mM 2-mercaptoethanol, and 10 mM sodium pyrophosphate. The peak activity tubes are pooled, concentrated, dialyzed, and applied to a second phosphocellulose column (8–10 ml bed volume) as above. The column is washed as above with the addition of 1.0% Triton X-100 to the buffer. Triton X-100 removes trace contaminating proteins but itself needs to be washed from the column with 3 bed volumes of additional buffer. AMP deaminase is eluted from the column with 0.45 M NaCl, 10 mM imidazole-HCl (pH 6.5), 0.1 mM 2-mercaptoethanol, and a linear sodium pyrophosphate gradient between 0–10 mM. Protein in the peak activity fractions is precipitated by addition of ammonium sulfate to 55% saturation, and the resuspended enzyme is dialyzed and stored in 10 mM imidazole-HCl (pH 6.5), 0.3 M NaCl, and 0.1 mM 2-mercaptoethanol.

AMP deaminase may be partially purified using AMP sepharose (8-[6-aminohexyl]-amino-AMP, prepared by the method of Lee et al.[5]). The enzyme binds readily to the affinity ligand and is recovered in high yield from the column after elution with 5′-AMP, or 2,3-DPG. This enzyme has lower specific activity than enzyme purified as above but does behave as a single protein species in standard disc gel electrophoresis and in gel filtration studies. Gel electrophoresis in sodium dodecyl sulfate demonstrates multiple protein bands and thus provides evidence that the product of AMP-Sepharose affinity chromatography is an aggregate of multiple protein species.

[5] C. Y. Lee, D. A. Loppi, B. Wermuth, J. Everse, and N. O. Kaplan, *Arch. Biochem. Biophys.* **163**, 561 (1974).

AMP DEAMINASE FROM HUMAN ERYTHROCYTES

Procedure	Volume	Total activity (units)	Specific activity (units/mg)	Yield (%)	Purification
Hemolysate	4400	6160	0.0091	100	1
DEAE-cellulose	4557	5298	0.063	86	6.9
$(NH_4)_2SO_4$ precipitation	647	4933	0.38	80.1	42
First phosphocellulose column	60	1327	197.5	21.6	21,708
Second phosphocellulose column	6	709	435.9	11.5	47,901

The results of a standard purification of human erythrocyte AMP deaminase are presented in the table.

Properties

Stability. The crude enzyme is stable for long periods of time if stored at 4°. Freezing the enzyme results in significant loss of activity. Sulfhydryl reagents stabilize the enzyme and are required for purification and storage. Monovalent cations in high concentrations also stabilize the enzyme during storage. Extensive dialysis of the enzyme against EDTA results in loss of catalytic activity. This loss in activity may be partially recovered by subsequent incubation of the enzyme with 1 mM $ZnCl_2$.

Physical Properties. The enzyme migrates as an essentially single band in gel electrophoresis in sodium dodecyl sulfate and has an estimated subunit molecule weight of 74,000. The sedimentation coefficient in 0.3 M NaCl is 11.0. The protein comigrates with band 4.2 when erythrocyte membranes are dissociated and subjected to electrophoresis in sodium dodecyl sulfate, according to the techniques of Steck.[6]

Specificity. AMP deaminase deaminates 5'-AMP but not 2'-AMP, 3'-AMP, adenosine, adenine, ADP, or ATP.

Kinetic Properties. The pH optimum is 7.0. ATP and dATP activate the enzyme. In order of increasing effectiveness, Li^+, Na^+, and K^+ activate the enzyme when ATP is not present during assay; the activation by monovalent cations is much less pronounced when ATP is present. The substrate saturation curve for 5'-AMP is sigmoidal but is converted to a hyperbolic curve when ATP (2 mM) and KCl (100 mM)

[6] T. L. Steck, *J. Cell Biol.* **62**, 1 (1974).

are included in the assay. The apparent K_m for AMP is ≈ 1 mM under standard assay conditions. Inorganic phosphate, inorganic pyrophosphate, 2,3-DPG, and GTP are inhibitors of catalytic activity. Similar results have been reported by Lian and Harkness[7] and by Yung.[8]

Relationship to the Erythrocyte Membrane. Partially purified AMP deaminase binds readily and reversibly to the inner (cytoplasmic) surface of isolated human erythrocyte membranes. The binding to the erythrocyte membrane is a hyperbolic function of the amount of enzyme added. Little if any binding to the outer membrane surface occurs. 2,3-DPG, 5'-AMP, and 5'-ATP markedly reduce the amount of enzyme that binds to the membrane. Membrane-bound enzyme is less than 20% as active as enzyme fully dissociated from the membrane. These relationships have not been investigated with the fully purified enzyme preparations.

[7] C. Y. Lian and D. Harkness, *Biochim. Biophys. Acta* **341**, 27 (1974).
[8] S. L. Yung, *Fed. Proc., Fed. Am. Soc. Exp. Biol.* **35**, 1604 (1976).

[67] Adenosine Deaminase from Human Erythrocytes

By R. P. AGARWAL and R. E. PARKS, JR.

$$\text{Adenosine} + H_2O \rightarrow \text{Inosine} + NH_3$$

Assay Method

Among various methods available, the direct spectrophotometric method is most commonly employed for the assay of adenosine deaminase.

Principle. The activity is determined by measuring the rate of decrease in absorbancy at 265 nm resulting from the conversion of adenosine to inosine.[1] Molar absorption changes and optimal wavelengths of various adenosine analogs that may be employed as substrates for the enzymic assay have been described elsewhere.[2,3]

[1] H. M. Kalckar, *J. Biol. Chem.* **167**, 461 (1947).
[2] R. P. Agarwal, S. M. Sagar, and R. E. Parks, Jr., *Biochem. Pharmacol.* **24**, 693 (1975).
[3] C. L. Zielke and C. H. Suelter, *in* "The Enzymes" (P. D. Boyer, ed.), 3rd ed., Vol. 4, p. 47. Academic Press, New York, 1971.

Reagents

Potassium phosphate buffer, 50 mM, pH 7.4
Adenosine, 10 mM

Procedure.[2] The reaction mixture (in a 1-ml cuvette; diam = 1.0 cm) contains 0.9 ml buffer, 10 μl of adenosine (0.1 μmole), the enzyme, and water to make 1 ml. The decrease in absorbancy is followed at 265 nm at 30°.[2]

One unit of adenosine deaminase is the amount of enzyme that catalyzes the deamination of 1 μmole of adenosine per minute ($-\Delta A$ = 8.6 min^{-1} ml^{-1}) under the conditions of the assay. Specific activity is expressed as units per milligram of protein.

Protein concentrations may be determined either by methods such as that of Lowry *et al.*[4] or by UV absorption at 280 nm.[5]

Other assay methods include: (1) measurement of inosine production by coupling the adenosine deaminase reaction to the purine nucleoside phosphorylase and xanthine oxidase reactions;[6] (2) measurement of ammonia production by reaction with α-ketoglutarate in the presence of glutamic acid dehydrogenase and NADH[2] or by a microdiffusion procedure.[7-9] The latter method is useful for studies of the enzyme in crude suspensions, intact cells, or with adenosine analogs where spectrophotometric assays are inconvenient or inaccurate.[9-12]

Purification Procedure[2]

Step 1. Preparation of Hemolysate. The preparation of hemolysates is described by Agarwal *et al.*[13]

Step 2. Calcium Phosphate Gel—Negative Adsorption.[2,13] This procedure is identical to that described by Agarwal *et al.*[13] Although this

[4] O. H. Lowry, N. J. Rosebrough, A. L. Farr, and R. J. Randall, *J. Biol. Chem.* **193**, 265 (1951).
[5] O. Warburg and W. Christian, *Biochem. Z.* **310**, 384 (1941); also see Vol. 3 [73].
[6] D. A. Hopkinson, P. J. L. Cook, and H. Harris, *Ann. Hum. Genet.* **32**, 361 (1969).
[7] D. Seligson and H. Seligson, *J. Lab. Clin. Med.* **38**, 324 (1951).
[8] A. L. Chaney and E. P. Marbach, *Clin. Chem.* **8**, 130 (1962).
[9] R. P. Agarwal, G. W. Crabtree, R. E. Parks, Jr., J. A. Nelson, R. Keightley, R. Parkman, F. S. Rosen, R. C. Stern, and S. H. Polmar, *J. Clin. Invest.* **57**, 1025 (1976).
[10] R. P. Agarwal and R. E. Parks, Jr., *Biochem. Pharmacol.* **24**, 547 (1975).
[11] R. E. Parks, Jr., G. W. Crabtree, C. M. Kong, R. P. Agarwal, K. C. Agarwal, and E. M. Scholar, *Ann. N. Y. Acad. Sci.* **255**, 412 (1975).
[12] R. P. Agarwal, G. W. Crabtree, K. C. Agarwal, R. E. Parks, Jr., and L. B. Townsend, *Proc. Am. Assoc. Cancer Res.* **17**, 214 (1976).
[13] R. P. Agarwal, K. C. Agarwal, and R. E. Parks, Jr., this volume [79].

step does not significantly change the specific activity of the enzyme, it is included primarily to remove most of the purine nucleoside phosphorylase, the activity of which, in human erythrocytes, is at least 50 times greater than the adenosine deaminase activity. Since purine nucleoside phosphorylase reacts with certain products of the adenosine deaminase reaction, e.g., inosine and deoxyinosine, it is desirable to remove this contaminant early during purification.

Step 3. DEAE-Cellulose Column Chromatography.[2,13] The supernatant fluid from step 2 is poured onto a column (4.5 × 36 cm) of DEAE-cellulose (phosphate form) which is prepared by equilibration with 0.01 M potassium phosphate buffer containing 0.05 M NaCl, pH 7.4. The enzyme is eluted with a linear gradient of 0.05–0.2 M NaCl solution (1 liter each) in 0.01 M potassium phosphate buffer, pH 7.4. The enzyme emerges at approximately 0.12–0.16 M NaCl. Successive fractions are collected and those containing most of the enzymic activity are pooled.

When performing purification of adenosine deaminase from large quantities of human blood by the "General Method" described by Agarwal et al.,[13] this enzyme is found in the ammonium sulfate precipitate of fraction 8.[13] This precipitate may be dissolved in a small volume of 0.01 M potassium phosphate, pH 7.4, containing 0.05 M NaCl and dialyzed thoroughly against the same solution. This material may then be subjected to DEAE-cellulose chromatography as described above.

Step 4. Ammonium Sulfate Fractionation. To the pooled fractions from step 3, finely dispersed solid ammonium sulfate (231 g/liter) is added slowly by sifting and gentle stirring to final concentration of 40% saturation. The mixture is stirred gently for 10–12 hr at 4°, and the precipitate is removed by centrifugation at 16,000 g for 60 min. More solid ammonium sulfate (186 g/liter) is added to the supernatant fluid to bring the saturation to 70%. After stirring for 6 hr at 4°, the precipitate is collected by centrifugation at 16,000 g for 60 min, and the pellet is dissolved in a small volume of 0.1 M Tris·HCl buffer, pH 7.5.

Step 5. Molecular Sieving. The enzyme solution from step 4 is added to a Bio-Gel P-60 column (2 × 60 cm) equilibrated with 0.1 M potassium phosphate buffer, pH 7.4. The enzyme is eluted with the same buffer, and the fractions containing the enzyme are pooled. The pooled enzyme is concentrated by ultrafiltration in an Amicon cell by use of an XM50 membrane under nitrogen pressure (50 psi).

Table I summarizes a typical purification procedure of adenosine deaminase from about 100 ml of packed human erythrocytes.

Several other methods of purification of erythrocytic adenosine deaminase have been described recently that employ affinity chromatog-

TABLE I
PURIFICATION OF ADENOSINE DEAMINASE FROM HUMAN ERYTHROCYTES[a]

Purification steps	Total activity (units)	Specific activity (units/mg protein)	Recovery (%)	Purification (fold)
1. Hemolysate	21.25	2.3×10^{-4}	100	1
2. Calcium phosphate gel treatment	17.0	1.4×10^{-4}	80	—
3. DEAE-cellulose column	14.0	3.7×10^{-2}	66	161
4. Ammonium sulfate fractionation (40–70% saturation)	7.4	2.7×10^{-1}	35	1174
5. Molecular sieving	3.2	7.8×10^{-1}	15	3087

[a] Agarwal et al.[2]

raphy, e.g., adenosine attached to Sepharose,[14] 9-(p-aminoben-zyl)adenine attached to agarose,[15] and adenosine deaminase specific rabbit antibody attached affinity columns.[16] The latter method has yielded enzyme that appears homogeneous.

Properties

Stability. The partially purified adenosine deaminase preparation from step 5 above is stable for several weeks at 4° and for several years if frozen.[17] This preparation retained 100% and 36% of activity after heating for 5 min at 65° and 70°, respectively.[17] Treatment of the enzyme with 1 mM p-chloromercuribenzoate at 25° for 30 min resulted only in 30% loss of activity.[17] The enzyme has a broad pH optimum from pH 6–8.[2]

Physical Properties. Molecular weight values of 30,000–38,000 have been reported for human erythrocytic adenosine deaminase.[2,14,16] Studies with an apparently homogeneous preparation have shown that the enzyme has a Stokes radius of 24 Å, and $s_{20,w}$ of 3.8×10^{-13}, a partial specific volume of 0.729 cm^3/g, and a frictional ratio of 1.077.[16] Evidence to date indicates that the enzyme consists of a single polypeptide chain.

Specificity and Kinetics.[2] Although the enzyme has a broad specificity for adenosine analogs, alterations in the purine ring or sugar moiety

[14] W. P. Schrader, A. R. Stacy, and B. Pollara, *J. Biol. Chem.* **251**, 4026 (1976).
[15] C. A. Rossi, A. Lucacchini, U. Montali, and G. Ronca, *Int. J. Pept. Protein Res.* **7**, 81 (1975).
[16] P. E. Daddona and W. N. Kelley, *J. Biol. Chem.* **252**, 110 (1977).
[17] Our unpublished results.

result in striking changes in substrate activity.[2] Adenosine analogs modified in the imidazole ring, e.g., formycin A and 8-azaadenosine, have both higher K_m and V_{max} values, whereas 7-deaza compounds such as toyocamycin and tubercidin are totally devoid of activity either as substrates or inhibitors.[2] The enzyme is inhibited by many compounds, including the products of the reaction and various adenosine analogs.[2,18−20] Among the most potent inhibitors of this enzyme discovered to date are coformycin (K_i, 10^{-11} M) and deoxycoformycin (K_i, 2.5×10^{-12} M).[18−20] In fact, these compounds are tight-binding inhibitors and might be regarded as "transition-state" analogs.[21] Kinetic parameters of various analogs are provided in Table II.

Compounds such as AMP, dAMP, cAMP, dADP, ATP, cytosine, cytidine, CMP, dCMP, CDP, dCDP, and CTP did not show any inhibition, whereas 4-amino-5-imidazole carboxamide ribonucleoside, 4-amino-5-imidazole carboxamide-HCl, 2,6-diaminopurine sulfate, 6-chloropurine, iodopurine, and adenine gave 15–60% inhibition.[16] No cofactor requirement has been established for the enzyme.

Comments

Several recent developments have greatly increased the interest in adenosine deaminase: (1) a hereditary deficiency in this enzyme has been associated with a severe combined immunodeficiency syndrome in which the children so affected lack both T- and B-lymphocyte function and often have bone abnormalities;[9,22,23] (2) many analogs of adenosine with chemotherapeutic potential are degraded to inactive compounds by this enzyme, e.g., arabinosyl adenine, formycin, cordycepin, etc.[2] Administration of these analogs with a potent adenosine deaminase inhibitor such as deoxycoformycin has resulted in marked synergy in antitumor activity in experimental animals.[24] These observations point to an important role played by adenosine deaminase, both in immunosuppression and chemotherapy. Furthermore, cells in which the adenosine deaminase is absent or inhibited may accumulate remarkably high

[18] S. Cha, R. P. Agarwal, and R. E. Parks, Jr., *Biochem. Pharmacol.* 24, 2187 (1975).

[19] R. P. Agarwal, T. Spector, and R. E. Parks, Jr., *Biochem. Pharmacol.* 26, 359 (1977).

[20] R. P. Agarwal, S. Cha, G. W. Crabtree, and R. E. Parks, Jr., *Adv. Chem. Ser.* (in press).

[21] L. Pauling, *Am. Sci.* 36, 58 (1948).

[22] E. R. Giblett, J. E. Anderson, F. Cohen, B. Pollara, and H. J. Meuwissen, *Lancet* 2, 1067 (1972).

[23] H. J. Meuwissen, R. J. Pickering, B. Pollara and I. H. Porter, eds., "Combined Immunodeficiency Disease and Adenosine Deaminase Deficiency: A Molecular Defect." Academic Press, New York, 1975.

[24] D. G. Johns and R. H. Adamson, *Biochem. Pharmacol.* 25, 1441 (1976).

TABLE II
KINETIC PARAMETERS OF HUMAN ERYTHROCYTIC ADENOSINE DEAMINASE[a,b,c]

Substrates or inhibitors	Relative V_{max}	K_m (M)	K_i (M)
Adenosine	100	2.5×10^{-5}	—
Formycin A	750–800	1.0×10^{-3}	—
8-Azaadenosine	310	1.3×10^{-4}	—
6-Chloropurine ribonucleoside	91	1.0×10^{-3}	—
2,6-Diaminopurine ribonucleoside	91	7.4×10^{-5}	—
6-Methylselenopurine ribonucleoside	88	2.7×10^{-5}	—
2'-Deoxyadenosine	60	7.0×10^{-6}	—
Xylosyl adenine	60	3.3×10^{-5}	—
Arabinosyl adenine	47	1.0×10^{-4}	—
3'-Amino-3'-deoxyadenosine	89	1.3×10^{-4}	—
4'-Thioadenosine	43	1.3×10^{-5}	—
Inosine	—	—	1.2×10^{-4}
2'-Deoxyinosine	—	—	6.0×10^{-5}
Guanosine	—	—	1.4×10^{-4}
2-Fluoroadenosine	—	—	6.0×10^{-5}
2-Fluorodeoxyadenosine	—	—	1.9×10^{-5}
N^6-methyladenosine	—	—	1.7×10^{-5}
N^1-methyladenosine	—	—	2.8×10^{-4}
6-Thioguanosine	—	—	9.2×10^{-5}
6-Thioinosine	—	—	3.3×10^{-4}
6-Methylthioinosine	—	—	2.7×10^{-4}
Arabinosyl-6-thiopurine	—	—	3.6×10^{-4}
1,6-Dihydro-6-hydroxymethylpurine ribonucleoside	—	—	1.3×10^{-6}
Erythro-9-(2-hydroxy-3-nonyl)adenine	—	—	1.6×10^{-9}
Coformycin	—	—	1.0×10^{-11}
Deoxycoformycin	—	—	2.5×10^{-12}

[a] Agarwal et al.[2]
[b] Cha et al.[18]
[c] Agarwal et al.[19]

concentrations of adenine nucleotides, e.g., ATP, when incubated with adenosine.[9,11,20,25] It is possible that the cytotoxicity of adenosine with certain tissues may result from these abnormally elevated levels of adenine nucleotides.[9,26]

[25] L. M. Rose and R. W. Brockman, *J. Chromatogr.* **133**, 335 (1977).
[26] S. H. Polmar, R. C. Stern, A. L. Schwartz, E. M. Wetzler, P. A. Chase, and R. Hirschhorn, *N. Engl. J. Med.* **295**, 1337 (1976).

[68] Adenosine Deaminase from *Escherichia coli*

By PER NYGAARD

$$\text{Adenosine} + H_2O \rightarrow \text{inosine} + NH_3$$

Adenosine deaminase (EC 3.5.4.4) is an inducible enzyme that has been shown to function in a pathway which converts adenine, adenosine, and deoxyadenosine to guanine nucleotides.[1] Addition of adenine, hypoxanthine, and their nucleosides to cultures of *E. coli* induces the synthesis of adenosine deaminase.[2,3]

Assay Method

Spectrophotometric Assay

Principle. The continous spectrophotometric assay is based on the decrease in optical density at 265 nm due to the removal of adenosine and the formation of inosine. The amount of product formed was calculated, assuming a decrease in the molecular extinction coefficient of 8.5×10^3.

Procedure. In a quartz cuvette of 1-cm light path are added 980 μl of 25 mM Tris-succinate, pH 7.6, and 10 μl of 15 mM adenosine (the cuvette compartment should be thermostated at 37°). The reaction is started by the addition of 10 μl of enzyme. The decrease in absorbancy at 265 nm is read in an automatic-recording spectrophotometer.

Radioactive Assay

Principle. In this assay the adenosine deaminase activity is measured as the amount of [14C]inosine formed from [14C]adenosine with time. The assay is based on a chromatographic separation of adenosine from inosine. Since the chromatographic system also separates adenine and hypoxanthine from adenosine and inosine, a possible cleavage of adenosine and inosine by purine nucleoside phosphorylase[4] can be determined. The radioactive assay also allows the determination of adenosine

[1] B. Jochimsen, P. Nygaard, and T. Vestergaard, *Mol. Gen. Genet.* **143**, 85 (1975).
[2] A. L. Koch and G. Vallee, *J. Biol. Chem.* **234**, 1213 (1959).
[3] F. Olsen and P. Nygaard, *Proc. Int. Congr. Biochem., 10th, 1976* Abstr. 01-08-018 (1976).
[4] K. F. Jensen and P. Nygaard, *Eur. J. Biochem.* **51**, 253 (1975).

METHODS IN ENZYMOLOGY, VOL. LI

deaminase activity in the presence of compounds which absorb at 265 nm.

Procedure. A reaction mixture of 100 μl is made up as follows: 80 μl of 25 mM Tris-succinate, pH 7.6, and 10 μl of 1.5 mM [8-^{14}C]adenosine (1 μCi/μmole) are mixed and incubated at 37°. The reaction is started by the addition of 10 μl of enzyme. At 2-min intervals 10-μl samples are withdrawn and applied to polyethyleneimine-impregnated cellulose plates.[5] Prior to application 10 nmoles of adenosine and inosine are applied to each start spot. The chromatographic plates are developed in methanol:*n*-butanol:lithium borate 4% in water (20:60:20). After drying, the spots corresponding to adenosine and inosine are cut out and counted in a liquid scintillation spectrophotometer using a toluene solution containing 5.5 g scintillator (Permablend III, Packard) per liter.

Units. One unit is defined as the amount of enzyme that catalyzes the deamination of 1 μmole of adenosine per minute at pH 7.6 and 37°. Protein is determined by the method of Lowry *et al.*[6]

Preparation of Cell-free Extracts. For routine analysis on crude extracts cells are homogenized at 4° in 0.1 M Tris-HCl, pH 8.2, and 20% ethylene glycol with an ultrasonic drill for 1 min. After removal of cell debri by centrifugation for 5 min at 6000 g in the cold, the supernatant fluid is dialyzed against 10 mM Tris·HCl (pH 8.2), 20% ethylene glycol, 2 mM CaCl₂, and 50 μM EDTA.

Purification Procedure

Growth Medium. The medium contains, in each liter: 2 g of (NH₄)₂SO₄, 6 g of Na₂HPO₄, 3 g of KH₂PO₄, 3 g of NaCl, 0.4 g of MgCl₂·6 H₂O, 15 mg of CaCl₂·2 H₂O, 1 mg of FeCl₃·6 H₂O, 20 g of casein hydrolyzate, 20 g of glucose, 0.4 g of methionine, 7 g of yeast extract, and 0.14 g of adenine, pH 7.0.

General. Unless otherwise stated all operations were performed at 0–4°. Standard Tris-buffer: 10 mM Tris·HCl (pH 8.8), 20% ethylene glycol, 2 mM CaCl₂, and 50 μM EDTA. Standard phosphate buffer: 10 mM potassium phosphate (pH 8.2), 20% ethylene glycol, 50 mM NaCl, and 50 μM EDTA. A summary of the purification procedure is shown in the table.

Growth of Organism. Escherichia coli K12 *met*[4] is inoculated into 1 liter of medium in an Erlenmeyer flask and grown on a rotary shaker at

[5] K. Randerath and E. Randerath, *Anal. Biochem.* **13**, 575 (1965).
[6] O. H. Lowry, N. J. Rosebrough, A. L. Farr, and R. Randall, *J. Biol. Chem.* **193**, 265 (1951).

PURIFICATION OF ADENOSINE DEAMINASE FROM *Escherichia coli*

Fraction	Volume (ml)	Protein (mg)	Specific activity (units/mg)	Recovery (%)
Crude extract	79	5216	0.142	100
Polyethylene glycol fraction	107	337	2.06	93
DEAE-cellulose, batchwise	16	67	3.77	34
Sephadex G-100, chromatography	44	30	5.74	23
Hydroxylapatite chromatography	12	0.48	277	18

37° for 16 hr. The culture is then poured into 9 liters of medium in a container in a 37° bath. The culture is aerated during growth by bubbling filtered compressed air through the medium; pH is controlled at 6.5–7.0 by addition of 5 M NH$_4$OH. After 5 hr of growth the culture is transferred to a chilled reservoir from which the cells are harvested using a continous-flow centrifuge. The usual yield was about 35–55 g wet weight of cells.

Preparation of Cell Extracts. Frozen cells, 39.5 g of wet weight, are suspended in 40 ml of 0.1 M Tris·HCl (pH 8.2), 20% ethylene glycol, 2 mM CaCl$_2$, and 50 μM EDTA and allowed to stand with occasional stirring until thawed. Cells are broken in a French pressure cell at 6000 psi.

Polyethylene Glycol Fractionation. The homogenate (79 ml) is mixed with 79 ml of 50% polyethylene glycol 6000 in water and stirred for 4 hr. Initially the mixture constitutes a gelatinous mass which with time is turned into a thick, less-viscous suspension. The heavy precipitate which is formed is removed by centrifugation at 123,000 g for 150 min. The supernatant solution, 107 ml, contains the enzyme.

DEAE-Cellulose Adsorption and Elution. The supernatant fluid was further diluted with 214 ml of standard Tris buffer containing 50 mM NaCl. To this was added 6 g of DEAE-cellulose (Whatman DE-52). This mixture was stirred for 7 hr and then centrifuged for 10 min at 6000 g. The sediment plus 12 ml of standard Tris buffer containing 50 mM NaCl was poured into a column, 1.5 × 15 cm. The DEAE-cellulose was allowed to pack, and the loading buffer was run through the column. The column was eluted with standard Tris buffer containing 250 mM NaCl. The flow rate was 32 ml/hr, and 2.7-ml fractions were collected. The fractions containing the bulk of the adenosine deaminase activity were pooled (16 ml).

Sephadex G-100 Chromatography. The above fraction was placed on a column (2.5 × 90 cm) of Sephadex G-100 equilibrated with the standard Tris buffer containing 50 m*M* NaCl and was eluted with the same buffer. The flow rate was 17.4 ml/hr, and 5.8 ml fractions were collected. Those with most activity were combined (43 ml).

Hydroxylapatite Chromatography. The combined fractions were concentrated before application; they were allowed to sink slowly into a small DEAE-cellulose column, 1 × 1.5 cm, equilibrated with standard Tris buffer. After application of the sample the column was eluted with standard phosphate buffer containing 250 m*M* NaCl. Fractions of 1.7 ml were collected; the flow rate was about 20 ml/hr. The active fractions were pooled (3.8 ml) and applied to a hydroxylapatite column (BioRad Labs.), 2.5 × 25 cm, previously equilibrated with standard phosphate buffer. This buffer was also used for the elution. The flow rate was 24 ml/hr, and 6-ml fractions were collected. The fractions containing the bulk of the enzymic activity were pooled and finally dialyzed against the standard Tris buffer (12 ml).

Properties

Purity. The adenosine deaminase preparation obtained by this procedure is about 90% pure as judged by analytical polyacrylamide gel electrophoresis.

Substrate Specificity. Nucleosides that serve as substrates are: adenosine, 2-deoxyadenosine, 6-methylaminopurine ribonucleoside, adenine arabinoside, 3-deoxyadenosine, and 2.3-dideoxyadenosine. Guanosine, cytidine, and phosphate derivatives of adenosine are not substrates. K_m values for adenosine and 2-deoxyadenosine, which are considered to be the natural substrates, are 75 μM and 40 μM, respectively.

Inhibitors. The enzyme is inhibited by reagents such as *p*-chloromercuribenzoate, and this inhibition is partly reversed by sulfhydryl reagents. Coformycin, a structural analogue of inosine, inhibits adenosine deaminase; $K_i = 0.03$ μM.

pH Optimum. Adenosine deaminase displays a broad pH optimum between pH 6.9 and 8.5 for both adenosine and 2-deoxyadenosine.

Estimated Molecular Weight. The molecular weight of adenosine deaminase, in crude extracts and with purified enzyme, as determined by gel filtration on Sephadex G-100, is about 29,000.

Stability. The purified enzyme is stable for several months when stored in 10 mM Tris·HCl (pH 8.2), 20% ethylene glycol, 2 mM CaCl$_2$, and 50 μM EDTA at 4°. Stability is greatly diminished in the absence of ethylene glycol.

[69] Guanine Deaminase from Rabbit Liver

By MORTON D. GLANTZ and ARTHUR S. LEWIS

Guanine + H$_2$O → xanthine + NH$_3$

Guanine deaminase (guanine aminohydrolase, EC 3.5.4.3) catalyzes the hydrolytic deamination of guanine to xanthine. Partially purified preparations have been reported and the literature briefly reviewed.[1] A procedure is described for the purification of guanine deaminase by affinity chromatography using 9-(p-β-aminoethoxyphenyl)guanine as the ligand.[2] The following describes the purification of the enzyme to homogeneity and its chemical, physical, and kinetic properties.[3,4]

Assay Method

Principle. The assay measures the disappearance of guanine spectrophotometrically.

Reagents

Phosphate buffer, 0.2 M, pH 7.0
Guanine, 2.0 × 10^{-4} M in 0.2 M phosphate buffer, pH 7.0
Enzyme solutions of aliquots of 10–25 μl

Procedure. The enzyme solution is pipetted into a 3.0-ml cuvette, which is previously incubated at 40° in the thermostated compartment of

[1] C. L. Zielke and C. H. Suelter, *in* "The Enzymes" (P. D. Boyer, ed.), 3rd ed., Vol. 4, p. 76. Academic Press, New York, 1971.
[2] H. Siebeneick and B. R. Baker, *in* "Methods in Enzymology" (W. B. Jakoby and M. Wilchek, eds.), Vol. 34, Part B, p. 523. Academic Press, New York, 1974.
[3] A. S. Lewis and M. D. Glantz, *J. Biol. Chem.* **250**, 8220 (1975).
[4] A. S. Lewis and M. D. Glantz, *J. Biol. Chem.* **249**, 3862 (1974).

a Beckman DU spectrophotometer. An aliquot of substrate (2.5 ml of guanine, 2.0 × 10⁻⁴ M, in 0.2 M phosphate buffer, pH 7.0) at 40° is added and mixed. The disappearance of substrate (guanine → xanthine) is monitored at 246 nm, and recorded as ΔA per minute per milligram of protein. The results are converted to μmoles of guanine deaminated per minute per milligram of protein by dividing the ΔA per minute per milligram of protein by the difference in extinction coefficients at 246 nm, of the substrate guanine and the product xanthine, for a concentration of 1.0 μmole/ml, in 0.2 M phosphate buffer (pH 7.0), and multiplying by the total volume of the assay mixture.

Total protein is determined by the method of Lowry et al.[5] Bovine serum albumin is utilized as standard.

Purification Procedure

All procedures are performed at 0–4° unless otherwise noted.

Step 1. First Ammonium Sulfate Fractionation. Frozen rabbit livers (2.4 kg, Type 2, obtained from Pel Freez Biologicals, Inc., Rogers, Arkansas) are partially thawed, cut into small pieces, and homogenized in a Waring Blendor in 10 liters of 0.2 M phosphate buffer (pH 7.0) at 4°, containing 20 mM 2-mercaptoethanol. The homogenate is centrifuged at 20,000 g for 30 min at 0°. The clear supernatant solution is removed and brought to 35% saturation with solid ammonium sulfate near pH 7.0. The precipitated proteins are removed by centrifugation, and the clear supernatant fluid is brought to 70% saturation with solid ammonium sulfate. The precipitated proteins are recovered by centrifugation, redissolved in 0.05 M phosphate buffer (pH 7.5) containing 20 mM 2-mercaptoethanol, and dialyzed overnight under running tap water, near 7°. The precipitate formed upon dialysis is removed by centrifugation, and 2-mercaptoethanol is added to a concentration of 10 mM. The pH of the solution is adjusted to pH 8.0 with a dilute solution of Tris base. A 2-to 3-fold purification is achieved in this step.

Step 2. Preparative Chromatography on DEAE-Cellulose. The enzyme solution (152 g protein in 3 liters) is adsorbed onto 3 columns of DEAE-cellulose with bed dimensions of 5.0 × 20 cm. The flow rate is 250 ml/hr. The columns are each washed with 3 column volumes of 0.05 M phosphate buffer, pH 8.0, followed by 3 volumes of 0.05 M phosphate buffer, pH 7.5. The resin-bound enzyme is then eluted from the column with 0.05 M citrate buffer, pH 6.5. All buffers contain 2-

[5] O. H. Lowry, N. J. Rosebrough, A. L. Farr, and R. J. Randall, *J. Biol. Chem.* **193**, 265 (1951).

mercaptoethanol at a concentration of 20 mM. Fractions with activity greater than the starting material are pooled and yield 13.1 g protein with a specific activity of 0.07 μmole/min/mg protein. A 23-fold purification of the enzyme is achieved.

Step 3. Second Ammonium Sulfate Fractionation. The enzyme solution from the preceding step is brought to 45% saturation with solid ammonium sulfate at 4° and pH 7.0. After centrifugation, the precipitate is discarded and the supernatant fluid brought to 70% saturation with ammonium sulfate. The precipitated enzyme is collected by centrifugation at 0°, redissolved in 0.02 M phosphate buffer (pH 7.5) containing 20 mM 2-mercaptoethanol, and dialyzed under running tap water near 7°. Dialysis is continued for 6 hr at 4° in two changes of 4 liters each of 0.01 M phosphate buffer (pH 7.5), containing 10 mM 2-mercaptoethanol.

Step 4. Second DEAE-Cellulose Chromatography. The dialyzed enzyme (5.3 g in 282 ml) is adsorbed on a column of DEAE-cellulose with bed dimensions of 3.5 × 20 cm which was equilibrated with 0.01 M phosphate buffer (pH 7.5). After washing with 4 column volumes of 0.05 M phosphate buffer (pH 7.5) containing 10 mM 2-mercaptoethanol, the enzyme is slowly eluted with 0.025 M citrate buffer (pH 6.8). Fractions of 15 ml are collected. The enzyme is eluted within a single sharp peak of protein, between 4.5–5.5 column volumes of the citrate buffer. Fractions with activity greater than the adsorbed material are pooled and precipitated at 80% saturation with solid ammonium sulfate near pH 7.0. The precipitated enzyme is recovered by centrifugation at 0° and redissolved in 0.025 M phosphate buffer (pH 7.5).

Step 5. Third DEAE-Cellulose Chromatography. The enzyme is dialyzed 6 hr against 4 liters of 0.01 M phosphate buffer (pH 7.5), containing 10 mM 2-mercaptoethanol, and then adsorbed on a column of DEAE-cellulose with bed dimensions of 2.6 × 15 cm. The total protein is 1.9 g in 118 ml of solution. The elution pattern is identical to that of the preceding step. Fractions of 7.5 ml are collected, and those fractions with activity greater than the adsorbed material are pooled and precipitated at 80% saturation with solid ammonium sulfate near pH 7.0. The precipitate is recovered by centrifugation at 0° and redissolved in 0.025 M phosphate buffer (pH 7.5). This showed a 288-fold purification.

Step 6. Chromatography on Hydroxylapatite. The enzyme is dialyzed as in step 5, and 35 ml of enzyme solution containing 740 mg protein are adsorbed on a column of hydroxylapatite (2.5 × 15 cm), previously equilibrated with 0.01 M phosphate buffer (pH 7.5). The column is eluted with 2 column volumes of 0.01 M phosphate buffer (pH

7.5), after which the enzyme is eluted with 0.04 M phosphate buffer (pH 7.5). Fractions of 4.0 ml are collected, and fractions with activity 5-fold greater than the adsorbed enzyme are pooled and precipitated at 80% saturation with ammonium sulfate near pH 7.0. The precipitate is collected by centrifugation, dissolved in 0.025 M phosphate buffer (pH 7.5), and dialyzed against the same buffer containing 10 mM 2-mercaptoethanol. Assay showed a purification of 3900-fold.

Step 7. Fourth DEAE-Cellulose Chromatography. The dialyzed enzyme is adsorbed on a column of DEAE-cellulose (1.8 × 10 cm) and eluted with 0.025 M citrate buffer (pH 6.8). Fractions of activity greater than 22.0 μmole/min/mg protein are pooled to give a total of 8.78 mg protein with an 8250-fold purification. Polyacrylamide disc electrophoresis indicated a small trace of protein contamination.

Step 8. Selective Precipitation with Ammonium Sulfate Solution. The pooled fractions from the preceding step are precipitated at 80% saturation with ammonium sulfate near pH 7.0 and redissolved in 5.0 ml of 0.1 M phosphate buffer (pH 7.5) containing 10 mM 2-mercaptoethanol. The enzyme solution is brought to a slight turbidity with saturated ammonium sulfate (pH 7.0) at 0° and centrifuged. The supernatant fluid is removed, brought to a definite turbidity with saturated ammonium sulfate solution, and allowed to stand overnight at 4°. The precipitated enzyme is recovered by centrifugation and redissolved in 0.1 M phosphate buffer (pH 7.5) containing 10 mM 2-mercaptoethanol. This final preparation contained 4.56 mg protein at a specific activity of 25 μmole/min/mg protein, and a 9000-fold purification.

A summary of the purification procedure is presented in the table.

Properties

Purity. The enzyme is shown to be homogeneous by polyacrylamide disc electrophoresis both in the absence and presence of sodium dodecyl sulfate. Homogeneity was also established by sedimentation velocity study.

pH Activity. The deamination of guanine is optimal at pH 6.8. The pH optimum is independent of the buffer system as observed in phosphate, Tris, and citrate buffers of the same molarity. A sharp pH optimum at pH 6.0 is obtained with 8-azaguanine as substrate. Analysis of plots of kinetic parameters versus pH indicate catalytic roles for groups at the active site of the enzyme, with pK_a values at pH 6.1, 7.5, and 7.8. The ionization of histidine with pK_a 6.1, and the early ionization

PURIFICATION OF GUANINE DEAMINASE FROM RABBIT LIVER

Purification step	Total protein (mg)	Specific activity (μmole/ min/mg)	Total activity (%)	Purification (fold)
Supernate after centrifugation of homogenate	411,200	0.003	100	1
1. First ammonium sulfate fractionation	152,000	0.007	98.4	2.6
2. Preparative chromatography on DEAE-cellulose	13,100	0.07	74.5	23
3. Second ammonium sulfate fractionation	5,300	0.14	67.9	53
4. Second DEAE-cellulose chromatography	1,900	0.3379	56.6	121
5. Third DEAE-cellulose chromatography	740	0.8024	51.9	288
6. Chromatography on hydroxylapatite	23.5	10.8	17.6	3900
7. Fourth DEAE-cellulose chromatography	8.78	23	17.6	8250
8. Selective precipitation with saturated ammonium sulfate	4.56	25.2	10.0	9000

of cystein-SH (pK_a 7.5 and 7.8) in the formation and dissociation of the enzyme–substrate complex, are strongly implicated.

Kinetic Constants. The apparent K_m values determined for substrates of the enzyme, in 0.2 M phosphate buffer at 40°, were 1.25 × 10^{-5} M for guanine at pH 7.0, 3.33 × 10^{-4} M for 8-azaguanine at pH 6.0, and 8.0 × 10^{-4} M for 6-thioguanine at pH 7.0.

Inhibition Studies. Irreversible inactivation was observed upon incubation with p-chloromercuribenzoate, at a concentration of 1.0 × 10^{-4} M. Iodoacetic acid and iodoacetamide were mildly inhibitory at a concentration of 1.0 × 10^{-4} M. The purine precursor 5-aminoimidazole-4-carboxamide was a strong competitive inhibitor, with an apparent K_i of 3.05 × 10^{-5} M. Its ribonucleoside derivative (5-aminoimidazole-4-carboxamide ribonucleoside) and tetrahydrofolic acid were also observed to be competitive inhibitors of the enzyme. Their respective apparent K_i values were 2.58 × 10^{-4} M for tetrahydrofolic acid and 5.68 × 10^{-4} M for 5-aminoimidazole-4-carboxamide ribonucleoside.

Stability. The enzyme is relatively stable between the temperatures of 40° and 50°. A substantial loss in the maximal catalytic activity of the enzyme is seen upon incubation at 55° for 20 min. Rapid and irreversible denaturation is observed upon incubation at 60°, with complete inactivation after 30 min of incubation at 60°. An inactivation energy of 4.94 kcal/mole, which is linear between temperatures of 20° and 50°, is determined for the enzyme. The enzyme is stabilized by 2-mercaptoethanol during the purification.

Molecular Weight. A molecular weight of 56,000 was calculated from gel filtration, and a molecular weight of 54,000 was determined from sodium dodecyl sulfate–polyacrylamide disc gel electrophoresis. An S value of 4.32 was calculated from sedimentation velocity study.

Mechanistic Analysis. Deuterium isotope studies have implicated the participation of water in the catalytic process. A mole of water is apparently consumed in the formation of products. An ordered mechanism, involving two substrates, is proposed, whereby a proton is shuttled from the solvent (water), by way of histidine and cysteine, for protonation of the primary amine of the substrate (guanine) in the initial step. Concerted protonation and alkylation by cysteine at the susceptible carbon of the substrate result in the release of the first product (NH_3) and the formation of an enzyme–purinyl intermediate. The dissociation of this intermediate is mediated by nucleophilic displacement by solvent hydroxyl to liberate xanthine as the second product.[3]

[70] Purine Nucleoside Phosphorylase from *Salmonella typhimurium* and Rat Liver

By PATRICIA A. HOFFEE, REYAD MAY, and B. C. ROBERTSON

Inosine + P_i ⇌ hypoxanthine + α-D-ribose-1-phosphate

When *Salmonella typhimurium* is grown in the presence of a purine ribonucleoside, purine nucleoside phosphorylase and phosphodeoxyribomutase are coordinately induced. But when cells are grown in the presence of deoxyribose or deoxyribonucleosides, four enzymes are induced. These include thymidine phosphorylase, purine nucleoside phosphorylase, phosphodeoxyribomutase, and deoxyribose-5-phosphate aldolase.[1] The demonstration that purified purine nucleoside phosphorylase has specificity for both purine ribo- and deoxyribonucleosides is consistent with the hypothesis that a single gene codes for the enzyme but that the gene is regulated by more than one effector, i.e., ribo- and deoxyribo- derivatives.

Purine nucleoside phosphorylase activity in rat liver extract was

---

[1] B. C. Robertson, P. Jargiello, J. Blank, and P. Hoffee, *J. Bacteriol.* **102**, 628 (1970).

METHODS IN ENZYMOLOGY, VOL. LI

originally described by Kalckar.[2] The phosphorylase is present in a relatively high concentration in rat liver tissue accounting for about 0.2% of the soluble protein. Although isozymes of purine nucleoside phosphorylase have been demonstrated in erythrocytes,[3] only a single species of the enzyme has been found in rat liver tissue.

Assay Method[2]

Principle. Enzyme activity is measured in the nucleoside breakdown direction by a coupled spectrophotometric assay developed by Kalckar.[2] The breakdown of inosine is followed by an increase in absorbance at 293 nm produced by uric acid in a coupled assay system with xanthine oxidase.

Reagents

Potassium phosphate buffer, 0.05 M, pH 7.5
Inosine, 0.05 M
Xanthine oxidase, milk (0.4 μg–10 mg/ml) (The purified enzyme preparation is obtained from Boehringer Mannheim Biochemicals.)

Procedure. To 0.98 ml of phosphate buffer in a 1-ml quartz cuvette add 0.01 ml of inosine, 0.005 ml xanthine oxidase, and purine nucleoside phosphorylase solution that gives a change in absorbance of 0.01–0.05 per minute. The absorbance is measured at 293 nm at 25° for the bacterial enzyme and at 37° for the rat liver enzyme. A control cuvette is run without enzyme solution to correct for any nonenzymic breakdown of substrate.

Definition of Units. One unit of activity cleaves 1 μmole of inosine per minute. The number of units is calculated using an extinction coefficient of 12.5. Protein is determined by the method of Lowry *et al.*[4] before ammonium fractionation (see *Purification*) and by the method of Bücher[5] or Waddel[6] after ammonium sulfate fractionation I.

[2] H. M. Kalckar, *J. Biol. Chem.* **167**, 429 (1947).
[3] K. C. Agarwal, R. P. Agarwal, J. D. Stoeckler, and R. E. Parks, Jr., *Biochemistry* **14**, 79 (1975).
[4] O. H. Lowry, N. J. Rosebrough, A. L. Farr, and R. J. Randall, *J. Biol. Chem.* **193**, 265 (1951).
[5] T. Bücher, *Biochim. Biophys. Acta* **1**, 292 (1947).
[6] W. J. Waddel, *J. Lab. Clin. Med.* **48**, 311 (1956).

Purification of Enzyme from *Salmonella typhimurium*[7]

Growth of Cells. Wild-type *Salmonella typhimurium* (ATCC 15277): Inoculate 2-liter flasks containing 1 liter of Casamino acid medium, buffered at pH 7.0 with phosphate, with 25 ml of an overnight culture grown on the same medium. Incubate the flasks with shaking at 37° for 4 hr or until cells are in mid-log phase. Add sterile 2-deoxyribose to the cells to a final concentration of 0.2%. Continue incubation of the cells at 37° with aeration for an additional 90 min and then rapidly chill and harvest. Wash the cell paste in 1 m*M* EDTA and store frozen at −10°. *Salmonella* strains constitutive for purine nucleoside phosphorylase: Inoculate 2-liter flasks containing 1 liter of Casamino acid medium, buffered at pH 7.0 with phosphate, and containing 10 μg/ml of thymine, with 25 ml of an overnight culture grown on the same medium. Incubate the flasks with shaking at 37° overnight and harvest in the morning. Wash the cell paste as above and store frozen at −10°. The use of the constitutive strains is much more convenient since it does not require induction with 2-deoxyribose and the amount of enzyme obtained is about 1.5–2 times the amount obtained from the same number of wild-type cells. The purification procedure described below can be used for both wild-type induced cells and constitutive strains.

Preparation of Cell Extract. Grind 25 g of frozen cells in a cold mortar with 50 g of Alumina A-301 (Bacteriological grade, Alcoa Chemicals, Bauxite, Arkansas). Take up the paste in 100 ml of 400 m*M* triethanolamine-HCl at pH 7.6, 1 m*M* in EDTA, and centrifuge the mixture in a refrigerated centrifuge at 27,000 *g* for 45 min. Assay the supernatant fluid for activity and protein.

Protamine Sulfate Treatment. Make a 1% solution of protamine sulfate (Eli Lilly). To each milliliter of extract, add 0.25 ml of protamine sulfate solution slowly with constant stirring. After 5 min at 4°, remove the precipitate by centrifugation at 12,000 *g* for 10 min and assay the supernatant fluid for activity and protein.

Ammonium Sulfate Fractionation 1. Bring the supernatant fluid from the protamine sulfate fraction to 45% ammonium sulfate saturation by adding 254 mg/ml of solid ammonium sulfate. After 30 min at 4°, centrifuge the suspension. Dissolve the precipitate in 5 m*M* potassium phosphate buffer, pH 7.5, and dialyze the solution overnight against 500 volumes of the same buffer.

[7] Procedure based on work previously published: B. C. Robertson and P. Hoffee. *J. Biol. Chem.* **248**, 2040 (1973).

Calcium Phosphate Gel Treatment. Add to the dialyzed fraction from above, calcium phosphate gel (Bio-Rad Laboratories) suspended in 5 mM potassium phosphate, pH 7.5, until all purine nucleoside phosphorylase activity is absorbed. Since different preparations of calcium phosphate gel vary in absorption ability, a series of test tubes containing a known amount of enzyme activity are titrated with increasing amounts of calcium phosphate gel to determine the amount necessary to absorb 95% of the enzyme. The suspension is centrifuged and the supernatant fluid discarded. The pellets are washed 4 times or until no more enzyme is recovered with 10 mM potassium phosphate, pH 7.5. The washes which contain the highest specific activity of purine nucleoside phosphorylase are brought to 90% ammonium sulfate saturation by the addition of 630 mg/ml of solid ammonium sulfate. After 30 min at 4°, centrifuge the suspension. Dissolve the precipitate in 10 ml of 5 mM potassium phosphate, pH 7.5, and dialyze overnight against 1000 volumes of the same buffer.

DEAE-Cellulose Chromatography. Apply the dialyzed calcium phosphate gel eluate to a DEAE-cellulose column (1.5 × 20 cm) equilibrated with 5 mM potassium phosphate, pH 7.5. Wash the column with the same buffer until no more 280-nm-absorbing material is eluted. Purine nucleoside phosphorylase is then eluted with a linear gradient of potassium phosphate, 5 mM, pH 7.5 to 175 mM, pH 8.0 in 700 cc. Combine fractions containing enzyme activity and bring to 90% ammonium sulfate saturation by the addition of 630 mg/ml of solid ammonium sulfate. After 30 min centrifuge the suspension and dissolve the precipitate in 4.5 ml of 5 mM potassium phosphate, pH 7.5.

Ammonium Sulfate Fractionation 2. Clarify the suspension from above by centrifugation. Slowly add solid ammonium sulfate to the clear supernatant fluid until the solution is slightly turbid (about 50% saturation). Allow the suspension to stand at 4° for 30 min and then centrifuge. Resuspend the pellet in 5 mM potassium phosphate, pH 7.5, and repeat the precipitation a second time. Resuspend the second precipitate in 5 mM potassium phosphate, pH 7.5.

The purification procedure is summarized in Table I.[1,8]

Properties[6]

Molecular Weight. Purine nucleoside phosphorylase isolated from *S. typhimurium* has a molecular weight of 141,000 and is composed of six subunits of molecular weight 23,500.

[8] J. Blank and P. Hoffee, *Mol. Gen. Genet.* **116,** 291 (1972).

TABLE I

PURIFICATION PROCEDURE FOR PURINE NUCLEOSIDE PHOSPHORYLASE FROM *Salmonella typhimurium*[a]

Fraction	Total volume (ml)	Total units[b]	Protein (mg/ml)	Specific activity (units/mg)	Purification (fold)	Recovery (%)
Crude extract	91	1670	26	0.7	1	100
Protamine sulfate	109	1820	11.9	1.4	2	100
Ammonium sulfate fraction 1	19	1030	16.4	3.3	4.7	61.5
CaPO₄ gel	20	550	1.53	18.0	26	33
DEAE-cellulose	4.5	545	1.57	77.0	110	32.5
Ammonium sulfate fraction 2	4.5	360	0.5	160.0	229	21

[a] Data given for a constitutive mutant pH 4021. Constitutive strains are available from the author or can be selected by the procedures described by Blank and Hoffee[8] or Robertson *et al.*[1]

[b] Units given as μmoles per minute.

Substrate Specificity. Purified purine nucleoside phosphorylase can catalyze the conversion of inosine, deoxyinosine, guanosine, deoxyguanosine, adenosine, and deoxyadenosine to their respective bases and pentose-1-phosphate. No activity is seen with pyrimidine nucleosides. The apparent K_m values for inosine and deoxyinosine are 50 and 47 μM and for phosphate 0.37 mM.

Stability. The enzyme is stable when stored at 4° in ammonium sulfate solution. It is not stable when frozen.

Effect of pH. Optimum activity is observed at pH 7.5 with a rapid decrease in activity both above and below this value.

Purification of Enzyme from Rat Liver

Preparation of Liver Homogenate. Cut 500 g of fresh rat livers into small pieces (0.5 cm²) and rinse 3 times with phosphate-buffered saline. Homogenize the tissue in 2 volumes of 50 mM Tris·HCl–1 mM EDTA–0.1 mM phenylmethyl sulfonyl fluoride (pH 7.5) (buffer A) in a motorized Thomas Model C Teflon–glass homogenizer using 10 passes. Centrifuge the homogenate at 27,000 g for 45 min. Assay the supernatant fluid for activity and protein. Note: All buffers used in the purification procedures, except the electrophoresis buffers, contain 10 mM β-mercaptoethanol.

Ammonium Sulfate Fractionation. Bring the supernatant fluid from the previous step to 35% ammonium sulfate saturation by adding 199 mg/ml of solid ammonium sulfate. After 30 min at 4°, centrifuge the suspension and remove the supernatant fluid. Bring the supernatant fluid to 65% saturation by adding 190 mg/ml of solid ammonium sulfate. After 30 min at 4°, centrifuge the suspension. Dissolve the precipitate in buffer A containing 15% glycerol and dialyze the solution overnight against 50 volumes of 10 mM potassium phosphate buffer, pH 6.0 (buffer B).

Heat Treatment. Centrifuge the dialyzed solution at 27,000 g for 20 min. Add solid inosine to yield a final concentration of 20 mM. Place the solution in a circulating water bath at 65°, and stir the solution with a magnetic stirrer for 15 min. (The protein solution takes about 10 min to reach 65°.) Place the heated suspension in an ice bath, and stir until the temperature reaches 4°. Remove the precipitate by centrifugation at 27,000 g for 20 min. Dialyze the supernatant fluid overnight against 50 volumes of buffer B.

CM-Sephadex Chromatography. Apply the dialyzed solution to a CM-Sephadex A-25 column (2.5 × 40 cm) equilibrated with buffer B. Collect all of the eluate containing enzyme activity, and adjust the solution to 30 mM potassium phosphate by adding 0.02 volume of 1 M potassium phosphate, pH 7.0.

QAE-Sephadex Chromatography. Apply the CM-Sephadex eluate to a QAE-Sephadex A-50 column (2.5 × 40 cm) equilibrated with 20 mM potassium phosphate, pH 7.0 (buffer C). Wash the column with buffer C until no more 280-nm-absorbing material is eluted. Purine nucleoside phosphorylase is then eluted with a linear gradient of NaCl, 0–0.2 M in 1 liter of buffer C. Combine the fractions containing enzyme activity with a specific activity greater than 6-fold of the starting material.

Back Extraction with Ammonium Sulfate. Bring the eluate from the previous step to 80% ammonium sulfate saturation by the addition of 534 mg/ml of solid ammonium sulfate. After 30 min at 4°, collect the precipitate by centrifugation. Discard the supernatant fluid, and extract the precipitate with 20 ml of 55% saturated ammonium sulfate solution. Extract the precipitate for 5 min using a glass rod to effect resuspension of the solid; then allow the suspension to stand for 10 min at 4°. Centrifuge the suspension at 12,000 g for 10 min, and remove the supernatant fluid. Repeat the extraction procedure twice using 10 ml of 48% saturated ammonium sulfate solution. Pool the two 48% ammonium sulfate supernatant fluids, and let the solution sit at room temperature for 30 min. Centrifuge the suspension at 27,000 g for 10 min at 20°. Bring

TABLE II
PURIFICATION PROCEDURE FOR PURINE NUCLEOSIDE PHOSPHORYLASE FROM RAT LIVER

Fraction	Total volume (ml)	Total units (μmoles/ min)	Protein (mg/ml)	Specific activity (units/mg)	Purifica- tion (fold)	Recovery (%)
High-speed supernatant	1430	18,000	35.4	0.36	1.0	100.0
Ammonium sulfate	405	15,870	74.0	0.53	1.5	88.0
Heat treatment	540	11,250	4.3	4.8	13.4	62.5
CM-Sephadex chromatography	600	9,500	1.6	9.9	27.5	53.0
QAE-Sephadex chromatography	220	7,330	0.4	83.3	231.4	40.7
Back extraction	20	6,250	2.8	112.0	311.0	34.7
Ultrogel chromatography	37	4,150	0.9	125.0	347.0	23.0
Polyacrylamide gel electrophoresis	2.6	2,830	7.6	143.0	397.0	15.7

this supernatant fluid to 80% ammonium sulfate saturation by the addition of 220 mg/ml of solid ammonium sulfate, and collect the precipitate by centrifugation. Dissolve the precipitate in 3 ml of 0.2 M potassium phosphate, pH 6.0 (buffer D). Note: The 50% and 48% ammonium sulfate solutions used in the extraction also contain 10 mM inosine and 10 mM β-mercaptoethanol.

Gel Filtration on Ultrogel AcA 34. Apply the solution from the previous step to an Ultrogel AcA 34 (LKB-Produkter, Bromma, Sweden) column (2.5 × 100 cm) equilibrated with buffer D. Purine nucleoside phosphorylase is eluted with buffer D at a flow rate of 22 ml/hr. Pool the fractions containing enzyme activity, and concentrate the solution to 2 ml using a Millipore Pellicon membrane type PTGC.

Preparative Polyacrylamide Gel Electrophoresis. Add sucrose to the enzyme solution from the previous step to yield a 30% solution. Add one small crystal of Bromophenol blue and bring to 10 mM with β-mercaptopropionic acid. Layer the sucrose solution on top of a discontinuous preparative polyacrylamide gel prepared as described below. A Shandon-Southern Unit Model number 2782[9] is used following the manufacturer's operating instructions. A 10-cm, 40-ml separating gel of 7.5% polyacrylamide and a 4-ml stacking gel of 2.5% polyacrylamide are

[9] P. Hauschild-Rogat and I. Smith, *in* "Chromatographic and Electrophoretic Techniques" (I. Smith, ed.). Vol. II, p. 475. Wiley (Interscience), New York, 1968.

used. The lower cone is a 10% polyacrylamide gel in lower tank buffer. The following buffers are used:

1. Stacking gel buffer: 46 mM Tris-phosphate, pH 6.7
2. Separating gel buffer: 0.375 M Tris·HCl, pH 9.1
3. Upper tank buffer: 43 mM Tris-glycine, pH 8.9, 5 mM β-mercaptopropionic acid
4. Lower tank buffer: 0.12 M Tris·HCl, pH 8.1
5. Elution buffer: 0.12 M Tris·HCl, pH 8.1, 10 mM β-mercaptoethanol, 10 mM inosine

After adding the sample to the top of the stacking gel, apply a current of 20 mA until the tracking dye enters the separating gel; then increase the current to 40 mA for the remainder of the run. When the tracking dye reaches the elution chamber, start pumping elution buffer through the chamber with a peristaltic pump at 20 ml/hr and collect fractions of 2 ml each. Pool the fractions containing enzyme activity, and bring to 90% ammonium sulfate saturation using 630 mg/ml solid ammonium sulfate. Resuspend the precipitate in buffer D. The enzyme can be stored at −20° in 20% glycerol and 10 mM inosine.

The purification procedure is summarized in Table II. The purified enzyme appears homogeneous and shows only one band on polyacrylamide gels and polyacrylamide gels containing SDS.

[71] Purine Nucleoside Phosphorylase from Rabbit Liver

By MORTON D. GLANTZ and ARTHUR S. LEWIS

Guanine or hypoxanthine + ribose-1-phosphate $\rightleftharpoons$ guanosine or inosine + phosphate

Purine nucleoside phosphorylase (purine nucleoside: orthophosphate ribosyltransferase, EC 2.4.2.1) catalyzes the phosphorolysis and synthesis of nucleosides with almost equal efficiency. The enzyme from rabbit liver[1] and bovine brain[2] was purified to homogeneity and partially characterized. The literature on the enzyme has been reviewed.[3,4]

[1] A. S. Lewis and M. D. Glantz, *J. Biol. Chem.* **251**, 407 (1976).
[2] A. S. Lewis and M. D. Glantz, *Biochemistry* **15**, 4451 (1976).
[3] M. Friedkin and H. Kalckar, *in* "The Enzymes" (P. D. Boyer, H. Lardy, and K. Myrbück, eds.), 2nd rev. ed., Vol. 5, p. 237. Academic Press, New York, 1961.
[4] R. E. Parks, Jr. and R. P. Agarwal, *in* "The Enzymes" (P. D. Boyer, ed.), 3rd ed., Vol. 7, p. 483. Academic Press, New York, 1972.

Recent studies on the enzyme have been reported from human and rat erythrocytes,[5,6] bovine spleen,[7] *E. coli*[8] and *S. typhimurium*,[8,9] and from Chinese hamster.[10]

Assay Methods

Method 1

Principle. The assay measures the disappearance of guanosine, in the reverse reaction by the phosphorolysis of guanosine, which can be determined spectrophotometrically.

Reagents

Phosphate buffer, 0.2 M, pH 6.3, containing 10 mM 2-mercaptoethanol

Purine nucleoside, guanosine or inosine, 2.0 × 10^{-4} M in 0.2 M phosphate buffer, pH 6.3

Enzyme solutions of aliquots of 1–50 μl

Procedure. The enzyme solution is pipetted into a 3.0-ml cuvette, which is previously incubated at 40° in the thermostated cuvette compartment of a Beckman DU spectrophotometer. An aliquot of substrate (2.5 ml of purine nucleoside, 2.0 × 10^{-4} M, in 0.2 M phosphate buffer, pH 6.3) at 40° is added and mixed. The disappearance of the substrate is recorded as ΔA/min/mg of protein at the absorption maximum of the substrate. The results are converted to μmole of substrate hydrolyzed per minute per milligram of protein by the following:

$$[\Delta A/\text{min/mg of protein}/(E_s - E_p) \times \text{total ml of assay mixture}]$$

where E_s and E_p are the calculated extinctions of substrate (E_s) and product (E_p) at 1.0 μmole/ml, in 0.2 M phosphate buffer (pH 6.3), at the absorption maximum of the substrate only.

The absorption maxima for the substrates of the enzyme in 0.2 M phosphate buffer (pH 6.3) are guanosine and deoxyguanosine (252 nm), inosine (249 nm), xanthosine (248 nm), and guanine (246 nm).

[5] K. C. Agarwal, R. P. Agarwal, R. E. Parks, Jr., and M. G. Baldini, *Fed. Proc., Fed. Am. Soc. Exp. Biol.* **32**, Abstr. No. 581 (1973).

[6] K. C. Agarwal, R. P. Agarwal, J. D. Stoeckler, and R. E. Parks, Jr., *Biochemistry* **14**, 79 (1975).

[7] Y. H. Edwards, P. A. Edwards, and D. A. Hopkinson, *FEBS Lett.* **32**, 235 (1973).

[8] K. F. Jensen and P. Nygaard, *Eur. J. Biochem.* **51**, 253 (1975).

[9] B. C. Robertson and P. Hoffee, *J. Biol. Chem.* **248**, 2040 (1973).

[10] G. Milman, D. L. Anton, and J. L. Weber, *Biochemistry* **15**, 4967 (1976).

Method 2

Principle. The synthesis of the nucleoside by the ribosylation of purines is measured in the forward reaction. This is determined spectrophotometrically by recording the increase in absorbance of the nucleoside product.

Reagents

Citrate buffer, 0.2 M, pH 6.3
Ribose 1-phosphate
Purine substrate, guanine or hypoxanthine, 2.0 × 10^{-4} M in 0.2 M citrate buffer, pH 6.3
Enzyme solutions of aliquots of 5–25 μl

Procedure. Enzyme preparations are dialyzed against 0.1 M citrate buffer (pH 6.3), and aliquots of 5–25 μl of enzyme of known protein concentration are preincubated with 100 μl (1.0 μmole) of ribose 1-phosphate solution in distilled water. Purine substrate (2.5 ml of 2 × 10^{-4} M) in 0.2 M citrate buffer (pH 6.3) at 40° is added and mixed. The increase in absorbance of the product is recorded as ΔA/min/mg of protein, at the absorption maximum of the product. The results are converted to μmoles of nucleoside synthesized per minute per milligram of protein, as in method 1, with $E_p - E_s$ substituted for $E_s - E_p$.
The protein is determined after the method of Zamenhof.[11]

Other Assay Methods

Various other assay methods for this enzyme have been reported. The phosphorolysis is determined spectrophotometrically by measuring the increase in absorbance at 293 nm in a coupled system containing a phosphate buffer, nucleoside, xanthine oxidase, and purine nucleoside phosphorylase.[12] The phosphorolysis of deoxyribonucleosides has also been measured by a colorimetric determination with thiobarbituric acid.[13,14] Nucleoside synthesis has been determined in a radioactive assay by following the formation of ^{14}C-labeled nucleoside from the ^{14}C-labeled purine.[8,10] Nucleoside phosphorolysis has been measured by the formation of [14C]ribose 1-phosphate from [U-14C]guanosine or replacing guanosine with inosine and [U-14C]guanosine with [U-14C]inosine.[10]

[11] S. Zamenhof, *in* "Methods in Enzymology" (S. P. Colowick and N. O. Kaplan, eds.), Vol. 3, p. 702. Academic Press, New York, 1957.
[12] H. M. Kalckar, *J. Biol. Chem.* **167**, 429 (1947).
[13] K. Hammer-Jespersen and A. Munch-Petersen, *Eur. J. Biochem.* **17**, 397 (1970).
[14] L. D. Saslaw and V. S. Waravdekar, *in* "Methods in Enzymology" (L. Grossman and K. Moldave, eds.), Vol. 12, p. 108. Academic Press, New York, 1967.

Purification Procedure

Isolation and purification procedures, and storage of the enzyme, are done between 4–10°, unless otherwise stated.

Step 1. Homogenate Extract. Thawed rabbit livers (1.4 kg) are homogenized (20 g%) in 0.1 *M* phosphate buffer (pH 7.0) at 4°, containing 20 m*M* 2-mercaptoethanol, in a Waring Blendor. The homogenate is centrifuged (20,000 *g* for 30 min at 0°), and the supernatant solution (5 liters containing 311 g of protein) is removed for ammonium sulfate fractionation.

Step 2. First Ammonium Sulfate Fractionation. The supernatant fluid is fractionated with solid ammonium sulfate near pH 7.0, and almost total recovery of the enzyme is obtained in the fraction precipitated between 35–70% saturation. After dialysis, the precipitated substances are removed by centrifugation and discarded. Mercaptoethanol is added drop by drop, with stirring, to a concentration of 10 m*M*, and the pH is adjusted to pH 7.5 with a saturated solution of Tris base.

Step 3. First DEAE-Cellulose Chromatography. The enzyme solution (1740 ml containing 63.4 g of protein) is adsorbed on a column of DEAE-cellulose (3.6 × 36 cm). Approximately 85% of the adsorbed proteins are eluted from the column within 2 liters of 0.05 *M* phosphate buffer (pH 7.5), with no observable enzyme activity. The enzyme is removed from the column by elution with 0.05 *M* citrate buffer (pH 6.8), and fractions of 15.0 ml are collected.

Step 4. Second Ammonium Sulfate Fractionation. The fractions from step 3 with activity greater than the starting material are pooled (835 ml containing 5.5 g of protein) and fractionated between 45–70% saturation with ammonium sulfate. The precipitated enzyme is recovered by centrifugation, redissolved in 0.05 *M* phosphate buffer (pH 7.5), and dialyzed for 2 hr against running tap water. Dialysis is then continued for 6 hr at 4°, in two changes of 4 liters each of 0.01 *M* phosphate buffer (pH 7.5) containing 10 m*M* 2-mercaptoethanol.

Step 5. Chromatography on Hydroxylapatite. The dialyzed enzyme (2 g in 75 ml) is adjusted to pH 7.5 and adsorbed on a column of hydroxylapatite (2.6 × 30 cm). A peak of protein containing no enzyme activity is completely eluted within 2 column volumes of 0.04 *M* phosphate buffer (pH 7.5). The enzyme is then eluted from the column with 0.06 *M* phosphate buffer (pH 7.5), and fractions of 4.0 ml are collected. Fractions with specific activity 10-fold greater than the starting material are pooled and precipitated at 80% saturation with ammonium sulfate. The precipitated protein is recovered by centrifugation and redissolved to a volume of 30 ml (containing 74 mg of protein) with 0.2 M phosphate buffer (pH 6.25) and 20 mM 2-mercaptoethanol.

Step 6. Heat Treatment of Enzyme. The enzyme solution from the previous step is raised to 55° in a water bath and maintained at that temperature for 15 min with gentle agitation. After chilling, denatured protein is removed by centrifugation, and the enzyme is precipitated at 80% saturation with ammonium sulfate. The enzyme is recovered by centrifugation and dissolved in 2.0 ml of 0.1 M phosphate buffer (pH 6.25).

Step 7. Filtration on Sephadex G-150. The enzyme solution (2.0 ml containing 35 mg of protein) is filtered through a column of Sephadex G-150 (1.9 × 42 cm). The column is eluted with 0.025 M phosphate buffer (pH 7.0), and 3.0-ml fractions are collected. Fractions with activity greater than the starting material are pooled, precipitated at 80% saturation with ammonium sulfate, and recovered by centrifugation.

Step 8. Second DEAE-Cellulose Chromatography. The dialyzed enzyme (5.0 ml containing 13.1 mg of protein) is adsorbed on a column of DEAE-cellulose (1.6 × 10.0 cm) and eluted with a 0–0.1 M NaCl linear gradient in 0.025 M phosphate buffer (pH 7.4). Fractions of 3.0 ml are collected, and fractions with specific activity greater than the starting material are pooled and precipitated at 80% saturation with ammonium sulfate. The protein is removed by centrifugation and redissolved.

Step 9. Selective Precipitation with Saturated Ammonium Sulfate Solution. The enzyme solution (2.0 ml containing 5.1 mg of protein) is brought to a slight turbidity with saturated ammonium sulfate (pH 7.0) and centrifuged. The clear supernatant solution is then brought to a definite turbidity with saturated ammonium sulfate solution and allowed to stand overnight at 4°. The precipitate is recovered by centrifugation. The resulting homogeneous preparation of the enzyme (2.0 mg) has a specific activity of 47 μmole/min/mg of protein at 4100-fold purification.

Alternate Step 9 Preparative Slab Gel Polyacrylamide Electrophoresis. The active fractions, after the second DEAE-cellulose chromatography, are pooled, dialyzed against 0.01 M phosphate buffer (pH 7.5) containing 10 mM 2-mercaptoethanol, and concentrated on a Millipore 90-mm Hi-Flux Cell. The concentrated enzyme (2 ml containing up to 5 mg protein) is layered on top of a 7.5% acrylamide gel in a water-cooled Bio-Rad 220 Slab Gel Preparative Electrophoresis unit. The proteins migrate in a constant current of 30 mA and are eluted from the gel into a fraction collector with the Tris-glycine running buffer at pH 8.6. A homogeneous protein is obtained, showing a single band in polyacrylam-

PURIFICATION OF PURINE NUCLEOSIDE PHOSPHORYLASE FROM RABBIT LIVER

Purification step	Total protein (mg)	Specific activity[a] (μmole/ min/mg)	Total activity (%)	Purification (fold)
1. Supernatant after centrifugation of homogenate	311,150	0.01	100	1
2. First ammonium sulfate fractionation	63,400	0.05	95.6	4.5
3. First DEAE-cellulose chromatography	5,500	0.35	59.4	33
4. Second ammonium sulfate fractionation	2,000	0.60	37.5	57
5. Chromatography on hydroxylapatite	74	7.45	16.8	705
6. Heat treatment 55° for 15 min	35	13.20	12.7	1250
7. Filtration on Sephadex G-150	13.1	27.80	11.1	2631
8. Second DEAE-cellulose chromatography	5.1	31.00	5.7	2954
9. Selective precipitation with saturated ammonium sulfate	2.0	47.12	3.6	4100

[a] Assayed by method 1 with guanosine as substrate.

ide disc electrophoresis. Step 7, filtration on Sephadex G-150, may be omitted if alternate step 9 is followed.

A summary of the purification procedure is presented in the table.

Properties

Purity. The enzyme is shown to be homogeneous by polyacrylamide disc gel electrophoresis, both in the presence and absence of sodium dodecyl sulfate and by isoelectric focusing. The enzyme forms a single band at a pI of 5.1.

pH Activity. A broad pH activity profile is observed, with an almost flat maximum between pH 6.0 and pH 8.0. A small peak of optimal activity is obtained at pH 6.25. The plots of pK_m versus pH and log V_{max} versus pH, as outlined by Dixon,[15] showed ionizations both in the free enzyme and in the enzyme–substrate complex, with pK_a values near pH 5.5 and pH 8.5, suggestive of the presence of histidine and cysteine at the active site.

Specificity and Kinetic Constants. Guanosine (K_m 5.00×10^{-5} M), deoxyguanosine (K_m 1.00×10^{-4} M), and inosine (K_m 1.33×10^{-4} M) are substrates for enzymic phosphorolysis. Inorganic orthophosphate (K_m 1.54×10^{-2} M) is an obligatory anion requirement, with arsenate

[15] M. Dixon, *Biochem. J.* **55**, 161 (1953).

substituting for phosphate with comparable results. Xanthosine is an extremely poor substrate, and adenosine is not phosphorylyzed at 20-fold excess of the homogeneous enzyme.

Inhibitors. Product inhibition is observed with guanine (K_i 1.25 × 10^{-5} M) and hypoxanthine (2.5 × 10^{-5} M) for the enzymic phosphorolysis of guanosine. Ribose 1-phosphate, another product of the reaction, gave noncompetitive inhibition (K_i 3.61 × 10^{-4} M) with guanosine as the variable substrate. Noncompetitive inhibition was observed with p-chloromercuribenzoate (K_i 5.68 × 10^{-6} M). The inactivated enzyme was completely reactivated upon the addition of an excess of 2-mercaptoethanol.

Photooxidation. The enzyme is highly susceptible to photooxidation in the presence of methylene blue. A pH dependence of photoinactivation is observed with near maximal photoinactivation obtained near pH 8.5.

Stability. The enzyme shows no significant changes in catalytic property when incubated in phosphate buffer between pH 5.5 and pH 9.0 for 10 min at 40° and assayed at pH 7.0. The stability of the enzyme at relatively higher temperatures is an asset to its purification. The enzyme is stabilized by 2-mercaptoethanol during the purification.

Molecular Weight. Recent studies show a molecular weight value of 30,500 by sodium dodecyl sulfate electrophoresis and 61,000 by Sephadex G-100 gel filtration. Cross-linking studies with dimethyl suberimidate show two bands indicating a dimer.[16]

[16] T. Treuman and M. D. Glantz, *Fed. Proc., Fed. Am. Soc. Exp. Biol.* 873, Abstr. No. 3176 (1977).

[72] Purine Nucleoside Phosphorylase[1] from Human Erythrocytes

By J. D. STOECKLER, R. P. AGARWAL, K. C. AGARWAL, and R. E. PARKS, JR.

Inosine (deoxyinosine) + P_i ⇌ hypoxanthine

+ ribose-1-phosphate (deoxyribose-1-phosphate)

Guanosine (deoxyguanosine) + P_i ⇌ guanine

+ ribose-1-phosphate (deoxyribose-1-phosphate)

[1] Purine nucleoside:orthophosphate ribosyltransferase, EC 2.4.2.1.

Purine nucleoside phosphorylase (PNPase) has been crystallized[2] from human erythrocytes where it is present in relatively high concentration (about 13 units/ml packed cells).[3] PNPase is essential for the reutilization of purine ribo- and deoxyribonucleosides in these cells which lack the *de novo* pathway for nucleotide synthesis. The enzyme has been found deficient in erythrocytes and lymphocytes of individuals who also lack T-lymphocyte functions.[4] These subjects have hyperactive *de novo* purine biosynthesis, hypouricemia, and hypouricosuria but markedly elevated blood and urine concentrations of nucleosides of guanine and hypoxanthine.[5]

Assay Method[6]

Principle. Hypoxanthine formed in the reaction is oxidized to uric acid by xanthine oxidase. The uric acid is measured by following the increase in absorbancy at 293 nm.[7]

Reagents[8]

Potassium phosphate buffer, 0.5 M, pH 7.5
Inosine, 5 mM
Xanthine oxidase from milk (commercially available ammonium sulfate ppt.), 0.2 units/ml water
Solutions containing PNPase activity are diluted in Tris or phosphate buffer, pH 7.5, so that an aliquot gives a 0.005–0.040 absorbancy change per minute.

Procedure. In a 1.5-ml quartz cuvette with a 1-cm light path, 0.1 ml each of the buffer, inosine, and xanthine oxidase are mixed with water to allow a final volume of 1.0 ml after enzyme addition. The reaction mixture is preincubated for several minutes at 30° to permit oxidation of any hypoxanthine or xanthine that might be present as contaminants in inosine. The reaction is initiated by the addition of PNPase, and activity

[2] R. P. Agarwal and R. E. Parks, Jr., *J. Biol. Chem.* **244**, 644 (1969).
[3] R. E. Parks, Jr. and R. P. Agarwal, *in* "The Enzymes" (P. D. Boyer, ed.), 3rd ed., Vol. 7, p. 483. Academic Press, New York, 1971.
[4] E. R. Giblett, A. J. Amman, R. Sandman, D. W. Wara, and L. K. Diamond, *Lancet* **1**, 1010 (1975).
[5] A. Cohen, D. Doyle, D. W. Martin, Jr., and A. J. Amman, *N. Engl. J. Med.* **295**, 1449 (1976).
[6] B. K. Kim, S. Cha, and R. E. Parks, Jr., *J. Biol. Chem.* **243**, 1763 (1968).
[7] H. M. Kalckar, *J. Biol. Chem.* **167**, 429 (1947).
[8] Phosphate and inosine stock solutions may be combined and are stable for many months if stored frozen. Diluted xanthine oxidase may be used for 1–2 days if refrigerated.

is calculated from the increase in absorbancy at 293 nm (ϵ_M = 1.25 $\times$ 10⁴).

Definition of Unit and Specific Activity. One unit of PNPase phosphorolyzes 1 μmole of inosine per minute under standard assay conditions. Specific activity is expressed as units per milligram of protein.

The protein concentration is determined from absorbancy at 280 nm during purification.[9]

Alternate Assay Procedures. (1) If spectrophotometric measurements at 293 nm are inconvenient or insensitive, e.g., in the presence of substances with high absorbancies near 293 nm, inosine may be replaced as the substrate by 6-thioinosine. The formation of 6-thiouric acid may be followed at 348 nm (ϵ_M = 2.45 $\times$ 10⁴).[10] (2) Direct spectrophotometric assay of either phosphorolysis or synthesis is possible with substrates that show significant absorbancy differences between the nucleoside and base (e.g., guanosine $\rightleftharpoons$ guanine, ϵ_M = 5.1 $\times$ 10³ at 252 nm, pH 7.5).[11] (3) A radioisotope assay has been described that is capable of detecting as little as 0.1 nmole product in either direction.[12] (4) PNPase activity may be demonstrated qualitatively in electrophoretic gels[13] and in spots of whole blood on DEAE-cellulose paper[14] by use of a tetrazolium dye. Various assay procedures are discussed in detail elsewhere.[3]

Purification Procedure[15]

All steps are carried out at 0–5°. The procedure described is for 100 ml of packed cells from fresh or stored heparinized or ACD human blood, but it may be scaled up or down if column sizes are adjusted accordingly.

Step 1. Preparation of Hemolysate. This step is described elsewhere in this volume.[16]

[9] O. Warburg and W. Christian, *Biochem. Z.* **310**, 384 (1941); see this series, Vol. III [73].

[10] M. R. Sheen, B. K. Kim, and R. E. Parks, Jr., *Mol. Pharmacol.* **4**, 293 (1968).

[11] A. F. Ross, K. C. Agarwal, S. H. Chu, and R. E. Parks, Jr., *Biochem. Pharmacol.* **22**, 141 (1973).

[12] G. Milman, D. L. Anton, and J. L. Weber, *Biochemistry* **15**, 4967 (1976).

[13] Y. H. Edwards, D. A. Hopkinson, and H. Harris, *Ann. Hum. Genet.* **34**, 395 (1971).

[14] M. Ansay, V. Baldewijns-Rouma, and J. E. Smith, *Anim. Blood Grps. Biochem. Genet.* **6**, 249 (1975).

[15] This method is basically similar to that reported earlier.[2,6] The initial purification steps were adapted from K. K. Tsuboi and P. B. Hudson, *J. Biol. Chem.* **224**, 879 (1957) and F. M. Huennekens, E. Nurk, and B. W. Gabrio, *ibid.* **221**, 971 (1956). Also see V. Zannis, D. Doyle, and D. W. Martin, *J. Biol. Chem.* **253**, 504 (1978) for a new affinity chromatographic method.

[16] R. P. Agarwal, K. C. Agarwal, and R. E. Parks, Jr., this volume [79].

Step 2. Adsorption on Calcium Phosphate Gel. The enzyme is adsorbed on calcium phosphate gel as described elsewhere in this volume.[16] In the procedure presented here, the enzyme is desorbed from the gel by eluting twice with 100 ml of 0.1 M potassium phosphate buffer, pH 7.5.[17]

Step 3. Ammonium Sulfate Fractionation. The eluate from step 2 is brought to 40% saturation by the gradual addition of solid ammonium sulfate (23.1 g/100 ml of initial volume) and stirred for at least 30 min. The precipitate is removed by centrifugation at 9000 g for 20 min. The supernatant fluid is brought to 60% saturation by slow addition of solid ammonium sulfate (12.3 g/100 ml of supernatant fluid). After about 1 hr, the suspension is centrifuged at 9000 g for 40 min, and the precipitate is dissolved in about 15 ml of 0.1 M Tris-acetate, pH 7.5. The solution is dialyzed in pretreated dialysis casings[18] against several changes of 4 liters of 0.03 M Tris-acetate, pH 7.5, for 12–16 hr.

Step 4. DEAE-Cellulose Column Chromatography. A 2.5 × 25 cm column of DEAE-cellulose (acetate form)[19] is equilibrated with 0.03 M Tris-acetate, pH 7.5, containing 1 mM dithiothreitol.[20] The dialyzed enzyme is adsorbed on the column, washed with about 50 ml of the same buffer, and eluted with a linear gradient (0.03–0.35 M; total volume, 600 ml) of Tris-acetate, pH 7.5. About 95% of the enzyme emerges in the range of 0.06–0.2 M Tris-acetate, and the fractions containing enzyme activity are pooled.

[17] Elution with 0.1 M phosphate, in contrast to the 20% ammonium sulfate eluant used by R. P. Agarwal *et al.*[16] permits the separation of purine nucleoside phosphorylase from many other proteins, e.g., nucleoside diphosphokinase, which desorb at higher phosphate concentrations. The procedure outlined in this chapter is specifically for the purification of purine nucleoside phosphorylase; however, if large quantities of erythrocytes are employed or purification of other enzymes is desired, it may be preferable to use the general method[16] where desorption with ammonium sulfate is followed by precipitation at 70% saturation and calcium phosphate gel chromatography. Fraction 5 of the general method,[16] which contains purine nucleoside phosphorylase at a specific activity of 2–5 (approximately equivalent to enzyme from step 3 above), may be dialyzed against 0.03 M Tris-acetate, pH 7.5, and purified further starting at step 4 above.

[18] P. McPhie, this series, Vol. 22, p. 25.

[19] DEAE-cellulose is washed free of fines by decantation with distilled water and converted to the -OH form by gentle stirring with 0.5 M NaOH for 1 hr followed by adequate rinsing with distilled water. The fluids may be removed by decantation, or more rapidly, by filtration on a Büchner funnel. The fibers are treated with 1 M sodium acetate until the filtrate pH is 7.5–8.0 and then equilibrated with 0.03 M Tris-acetate, pH 7.5.

[20] From this step onward, all solutions used in the purification should contain 5–10 mM mercaptoethanol or 1 mM dithiothreitol. It should be noted that the oxidized form of dithiothreitol may interfere with the measurement of absorbancy at 280 nm.

Step 5. Calcium Phosphate Gel–Cellulose Column Chromatography.
Calcium phosphate gel–cellulose is prepared as described elsewhere[16]
and poured into a column (2.5 × 25 cm). The column is equilibrated with
0.1 M Tris-acetate, pH 7.5. The pooled enzyme from step 4 is loaded
and eluted with a linear gradient of potassium phosphate (0–0.1 M in 0.1
M Tris-acetate, pH 7.5; total volume, 500 ml). Ten-milliliter fractions are
collected. The enzyme, which follows the colored proteins, emerges at
the phosphate concentration range of 0.05–0.07 M. The pooled enzyme
(almost colorless) is concentrated by addition of solid ammonium sulfate
to 65% saturation (40.7 g/100 ml). The precipitate is sedimented by
centrifugation at 9000 g for 40 min and then dissolved in 1–2 ml of 0.1 M
Tris-acetate buffer, pH 7.5.

Step 6. Gel Filtration. The enzyme solution from step 5 is applied to
a Sephadex G-100 column (1.9 × 50 cm). Either 0.1 M Tris-acetate or
0.05 M potassium phosphate buffer, pH 7.5, is employed to equilibrate
the column and filter the enzyme. The pooled fractions containing the
enzymic activity are concentrated by precipitation with ammonium
sulfate at 80% saturation (39.5 g/100 ml).

To achieve a specific activity of 80–95, it may be necessary to repeat
steps 5 or 6.

Step 7. Crystallization.[2] The enzyme solution from the last chromato-

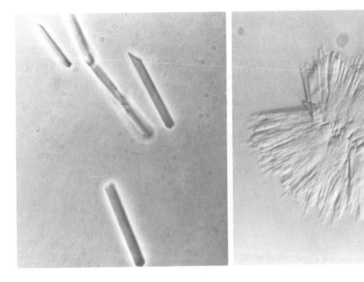

Fig. 1. Erythrocytic purine nucleoside phosphorylase.

PURIFICATION OF HUMAN ERYTHROCYTIC PURINE NUCLEOSIDE PHOSPHORYLASE

Purification steps	Total activity (units)	Specific activity	Purification	Recovery (%)
1. Hemolysate	14,500	0.01	1	100
2. Calcium phosphate gel adsorption	14,820	0.35	35	102
3. Ammonium sulfate fractionation	14,520	2.23	223	100
4. DEAE-cellulose chromatography	10,780	9.19	919	74
5. Calcium phosphate gel–cellulose chromatography	6,670	48.3	4833	46
6. Gel filtration[a]	5,120	69.6	6962	35
Repetition of steps 5 and 6, concentration and lyophilization[b]	2,750	93.1	9310	19
7. Crystallization Recrystallization	1,340	93.0	9300	9[a]

[a] If seed crystals are available, crystallization is possible after step 6 with an overall recovery of 15–20%.

[b] The enzyme was concentrated and dialyzed by DIAFLO ultrafiltration (Model 50, Amicon Corp., Lexington, Massachusetts) under 40 psi of N_2 pressure; however, other techniques that do not increase the salt concentration may be employed. A Collodion Bag Apparatus (Schleicher and Schuell, Inc., Keene, New Hampshire) is preferable for smaller volumes.

graphic step is concentrated in a Collodion Bag Apparatus (Schleicher & Schuell, Inc., Keene, New Hampshire) to a protein concentration of about 10 mg/ml. Alternatively, the enzyme may be concentrated by precipitation at 70% ammonium sulfate saturation. The precipitate is dissolved in 30 mM Tris-acetate buffer, pH 7.5, to give a protein concentration of about 10 mg/ml. The enzyme solution is adjusted very slowly and with continuous mixing to about 35% saturation with saturated ammonium sulfate (recrystallized) solution. The turbid mixture is centrifuged at 9000 g for 30 min to remove the amorphous precipitate. Saturated ammonium sulfate is added to the supernatant fluid until slight turbidity develops (at about 40% saturation). Upon overnight storage at 4°, about 50% of the enzyme crystallizes as fine needles and bundles of needles (Fig. 1).

The crystals are harvested by centrifugation at 9000 g for 10 min. A few crystals are reserved for seeding, and the remainder are dissolved in 30 mM Tris-acetate, pH 7.5, and brought to 40% saturation with saturated ammonium sulfate. Seed crystals are added, and 95% of the enzyme crystallizes within 12 hr. The table summarizes the purification procedure from approximately 2 liters of packed erythrocytes.

Properties

Stability. The crude enzyme is stable for several days, even at room temperature, in concentrated solutions, e.g. 1% in 0.02% sodium azide. Purified enzyme is stable as an ammonium sulfate precipitate at 4° or frozen for long periods of time.[21] Enzyme preparations of high specific activity (70–96) can decrease in activity rapidly in the absence of sulfhydryl compounds, e.g., dithiothreitol, 1 mM. Inactivation by the sulfhydryl reagents, PCMB and 5,5'-dithio(2-nitrobenzoic acid), can be reversed by dithiothreitol.[6] Freezing the enzyme in the presence of dithiothreitol results in greater loss of activity than in the absence of a sulfhydryl reagent.[21] Homogeneous enzyme loses about 50% of its activity upon incubation at 57° for 15 min and is also rapidly inactivated at pH values below 6.2 and above 10.0.[2]

Molecular Properties. Multiple peaks of enzymic activity with pI values from 5.85–6.25 are detectable after column isoelectric focusing of the crystalline enzyme (Fig. 2).[22] Starch gel electrophoresis of hemolysates of washed erythrocytes also reveals multiple bands of enzymic activity, and the presence of the more acidic variants is related to senescence of the cells.[23]

Molecular weights have been estimated at 80,000–92,000, based on gel filtration and sedimentation equilibrium analysis.[2,24,25] Other physical parameters are: Stokes radius, 38 Å (gel filtration); $s_{20,w}$, 5.4 and 5.5 (from sedimentation velocity and sucrose density gradient centrifugation, respectively); $D_{20,w}$, 5.7 × 10^{-7} cm^2 sec^{-1}; frictional ratio, 1.29; partial specific volume, 0.73 cm^3 g^{-1}.[25]

Three moles of hypoxanthine bind per mole of enzyme,[2] and a single protein band of molecular weight 30,000 ± 500 is observed on SDS gel electrophoresis.[25] Therefore the enzyme appears to be a homologous trimer.

The CD spectrum of the enzyme indicates approximately 65% random coil structure and a very low α-helix content. Tryptophan, cysteine, methionine, histidine, and isoleucine are present in lowest concentration. $E_{1cm}^{1\%} = 9.6$ at 280 nm.[25]

Catalytic Properties. The enzyme catalyzes the reversible phospho-

---

[21] Our unpublished observations.

[22] K. C. Agarwal, R. P. Agarwal, J. D. Stoeckler, and R. E. Parks, Jr., *Biochemistry* **14**, 79 (1975).

[23] B. M. Turner, R. A. Fisher, and H. Harris, *Eur. J. Biochem.* **24**, 288 (1971).

[24] Y. H. Edwards, P. A. Edwards, and D. A. Hopkinson, *FEBS Lett.* **32**, 235 (1973).

[25] J. D. Stoeckler, R. P. Agarwal, K. C. Agarwal, K. Schmid, and R. E. Parks, Jr., *Biochemistry* **17**, 278 (1978).

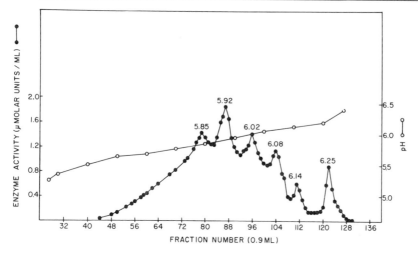

FIG. 2.

rolysis of the ribo- and deoxyribonucleosides of guanine, hypoxanthine, and xanthine. The K_m values of some substrates are: hypoxanthine, $1.9 \times 10^{-5} M$;[26] inosine, $4.8 \times 10^{-5} M$;[21] deoxyinosine, $6.6 \times 10^{-5} M$;[6] guanosine, $4.7 \times 10^{-5} M$;[21] P_i, $3.2 \times 10^{-5} M$.[6] Compared to hypoxanthine, adenine is a very poor substrate (V_{max}, 0.6%; K_m, $4.1 \times 10^{-4} M$).[26] Formycin B inhibits the enzyme with a K_i value of $1 \times 10^{-4} M$.[10] Many other analogs have been tested for substrate or inhibitory activity.[27]

The reaction follows an ordered bi-bi mechanism with nucleoside being the first substrate to add and base the last product to leave.[6] No cofactors or metal requirements are known. No evidence has been found of a ribosylated or phosphorylated intermediate, and the existence of a ribosyl transfer reaction from nucleoside to base in the absence of P_i remains doubtful.[2,6]

[26] T. P. Zimmerman, N. Gersten, A. F. Ross, and R. P. Miech, *Can. J. Biochem.* **49**, 1050 (1971).

[27] 6-Mercaptopurine ribonucleoside, 6-selenoguanosine, 6-thioguanosine, and 8-azaguanosine are readily phosphorolyzed.[10,11] Allopurinol is ribosylated [T. A. Krenitsky, G. B. Elion, R. A. Strelitz, and G. H. Hitchings, *J. Biol. Chem.* **242**, 2675 (1967)]. Arabinosylhypoxanthine, 3'-aminoinosine, N^2-methylguanosine, and 3-deazaguanosine have low substrate activities.[21] 7-Deazainosine and 7-deazathioinosine are competitive inhibitors [M. R. Sheen, H. F. Martin, and R. E. Parks, Jr., *Mol. Pharmacol.* **6**, 255 (1970)] as are also 2,6-diaminopurine and numerous methyl and methoxy purine derivatives [T. A. Krenitsky, G. B. Elion, A. M. Henderson, and G. H. Hitchings, *J. Biol. Chem.* **243**, 2876 (1968)]. Arsenate is a competitive inhibitor of phosphate[6]; contrary to an earlier report [F. M. Huennekens, E. Nurk, and B. W. Gabrio, *J. Biol. Chem.* **221**, 971 (1956)], pyrophosphate has no effect on the reaction.[21]

Activation of enzymic activity is observed at high concentrations of nucleoside substrates. After electrophoretic fractionation of the enzyme, it is seen that activation is more pronounced in the acidic variants than in the basic variants.[22,23] Treatment with 5,5'-dithiobis(2-nitrobenzoic acid) results in a 60% decrease in activity and loss of substrate activation. Both the activity and the phenomenon of substrate activation are fully recovered after dithiothreitol treatment.[28]

Sulfhydryl groups play a key role in the enzymic behavior. With labeled PCMB it is possible to titrate 12 -SH groups per mole of enzyme. Activity is lost after the first 3–4 -SH are reacted but may be fully restored with dithiothreitol. As noted above, 5,5'-dithiobis(2-nitrobenzoic acid) reacts with only 2–4 sulfhydryl groups causing partial loss of activity; however, in the presence of SDS, approximately 12 -SH groups are titratable.[28] The effect of pH on kinetic parameters suggests an essential role for cysteine and histidine in catalysis.[2]

At pH values below 6 and above 8, substrate activation and inhibition, respectively, are observed with inorganic phosphate.[2]

The synthetic reaction is favored with a K_{eq} value of 54 for both inosine and deoxyinosine at pH 7.4, 24°.[29,30]

---

[28] R. P. Agarwal and R. E. Parks, Jr., *J. Biol. Chem.* **246**, 3763 (1971).
[29] H. M. Kalckar, *J. Biol. Chem.* **167**, 477 (1947).
[30] M. Friedkin, *J. Biol. Chem.* **184**, 449 (1950).

[73] Chinese Hamster Purine Nucleoside Phosphorylase[1]

By Gregory Milman

$$\text{(Hypoxanthine or guanine)} + \text{ribose-1-P} \underset{\text{"breakdown"}}{\overset{\text{"synthesis"}}{\rightleftharpoons}} \text{(inosine or guanosine)} + P_i$$

Assay Method

Principle. Purine nucleoside phosphorylase (EC 2.4.2.1; purine nucleoside:orthophosphate ribosyltransferase) in eukaryotes catalyzes the reversible conversion between the purine bases, hypoxanthine and guanine, and their corresponding nucleosides, inosine and guanosine.

Nucleoside synthesis assay. The formation of inosine or guanosine is measured in a radioisotope assay in which ^{14}C-labeled purine base

---

[1] Abbreviations used are: P and P_i, phosphate and inorganic phosphate, respectively; Tricine, *N*-Tris(hydroxmethyl)methylglycine.

is converted to labeled nucleoside. In 1 N NH$_4$OH, the labeled nucleoside passes through Cu^{2+} Chelex while the labeled purine base is bound.

NUCLEOSIDE PHOSPHORYLASE ASSAY. The conversion of nucleoside to purine base and ribose 1-phosphate is measured by the formation of [^{14}C]ribose 1-phosphate from U-^{14}C-labeled nucleoside. In water, the unreacted nucleoside and purine base produced during the enzyme reaction bind to Cu^{2+} Chelex, while the [^{14}C]ribose 1-P is eluted.

Reagents

NH$_4$OH, 1 N and 5 N

Cu^{2+} Chelex columns: Typically, 250 g of Chelex 100 (Bio-Rad) are allowed to stand overnight in 2 liters of 0.5 M CuCl$_2$, and then are washed with 10 liters of H$_2$O and 1 liter of 1 N NH$_4$OH and stored in 1 N NH$_4$OH. This preparation is used for the nucleoside synthesis assay. For the nucleoside phosphorylase assay, the Cu^{2+} Chelex is rinsed 10 times, each time with 2 liters of H$_2$O and stored in H$_2$O. Approximately 1–1.5 ml Cu^{2+} Chelex is placed into Pasteur pipet columns (3.5 × 0.5 cm) containing a glass-fiber plug. We usually use Cu^{2+} Chelex columns only once. However, they can be regenerated, first by washing with 25 ml of 5 N NH$_4$OH, and then with either 12 ml of 1 N NH$_4$OH for the nucleoside synthesis assay or with 25 ml of H$_2$O for the nucleoside phosphorylase assay. A convenient reservoir is a plastic syringe attached to the top of the column with a short piece of rubber tubing.

Enzyme buffer: Used to dilute enzyme samples, it consists of 20 mM Tris·HCl (pH 7.8), 6 mM MgCl$_2$, 100 mM KCl, 0.1 mM EDTA, and 0.5 mM dithiotreitol.

Assay buffer: Consists of 500 mM Tris·HCl (pH 7.8), 50 mM MgCl$_2$, 20 mM dithiotreitol

Ribose 1-phosphate, 11.5 mM

Guanine (or hypoxanthine), 0.8 mM

Guanosine (or inosine), 0.8 mM

Potassium phosphate (pH 7.8), 100 mM

[8-^{14}C]guanine (or [8-^{14}C]hypoxanthine), specific activity approximately 50 mCi/mmole. Background cpm in the absence of enzyme of less than 0.5% of the initial cpm in the assay are obtained by purifying the ^{14}C-labeled purine bases from the suppliers prior to use. ^{14}C-labeled purine base is absorbed to a standard Cu^{+2} Chelex column in 1 N NH$_4$OH, washed with 5 ml 1 N NH$_4$OH, eluted with 5 N NH$_4$OH, and lyophilized.

[U-^{14}C]guanosine (or [U-^{14}C]inosine), specific activity approximately 200–500 mCi/mmole.

Nucleoside synthesis assay reagent: An assay reagent for 500 assays is prepared by combining 5 ml assay buffer, 5 ml ribose 1-phosphate, 5 ml guanine (or hypoxanthine), approximately 5×10^7 cpm of [8-^{14}C]guanine (or hypoxanthine), and H_2O in a final volume of 20 ml.

Nucleoside phosphorylase assay reagent: An assay reagent for 500 assays is prepared by combining 5 ml assay buffer, 5 ml potassium phosphate, 5 ml guanosine (or inosine), approximately 2×10^7 cpm of [U-^{14}C]guanosine (or inosine), and H_2O in a final volume of 1 ml. The assay reagents are frozen in 1-ml portions (25 assays) and are stable for months in a liquid nitrogen freezer.

Scintillation fluid: Consists of 9.1 g of 2-5-diphenyloxazole (PPO), 0.61 g of 1,4-bis[2-(5-phenyloxazolyl)]benzene (POPOP), 2140 ml of toluene, and 1250 ml of Triton X-100.

Procedure

NUCLEOSIDE SYNTHESIS ASSAY. A reaction is initiated by adding 40 μl of nucleoside synthesis assay reagent to 60 μl of enzyme sample in enzyme buffer in a 10×75 mm test tube on ice, mixing for approximately 1 sec on a Vortex mixer, and placing the reaction mixture in a 37° water bath. In some cases 10 μl of 1 mg/ml bovine serum albumin are included in the assay to improve the stability of the more purified enzyme samples. After 15 min, the incubation is terminated by adding 150 μl of 1 N NH_4OH, mixing well, and placing the mixture on ice. A 200-μl sample of each reaction is applied to a Cu^{2+} Chelex column equilibrated in 1 N NH_4OH. Nucleoside is eluted with 1.5 ml of 1 N NH_4OH, collected directly into a scintillation vial, and counted in 10 ml of scintillation fluid.

NUCLEOSIDE PHOSPHORYLASE ASSAY. A reaction is initiated by adding 40 μl of nucleoside phosphorylase assay reagent to 60 μl of enzyme sample in enzyme buffer. After 15 min, the incubation is terminated by adding 150 μl of H_2O and immediately placing the mixture in an 80° water bath for 10 min. A 200-μl sample of each cooled reaction mixture is applied to a Cu^{2+} Chelex column equilibrated in H_2O. Ribose 1-phosphate is eluted with 1.5 ml of H_2O, collected directly into scintillation vials, and counted in 10 ml of scintillation fluid.

Definition of Unit and Specific Activity. A unit is defined as the amount of enzyme which catalyzed the formation of 1 μmole of guanosine per minute at 37° in the nucleoside synthesis assay. The number of nanomoles of substrate in a standard assay is determined from the amount of unlabeled and radioisotope-labeled purine base. The specific activity of the purine base is determined from the number of cpm in an assay counted in the same manner as the assay samples. The nanomoles of product formed in the reaction is 1.25 times the product eluted from the Cu^{2+} Chelex columns. Specific activity is expressed as units of enzyme per milligram of protein determined by the method of Lowry et al.[2] with crystalline bovine serum albumin as a standard.

Purification Procedure

Purification from Chinese Hamster V79 Tissue Culture Cells. All procedures are performed at 0–5° unless otherwise specified. All enzyme fractions are stored in a liquid-nitrogen freezer. Enzyme activity is measured by the formation of guanosine from guanine and ribose 1-phosphate in the nucleoside synthesis assay described above. A summary of a sample purification is presented in the table.

Crude Extract. Chinese hamster V79 cells[3] are grown in suspension to a density of 0.5–1 × 10⁶/ml in Joklik's minimum essential media supplemented with nonessential amino acids, 1 mM sodium pyruvate, 50 mM Tricine (pH 7.6), and 5% fetal bovine serum. The cells are harvested by centrifugation and resuspended in 2 volumes of enzyme buffer (assay reagent). The cells are lysed by freezing and stored at −70°. The lysate from 25 ml of packed cells is thawed and centrifuged at 20,000 *g* for 20 min, and the supernatant fluid is saved. The pellet is resuspended in an additional 10 ml of the same buffer and centrifuged as

PURIFICATION OF CHINESE HAMSTER V79 TISSUE CULTURE CELL PURINE NUCLEOSIDE PHOSPHORYLASE

Fraction	Volume (ml)	Protein (mg)	Specific activity (units/mg)	Cumulative recovery (%)	Purification (fold)
Crude extract	76	1626	0.14	100	1
High-speed supernatant	73	920	0.16	65	1.1
DEAE-Sephadex	2.0	39	1.59	27	11
G-100 Sephadex	1.4	5.3	11.8[a]	27	84
Isoelectric focusing	6.0	0.5[b]	60.0[a]	13	430

[a] Assay includes 0.1 mg/ml of bovine serum albumin.
[b] Protein determined from intensity of enzyme band on sodium dodecyl sulfate–polyacrylamide gel in comparison with protein standards.

[2] O. H. Lowry, N. J. Rosebrough, A. L. Farr, and R. J. Randall, *J. Biol. Chem.* **193**, 265 (1951).
[3] D. K. Ford and G. Yerganion, *J. Natl. Cancer Inst.* **21**, (1958).

described above. The two supernatant fluids are combined to form the crude extract.

High-Speed Supernatant Fraction. The crude extract is centrifuged for 2.5 hr at 60,000 g in a Spinco 40 rotor. The supernatant fluid forms the high-speed supernatant fraction.

DEAE-Sephadex Fraction. The high-speed supernatant fraction is applied to a column (0.9 × 20 cm) containing DEAE-Sephadex A-50 equilibrated in enzyme buffer. The column is washed with 40 ml of the same buffer, and then the enzyme is eluted in a 400-ml linear KCl gradient (100–250 mM) in enzyme buffer. Fractions of 4 ml are collected and assayed for enzyme activity which elutes in an irregular peak from 140–180 mM KCl. Only the most active fractions containing peak enzyme activity, approximately 80 ml, are pooled and concentrated to about 2 ml using an Amicon ultrafiltration cell with a PM-10 membrane.

G-100 Sephadex Fraction. DEAE-Sephadex fraction enzyme is applied to a column (1.5 × 90 cm) containing G-100 Sephadex equilibrated in enzyme buffer. The column is developed with the same buffer, and 2-ml fractions are collected and assayed for enzyme activity. Fractions containing enzyme activity (approximately 16 ml) are pooled and concentrated to about 1.5 ml using an Amicon ultrafiltration cell with a PM-10 membrane.

Isoelectric Focusing Fraction. A 110-ml, 0–46% linear sucrose gradient with 1% pH 5–7 ampholytes and 2 mM dithiothreitol is prepared manually in an LKB Model 8180 ampholine column following the procedure in the LKB manual. G-100 Sephadex fraction enzyme is applied to the middle of the gradient replacing the "light" solution in that region. Electrofocusing is conducted for 3–5 days at 3–5°. During that time, the voltage is increased from 300 to 800 V. One-milliliter fractions are collected, and the pH and enzyme activity are measured. Fractions containing peak enzyme activity (2–3 fractions) are pooled.

Properties[4]

Chinese hamster purine nucleoside phosphorylase from both V79 tissue culture cells and Chinese hamster organs appears to have identical structural and catalytic properties. At 37° the purified enzyme converts 60 μmoles of guanine to guanosine per minute per milligram of protein. Electrophoresis in sodium dodecyl sulfate–polyacrylamide gels indicates

[4] Reprinted with permission from G. Milman, D. L. Anton, and J. L. Weber, *Biochemistry* **15**, 4967 (1976). Copyright by the American Chemical Society.

that the enzyme is composed of identical subunits of 30,000 molecular weight. The native enzyme behaves as a mixture of dimers of 68,000 molecular weight and trimers of 89,000 molecular weight during Sephadex G-100 chromatography. Sucrose gradient centrifugation indicates that the enzyme has a sedimentation coefficient of 5.4 S, which corresponds to a molecular weight of 94,000 and suggests a trimer structure. The enzyme displays Michaelis–Menten kinetics with apparent Michaelis constants of 20 μM for both hypoxanthine and guanine, 35 μM for guanosine, 50 μM for inosine, and 200 μM for both ribose 1-P and P_i. During electrofocusing, the enzyme forms a single major band at a pI of 5.25.

[74] Hypoxanthine Phosphoribosyltransferase from Chinese Hamster Brain and Human Erythrocytes[1]

By ANNE S. OLSEN and GREGORY MILMAN

$$\text{Hypoxanthine or guanine} + \text{PRPP} \xrightarrow{\text{Mg}^{2+}} \text{IMP or GMP} + \text{PP}_i$$

Assay Method

Principle. Hypoxanthine phosphoribosyltransferase (IMP: pyrophosphate phosphoribosyltransferase, EC 2.4.2.8) is a salvage enzyme responsible for the conversion of the purine bases hypoxanthine or guanine to the corresponding 5′-ribonucleotides IMP or GMP. The formation of IMP or GMP is measured in a radioisotope assay in which ^{14}C-labeled purine base is converted to the labeled nucleotide. The uncharged purine base is separated from the negatively charged nucleotide on DEAE-cellulose.

Reagents

Enzyme buffer, used to dilute enzyme samples, consisting of 20 mM Tris·HCl (pH 7.8), 20 mM KCl, 6 mM MgCl$_2$, 0.1 mM EDTA, and 0.5 mM dithiothreitol

Assay buffer, consisting of 500 mM Tris·HCl (pH 7.8), 5 mM MgCl$_2$, and 20 mM dithiothreitol

PRPP (sodium salt), 10 mM

Hypoxanthine (or guanine), 0.6 mM; add HCl to dissolve guanine

---

[1] The abbreviations used are: PRPP, 5-phosphoribosyl 1-pyrophosphate; PP$_i$, inorganic pyrophosphate; HPRT, hypoxanthine phosphoribosyltransferase.

METHODS IN ENZYMOLOGY, VOL. LI

[^{14}C]hypoxanthine (or guanine), specific activity approximately 50 mCi/mmole

HPRT assay reagent: An assay reagent for 500 assays is prepared by combining 5 ml assay buffer, 5 ml PRPP, 5 ml hypoxanthine (or guanine), approximately 1.5×10^7 cpm of [^{14}C]hypoxanthine (or guanine), and water in a final volume of 20 ml. The assay reagent is frozen in 1-ml portions (25 assays) and is stable for months in a liquid-nitrogen freezer.

Bovine serum albumin, 1 mg/ml

Tris·HCl buffer, 50 mM, pH 8

HCl, 1 N

DEAE-cellulose (Bio-Rad Cellex-D, capacity 0.7 meq/g, fines removed) suspended in 50 mM Tris·HCl, pH 8

Scintillation fluid, consisting of 9.1 g Omnifluor (New England Nuclear) in 2000 ml of toluene and 1200 ml of Triton X-100

Procedure. A reaction is initiated by adding 40 μl of HPRT assay reagent to 60 μl of enzyme sample in enzyme buffer in a 10 × 75 mm test tube on ice, mixing for approximately 1 sec on a Vortex mixer, and placing the reaction mixture in a 37° water bath. In some cases, 10 μl of 1 mg/ml bovine serum albumin are included to improve the stability of purified enzyme samples. After 15 min, the incubation is terminated by adding 150 μl of ice-cold Tris·HCl buffer, mixing well, and placing the mixture on ice. A 200-μl sample of each reaction is removed and added to a Pasteur pipette column filled with 1 ml of packed DEAE-cellulose. The column is then washed with 5 ml of Tris·HCl buffer. A 5-ml plastic syringe attached to the top of the column with a short piece of rubber tubing serves as a convenient reservoir. The nucleotide is eluted with 1.5 ml of HCl, collected directly into a scintillation vial, and counted with 10 ml of scintillation fluid.

Definition of Unit and Specific Activity. A unit is defined as the amount of enzyme which catalyzes the formation of 1 μmole of IMP per minute at 37°. The number of nanomoles of substrate in a standard assay is determined from the amount of unlabeled and radioisotope-labeled purine base. The specific activity of the purine base is determined from the number of counts per minute in an assay reaction counted in the same manner as the assay samples. The nanomoles of product formed in the reaction is 1.25 times the product eluted from the DEAE-cellulose columns. Specific activity is expressed as units of enzyme activity per milligram of protein determined by the method of Lowry *et al.*,[2] with crystalline bovine serum albumin as a standard.

[2] O. H. Lowry, N. J. Rosebrough, A. L. Farr, and R. J. Randall, *J. Biol. Chem.* 193, 265 (1951).

Purification Procedures

Purification from Chinese Hamster Brain[3]

HPRT is purified from brain because the specific activity of HPRT is higher in brain extracts than in other tissues. All steps are performed at 0–5° unless otherwise specified. Enzyme samples are stored in a liquid-nitrogen freezer. A summary of a sample purification is presented in Table I.

Crude Extract. Brains (15 g) from 24 adult male Chinese hamsters are homogenized in 15 ml of 50 mM Tris·HCl (pH 7.8), 50 mM KCl, 6 mM MgCl$_2$, 0.1 mM EDTA, and 2 mM dithiothreitol. The homogenate is centrifuged at 20,000 g for 20 min, and the supernatant fluid is saved. The pellet is homogenized in an additional 15 ml of the same buffer, centrifuged, and the supernatant fluids are combined.

pH 4.5 Supernate. The crude extract is brought to pH 4.5 by the addition of a 0.2 volume of 1 M sodium acetate, pH 4.5. After 10 min, the precipitate is removed by centrifugation at 20,000 g for 15 min, and the supernatant fluid is adjusted to pH 7.8 by the addition of a 0.2 volume of 2 M Tris·HCl, pH 8.0.

65° Supernate. To the pH 4.5 supernate is added 0.01 volume of 10 mM PRPP. The solution is heated at 65° for 10 min, quickly chilled, and centrifuged at 20,000 g for 15 min.

TABLE I

PURIFICATION OF CHINESE HAMSTER BRAIN HYPOXANTHINE
PHOSPHORIBOSYLTRANSFERASE

Fraction	Volume (ml)	Total protein (mg)	Specific activity (units/mg)	Cumulative recovery (%)	Purification (fold)
Crude extract	24.7	361	0.017	100.0	1.0
pH 4.5 Supernate	34.2	140	0.023	52.7	1.4
65° Supernate	30.2	30.2	0.054	26.7	3.2
40–70% Ammonium sulfate	3.6	12.6	0.232	48.3	14
DEAE-Sephadex A-50	3.3	0.7	1.23[a]	14.8	75
85° Supernate	3.3	0.07[b]	9[a]	10.7	540

[a] Assay included 1 mg/ml bovine serum albumin.
[b] Protein concentration determined from intensity of enzyme band on sodium dodecyl sulfate–polyacrylamide gel.

[3] A. S. Olsen and G. Milman, *J. Biol. Chem.* **249**, 4030 (1974).

40–70% Ammonium Sulfate Fraction. The 65° supernate is brought to 40% saturation in ammonium sulfate by the addition of a 0.67 volume of saturated ammonium sulfate solution. After 20 min, the precipitate is removed by centrifugation, and the supernatant fluid is brought to 70% saturation by the addition of 1 volume of saturated ammonium sulfate. After 20 min, the suspension is centrifuged, and the supernatant fluid is discarded. The precipitate is dissolved in a small volume of enzyme buffer (assay reagent).

DEAE-Sephadex Fraction. The ammonium sulfate fraction is dialyzed against enzyme buffer and applied to a column (0.9 × 20 cm) filled with DEAE-Sephadex A-50 equilibrated with enzyme buffer. The column is washed with 40 ml of enzyme buffer, and the enzyme is eluted with a 200-ml linear KCl gradient (20–150 mM) in enzyme buffer. Fractions of 2 ml are collected and assayed for enzyme activity which elutes in an irregular peak from 30–70 mM KCl. Only the most active fractions containing peak enzyme activity (approximately 16 ml) are pooled and concentrated in an Amicon ultrafiltration cell with a PM-10 membrane.

85° Supernate. PRPP is added to the DEAE-fraction to a concentration of 1 mM, and the enzyme is incubated 15 min at 37° and then heated at 85° for 10 min. The solution is quickly cooled, and then centrifuged for 30 min at 50,000 g to remove denatured contaminants.

Purification from Human Erythrocytes [4]

Although the specific activity of HPRT in human erythrocytes (1.3 munits/mg) is low compared to that in Chinese hamster brain (17 munits/mg), the major contaminant in the erythrocytes is hemoglobin which is easily removed by adsorption of HPRT to DEAE-cellulose, resulting in an 800-fold increase in specific activity. The availability of large quantities of erythrocytes in outdated blood from blood banks makes this an excellent source for purification of milligram quantities of the human enzyme.

All purification steps are performed at 0–5° unless otherwise specified. A summary of a sample purification is given in Table II.

Crude Lysate. Erythrocytes from 10 pints of outdated blood are sedimented by centrifugation at 4000 g for 10 min, and the serum is discarded. The erythrocytes are washed twice with an equal volume of 0.9% NaCl, then frozen and stored at −20°. The lysed cells are thawed at room temperature and diluted 5-fold with enzyme buffer (assay reagent).

[4] Reprinted with permission from A. S. Olsen and G. Milman, *Biochemistry* **16**, 2501 (1976). Copyright by the American Chemical Society.

TABLE II
PURIFICATION OF HUMAN ERYTHROCYTE HYPOXANTHINE PHOSPHORIBOSYLTRANSFERASE

Fraction	Volume (ml)	Protein (mg)	Specific activity (units/mg)	Cumulative recovery (%)	Purification (fold)
Crude lysate	10,125	527,000	.00134	100	1.0
65° Supernate	8,575	394,000	.00125	70	0.9
DEAE-cellulose	105	336	1.11	53	830
40–70% Ammonium sulfate	25	195	2.00	56	1,500
85° Supernate	26	77.5	4.50	49	3,300
DEAE-Sephadex	7.5	12.7	12.5	23	9,300
Second 85° supernate	7.5	6.7	17.5	17	13,000

$65°$ *Supernate*. The crude lysate, in 200-ml portions, is placed in a $65°$ water bath and heated to $62°$ (about 15 min). It is maintained at $62–65°$ for 10 min, quickly cooled, and centrifuged at 4000 g. Although the $65°$ heat step yields no net purification, it greatly improves the purification achieved by the subsequent DEAE-cellulose column chromatography step.

DEAE-Cellulose Fraction. DEAE-cellulose (Bio-Rad Cellex-D, capacity 0.6 meq/g) is washed with 0.25 M NaOH containing 0.25 M NaCl, rinsed with distilled water, then washed with 0.25 M HCl, followed by extensive washing with distilled water. The DEAE-cellulose is then suspended in enzyme buffer, allowed to settle, and the supernatant fluid containing fine DEAE-cellulose particles is decanted. The settling procedure is repeated several times, and the suspension is then poured into a column (6.5 × 45 cm). The $65°$ supernate (about 8.5 liters) is applied to the column which is then washed with 3 liters of enzyme buffer. The enzyme is eluted with a 8 liters of linear KCl gradient (20–150 mM) in enzyme buffer. Fractions containing the HPRT peak (about 2 liters) are pooled and concentrated to approximately 100 ml in an Amicon ultrafiltration cells with a PM-10 membrane.

40–70% Ammonium Sulfate Fraction. The DEAE-cellulose fraction is brought to 40% saturation in ammonium sulfate by addition of solid ammonium sulfate (22.6 g/100 ml). After 30 min, the sample is centrifuged at 12,000 g for 15 min. The 40% supernatant fluid is brought to 70% saturation by addition of solid ammonium sulfate (18.2 g/100 ml). After 1 hr, the suspension is centrifuged for 15 min at 12,000 g. The pellet is resuspended in enzyme buffer and dialyzed for 18 hr against enzyme buffer (3 changes of 1 liter each). Dialysis of the ammonium sulfate fraction prior to the 85° heat step is extremely important, since the enzyme is completely inactivated at 85° if ammonium sulfate is present at greater than 2–5% saturation.

85° Supernate. PRPP in enzyme buffer is added to the dialyzed ammonium sulfate fraction to a final concentration of 1 mM. The sample is placed in a 37° water bath for 15 min and then heated in an 85° water bath for 10 min. The enzyme solution is quickly cooled on ice and centrifuged at 12,000 g for 15 min.

DEAE-Sephadex Fraction. The 85° supernate is applied to a column (2.5 × 40 cm) of DEAE-Sephadex A-50 equilibrated with enzyme buffer. The column is washed with 400 ml of enzyme buffer, and the enzyme is eluted with a 4-liter linear KCl gradient (20–150 mM) in enzyme buffer. Fractions containing the HPRT peak (about 800 ml) are pooled and concentrated to 5–10 ml in an Amicon ultrafiltration cell with a PM-10 membrane.

Second 85° Supernate. To the DEAE-Sephadex fraction is added 0.1 volume of 10 mM PRPP. The sample is placed in a 37° water bath for 15 min and then heated in an 85° water bath for 10 min. The solution is quickly chilled and then centrifuged for 30 min at 41,000 g to remove denatured contaminant proteins.

Properties

The Chinese hamster brain HPRT[3] has a native molecular weight of 78,000–85,000 determined by Sephadex G-100 column chromatography and polyacrylamide gel electrophoresis. The enzyme appears to consist of subunits of molecular weight 25,000 determined by sodium dodecyl sulfate–polyacrylamide gel electrophoresis. Electrofocusing and polyacrylamide gel electrophoresis demonstrate the presence of at least three isozymes at pH 6.2, 6.4, and 6.6. The enzyme is remarkably stable at 85° if first incubated in 1 mM PRPP. The enzyme is active from pH 5.5–11 with maximum activity at pH 10. The enzyme displays Michaelis–Menten kinetics with apparent Michaelis constants for hypoxanthine, guanine, and PRPP of 0.52, 1.1, and 5.3 μM, respectively.

Human HPRT[4,5] has a sedimentation coefficient of 5.9 S, determined by analytical ultracentrifugation, and a molecular weight of 81–83,000, determined by sedimentation equilibrium centrifugation. Sodium dodecyl sulfate–polyacrylamide gel electrophoresis indicates a subunit molecular weight of 26,000. Isoelectric focusing resolves three peaks of

[5] A. S. Olsen and G. Milman, *J. Biol. Chem.* **249**, 4038 (1974).

enzyme activity at pH 5.6, 5.7, and 5.9. The amino acid composition of human HPRT is 17 Lys, 5 His, 12 Arg, 0 Trp, 31 Asx, 12 Thr, 14 Ser, 16 Glx, 14 Pro, 19 Gly, 12 Ala, 5 Cys, 18 Val, 5 Met, 11 Ile, 20 Leu, 10 Tyr, and 9 Phe. The enzyme appears to have a blocked N-terminus.

Immunoprecipitated HPRT[6] from hemolysates displays two major spots after two-dimensional polyacrylamide gel electrophoresis (isoelectric focusing—SDS–polyacrylamide gel electrophoresis). HeLa cells or human lymphoblasts display only a single HPRT spot located at the same position as the most basic of the hemolysate HPRT spots.[7] This suggests that the most basic spot is the form initially synthesized and the more acidic hemolysate HPRT spot (a pseudo-isozyme) is probably derived from the first by an age-related modification (e.g., deamidation).

[6] G. S. Ghangas and G. Milman, *Science* **196**, 1119 (1977).
[7] G. Milman, E. Lee, G. S. Ghangas, J. R. McLaughlin, and M. George, Jr., *Proc. Natl. Acad. Sci. U.S.A.* **73**, 4589 (1976).

[75] Hypoxanthine Phosphoribosyltransferase and Guanine Phosphoribosyltransferase from Enteric Bacteria

By Joy Hochstadt

Hypoxanthine + 5' phosphoribosyl-α-1-pyrophosphate $\rightarrow$ IMP + inorganic pyrophosphate

Guanine + 5' phosphoribosylpyrophosphate $\rightarrow$ GMP + inorganic pyrophosphate

Phosphoribosyltransferases (sometimes referred to as PRTs) for hypoxanthine and guanine are separate activities representing the gene products of only distantly linked genes on the chromosome of *Escherichia coli* or *Salmonella typhimurium*. Mutants lacking functional product of one or the other gene cannot utilize hypoxanthine[1] or guanine,[2,3] respectively. Membrane vesicles prepared from such mutants exhibit substrate specificity both with respect to uptake of the respective bases and with regard to 6-OH purine phosphoribosyltransferase activity. When protein is removed from the membrane, however, or when intact cells are ruptured, the activity that is released has specificity for both guanine and hypoxanthine. Thus the two enzymes have restricted

[1] J. Y. Chou and R. G. Martin, *J. Bacteriol.* **112**, 1010 (1972).
[2] J. S. Gots, C. E. Benson, and S. R. Shumas, *J. Bacteriol.* **112**, 910 (1972).
[3] C. E. Benson and J. S. Gots, *J. Bacteriol.* **121**, 77 (1975).

activity in intact cells and while still in isolated membranes, but lose strict substrate specificity upon entry into an aqueous environment where overlapping specificity is observed. The two enzymes are quite similar[4] but can be resolved chromatographically;[3-6] each retains overlapping specificity in the aqueous environment. Both are inducible[4] and readily released from the periplasm upon osmotic shock.[7] The enzymes have been shown to mediate group translocation of 6-OH purines across the bacterial membrane[7,8] while a single enzyme with combined activity mediates transport of the two purines across plasma membrane of mouse fibroblasts.[9] Enzyme and uptake activity are regulated in enteric bacteria by a wide variety of 5′ nucleotides.[8,10]

Assay Method

Principle. The assay depends on the chromatographic separation of the radioactively labeled product IMP or GMP from the radioactively labeled substrate hypoxanthine or guanine, respectively.

Reagents

$MgCl_2$, 0.1 M
Tris·HCl, pH 7.9, 1 M[11]
Tris·HCl, pH 8.4, 1 M[12]
[8-^{14}C]Guanine, 20 mM (specific radioactivity adjusted to 5–10 mCi/mmole)[11]
[8-^{14}C]Hypoxanthine, 20 mM (specific radioactivity adjusted to 5–10 mCi/mmole)[12]
K^+Mg^{2+} EDTA Titriplex, 1 M
Mg_2PRPP, 20 mM
Hypoxanthine, 20 mM[12]
IMP, 20 mM[12]
Guanine, 20 mM[11]
GMP, 20 mM[11]
[K][EDTA], pH 7.0

[4] J. Hochstadt, *Crit. Rev. Biochem.* **2**, 259 (1974).
[5] J. A. Holden, J. D. Wall, and P. I. Harriman, *J. Bacteriol.* **126**, 1141 (1976).
[6] T. A. Krenitsky, S. M. Neil, and R. L. Miller, *J. Biol. Chem.* **245**, 2605 (1970).
[7] J. Hochstadt-Ozer and E. R. Stadtman, *J. Biol. Chem.* **246**, 5312 (1971).
[8] L. Jackman and J. Hochstadt, *J. Bacteriol.* **126**, 312 (1976).
[9] J. Hochstadt and D. C. Quinlan, *J. Cell. Physiol.* **89**, 839 (1976).
[10] J. Hochstadt-Ozer, *J. Biol. Chem.* **247**, 2419 (1972).
[11] For assay of guanine phosphoribosyltransferase only.
[12] For assay of hypoxanthine phosphoribosyltransferase only.

Ammonium acetate, 1 M[11]
Solution A (1246 ml butanol and 84 ml distilled water)[12]
Solution B (640 ml propionic acid and 790 ml of distilled water)[12]
Cellulose thin-layer sheets (Eastman chromagram #6065 with fluorescent indicator)

Procedure. The following are added to 12 × 75 mm glass test tubes using Hamilton PB600-1 repeating dispensors: 5 μl [14C]guanine[11] (or [14C]hypoxanthine[12]), 5 μl Tris buffer, 0.5 μl $MgSO_4$, 0.5 μl K^+Mg^+ Titriplex, and enzyme sample in a volume 0.5–10 μl (containing between 5 ng to 5 μg of protein according to relative purity). Distilled water is added to make a total volume of 45 μl. Reactions are initiated with the addition of 5 μl of Mg_2PRPP. A control tube is prepared to which 5 μl of distilled water are added instead of Mg_2PRPP. Reaction mixtures are incubated at 37° with shaking usually for 10 min. The reaction is terminated by addition of 5 μl [K][EDTA]. Cellulose thin layers (plastic-backed) are cut to strips 7 cm high and 1.5 times the number of samples wide to a maximum of 20 cm wide. Channels, 6 × 1.5 cm, are ruled on the thin layer with a ruler and pencil. At a point 1 cm from the edge, 0.5 μl GMP[11] carrier solution (or IMP[12]) and 0.5 μl of guanine[11] carrier solution (or hypoxanthine[12]) are spotted in the middle of the 1.5-cm channel; 3 μl of the reaction mixture are then spotted on top of the carrier GMP/guanine[11] (IMP/hypoxanthine)[12] spot. The spots are dried,[12] or the thin layer is immediately placed in the chromatography tank without drying[11] depending upon the substrate to be assayed. The chromatograms are then developed in either 1 M ammonium acetate[11] or a solution containing equal parts of solution A and solution B (miscible with each other at temperatures above 22°)[12] in a glass chromatography tank. When the ascending solvent front reaches the upper edge of the thin layers, they are removed and thoroughly dried with a hair dryer. The spots (R_fs:GMP, 0.8; guanine, 0.4; IMP, 0.0; hypoxanthine, 0.4) are marked with a pencil after having been localized under UV light, and each nucleoside-monophosphate-containing spot is cut out and placed in a scintillation vial. Toluene-based fluor solution is added to the vial; a few milliliters are sufficient, for the entire thin-layer strip need only be saturated with the fluor solution—it need not be totally submerged in the liquid. The samples are counted in a liquid scintillation counter; with an appropriate fluor solution the counting efficiency for 14C on the cellulose-backed thin layers is 75–80%.

Linearity. The reaction goes to completion and is often linear until a majority of the substrate (60–70%) has been utilized. For kinetic studies, concentration of enzyme and incubation times are adjusted, however, to utilize no more than 15–25% of the substrate.

Standardization of Substrates. [^{14}C]guanine and [^{14}C]hypoxanthine are standardized by spectral assay using molar extinction coefficients of 10,700 at 246 nm and 8150 at 276 nm for guanine at pH 7 and 10,700 at 249.5 nm for hypoxanthine at pH 6. One to three microliters pipetted directly into a scintillation vial, dried completely, and covered with 10 ml fluor solution gave a counting efficiency of approximately 97% and were used to confirm radiospecific activity assigned by the manufacturer.

Mg$_2$PRPP solutions are standardized by allowing the enzyme reaction to proceed to completion using a limiting amount of Mg$_2$PRPP and an excess of [^{14}C]purine standardized as above. The amount of nucleoside monophosphate formed under such conditions is taken as a measure of the Mg$_2$PRPP present since the reaction does go to virtual completion. Commercially purchased PRPP salts are often found to be 50–60% impure.

Purification of Substrate PRPP. In order to assay a variety of PRPP salts in addition to those commercially available, as well as to remove contaminants, the following procedure[13] is employed: 100 μmoles of Mg$_2$PRPP (in 1.5 ml volume) are passed over a 7.5 × 0.5 cm bed Dowex 50 which has been well washed with water after preliminary treatment with LiCl or KCl (for preparing Li$_4$PRPP or K$_4$PRPP, respectively, for example). Water (3.5 ml) is added to wash through the PRPP in 0.5-ml portions, and the effluent fractions are pooled and stored at −79° in 0.5-ml portions. All PRPP activity is standardly recovered from the Dowex 50. The quantity and concentration of the PRPP are then determined by the endpoint assay described above.

Definition of Unit and Specific Activity. One unit is defined as the amount of enzyme that catalyzes the phosphoribosylation of 1 μmole of guanine (or hypoxanthine) to form 1 μmole of GMP (or IMP) per minute. Specific activity is expressed in terms of units per milligram of protein. Protein is determined by the Biuret[14] procedure during initial steps in the purification and thereafter by the method of Lowry *et al.*[15]

Alternate Assay Procedure. When even greater sensitivity is required, assay of enzyme activity is performed in 20 μl of reaction volume in glass tubes 6 × 50 mm, containing 0.2 m*M* [^{14}C]guanine (or [^{14}C]hypoxanthine) (specific radioactivity >50.0 mCi/mmole), 2 m*M* Mg$_2$PRPP, 100 m*M* Tris·HCl, and 1–10 ng purified enzyme. The surface

[13] J. Hochstadt-Ozer and E. R. Stadtman, *J. Biol. Chem.* **246**, 5294 (1971).

[14] M. Dittebrandt, *Am. J. Clin. Pathol.* **18**, 439 (1978).

[15] O. H. Lowry, N. J. Rosebrough, A. L. Farr, and R. J. Randall, *J. Biol. Chem.* **193**, 265 (1954).

tension properties of the reaction mixtures in tubes of this geometry, however, required vortexing each tube for a second, once each minute during the incubation to ensure adequate mixing during the reaction. Other aspects of the assay are identical.

Application to Measurements in Crude Extracts. Because nucleotide pool constituents are potent inhibitors of enzyme activity, all crude extracts should be dialyzed or otherwise separated from cellular metabolites prior to assay of the protein fraction for activity. Though the nucleotides are competitive with PRPP, simple increase in PRPP concentrations is not an adequate means of dealing with the situation since, as purchased, the PRPP is contaminated with inorganic pyrophosphate and possibly other inhibitors of the enzyme. Other interfering contaminants (e.g., proteases) of the crude extract can be assessed after enzyme purification by reconstructive mixing of a sample of purified enzyme with a sample of crude extract each of known activity alone and assay of the combined activity observed.

Purification Procedure

Step 1. Growth and Harvest of Organism. As much as a 100-fold enrichment of these two phosphoribosyltransferases can be obtained by growth under conditions which inhibit *de novo* purine synthesis and render the organism dependent on exogenous purine and phosphoribosyltransferase activity.[4] The medium designed for this purpose, PAT medium,[4,13] contains a mixture of "P"urines each at 10^{-4} M, "A"minopterin or amethopterin (to inhibit *de novo* purine biosynthesis; it also inhibits thymidine synthesis) 3×10^{-7} M, and compensatory "T"hymine or thymidine 10^{-5} M. These "PAT" additions are made to the basic Vogel-Bonner-Citrate medium as described by Korn and Weissbach.[16] *Escherichia coli* K_{12} or *Salmonella typhimurium* LT-2 *Pro*AB47' or Hpt1[1] are grown to the end of exponential growth (turbidity monitored) in either 20-liter glass bottles with aeration or in large-scale fermentation apparatus (e.g., 60–300 liters) at 37°, harvested by centrifugation, frozen in liquid N_2, and stored at $-79°$.

Step 2. Preparation of Cell Extracts. Frozen cells are thawed at room temperature, suspended in 5–10 volumes of 50 mM potassium phosphate buffer (pH 7.5), and sonically disrupted at 0° by six 30-sec bursts (at maximum output with Heat Systems–Ultrasonic, Inc., Model 185W Sonifier Cell Disruptor) with intervening cooling at 0°; cell debris is removed by centrifugation at 10,000 rpm for 20 min.

---

[16] D. Korn and A. Weissbach, *Biochim. Biophys. Acta* **61**, 775 (1962).

Step 3. Streptomycin Preparation. Streptomycin sulfate at 4°, 10% by volume of a 10% solution, is added to the supernatant solution also at 4° (20–40 mg of protein/ml) obtained from step 2. After stirring for 10 min in the cold, the mixture is centrifuged.

Step 4. Ammonium Sulfate Precipitation. The supernatant solution from step 3 is adjusted to a protein concentration of 10 mg/ml, pH 7.8, with Tris·HCl (50 mM final), and a saturated solution of $(NH_4)_2SO_4$ is added to a final concentration of 35% saturation by volume. The treatment is at room temperature and equilibration is for 15 min with stirring. The precipitate is collected by centrifugation in the cold for 10 min at 10,000 rpm. The pellet is discarded. Saturated $(NH_4)_2SO_4$ solution is added to that supernatant solution to 42% saturation by volume. Equilibration and collection of the precipitate are as before; the precipitate is resuspended in 20 mM Tris·HCl (pH 8.0). The fraction precipitating between 42% and 49% saturated ammonium sulfate is then collected and tested for adenine, guanine, hypoxanthine, and xanthine phosphoribosyltransferase activities; it is either pooled with the 35–42% fraction if it has considerable 6-OH purine transferase activity, used to prepare the adenine phosphoribosyltransferase if that activity predominates, or discarded.

Step 5. Precipitation with Acetone. The protein fraction salting out between 35–42% saturated $(NH_4)_2SO_4$ (or 35–49%, see above) is adjusted to 10 mg/ml protein and 20 mM Tris·HCl (pH 8.0). One volume of acetone at −7° is added with stirring to 4 volumes of protein solution cooled in an ice–salt bath. The mixture is centrifuged and the supernatant fluid brought to −10° with an ice–salt bath; 4 additional volumes of acetone at −10° are added with stirring, and the mixture is centrifuged at −15°. The pellet is collected, drained, resuspended in 20 mM Tris·HCl (pH 8.0) at 0° (final volume—one-half that of the protein solution originally treated with acetone), and dialyzed against 1 liter of 20 mM Tris·HCl at 0° for 2 hr.

Step 6. Treatment with Cγ gel. The solution from step 5 is adjusted to pH 6.0 with 50 mM potassium phosphate buffer and mixed with a suspension of aged Cγ gel (obtainable from Bio-Rad) in 50 mM potassium phosphate buffer. Elution is at pH 7.5, and in some batches an additional pH 8.0 elution step was performed when recoveries at pH 7.5 are low.

Step 7. Chromatography on Bio-Gel P150. The 6-OH purine enzyme activities are recovered in the void volume after application to a Bio-Gel P150 column equilibrated with 75 mM potassium phosphate buffer (pH 1.5).

Step 8. Chromatography of Bio-Gel A 0.5. Chromatography is performed under conditions employed for the other Bio-Gel column in step 7. The enzyme activities for all 6-OH purines cochromatograph on the A 0.5 column with the exception that a second lower molecular weight peak (activity unstable) is observed for the hypoxanthine to IMP reactions in some, but not all, preparations. The peak fractions are concentrated by precipitation with ammonium sulfate and suspended in 9 mM Tris·HCl–0.1 mM MgSO$_4$ (pH 7.7).

Step 9. Chromatography on Ecteola-Cellulose. Approximately 5 ml of enzyme solution (containing approximately 20 mg of protein) [dialyzed against the Tris-Mg buffer the (NH$_4$)$_2$SO$_4$ precipitate is suspended in, in order to remove the (NH$_4$)$_2$SO$_4$] were applied to a 1 × 22 cm column of Ecteola-cellulose equilibrated with the same buffer at 4°.[6] The two phosphoribosyltransfer enzymes are resolved from each other by elution with a linear gradient of KCl 0–0.15 M in 9 mM Tris·HCl–0.1 mM MgSO$_4$ (pH 7.7) at a flow rate of about 15 ml/hr. The hypoxanthine PRT activity peak coincided with a KCl concentration of approximately 0.049 M while the guanine PRT activity eluted at approximately 0.073 M. Pooled peak fractions were frozen at −20° and studied within the next 3 weeks.

Summary. The table shows the results of the purification with respect to specific activity and yields at each step.

Properties

Stability. The early purification steps yield preparations that are quite stable. Later steps require addition of 2 mM β-SH ethanol to storage solutions. Ammonium sulfate suspensions are also advised for storage. Dilute aqueous solutions of the final preparation (even with β-SH ethanol) stored at −20° degenerate within a few weeks. PRPP helps to stabilize the enzyme.

Homogeneity. A single-band protein coincident with enzyme activity is detected when the pooled peak fractions (each of the two peaks pooled separately) from step 9 of the purification are subjected to acrylamide gel electrophoresis at pH 7.5 in the presence of 5 mM MgSO$_4$.

Stoichiometry. For each mole of guanine and PRPP utilized, 1 mole each of PP$_i$ and GMP are generated by GPRT; and for each mole of hypoxanthine and PRPP utilized, 1 mole each of PP$_i$ and IMP are generated by HPRT.

SUMMARY OF PURIFICATION OF 6-OH PURINE PHOSPHORIBOSYLTRANSFERASES

Step	Specific activity for		Recovery (%)	Comments
	$G \rightarrow GMP$	$H \rightarrow IMP$		
1. Growth on standard media: crude sonicate 10,000 g supernatant	$\leq 5 \times 10^{-4}$	$\leq 5 \times 10^{-4}$	<100	
2. Growth on 3×10^{-7} aminopterin, $10^{-4}\,M$ purines (AGHX), $10^{-5}\,M$ thymine	$\sim 3.5 \times 10^{-3}$	3.3×10^{-3}	<100	Interference by guanine deaminase, xanthine oxidase, and protease observed
3. 1% streptomycin supernatant	4.5×10^{-3}	3×10^{-3}	100	
4. 35–42% saturated $(NH_4)_2SO_4$ precipitate	4.2×10^{-2}	4.8×10^{-2}	70	
5. 44.5–54.4% acetone precipitate	0.288	0.207	47	
6. Adsorption and elution from C_γ gel	1.17	1.09	40	
7. Bio-Gel P 150 column void	2.45	2.2	33	
8. Bio-Gel A 0.5 column chromatography	8.0	7.0	23	
9. Ecteola-cellulose chromatography	$11.4(6.7)^a$	$17.3(8.4)^a$	16^b 12^c	

[a] Numbers in parenthesis indicate specific activity for that reaction by the opposite gene product (e.g., GMP formation by the HPRT'ase).

[b,c] Separate peaks for hypoxanthine PRT'ase ([c]) and guanine PRT'ase ([b]) (which peak was which gene product—since both had dual activity—was determined by peak position in the ProAB47 deletion lacking guanine PRTase activity *in situ*).

Substrate specificity. The enzymes exist on the membrane *in situ* where their substrate specificity is limited to either HPRT activity or GPRT activity individually as distinct enzymes. When either enzyme is released from the membrane it acquires the ability to catalyze the reaction with either substrate thus losing the strict specificity in its membrane-localized state.[8] Thus mutants mapping at the HPRT locus[1,5] are unable to use hypoxanthine to satisfy purine requirement if they are also purine auxotrophs while they are able to use guanine for growth. The reverse is true of GPRT mutants mapping in the *pro* AB region.[2,3] If extracts are prepared from such mutants both GPRT and HPRT activity are observed in each mutant.[1–3,5] If membrane vesicles are prepared from such mutants, the vesicles take up only the substrate the mutant was capable of growth on and catalyze the enzyme reaction with only the base for which a mapable PRT mutation does not exist, i.e., the membrane enzyme has the same strict substrate specificity that uptake has. If the purified membranes are dissociated, the single-specificity enzyme is solubilized to a multispecificity activity.[8] In addition, on- or off-the-membrane xanthine and 8-azaguanine can serve as substrate for the wild-type guanine enzyme, and 6-mercaptopurine can serve as substrate for the wild-type hypoxanthine enzyme.[2] Further, selection of deletion mutants in the guanine enzyme for the ability to utilize guanine led to alteration of the hypoxanthine enzyme or of the membrane such that the hypoxanthine enzyme could utilize guanine *in situ* as well as in solution.[2]

Kinetic Constants. The K_m for guanine is 2.5 μM and for PRPP is 100 μM by GPRT,[4] while the K_m for hypoxanthine is 120 μM and for PRPP 200 μM for HPRT.

Metal Ion Requirement. An absolute requirement for either Mn^{2+} or Mg^{2+} was found; the acetate, chloride, or sulfate salts were all found satisfactory with Mg acetate giving slightly higher values. Mg_2PRPP at a 2:1 ratio of Mg to PRPP was optimal. Ba^{2+}, Ca^{2+}, and Zn^{2+} are highly inhibitory and $K^+Mg^{2+}EDTA$ Titraplex is used in the assay to chelate traces of those metals.

pH Optimum. Guanine PRT has a pH optimum of 7.9[4] while hypoxanthine PRT has a pH optimum of 8.4. The xanthine PRT activity associated with GPRT has a pH optimum of 8.6.

Effectors of Enzyme Activity. All free 5′ nucleotides affect enzyme activity. 6-OH purine nucleotides are most inhibitory, while 6 NH_2 purine nucleotides are required at higher concentrations to achieve the same level of inhibition.[4,10] Often there is slight stimulation by nucleotide effectors when present at the 100–200 μM range. ppGpp is an especially potent inhibitor and at physiological concentrations can account for the drop in guanine nucleotides in the pool during the stringent response to amino acid control of nucleic acid synthesis.[17]

Physiological Function. In enteric bacteria the 6-OH purine phosphoribosyltransferases reside in the periplasm at the membrane[7,8] and there mediate the uptake of 6-OH purines. The specificity of the enzymes on the isolated membrane vesicles reflects the metabolic capacities of the cells even though broader substrate specificity is observed upon solubilization.[8]

Alternative Purification Scheme. An alternative method of enzyme purification is available which reduces the preparation to two steps. It is not as useful, however, for preparing very large amounts of enzyme. Based on the observation that much of the enzyme is released from the periplasm upon osmotic shock,[7] it can be obtained in the following manner. Cells are grown as described, but must be harvested at midexponential growth. Osmotic shock fluid is prepared by dilution of cells in 20% sucrose solution, 1:100 into 20 μM $MgSO_4$. The cells are removed by centrifugation. The supernatant "shock fluid" containing the enzyme is filtered through a 0.45 μm nitrocellulose filter and concentrated by ultrafiltration to a protein concentration of 2–4 mg/ml. An $(NH_4)_2SO_4$ fraction, 35–50% saturation, is prepared (see step 4 of the purification procedure above) and applied directly to the Ecteolacellulose column and eluted as described in step 9 of the purification procedure. Though the peak fractions from the column are homogeneous after only these two steps, in addition to the preparation of osmotic shock fluid, cells, subjected to osmotic shock immediately upon harvest from mid-exponential growth and never frozen, must be used.

[17] J. Hochstadt-Ozer and M. Cashel, *J. Biol. Chem.* **247**, 1067 (1972).

[76] Adenine Phosphoribosyltransferase from *Escherichia coli*

By Joy Hochstadt

Adenine + 5' phosphoribosyl-α-1 pyrophosphate → 5' AMP + inorganic pyrophosphate

Adenine phosphoribosyltransferase is an inducible enzyme in *Escherichia coli*[1] that though readily solubilized into aqueous extracts upon

[1] J. Hochstadt-Ozer and E. R. Stadtman, *J. Biol. Chem.* **246**, 5294 (1971).

cell rupture appears to be localized on the cell membrane *in situ*[2] where it mediates the translocation of adenine into the cell as AMP.[2] Its location at the cell surface is further indicated by its release from the periplasm upon osmotic shock treatment.[3] Enzyme activity is regulated by nucleotide pool constituents *in situ*;[3] experimentally it is quite sensitive to competitive inhibition by a very wide variety of 5′ nucleotides.[1]

Assay Method

Principle. The assay depends on the chromatographic separation of the radioactively labeled product, AMP, from the radioactively labeled substrate adenine.

Reagents
$MgCl_2$, 0.1 M
Tris·HCl, pH 7.8, 1 M
[8-^{14}C]Adenine, 20 mM (specific radioactivity adjusted to 5–10 mCi/mmole)
K^+Mg^{2+}EDTA Titriplex, 1 M
Mg_2PRPP, 20 mM
Adenine, 20 mM
5′ AMP, 20 mM
[K][EDTA], pH 7.0, 1 M
Eastman Chromagram with fluorescent indicator #6065
Ammonium acetate, 1 M

Procedure. The following are added to 12 × 75 mm glass test tubes using Hamilton syringes and Hamilton PB600-1 repeating dispensers: 5 μl [^{14}C]adenine, 5 μl Tris buffer, 0.5 μl $MgCl_2$, 0.5 μl K^+Mg^{2+} Titriplex, and enzyme sample in a volume 0.5–10 μl (containing between 5 ng to 5 μg of protein according to relative purity). Distilled water is added to make a total volume of 45 μl. Reactions are initiated with the addition of 5 μl of Mg_2PRPP. A control tube is prepared to which 5μl of distilled water is added instead of Mg_2PRPP. Reaction mixtures are incubated at 37° with shaking usually for 10 min. The reaction is terminated by addition of 5 μl [K][EDTA]. Cellulose thin layers (plastic-backed) are cut to strips 7-cm high and 1.5 cm times the number of samples wide to a maximum of 20 cm wide. Channels, 6 × 1.5 cm, are ruled on the thin layer with a ruler and pencil. At a point 1 cm from the edge, 0.5 μl of AMP carrier solution and 0.5 μl of adenine carrier solution are spotted in

[2] J. Hochstadt-Ozer and E. R. Stadtman, *J. Biol. Chem.* **246**, 5304 (1971).
[3] J. Hochstadt-Ozer and E. R. Stadtman, *J. Biol. Chem.* **246**, 5312 (1971).

the middle of the 1.5-cm channel. Then 3 μl of the reaction mixture are spotted on top of the carrier AMP/adenine spot. No effort is made to dry the spots prior to placement of the thin layer in the chromatography tank. The chromatograms are then developed in 1 mM ammonium acetate in a glass chromatography tank until the ascending solvent front has reached the upper edge of the 7-cm-high thin layer. The thin layers are removed and dried with a hair dryer. The AMP (R_f = 0.7) and adenine (R_f = 0.4) spots are localized under UV light and marked with a pencil; each AMP spot is cut out and placed in a scintillation vial. Toluene-based fluor solution is added to the vial; a few ml are sufficient for the entire thin-layer strip need only be saturated with the fluor solution—it need not be totally submerged in the liquid. The samples are counted in a liquid scintillation counter; with an appropriate fluor solution the counting efficiency for ^{14}C on the cellulose-backed thin layers is 75–80%.

Linearity. The reaction goes to completion and often is linear until a majority of the substrate (60–70%) has been utilized. For kinetic studies, concentration of enzyme and incubation times are adjusted, however, to utilize no more than 15–25% of the substrate.

Standardization of Substrates. [^{14}C]adenine is standardized by spectral assay using a molar extinction coefficient of 13,400 at 260.5 nm at pH 7. One to three microliters pipetted directly into a scintillation vial, dried completely, and covered with 10 ml of fluor solution gave a counting efficiency of approximately 97%; this was used to confirm radiospecific activity assigned by the manufacturer.

Mg$_2$PRPP solutions are standardized by allowing the enzyme reaction to proceed to completion using a limiting amount of Mg$_2$PRPP and an excess of [^{14}C]adenine standardized as above. The amount of AMP formed under such conditions is taken as a measure of the Mg$_2$PRPP present since the reaction does go to virtual completion. Commercially purchased PRPP salts are often found to be 50–60% impure.

Purification of substrate PRPP. In order to assay a variety of PRPP salts in addition to those commercially available, as well as to remove contaminants, the following procedure[1] is employed: 100 μmole of Mg$_2$PRPP (in a 1.5 ml volume) are passed over a 7.5 × 0.5 cm bed of Dowex 50 which has been well washed with water after preliminary treatment with LiCl or KCl (for preparing Li$_4$PRPP or K$_4$PRPP, respectively, for example). Water (3.5 ml) is added to wash through the PRPP in 0.5-ml portions, and the effluent fractions are pooled and stored at $-79°$ in 0.5-ml portions. All PRPP activity is standardly recovered from the Dowex 50. The quantity and concentration of the PRPP are then determined by the endpoint assay described above.

Definition of Unit and Specific Activity. One unit is defined as the amount of enzyme that catalyses the phosphoribosylation of 1 μmole of adenine to form 1 μmole of AMP per minute. Specific activity is expressed in terms of units per milligram of protein. Protein is determined by the Biuret[4] procedure during initial steps in the purification and thereafter by the method of Lowry *et al.*[5]

Alternate Assay Procedure. When even greater sensitivity is required, assay of enzyme activity is performed in 20 μl of reaction volume in glass tubes, 6 × 50 mm, containing 0.2 mM [^{14}C]adenine (specific radioactivity >50.0 mCi/mmole), 2 mM Mg$_2$PRPP, 100 mM Tris·HCl, and 1–10 ng of purified enzyme. The surface tension properties of the reaction mixtures in tubes of this geometry required vortexing each tube for a second, once each minute during the incubation, to ensure adequate mixing during the reaction. Other aspects of the assay are identical.

Application to Measurements in Crude Extracts. Because nucleotide pool constituents are potent inhibitors of enzyme activity, all crude extracts should be dialyzed or otherwise separated from cellular metabolites prior to assay of the protein fraction for activity. Though the nucleotides are competitive with PRPP, simple increase in PRPP concentrations is not an adequate means of dealing with the situation since, as purchased, the PRPP is contaminated with inorganic pyrophosphate and possibly other inhibitors of the enzyme. Other interfering contaminants (e.g., proteases) of the crude extract can be assessed after enzyme purification by reconstructive mixing of a sample of purified enzyme with a sample of crude extract, each of known activity alone, and determination of the combined activity in the mixture.

Purification Procedure

Growth and Harvest of Organism. As much as a 100-fold enrichment of adenine phosphoribosyltransferase can be obtained by growth under conditions which inhibit *de novo* purine synthesis and render the organism dependent on exogenous purine and phosphoribosyltransferase

[4] M. Dittebrandt, *Am. J. Clin. Pathol.* **18**, 439 (1948).
[5] O. H. Lowry, N. J. Rosebough, A. L. Farr, and R. J. Randall, *J. Biol. Chem.* **193**, 265 (1951).

activity.[1] The medium designed for this purpose, PAT medium,[1] contains a mixture of "P"urines each at 10^{-4} M, "A"minopterin or amethopterin (to inhibit *de novo* purine biosynthesis; it also inhibits thymidine synthesis) 3×10^{-7} M, and compensatory "T"hymine or thymidine 10^{-5} M. These "PAT" additions are made to the basic Vogel-Bonner-Citrate medium as described by Korn and Weisbach.[6] *Escherichia coli* K_{12} is grown to the end of exponential growth (turbidity monitored) in either 20-liter glass bottles with aeration or in large-scale fermentation apparatus (e.g., 60–300 liters) at 37°, harvested by centrifugation, frozen in liquid N_2, and stored at $-79°$.

Step 1. Preparation of Cell Extracts. Frozen cells are thawed at room temperature, suspended in 5–10 volumes of 50 mM potassium phosphate buffer (pH 7.5), and sonically disrupted at 0° by six 30-sec bursts (at maximum output with Heat Systems–Ultrasonic, Inc., model 185W Sonifier Cell Disruptor) with intervening cooling at 0°. Debris is removed from this homogenate by centrifugation at 10,000 rpm for 20 min.

Step 2. Streptomycin Precipitation. Streptomycin sulfate, 10% by volume, of 10% solution, is added to the supernatant solution (20–40 mg of protein per ml) obtained in step 1. After stirring for 10 min in the cold the mixture is centrifuged.

Step 3. Ammonium Sulfate Precipitation P5a. The supernatant solution from step 2 is adjusted to a protein concentration of 10 mg/ml and pH 7.8 with Tris·HCl (50 mM, final concentration) and a saturated solution of $(NH_4)_2SO_4$ is added, to a final concentration of 35% saturation by volume. The treatment is at room temperature and equilibration is for 15 min with stirring. The precipitate is collected by centrifugation at 4° for 10 min at 10,000 rpm (Sorvall SS-34 rotor). The pellet is discarded. Saturated $(NH_4)_2SO_4$ solution is added to that supernatant solution to make it 42% saturation by volume. Equilibration and collection of the precipitate are as before; the precipitate is resuspended in 20 mM Tris·HCl (pH 8.0) and saved for the preparation of 6-OH purine PRT. The supernatant fraction is adjusted to 49% saturated ammonium sulfate as above, and the precipitate is collected and resuspended as above. After testing this fraction for adenine and other purine PRT enzyme activities, it can be pooled with either the 35–42% fraction if rich in GPRT/HPRT, pooled with 49–56% fraction if rich in APRT, or discarded according to activities present. The supernatant fluid from this fraction is then made 56% with respect to $(NH_4)_2SO_4$, and the 56%

[6] D. Korn and A. Weissbach, *Biochim. Biophys. Acta* **61**, 775 (1962).

precipitate is collected as before to provide the peak APRT fraction for subsequent steps. If new growth conditions are being employed it may be possible that a shift in salting out may occur—an optional additional step is the preparation and APRT assay of a protein fraction salting out between 56–59% saturation. In about 35% of the fractionations, that fraction had the greatest APRT specific activity; however, it never contained more than 10–15% of the total activity.

Step 4. Precipitation with Acetone. The protein fraction salting out between 49–56% saturated $(NH_4)_2SO_4$ (or 42–56%, see above) is adjusted to 10 mg/ml of protein in 20 mM Tris·HCl (pH 8.0). One volume of acetone at $-7°$ is added to 4 volumes of protein solution and stirred in a NaCl–ice bath at $-10°$, and then 4 more volumes of acetone at $-10°$ are added with stirring. The mixture is centrifuged at $-15°$ after 1 more volume of acetone $(-15°)$ has been added. The pellet is then collected as before, drained, and resuspended in 20 mM Tris·HCl (pH 8.0) at $0°$. The final volume it is resuspended in is $\leq$ one-half that of the protein solution originally treated with acetone. This solution is then dialyzed against 1 liter of 20 mM Tris·HCl at $0°$ for 2 hr.

Step 5. Treatment with C_γ Gel. The solution from step 4 is adjusted to pH 6.0 with 50 mM potassium phosphate buffer and mixed with a suspension of aged C_γ gel in 50 mM potassium phosphate buffer. The ratio of gel to protein is 34.5 mg of gel to 1 mg of protein. After stirring for 15 min in the cold, the supernatant solution is collected by centrifugation.

Step 6. Treatment with Calcium Phosphate Gel. The supernatant solution from step 5 is treated with a suspension of calcium phosphate gel in 50 mM potassium phosphate buffer (pH 6.0). The ratio of calcium phosphate gel to protein is 17.6 mg of gel per milligram of protein. After stirring for 15 min in the cold, the supernatant fluid is collected and dialyzed twice in the cold, against 3 liters of 50 mM potassium phosphate buffer (pH 7.5) for 1 hr each and once against 3 liters of 10 mM potassium phosphate buffer (pH 7.5), and lyophilized.

Step 7. Chromatography on Bio-Gel P-150. The lyophilized powder is resuspended to one-tenth its original volume in distilled water and divided into 1-ml portions and stored at $-79°$. One- or 2-ml portions at a time are applied to a column (1.5 × 30 cm, resin bed) of Bio-Gel P-150 equilibrated with 75 mM potassium phosphate buffer (pH 7.5). Fractions (1–2 ml) are collected in the cold, automatically. The void volume containing contaminating protein is discarded. The enzyme fractions with activity greater than half-peak activity are pooled unless otherwise

Protein fraction[c]	Specific activity (μmoles/min/ mg protein)	Percentage recovery
1. Untreated extract	0.010	100.0 (54 observed)
2. Streptomycin supernatant solution	0.015	73.0
3. Ammonium sulfate precipitate	0.08	23.2
4. Acetone precipitate	0.70	22.2
5. C_γ gel supernatant	3.10	21.0
6. Calcium phosphate gel supernatant	7.20	14.0
7. Bio-Gel P 150 column peak	14.00	12.6

[a] Reprinted from Hochstadt-Ozer and Stadtman,[1] p. 5294.
[b] PAT medium is VBC medium plus purines (adenine, guanine, hypoxanthine, xanthine, 0.1 mM each); amethopterin, 0.3 μM; thymine, 0.1 mM.
[c] The numbers refer to steps described in the text.

noted. Fractions are stored at $-20°$ for 2 weeks or less and at $-79°$ for longer periods.

The table shows a summary of the purification with respect to specific activity and yields at each step. Recovery from the streptomycin step is higher than the total activity observed in the crude extract. The extent to which inhibitory conditions or inhibitors present in the crude extract assay contributes to the decreased activity observed is determined by adding a known amount of enzyme activity from step 7 to a sample of step 1 and noting the activity increment actually observed. It is from this latter figure that the total activity of the extract and subsequent yields are calculated.

Properties

Stability. The adenine phosphoribosyltransferase is reasonably stable to heat treatment of 5 min at 60° [crude extracts, however, require substrate to stabilize during heat treatment, presumably due to protease which is removed during $(NH_4)_2SO_4$ fractionation]. Activity is unaffected by 10 mM mercaptoethanol. After treatment with calcium phosphate gel (step 6 of purification), instability to storage at 4° (or $-20°$) is noted. Storage in 0.1 M potassium phosphate buffer (pH 7), however, was found to be a satisfactory means of preventing enzyme inactivation. The completely purified enzyme may be stored in a very dilute solution (i.e., 0.1 mg/ml at $-79°$ for several months), but considerable loss of activity was noted after subsequent storage of such solutions at $-20°$ for 2–3 weeks.

Homogeneity and Molecular Weight Estimations. Rechromatography of peak fractions from Bio-Gel P 150 leads to the isolation of a single protein peak, all portions of which exhibit the same specific enzyme activity. The homogeneity of this fraction is further demonstrated by the observation of a single band of protein coincident with enzyme activity after acrylamide gel electrophoresis at pH 7.2 in the presence of 5 mM MgSO$_4$. Under standard conditions of electrophoresis, however (i.e., pH 9.5 and without MgSO$_4$[7]), the enzyme dissociates into several catalytically active bands. The smallest active fraction (fastest mobility) on the gel is least stable, but activity can be partially and temporarily enhanced by reaggregation at $4°$ in buffer solutions or in the presence of magnesium or PRPP. Similar dissociation behavior is observed when the enzyme is subjected to gel filtration chromatography at alkaline pH and in the absence of divalent cation. Calibration of the column with protein standards of known molecular weight led to estimation of the active polypeptide moieties of the enzyme at about 20,000, 30,000, and 40,000.[1]

Stoichiometry. For each mole of adenine and PRPP utilized, 1 mole each of PP$_i$ and AMP are generated.

Specific Activity and Turnover Number. Based on an estimated molecular weight of 40,000[1] for the native enzyme and a specific activity of 14.00[1] μmoles of AMP generated per minute per milligram enzyme protein, an enzyme turnover number was calculated to be 5.6×10^2 reactions/min/40,000 MW moiety.

Substrate Specificity. The *E. coli* enzyme is specific for adenine or 2,6-diamino-purine. Numerous other tested purines,[1] including aminoimidazolicarboximideriboside previously reported to serve as substrate for partially purified bovine adenine phosphoribosyltransferase,[8] do not serve as substrates.

Kinetic Constants. Michaelis–Menton kinetics are observed with respect to both PRPP and adenine. K_m for PRPP in the presence of a 2-fold excess of Mg^{2+} is 125 μM (Mg^{2+}, though required for the reaction, is also competitive with PRPP[1]); K_m for adenine is approximately 20 μM.

[7] A. C. Chrambach, *Anal. Biochem.* **15**, 544 (1964).
[8] J. G. Flaks, M. J. Erwin, and J. M. Buchanan, *J. Biol. Chem.* **228**, 201 (1957).

Metal Ion Requirement. An absolute requirement for either Mn^{2+} or Mg^{2+} was found. The chloride salt of either metal was found satisfactory. A complex relationship between activity, metal ion concentration, and PRPP concentration exists[1,9] evidence suggests that alternate reaction mechanisms may be possible at varying cation concentrations depending on the relative concentration of PRPP as the free acid, MgPRPP or Mg_2PRPP[1,9], e.g., 2:1 ratio of Mg^{2+} to PRPP was found to be optimal. Other divalent cations such as Zn^{2+}, Ba^{2+}, and Ca^{2+} were highly inhibitory to the reaction. Based on this observation, $K^+Mg^{2+}EDTA$ Titraplex is included in the assay to bind continuous traces of such metals.

pH Optimum. Enzyme activity has a strikingly sharp pH optimum at pH 7.8.[1]

Reaction Mechanism. Under most experimental conditions a ping-pong mechanism is observed.

PRPP + enzyme → phosphoribosylenzyme + PP_i
$\qquad\qquad$ Phosphoribosylenzyme + adenine → AMP + enzyme

Effectors of Enzyme Activity. All nucleotides with free 5′ phosphate groups are effectors.[1] Stimulation of activity at low effector concentrations (especially at saturating PRPP concentrations) is noted for some purine nucleotides[1]. Higher concentrations of all 5′ nucleotides are competitive with PRPP. $6-NH_2$ purine nucleotides are most inhibitory, 6-OH purine nucleotides are moderately inhibitory, and pyrimidine nucleotides are least inhibitory; however, all inhibited activity greater than 50% when included in the reaction mixture (standard assay conditions) at 2 mM. The reaction products AMP and PP_i are also inhibitors of the reaction. AMP is competitive with PRPP[1] while PP_i competition with PRPP may be cooperative (Mg^{2+} varied with PRPP as in Mg_2PRPP) or anticooperative (at constant excess magnesium).[1]

Physiological Function. In *E. coli* cells which are actively producing purine nucleotides by the *de novo* biosynthetic pathway, approximately 50 enzyme molecules are found per bacterial cell; while in those cells wholly dependent on exogenous purines for growth, approximately 5000 enzyme molecules are found per cell.[1,3] The enzyme, though readily solubilized upon cell rupture as described above, appears to largely reside on the membrane in such derepressed cells where it mediates adenine translocation across the cell membrane.[2] Evidence for this comes both from studies with isolated vesicles[2] in which both

---

[9] T. A. Krenitsky, R. Papaioannou, and G. B. Elion, *J. Biol. Chem.* **244**, 1263 (1969).

enzyme activity and transport reaction have been studied[2] and from its recovery with other periplasmic constituents upon osmotic shock.[3] In noninduced cells this group translocation mechanism of adenine uptake may account for only a portion of the small amounts of adenine taken up,[10] while in the cells grown on the PAT medium it clearly accounts for virtually all of the uptake.[2] The enzyme also appears necessary for appropriate regulation of purine biosynthesis since its absence is associated with purine excretion.

Other Properties. The enzyme therefore exists *in situ* in an environment very different from the one in which it is purified, aqueous solution, and characterized. Certain properties which it exhibits when *in situ* in the membrane differ from what is observed in aqueous solution. These differences disappear when it is released from the membrane by detergent, freeze-thawing, or sonic oscillation; among them are greater sensitivity to effectors in aqueous solution and ability to carry out an exchange reaction between AMP and adenine only while membrane-bound. Both properties suggest that membrane-localized enzyme exists as phosphoribosyl-enzyme and that the phosphoribosyl moiety is discharged in aqueous solution. One approach to study of this enzyme in a homogeneous environment more reflective of the membrane milieu is to incorporate it into artificial phospholipid bilayers.

Alternative Purification Scheme. An alternative method of enzyme purification is available which reduces the preparation to two steps. It is not as useful, however, for preparing very large amounts of enzyme. Based on the observation that much of the enzyme is released from the periplasm upon osmotic shock,[3] it can be obtained in the following manner. Cells are grown as described, but must be harvested at mid-exponential growth. Osmotic shock fluid is prepared by dilution of cells in 20% sucrose solution, 1:100 into 20 μM MgSO$_4$. The cells are removed by centrifugation. The supernatant "shock fluid" containing the enzyme is filtered through a 0.45 μm pore size nitrocellulose filter and concentrated by ultrafiltration to a protein concentration of 2–4 mg/ml. An (NH$_4$)$_2$SO$_4$ fraction, 45–55% saturation, is prepared (see step 3 purification procedure above) and applied directly to a Bio-Gel P 150 column and eluted as described in step 7 of the purification procedure. Though the peak fractions from the column are homogeneous after only two steps in addition to osmotic shock fluid preparation, cells—subjected to osmotic shock immediately upon harvest from mid-exponential growth and never frozen—must be used for this method.

[10] S. Roy-Berman and D. W. Visser, *J. Biol. Chem.* **250**, 9270 (1975).

[77] Adenine Phosphoribosyltransferase

By WILLIAM J. ARNOLD and WILLIAM N. KELLEY

Human adenine phosphoribosyltransferase (EC 2.4.2.7) (APRT) catalyzes the magnesium-dependent transfer of the ribose-5-phosphate moiety of 5-phosphoribosyl-1-pyrophosphate (PP-ribose-P) to the 9 position of the purine base adenine to form adenosine-5'-monophosphate (AMP). Human APRT activity is found exclusively in the cytoplasm, while in bacteria APRT activity appears to be loosely associated with the cell membrane in the periplasmic space.[1] Human APRT activity has been found in all tissues with the highest specific activity in nucleated cells.

Assay Method

Principle. Human APRT activity has been assayed by a variety of radiochemical techniques.[1-6] These techniques differ primarily by the procedure utilized to separate the radioactive purine substrate, adenine, from the product, AMP. Each technique may have advantages depending on the specific purpose for assaying the enzyme. We have found that separation by high-voltage electrophoresis on Whatman 3 MM chromatography paper is an accurate, rapid technique useful for assaying APRT activity during purification and for screening large numbers of samples for abnormalities of APRT activity.[7]

Preparation of Sample for Assay. Human APRT is most frequently assayed from hemolysate although with minor modifications this assay procedure is also applicable to the assay of APRT activity from virtually any tissue source. Five to ten milliliters of whole blood are collected by venipuncture in heparinized tubes (10 units/ml whole blood). After centrifugation at 1000 g for 10 min, the plasma and buffy coat are removed. Cold 0.85% sodium chloride is added in a volume equal to the

[1] J. Hochstadt-Ozer and E. R. Stadtman, *J. Biol. Chem.* **246**, 5312 (1971).
[2] M. Hori and J. F. Henderson, *J. Biol. Chem.* **241**, 1406 (1966).
[3] A. W. Murray and P. C. L. Wong, *Biochem. J.* **104**, 669 (1967).
[4] R. D. Berlin and E. R. Stadtman, *J. Biol. Chem.* **241**, 2679 (1966).
[5] B. Bakay, M. A. Telfer, and W. L. Nyhan, *Biochem. Med.* **3**, 230 (1969).
[6] H. Kizaki and T. Sakurada, *Anal. Biochem.* **72**, 49 (1976).
[7] C. B. Thomas, W. J. Arnold, and W. N. Kelley, *J. Biol. Chem.* **248**, 2529 (1973).

METHODS IN ENZYMOLOGY, VOL. LI

remaining erythrocytes. The erythrocytes are then suspended by gentle inversion and centrifuged as before. The sodium chloride is removed, and the wash is repeated twice. The thrice-washed erythrocytes are then lysed by freezing and thawing twice in a Dry Ice–acetone bath. The buffy coat-poor, washed, lysed erythrocytes (hemolysate) are then dialyzed for 2 hr against 1000 volumes of 5 mM Tris·HCl, pH 7.4. The dialyzed hemolysate may then be stored at $-35°$ until ready for assay. Human APRT activity in dialyzed hemolysate is stable for up to 6 months when stored at $-35°$.

Reaction Mixture. The APRT reaction is carried out in 12 × 75 mm plastic tubes. In a final volume of 100 μl the reaction mixture contains 0.03 μmole of [8-^{14}C]adenine (specific activity 6.83 mCi/mmole), 0.1 μmole of PP-ribose-P, 0.5 μmole magnesium chloride, 5 μmoles Tris·HCl (pH 7.4), and 100–200 μg of hemolysate protein. The reaction is initiated by the addition of [8-^{14}C]adenine after a preliminary incubation of 30 sec at 37°. The reaction mixture is then incubated at 37° in a shaking water bath for 20 min. The reaction is stopped by the addition of 4 μmoles of ethylenediamine tetraacetic acid (EDTA) and immersion in a Dry Ice–acetone bath.

After the reaction mixture has been allowed to thaw in an ice-water bath, a 20-μl aliquot is spotted on Whatman No. 3 MM chromatography paper with 20 μg of cold AMP as carrier. Fifteen spots may be placed on each sheet when a distance of 1 inch between spots is used.

The spots are then thoroughly dried with warm blown air and the sheet wetted with 25 mM sodium borate buffer, pH 8.5, containing 1 mM EDTA. Prior to electrophoresis the chromatography paper is blotted to remove puddles of buffer which may cause unequal migration and/or streaking. With 25 mM sodium borate–1 mM EDTA (pH 8.5) as both the anode and cathode buffer, electrophoresis is performed at 250 mA (4000–6000 V) for 15 min. After drying, the mononucleotide spot is visualized under ultraviolet light (254 nm) and cut out. With care taken to avoid touching the spot, it is then placed in a scintillation vial with 10 ml of Liquifluor scintillation fluid (New England Nuclear) and counted in a Packard Tricarb liquid scintillation spectrometer at 74% efficiency.

Using this assay procedure for human hemolysate, [8-^{14}C]AMP production is linear from 5–30 min of incubation at 37° and with 50–300 μg of hemolysate protein. Assay duplicates always agree within 10% with a day-to-day reproducibility on the same sample of 10%. The use of high-voltage electrophoresis to separate the radioactive purine base and substrate allows for the rapid processing of a large number of samples and the direct identification of the mononucleotide product by its comigration with added standard. This latter characteristic is of particular importance when the APRT assay is attempted in crude preparations from nucleated tissues. These crude preparations frequently also contain 5'-nucleotidase activity which will further metabolize the AMP product to adenosine.[8] Because the high-voltage electrophoretic technique adequately separates AMP, adenosine, and adenine (R_f values are 1.0, 0.43, and 0.18, respectively) the APRT activity can be determined by the sum of the radioactivity located in the AMP and adenosine spots. Alternatively, the addition of thymidine-5'-triphosphate (final concentration 3.3 mM) to the reaction mixture will inhibit 5'-nucleotidase activity by greater than 90%.[8]

Purification of Human APRT

Human APRT has been purified 33,000-fold from erythrocytes to a final specific activity of 9.58 IU/mg protein (Table I).[7] With increasing purity, the activity of human APRT becomes unstable under the usual assay conditions. The addition of bovine serum albumin to the reaction mixture at a final concentration of 0.5 mg/ml and shortening of the incubation time at 37° to 5 min are necessary to accurately assay APRT activity in purified preparations.

In addition, purified preparations of APRT are unstable during storage at either 4° or −70°. Virtually complete protection of the APRT activity for up to 2 months storage at −70° can be achieved by the addition of 5 mM $MgCl_2$ and 0.1 mM PP-ribose-P to the enzyme preparation.

Preparation of Hemolysate. Citrated whole blood (4000 ml) which would otherwise be discarded after 21–30 days by a blood bank may be

TABLE I
PURIFICATION OF HUMAN ERYTHROCYTE APRT

Step	Specific activity (IU/mg protein)	Recovery (% initial activity)	Purification (fold)
1. Hemolysate	0.000287	100	—
2. DEAE-eluate	0.0380	55	130
3. Ammonium sulfate precipitation	0.0375	17	130
4. Ethanol precipitation	0.0790	14	280
5. CM-cellulose column	9.58	9.5	33,000
6. Sephadex G-75	5.08	2.7	18,000

[8] W. N. Kelley and J. C. Meade, *J. Biol. Chem.* **246,** 2953 (1971).

used for the purification of APRT. The plasma and buffy coat are removed and the erythrocytes washed and lysed as described above. Unless otherwise stated, all further steps are carried out at 4°.

DEAE-Cellulose Chromatography. The procedure of Hennessy *et al.* is used to prepare the DEAE-cellulose (exchange capacity, 0.75 meq/g).[9] One hundred grams of an aqueous suspension of DEAE-cellulose are adjusted to pH 7.0 with 4 N HCl just prior to use. Hemolysate (562 ml) diluted with an equal volume of distilled water is then added, and the mixture is stirred for 2 hr. After filtration on a Büchner funnel the cellulose is resuspended in 2 liters of 2 mM sodium phosphate buffer, pH 7.0. After stirring for 30 min the cellulose is filtered as before, mixed with 2 liters of the same buffer, and poured into a glass column (5 × 50 cm). After washing with 2 liters of 20 mM sodium phosphate buffer, pH 7.0, the column is then developed with a 2-liter gradient, linear in KCl (0–0.3 M) and constant in sodium phosphate (20 mM, pH 7.0). Tenmilliliter fractions are collected, and the peak of APRT activity is pooled.

Ammonium Sulfate Precipitation. Next 209 g of crystalline ammonium sulfate are added to each liter of the pooled eluate. After being stirred for 1 hr, the precipitate is removed by centrifugation at 30,000 g for 15 min and 164 g of ammonium sulfate are added to each liter of the supernatant fluid and stirred and centrifuged as above. The final precipitate is resuspended to 5 times its volume in 50 mM Tris·HCl, pH 7.4. The resuspended precipitate is then clarified by centrifugation at 30,000 g for 20 min.

Ethanol Precipitation. The clarified, resuspended ammonium sulfate precipitate (1 volume) is stirred into 34 volumes of 50 mM sodium acetate buffer, pH 5.0, which had been chilled to 0°. Absolute ethanol (15 volumes) prechilled to −20° is then added immediately over a 60-sec period as the solution is chilled to −10°. The solution is stirred for 45 min. After centrifugation at 30,000 g for 15 min at −10° the precipitate is resuspended in approximately 5 times its volume of 50 mM Tris·HCl, pH 7.4, containing 5 mM MgCl$_2$ and 0.1 mM PP-ribose-P and clarified as above.

CM-Cellulose. Sixty grams wet weight of preswollen microgranular CM-cellulose (capacity, 1.0 ± 0.1 meq/g) are added to 300 ml of 10 mM sodium succinate buffer, pH 5.18, containing 5 mM MgCl$_2$. The suspension is adjusted to pH 5.18 with a solution of 10 mM succinic acid–5 mM

[9] M. A. Hennessey, A. M. Waltersdorph, F. M. Heunnekens, and B. W. Gabrio, *J. Clin. Invest.* **41**, 1257 (1962).

MgCl$_2$ and then filtered on a Büchner funnel. The cellulose is resuspended in the original succinate–magnesium buffer to a total volume of 300 ml. The suspension is placed in a graduated cylinder and the fines allowed to settle for a predetermined length of time (minutes) equal to twice the height (centimeters) of the slurry. At the proper time, the supernatant fluid was removed to leave a slurry volume equal to 120% of the wet settled volume. The succinate–Mg buffer is added to bring the total volume up to 150% of wet settled volume. The slurry is made 0.1 mM in PP-ribose-P, poured into a column (1.7 × 31.1 cm), and then equilibrated with sodium succinate–magnesium–PP-ribose-P, pH 5.18 (10 mM succinate, 5 mM MgCl$_2$, 0.1 mM PP-ribose-P) until buffer and effluent give identical pH and absorbance at 215 nm. The sample is dialyzed for 4 hr against this buffer and then applied to the column. APRT activity is eluted from the column with equilibrating buffer at a flow rate of 30 ml/hr. Two-milliliter fractions are collected, and those with peak APRT activity are pooled. The pooled fractions are concentrated to a 1-ml volume by dialysis against polyethylene glycol containing 25 mM Tris·HCl (pH 7.4), 5 mM MgCl$_2$, and 0.1 mM PP-ribose-P.

Sephadex Gel Filtration. The concentrated sample is applied to a Sephadex G-75 column (1.7 × 70 cm) and eluted with 50 mM Tris·HCl, pH 7.4, containing 5 mM MgCl$_2$ and 0.1 mM PP-ribose-P. Fractions of 1.5 ml are collected, and those containing the peak activity are pooled.

Results. This purification scheme reproducibly results in the isolation of a highly purified preparation of human APRT as judged by the presence of a single protein band in the step 6 enzyme preparation on polyacrylamide disc gel electrophoresis and sodium dodecyl sulfate–polyacrylamide electrophoresis. Attempts to change the ammonium sulfate precipitation step improved the recovery of APRT activity but resulted in a less pure APRT preparation. The Sephadex gel filtration step is necessary to achieve a homogeneous preparation of APRT protein, but it also results in a loss of specific activity despite the inclusion of MgCl$_2$ and PP-ribose-P.

Properties

The physical characteristics of human APRT are shown in Table II.[10]
A homogeneous preparation of human APRT enzyme demonstrated greatest activity with the 6NH$_2$ purine, adenine.[7] However, significant nucleotide formation from 2,6-diaminopurine, 4-amino-5-imidazolecarboxamide, and 6-mercaptopurine was also noted. No activity was

[10] L. M. Siegel and K. J. Monty, *Biochim. Biophys. Acta* **112**, 346 (1966).

TABLE II
PHYSICAL CHARACTERISTICS OF HUMAN APRT

1. Stokes radius	24.9 Å
2. $s_{20,w}$	3.35
3. Native molecular weight[a]	34,000 ± 3000
4. Isoelectric point	pH 4.85

[a] Determined by filtration on Sephadex G-75. Alternatively, assuming a partial specific volume of 0.725, the $s_{20,w}$ of 3.35 and Stokes radius of 24.9 yield a molecular weight of 34,400 by the method of Siegel and Monty.[10]

apparent with hypoxanthine, guanine, or adenosine as substrate. The apparent affinity constant (K_m) of the human APRT enzyme for adenine is $1.4 \times 10^{-4} M$.[11] PP-ribose--P is the only effective donor of the ribose-5 moiety for the APRT reaction studied to date.[12] The APRT enzyme is inactive with ribose-5-phosphate or ribose 1,5-diphosphate as substrate. Human APRT has a K_m of $6 \times 10^{-6} M$ for PP-ribose-P.[13] The enzyme activity of purified preparation of human APRT exhibits an absolute requirement for divalent cations. Magnesium has been shown to be the most effective metal cofactor; however, manganese (Mn^{2+}), calcium (Ca^{+2}), cobalt (Co^{+2}), nickel (Ni^{+2}), and zinc (Zn^{+2}) will also support enzyme activity in order of decreasing effectiveness.[2] Maximal APRT activity is achieved when magnesium is at approximately twice the concentration of PP-ribose-P.

Nucleotides, metal ions, and sulfhydryl binding agents have been demonstrated to effectively inhibit APRT enzyme activity.

The nucleotide mono-, di-, and triphosphates of adenine, guanine, and hypoxanthine are competitive inhibitors of APRT activity versus PP-ribose-P.[14–16] The monovalent cation sodium (Na^+) and the anions sulfate, succinate, and citrate are also inhibitors of APRT activity.[2,11] Depending on the source of APRT enzyme, divalent metal ions and sulfhydryl binding agents have been shown to inhibit APRT activity. Mercuric (Hg^{2+}) ion at low concentration ($10^{-5} M$) has been shown to produce essentially complete inhibition of APRT enzyme activity from *E. coli,* Ehrlich ascites cells, and human erythrocytes.[1,2,11] Sodium

[11] S. K. Srivastava and E. Beutler, *Arch. Biochem. Biophys.* **142**, 426 (1971).
[12] R. E. A. Gadd and J. F. Henderson, *J. Biol. Chem.* **245**, 2979 (1970).
[13] J. F. Henderson, *Fed. Proc., Fed. Am. Soc. Exp. Biol.* **27**, 1053 (1968).
[14] A. W. Murray, *Biochem. J.* **100**, 671 (1966).
[15] J. F. Henderson, R. E. A. Gadd, H. M. Palser, and M. Hori, *Can. J. Biochem.* **48**, 573 (1969).
[16] J. F. Henderson, M. Hori, H. M. Palser, and R. E. A. Gadd, *Biochim. Biophys. Acta* **268**, 70 (1972).

iodoacetate (0.1 mM) and p-hydroxymercuribenzoate (1.0 μM) are also effective inhibitors; however, p-CMB has no effect on the APRT enzyme from monkey liver.[17]

Highly purified human APRT has maximal activity over a broad pH range from 7.4–9.5.[7]

[17] T. A. Krenitsky, S. M. Neil, G. B. Elion, and G. H. Hitchings, *J. Biol. Chem.* **244**, 4779 (1969).

[78] Adenine Phosphoribosyltransferase from Rat Liver

By Donald P. Groth, Leona G. Young, and James G. Kenimer

Adenine phosphoribosyltransferase (AMP:pyrophosphate phosphoribosyltransferase, EC 2.4.2.7) has been purified in reasonable yield from several mammalian tissues. The rat liver enzyme possesses unusual "initial burst" kinetics in that a rapid limited synthesis of AMP is followed by a sustained steady-state synthesis of AMP.[1,2] When the liver enzyme is purified to apparent homogeneity the two reaction phases of AMP synthesis are found to be catalyzed by the same protein molecule.

Assay Procedures

Adenine phosphoribosyltransferase converts adenine into AMP using PP-ribose-P as cosubstrate. Unreacted adenine is readily extracted by butanol from neutral solutions, leaving the product AMP in the water phase. If radioactive adenine is employed, the product AMP can be measured in a liquid scintillation counter with high sensitivity and precision.

Steady-State Assay. The assay method of Hori and Henderson[3] gives very reproducible results. The reagents required are:

1. Tris·HCl buffer, pH 7.5, 0.2 M containing 0.2 mg/ml bovine albumin
2. $MgCl_2$, 40 mM
3. PP-ribose-P, sodium salt, 2 mM
4. [8-^{14}C]adenine (40–60 Ci/mole), 0.5 mM

[1] J. G. Kenimer, L. G. Young, and D. P. Groth, *Biochim. Biophys. Acta* **384**, 87 (1975).
[2] D. P. Groth and L. G. Young, *Biochem. Biophys. Res. Commun.* **43**, 82 (1971).
[3] M. Hori and J. F. Henderson, *J. Biol. Chem.* **241**, 1406 (1966).

5. 1-Butanol saturated with water at room temperature
6. Diethyl ether saturated with water

It is usually necessary to purify commercial preparations of radioactive adenine to reduce the content of radiation self-decomposition products which are not extractable by butanol. One milliliter of reagent 4 is extracted 6 successive times with 5 ml of reagent 5. The combined butanol extracts (30 ml) are extracted twice with 1 ml of H_2O. The butanol extract is evaporated to dryness under reduced pressure at a temperature below 40°. The purified [8-^{14}C]adenine should contain less than 0.2% of radioactivity that is not extractable by 6 butanol washes under the standard assay conditions.

Incubation mixtures are prepared which contain the equivalent of 25 μl reagent 1, 5 μl reagent 2, 10 μl reagent 3, 10 μl reagent 4, and 50 μl of an adenine phosphoribosyltransferase preparation containing 0–1 munits of activity. The mixtures are incubated at 30° for 3 min. The reaction in each mixture is stopped by the addition with mixing of 2 ml of reagent 5. Water (0.4 ml) is added with mixing. After phase separation has occurred, the butanol is removed by aspiration. The butanol extraction is repeated 5 additional times. Residual butanol is removed by ether extraction and residual ether by aeration in a 30° water bath. The conversion of [8-^{14}C]adenine into butanol-insoluble [8-^{14}C]AMP is measured by liquid scintillation spectrometry.

"Initial Burst" Assay. The rapid phase of AMP synthesis catalyzed by adenine phosphoribosyltransferase is measured in the same incubation mixtures as with the steady-state assay, except that the mixtures are incubated at approximately 0° in an ice bath. The enzyme solution is added last and after only 10 sec of incubation, 2 ml of reagent 5 are added with mixing to stop the reaction as before. Under these conditions, the steady-state synthesis of AMP contributes less than 5% of the total reaction product obtained.

With crude tissue extracts, 50 mM potassium fluoride (final concentration) is included in the reaction mixtures.

One activity unit is defined as the conversion of one μmole of [8-^{14}C]adenine to AMP per minute.

Enzyme Isolation

All steps in the purification are carried out at 2–4°, and all solutions contain 5 mM mercaptoethanol, unless stated otherwise.

Step 1. Preparation of Crude Extract. Four to six rats (150–250 g) are fasted overnight and decapitated. The livers are removed, cut into small pieces, and homogenized with a Teflon homogenizer in 3 volumes of 30 mM Tris·HCl buffer, pH 7.5. The homogenate is centrifuged for 90 min at 30,000 rpm in a No. 30 rotor in the Spinco Model L ultracentrifuge. The supernatant solution contains essentially all of the activity in the homogenate.

Step 2. DEAE-Sephadex A-50 Chromatography. The supernatant fraction of the homogenate is adjusted to 45% saturation by the addition of saturated ammonium sulfate prepared in Tris·HCl buffer, 50 mM, pH 7.5. After several hours the precipitate is collected by centrifugation and discarded. The ammonium sulfate concentration is then raised to 75% of saturation, and the precipitate is collected by centrifugation and dissolved in a minimal amount (approx. 10 ml) of 1 mM phosphate buffer, pH 7.0. This solution is desalted by passage through a column (4 × 30 cm) of Sephadex G-25, which has been previously equilibrated with 1 mM phosphate buffer, pH 7.0. The desalted protein solution is percolated at a flow rate of 1–3 ml/min through a column of DEAE-Sephadex A-50 (4 × 5 cm) which has been equilibrated with 1 mM phosphate buffer, pH 7.0. The column is washed with 250 ml of 1 mM phosphate buffer, pH 7.0, containing 5 mM $(NH_4)_2SO_4$. The column is then eluted with 1 mM phosphate buffer, pH 7.0, containing 150 mM $(NH_4)_2SO_4$ at a flow rate of 1 ml/min. The first 50 ml of eluant contain little enzyme activity. Most of the activity is eluted from the column in ml 50–100 of eluant. After precipitation of the enzyme activity with $(NH_4)_2SO_4$ (75% saturation) the precipitated protein is dissolved in approximately 4 ml of 1 mM phosphate buffer, pH 7.0, containing 150 mM $(NH_4)_2SO_4$.

Step 3. Sephadex G-100 Chromatography. The enzyme solution above is layered on the top of a Sephadex G-100 column (2 × 140 cm) equilibrated with 1 mM phosphate buffer, pH 7.0, containing 150 mM $(NH_4)_2SO_4$. The flow rate of the column is maintained at 10–15 ml/hr (Fig. 1). The adenine phosphoribosyltransferase in ml 200–230 is concentrated by precipitation with $(NH_4)_2SO_4$ (75% of saturation). After dissolving the protein in 1–2 ml of 50 mM Tris·HCl buffer, pH 7.5, the solution is dialyzed for 2 hr against 2 liters of 20 mM Tris·HCl buffer, pH 7.5, containing 20 mM $(NH_4)_2SO_4$.

Step 4. Isoelectrofocusing in an Ampholyte pH Gradient. An LKB electrofocusing column (110-ml capacity) is siliconized before use with Siliclad. The cathode is put on the bottom of the column, 1.5% ampholine solution (pH range 5–8) in a sucrose gradient is used in the middle, and the anode is arranged at the top. Mercaptoethanol (5 mM) is included in the ampholine solution; otherwise the detailed directions in the instructional manual are followed in filling the column. Before adding the concentrated G-100 enzyme to the column, the ampholine

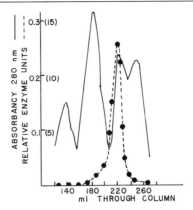

FIG. 1. Chromatography of rat liver adenine phosphoribosyltransferase on Sephadex G-100. The external volume of the column was approximately 140 ml as determined with Dextran blue. The effluent absorbance at 280 nm was measured in a recording spectrophotometer equipped with flow-through cuvettes (solid line). Assay for enzyme activity (dashed line) was by the steady-state method. Reproduced from Kenimer et al.,[1] with permission.

gradient is prefocused for 12 hr at 300 V and an additional 36 hr at 500 V. A piece of 0.5-mm polyethylene tubing connected to a 5-ml syringe is then carefully inserted through the gradient to a predetermined point at which the pH of the solution is 5.6–5.7. Two milliliters of the solution are drawn into the syringe. The polyethylene tubing is left inserted in the gradient, and the enzyme solution is combined with the ampholine solution in the syringe. Sufficient sucrose is added to the enzyme solution so that its density is equivalent to the ampholine solution. After adjusting the pH of the mixture to 5.6–5.7 with 0.1 M HCl, any precipitate which forms is removed by centrifugation. The mixture is then carefully reinserted into the ampholine gradient in the column. Electrofocusing of the column is then resumed for an additional 48 hr at 500 V. The contents of the isoelectrofocusing column are drained at a flow rate of 1 ml/min using a recording spectrophotometer with flow-through cuvettes to monitor the regions of the gradient for protein (Fig. 2). Fractions of 1.3 ml are collected. Fractions containing adenine phosphoribosyltransferase activity are stored at 2°. Several other methods of purification [including hydroxylapatite chromatography and $Ca_3(PO_4)_2$ gel absorption and elution of step 2, 3, and 4 material (see the table); preparative disc gel electrophoresis of step 3 material (as an alternative to isoelectrofocusing); and alumina absorption and elution of step 2, 3, and 4 material] are ineffective as means of improving overall yield of enzyme activity or increased specific activity. The overall yield of enzyme activity can be increased to approximately 10% by taking wider cuts of the eluates from the G-100 and isoelectrofocusing columns;

PURIFICATION OF RAT LIVER ADENINE PHOSPHORIBOSYLTRANSFERASE[a]

Step	Volume (ml)	Protein (mg)	Specific activity[b] (units/mg protein)	Yield (%)	Purification (fold)	Initial-burst[c] steady-state reaction
1. Crude supernatant fraction	114	2600	0.001	100	1.0	0.31
2. DEAE-Sephadex A-50	6	248	0.01	68	10.0	0.28
3. Sephadex G-100	30	28	0.04	33	40.0	0.26
4. Isoelectrofocused	5.2	0.63	1.10	4	1100.0	0.39
a. Peak tube (no. 49)[d]	1.3	0.31	1.12	2		
b. Side tubes (nos. 48, 50, 51)	3.9	0.32	1.08	2		

[a] Reproduced from Kenimer et al.[1] with permission.

[b] Assays carried out as described in the text using between 0.05–0.5 munit of enzyme per milliliter. Initial wet weight liver = 32.5 g.

[c] See text for explanation.

[d] Tube numbers refer to fractions obtained from isoelectrofocusing experiments presented in Fig. 2.

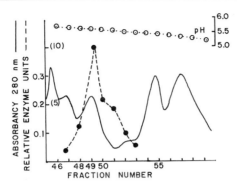

FIG. 2. Isoelectrofocusing of rat liver adenine phosphoribosyltransferase on a pH 5–8 sucrose gradient. The isoelectrofocusing was performed on an enzyme preparation after Sephadex G-100 chromatography. See the text for a complete description of the procedure employed. After isoelectrofocusing the column was drained from the bottom, and the contents were monitored at 280 nm as in Fig. 1 (solid line). Only the pH region (5.25–5.75) of the gradient in close proximity to the enzyme activity (dashed line) is presented. Reproduced from Kenimer *et al.*,[1] with permission.

however, preparations with lower specific activity and homogeneity result.

Properties

Enzyme Stability. Step 4 adenine phosphoribosyltransferase has an average half-life of 2–4 weeks when stored at 2–4° in the ampholine gradient solution. Both dialysis and dilution accelerated inactivation of the enzyme.

Homogeneity. Step 4 enzyme appears to be homogeneous when analyzed by disc gel electrophoresis either by the Davis system at pH 8.3 or on gels containing ampholines in the range of pH 5–8. A single sharp band is obtained at loads up to 0.2 mg protein.

Estimates of Molecular Weight. Step 4 enzyme has a sedimentation coefficient of 2.85 S when analyzed by zone sedimentation in a sucrose density gradient[4] using cytochrome c as an internal standard ($s_{20,w}$ = 1.9). The apparent molecular weight for the adenine phosphoribosyltransferase was calculated to be 22,000. Disc-gel electrophoresis in the presence of sodium dodecylsulfate yielded a molecular weight of 17,500 when the enzyme was compared to internal standards of trypsin, avidin, and lysozyme.[5]

[4] R. G. Martin and B. N. Ames, *J. Biol. Chem.* **236**, 1372 (1961).
[5] A. K. Dunker and R. R. Rueckert, *J. Biol. Chem.* **244**, 5074 (1969).

Amino-Terminal Analysis. The highly purified enzyme preparations contain a single amino terminal identifiable as the phenylthiohydantoin (PTH) derivative of phenylalanine. Authentic phenylthiocarbamyl-phenylalanine (PTC-phenylalanine) migrated on thin-layer plates[6] and on the gas chromatograph[7] with a mobility identical to that prepared from adenine phosphoribosyltransferase. PTC-phenylalanine accounted for a minimum of 85% of the total PTH derivatives after a single round of the Edman degradation.[8]

Apparent K_m Values. Kinetic constants for adenine and PP-ribose-P were determined at various stages of enzyme purification to be relatively constant. K_m (adenine) = 0.8–1.0 ± 0.3 μM. K_m (PP-ribose-P) = 5 ± 2 μM.

pH Optima. In the steady-state assay the enzyme was active over a broad range increasing progressively in activity from pH 5.5 to 10. The "initial burst" assay was relatively independent of pH in the same range.

"Initial Burst" Kinetics. A rapid phase of AMP synthesis is obtained with the enzyme at all stages of purification. In the table the numerical value of the ratio of the initial-burst activity (10 sec at 0°) to the steady-state activity (3 min at 30°) are presented for each step of the purification procedure. Similar fractional ratios were obtained at each step.

Sulfhydryl Content of the Purified Enzyme. Step 4 enzyme was freed of mercaptoethanol by dialysis and precipitation with ammonium sulfate. Sulfhydryl content as determined by the method of Boyer[9] was 3.65 moles of thiol per mole of enzyme. Sulfhydryl content measured as described by Ellman[10] was 3.6 moles of thiol per mole of enzyme (in the presence of 4.9 M urea). These values compare favorably with a value of 3.3 cysteic acid residues obtained by amino acid analysis after performic acid oxidation.[11] These results indicated the absence of any disulfide linkages in the enzyme.

Other Physical Constants. The catalytic constant for the enzyme was calculated to be approximately 20 at 37° and pH 7.5. The isoelectric point is 5.65 as determined in the isoelectrofocusing column.

[6] J. Jeppsson and J. Sjöquist, *Anal. Biochem.* **18,** 264 (1967).
[7] J. J. Pisano, T. J. Bronzert, and H. B. Brewer, Jr., *Anal. Biochem.* **45,** 43 (1972).
[8] J. B. Mills, S. C. Howard, S. Scapa, and A. E. Wilhelmi, *J. Biol. Chem.* **245,** 3407 (1970).
[9] P. D. Boyer, *J. Am. Chem. Soc.* **76,** 4331 (1954).
[10] G. L. Ellman, *Arch. Biochem. Biophys.* **82,** 70 (1959).
[11] S. Moore, *J. Biol. Chem.* **238,** 235 (1963).

[79] A General Method for the Isolation of Various Enzymes from Human Erythrocytes[1]

By R. P. AGARWAL, K. C. AGARWAL and R. E. PARKS, JR.

Although human erythrocytes are severely limited in metabolic function, they contain many enzymes, often in large quantities, of great interest to enzymologists and medical scientists. Furthermore, often blood banks are able to provide outdated blood or erythrocytes for research purposes, thus offering a source of substantial quantities of certain enzymes from humans. A problem faced by most investigators who wish to isolate a specific enzyme from erythrocytes is the fact that this tissue contains large amounts of hemoglobin (35 g/100 ml of packed cells). Therefore, an early step in most purification protocols involves the separation of the hemoglobin from the desired enzyme or enzymes. Certain of the procedures employed for this purpose cause protein denaturation or require the use of expensive reagents (see reference 1 for discussion). However, combinations of several relatively gentle, nondestructive procedures first described a number of years ago[2-4] have been adapted for this purpose.[1] These methods include adsorption on calcium phosphate gel and DEAE-cellulose (phosphate), pH 7.0–7.5.[5] Both adsorbents retain many enzymes but not hemoglobin. These two procedures and a calcium phosphate gel–cellulose column chromatographic method have proved successful for the partial purification in excellent yield of a number of human erythrocytic enzymes from quantities of blood ranging from a few milliliters to many liters.[6] Many of the purification procedures for human erythrocytic enzymes described elsewhere in this volume (Chapters 49, 64, 67, and 72) rely on this general method for the initial stages of purification. In addition, these methods may prove suitable for the initial stages of large-scale purifica-

[1] R. P. Agarwal, E. M. Scholar, K. C. Agarwal, and R. E. Parks, Jr., *Biochem. Pharmacol.* **20**, 1341 (1971).

[2] K. K. Tsuboi and P. B. Hudson, *J. Biol. Chem.* **224**, 879 (1957).

[3] M. A. Hennessey, A. M. Waltersdorph, F. M. Huennekens, and B. W. Gabrio, *J. Clin. Invest.* **41**, 1257 (1962).

[4] S. Cha, C.-J. M. Cha, and R. E. Parks, Jr., *J. Biol. Chem.* **242**, 2577 (1967).

[5] The purification procedure described here employed DEAE-cellulose (phosphate) at pH 7.5. However, one may choose a pH of 7.0–7.5, depending on the enzyme to be isolated.

[6] This general method has been performed successfully by the Enzyme Center at Tufts University School of Medicine, Boston, Massachusetts, at least four times on quantities of human blood in the range of 25–50 liters.

METHODS IN ENZYMOLOGY, VOL. LI

tions of various other human erythrocytic enzymes not discussed in this volume.[2,7] Modification of the quantities of reagents according to the relative proportions described below permits application of these methods to wide ranges of starting material. All steps in purification are performed at 4°. A flow diagram is presented in Fig. 1.

Reagents

Calcium phosphate gel[8]
DEAE-cellulose (phosphate)[9]
Tris-acetate, 0.05 M, pH 8.0 (Tris buffer)
Potassium phosphate, 1 M, pH 7.5 (phosphate buffer)

Methods

Step 1. Preparation of Hemolysate. The erythrocytes are sedimented by centrifugation at 2000 g for 10 min from human blood collected in an anticoagulant (both heparin and acid-citrate-dextrose, ACD, have proven satisfactory). The plasma and buffy coat are removed by

[7] C. M. Kong and R. E. Parks, Jr., *Mol. Pharmacol.* **10**, 648 (1974).

[8] The calcium phosphate gel and the calcium phosphate gel:cellulose columns are prepared essentially as described by Tsuboi and Hudson[2] and S. Cha *in* ["Methods in Enzymology" (J. M. Lowenstein, ed.), Vol. 13, p. 62. Academic Press, New York, 1969]. To 1 liter of 0.5 M Na_2HPO_4 is added 30 ml of concentrated ammonium hydroxide followed by 7.5 liters of 0.1 M $CaCl_2$, with vigorous stirring. The gel is allowed to settle overnight, and the supernatant fluid is siphoned off. The gel is washed repeatedly by resuspension in about 8 liters of distilled water until the supernatant fluid is free of chloride ions. After the final washing the excess supernatant fluid is siphoned off. The dry weight of the gel per unit volume is determined by heating a known volume of the gel suspension to dryness. The gel suspension is diluted to a concentration of 15–30 mg of gel, and may be stored at 4° for a long period of time. Commercially available hydroxylapatite gel may also be used.

For a calcium phosphate gel:cellulose column (5 × 50 cm), 130 gm of cellulose powder (coarse fibers, obtained from Reeve Angel & Co.) is suspended in 1 liter of the 0.05 M Tris buffer; then an appropriate volume of calcium phosphate gel slurry containing 5 g by dry weight is added to the suspension, mixed well, and kept at 4° overnight or longer. The whole suspension is packed into a column stoppered with glass wool at the bottom. The unsettled portion of the suspension is stirred frequently to prevent channeling and gross separation of the gel and the cellulose. When the gel and the cellulose have settled completely, to minimize channeling, 10 g of cellulose (preswollen by overnight soaking in about 200 ml of 0.05 M Tris buffer) are poured on the top of the gel:cellulose bed. The column is washed with 100–200 ml of the same buffer.

[9] About 100 g of DEAE-cellulose are repeatedly suspended in distilled water and the supernatant fluid is decanted to remove fines. Washed DEAE-cellulose is converted to the phosphate form by suspension in 1 M phosphate buffer. After thorough washing with distilled water, the DEAE-cellulose is suspended in 0.005 M phosphate buffer, pH 7.5.

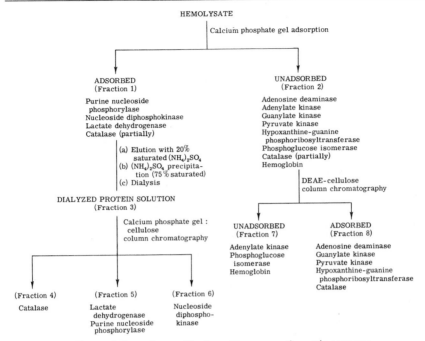

FIG. 1. Scheme for purification of human erythrocytic enzymes.

siphoning, and the erythrocytes are washed 3 times by resuspension in an equal volume of 0.9% NaCl and centrifugation at 2000 g for 10 min. The washed erythrocytes are lysed by addition of 1 volume of distilled water, with freezing and thawing followed by further addition of distilled water to an overall dilution of 10-fold. After thorough mixing, the erythrocytic stroma is removed by centrifugation.[10]

Step 2. Calcium Phosphate Gel Adsorption.[1] This procedure is a modification of a method described earlier by Tsuboi and Hudson.[2] Calcium phosphate gel slurry is added slowly with stirring to the hemolysate from step 1 in a ratio of gel to protein of 1:20 (dry weight). In order to achieve greater selectivity in enzymic adsorption, the gel to protein ratio may be modified as shown in Fig. 2. The mixture is stirred for 30 min after which the gel is sedimented by centrifugation (5000 g for 10 min) and washed 3 times by resuspension in water and centrifugation. The supernatant fluid and washings are pooled (fraction 2, Fig. 1) for

[10] At this stage, the hemolysate may be stored frozen. If the hemolysate is not frozen, it is advisable to proceed as soon as possible to the next step of purification to minimize the degradation of the enzymes by reactions such as proteolysis.

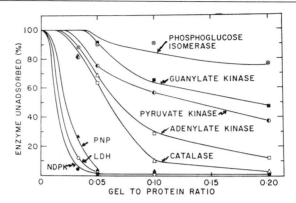

FIG. 2. Relative enzyme affinities for calcium phosphate gel. The gel and hemolysate were thoroughly mixed for 30 min and centrifuged. The gel was washed once by suspending in distilled water. The wash was combined with the supernatant, and residual enzymic activities were determined. From Agarwal et al.[1]

further treatment in step 4. The sedimented, washed calcium phosphate gel is dispersed thoroughly in approximately 5 volumes of 20% saturated ammonium sulfate solution (adjusted to pH 7.0 with NH₄OH). After gentle stirring for at least 30 min, the supernatant fluid is collected by centrifugation. This ammonium sulfate elution procedure is usually repeated 2 times (as found necessary by enzymic assays), and the supernatant fractions are pooled (fraction 1). Finely ground solid ammonium sulfate is added slowly with stirring to fraction 1 to a final concentration of 75% saturation. The pH is maintained at 7.0. The ammonium sulfate suspension is permitted to stand for at least 12 hr at 4°, after which the supernatant fluid is removed by centrifugation at 10,000 g for 60 min. The precipitate is dissolved in a minimal volume of the Tris buffer, containing 1.0 mM dithiothreitol or 10 mM mercaptoethanol. After thorough dialysis against the above buffer solution, the dissolved proteins (fraction 3) are subjected to calcium phosphate gel:cellulose chromatography as described in step 3.

Step 3. Calcium Phosphate Gel:Cellulose Column Chromatography.[1] The calcium phosphate gel:cellulose mixture is poured slowly into a column (diameter:height ratio about 1:10) with constant stirring to ensure uniform packing.[8] The column is equilibrated with the Tris buffer. Fraction 3 is added, and the column is washed with the same buffer until the eluate is essentially color-free. Usually 2 or 3 bed volumes of buffer suffice. Elution is then carried out with a linear gradient of phosphate (0–0.3 M in the Tris buffer, pH 8.0, containing 1 mM dithiothreitol). The total quantity of elution fluids is usually about

20 bed volumes. Successive fractions are collected, and the appropriate enzymic assays are performed. Fractions containing enzymic activities are pooled (fractions 4, 5, and 6; see Fig. 1), and the proteins are precipitated by addition of solid ammonium sulfate to a saturation of about 75%. These ammonium sulfate precipitates can be collected and stored frozen, often for prolonged periods without serious loss of enzymic activity, before further purification.

Step 4. DEAE-Cellulose Fractionations.[1] Many erythrocytic enzymes are adsorbed by DEAE-cellulose (phosphate) at pH values between 7.0–7.5,[5] whereas hemoglobin is not retained under these conditions. Both column and batch procedures have been employed successfully. For chromatography, DEAE-cellulose (phosphate) suspension is added to a column (relative dimensions 1 × 10) and equilibrated with the 0.005 M phosphate buffer. Fraction 2 is added, and the column is washed with equilibration buffer (usually 2–3 bed volumes) until the eluate is essentially color-free. The enzymes and remaining proteins may be eluted with a linear gradient of the phosphate buffer (0.005–0.5 M). Usually about 10 bed volumes of eluent solutions are employed. Successive fractions are collected, assayed for enzymic activities, and pooled as desired. These pooled fractions may be precipitated with ammonium sulfate and stored as described under step 3.

To perform a batch procedure, a slurry of DEAE-cellulose (phosphate)[9] is added slowly with stirring to fraction 2 (pH adjusted to 7.5). In order to determine the quantity of DEAE-cellulose needed for complete adsorption of the desired enzymic activity, assays are performed periodically during the addition of the DEAE-cellulose. After sufficient DEAE-cellulose has been added, the suspension is stirred for 30–60 min and is poured onto a Büchner funnel. Suction is applied until the filter cake is semisolid. Successive washings with the low-concentration buffer are carried out with suction until the filtrate is essentially colorless. The filter cake is then transferred to a suitable container and is suspended in about 5 bed volumes of 0.5 M phosphate buffer, pH 7.5. After stirring for about 1 hr, the suspension is again filtered on a Büchner funnel and the filter cake is washed 2 or 3 times by addition of the same buffer. The adequacy of the washing procedure may be tested by enzymic assay on the filtrate. The filtrate and washings are pooled and may be precipitated by the addition of ammonium sulfate to 75% saturation.

Comments

It is not always necessary or desirable to follow the sequence described above. It is usually preferred, however, to remove interfering enzymes as soon as possible. For example, adenosine deaminase, hypoxanthine-guanine phosphoribosyltransferase, and guanylate kinase are effectively adsorbed on DEAE-cellulose under the conditions described above, but poorly, if at all, by calcium phosphate gel. Therefore, if one is primarily interested in isolation of one of these enzymes, it might be preferable to carry out the DEAE-cellulose step before calcium phosphate gel addition. It may be desirable to include the calcium phosphate gel step as a negative adsorption procedure to remove interfering enzymes, e.g., purine nucleoside phosphorylase from adenosine deaminase.

Author Index

Numbers in parentheses are reference numbers and indicate that an author's work is referred to although his name is not cited in the text.

Subject Index

hibitor, of nucleoside phosphotransferase, 393
N-2,4-Dinitrophenyl-methionine, marker,
of total available column volume, 24
Dioxane
in CO_2 determination, 156
inhibitor, of deoxythymidine kinase, 365
in scintillation cocktail, 52
in Sepharose substitution reaction, 311
Diphenylamine assay, 442, 443
Diphenylamine reagent, preparation, 248
2,5-Diphenyloxazole, in scintillation fluid,
4, 30, 157, 292, 362, 409, 540
Diphosphatidylglycerol, activator, of dihydroorotate dehydrogenase, 62
2,3-Diphosphoglycerate, inhibitor, of adenylate deaminase, 502
Disodium p-nitrophenyl phosphate hexahydrate, in 5-fluoro-2'-deoxyuridine 5'-(p-
nitrophenylphosphate) synthesis, 98
5,5'-Dithiobis(2-nitrobenzoic acid)
inhibitor, of guanylate kinase, 489
of orotidylate decarboxylase, 79
of purine nucleoside phosphorylase,
536, 538
of thymidylate synthetase, 96
of uridine phosphorylase, 429
Dithioerythitol, reducing substrate, of ribonucleoside diphosphate reductase,
247
Dithiothreitol
activator, of cytidine deaminase, 407,
411, 412
of deoxycytidine kinase, 345
of GMP synthetase, 223
of pyrimidine nucleoside monophosphate kinase, 330
in dissociation, of carbamoyl-phosphate
synthetase, 24, 25
effect on orotate phosphoribosyltransferase activity, 152
in erythrocyte lysis, 282
inhibitor, of glutaminase activity, 25
interference with protein determination,
160
in PED buffer, 137
in pyrophosphate determination, 275
reducing substrate, of ribonucleoside triphosphate reductase, 247, 258

stabilizer, of adenosine monophosphate
nucleosidase, 264, 265
of adenylate kinase, 468, 471
of adenylosuccinate synthetase, 207,
208, 210
of carbamoyl-phosphate synthetase,
34, 112, 122, 124
of cytidine deaminase, 410
of deoxycytidine kinase, 338
of deoxythymidylate phosphohydrolase, 228
of erythrocyte enzymes, 584
of FGAM synthetase, 197
of GMP synthetase, 215, 216
of hypoxanthine phosphoribosyltransferase, 543, 545
of nucleoside diphosphokinase, 380,
381, 386
of nucleoside phosphotransferase, 393
of OPRTase-OMPdecase complex,
156, 162
of ortidylate decarboxylase, 158
of phosphoribosylpyrophosphate synthetase, 16
of protein B1, 228, 230
of purine nucleoside phosphorylase,
553, 536, 538, 539
of pyrimidine nucleoside monophosphate kinase, 322–325, 328
of pyrimidine nucleoside phosphorylase, 433, 434
of ribonucleoside diphosphate reductase, 229–233, 238–240
of thymidine kinase, 366–368
of uridine-cytidine kinase, 302
for storage, of deoxycytidine kinase, 341,
344
Dodecyl sulfate, inhibitor, of dihydroorotate dehydrogenase, 62
DON, see 6-Diazo-5-oxonorleucine
Dowex 50 column
in assay, of cytidine triphosphate synthetase, 85
of phosphoribosylglycinamide synthetase, 181
of ribonucleoside diphosphate reductase, 228
in preparation of derivatized Sepharose,
253

I

of pyrimidine nucleoside monophosphate kinase, 321, 329
of ribonucleoside diphosphate reductase, 227
of succino-AICAR synthetase, 193
of thymidine kinase, 358, 370
of UMP-CMP kinase, 331, 332
of uridine-cytidine kinase, 299, 305, 308, 314, 319
Magnesium sulfate
  in assay, of GMP synthetase, 219
  of phosphoribosylpyrophosphate synthetase, 13
  in purification, of phosphoribosylglycinamide synthetase, 184
Malate
  activator, of adenylate kinase, 465
  inhibitor, of rat liver acid nucleotidase, 274
Maleate buffer, interference by, in OPRTase-OMPdecase complex assay, 167
Manganous chloride
  in assay, of amidophosphoribosyltransferase, 172
  in purification, of adenylosuccinate AMP-lyase, 204
Manganous ion
  activator, of adenine phosphoribosyltransferase, 566, 573
  of cytidine deaminase, 411
  of dCTP deaminase, 422
  of deoxycytidine kinase, 345
  of deoxycytidylate deaminase, 417
  of guanine phosphoribosyltransferase, 557
  of guanylate kinase, 481, 489
  of hypoxanthine phosphoribosyltransferase, 557
  of nucleoside diphosphokinase, 385
  of phosphoribosylpyrophosphate synthetase, 11, 16
  of pyrimidine nucleoside monophosphate kinase, 329
  of succino-AICAR synthetase, 193
  of thymidine kinase, 358, 370
  of uridine-cytidine kinase, 319
  inhibitor, of deoxythymidine kinase, 365

of nucleoside phosphotransferase, 393
of uridine nucleosidase, 294
Mannitol, in fungal mitochondrial membrane preparation, 65
Manton-Gaulin mill, in bacterial cell extraction, 6, 32, 137, 414
Membrane particle preparation, 60–61
Menadione, electron acceptor, for dihydroorotate dehydrogenase, 64
Menaquinone, cofactor, for dihydroorotate dehydrogenase, 63
5-Mercaptodeoxyuridine, substrate, of thymidine kinase, 358
2-Mercaptoethanol
  activator, of orotidylate decarboxylase, 74, 75
  of pyrimidine nucleoside monophosphate kinase, 330–331
  in assay, of dCTP deaminase, 418, 419
  of cytidine triphosphate synthetase, 85
  of deoxycytidylate deaminase, 413
  of nucleoside diphosphokinase, 372
  of orotate phosphoribosyltransferase, 70
  of uridine phosphorylase, 425
  in dissociation of aspartate transcarbamoylase, 36
  in electrophoresis buffer, 342, 357
  in Enzyme Buffer, 14
  for enzyme storage, 82, 362
  inhibitor, of GMP synthetase, 223
  in (±)-L-methylenetetrahydrofolate preparation, 91
  in protein determination, 339
  in purification, of adenine phosphoribosyltransferase, 575, 576
  of adenylate deaminase, 492, 495, 499, 500
  of adenylate kinase, 461
  of adenylosuccinate AMP-lyase, 205
  of amidophosphoribosyltransferase, 174, 175, 176
  of aspartate carbamyltransferase, 45, 53
  of cytidine triphosphate synthetase, 81, 86
  of dCTP deaminase, 420, 421, 423
  of deoxycytidine kinase, 339, 340, 341

O

Octanol, in purification, of orotate phosphoribosyltransferase, 71

O-Octyl Sepharose, in purification, of *Lactobacillus* deoxynucleoside kinases, 351

Oleic acid, inhibitor, of adenylate kinase, 465

Omnifluor, in scintillation cocktail, 220, 544

OMPdecase, *see* Orotidylate decarboxylase

OPRTase, *see* Orotate phosphoribosyltransferase

L-Ornithine, activator, of carbamoyl-phosphate synthetase, 22, 27, 33, 34, 105, 112

Ornithine carbamoyltransferase, in assay, of carbamoyl-phosphate synthetase, 105, 112

Orotate
  in assay, of phosphoribosylpyrophosphate synthetase, 5
  chromatographic separation, 158–159
  inhibitor, of dihydroorotate dehydrogenase, 62, 68
  product, of dihydroorotate dehydrogenase, 63
  substrate, of OPRTase: OMPdecase complex, 135, 144, 145, 155
    of orotate phosphoribosyltransferase, 69–70

Orotate phosphoribosyltransferase, *see also* Orotate phosphoribosyltransferase: orotidylate decarboxylase complex
  activators, 74, 152
  animal sources, 72–73
  assay, 69–70, 135–136, 141, 143, 144–145, 156–157
  equilibrium constant, 164
  inhibitors, 74, 153, 163, 166
  kinetic properties, 74, 140, 152
  molecular weight, 73
  pH optimum, 73, 152
  purification, 70–73
  purity, 73
  pyrophosphorolysis by, 154
  stability, 73, 153–154

substrate specificity, 152
  from yeast, 69–74

Orotate phosphoribosyltransferase: orotidylate decarboxylase complex, 135–167
  assay, 135–136, 156
  dissociation, 141
  from Ehrlich ascites cells, 155–167
  from erythrocyte, 143–154
  kinetic properties, 140–141, 166–167
  molecular weight, 137, 138, 154, 164–166
  purification, 137–138, 146–151, 160–164
  purity, 167
  from *Serratia marcescens*, 135–143
  stability, 167

Orotidine, inhibitor, of orotate phosphoribosyltransferase, 166

Orotidylate
  in assay, of orotate phosphoribosyltransferase, 144
  chromatographic separation, 158–159
  extinction coefficient, 158
  inhibitor, of orotate phosphoribosyltransferase, 74
  product, of orotate phosphoribosyltransferase, 69
  substrate, of OPRTase-OMPdecase complex, 135, 155, 157
    of orotidylate decarboxylase, 74

Orotidylate decarboxylase, 74–79, *see also* Orotate phosphoribosyltransferase: orotidylate decarboxylase complex
  assay, 74–75, 136, 145
  inhibitors, 79, 153, 163, 166
  kinetics, 78–79, 140
  molecular weight, 78
  in orotate phosphoribosyltransferase assay, 69
  pH optimum, 78, 152
  properties, 78–79
  purification, 75–78
  purity, 78
  stability, 78, 153–154
  storage, 70
  from yeast, 74–79

Osmotic shock fluid, preparation, 558, 567

Oxalate, inhibitor, of rat liver acid nucleotidase, 274

Oxime reagent, 51